Association schemes were invented by statisticians on order to enable them to extend their range of block designs from 2-designs to partially balanced designs. However, they have intrinsic interest to combinatorialists and to group theorists as they include strongly regular graphs and schemes based on Abelian groups. They also have a wider use within statistics, forming the basic structures of many designed experiments. This book blends these topics in an accessible way without assuming any background in either group theory or designed experiments.

CAMBRIDGE STUDIES IN ADVANCED MATHEMATICS 84

ASSOCIATION SCHEMES
DESIGNED EXPERIMENTS, ALGEBRA AND COMBINATORICS

CAMBRIDGE STUDIES IN ADVANCED MATHEMATICS

Already published

17 W. Dicks & M. Dunwoody *Groups acting on graphs*
18 L.J. Corwin & F.P. Greenleaf *Representations of nilpotent Lie groups and their applications*
19 R. Fritsch & R. Piccinini *Cellular structures in topology*
20 H. Klingen *Introductory lectures on Siegel modular forms*
21 P. Koosis *The logarithmic integral II*
22 M.J. Collins *Representations and characters of finite groups*
24 H. Kunita *Stochastic flows and stochastic differential equations*
25 P. Wojtaszczyk *Banach spaces for analysts*
26 J.E. Gilbert & M.A.M. Murray *Clifford algebras and Dirac operators in harmonic analysis*
27 A. Frohlich & M.J. Taylor *Algebraic number theory*
28 K. Goebel & W.A. Kirk *Topics in metric fixed point theory*
29 J.F. Humphreys *Reflection groups and Coxeter groups*
30 D.J. Benson *Representations and cohomology I*
31 D.J. Benson *Representations and cohomology II*
32 C. Allday & V. Puppe *Cohomological methods in transformation groups*
33 C. Soule et al. *Lectures on Arakelov geometry*
34 A. Ambrosetti & G. Prodi *A primer of nonlinear analysis*
35 J. Palis & F. Takens *Hyperbolicity, stability and chaos at homoclinic bifurcations*
37 Y. Meyer *Wavelets and operators I*
38 C. Weibel *An introduction to homological algebra*
39 W. Bruns & J. Herzog *Cohen-Macaulary rings*
40 V. Snaith *Explict Brauer induction*
41 G. Laumon *Cohomology of Drinfeld modular varieties I*
42 E.B. Davies *Spectral theory and differential operators*
43 J. Diestel, H. Jarchow, & A. Tonge *Absolutely summing operators*
44 P. Mattila *Geometry of sets and measures in Euclidean spaces*
45 R. Pinsky *Positive harmonic functions and diffusion*
46 G. Tenenbaum *Introduction to analytic and probabilistic number theory*
47 C. Peskine *An algebraic introduction to complex projective geometry*
48 Y. Meyer & R. Coifman *Wavelets*
49 R. Stanley *Enumerative combinatorics I*
50 I. Porteous *Clifford algebras and the classical groups*
51 M. Audin *Spinning tops*
52 V. Jurdjevic *Geometric control theory*
53 H. Volklein *Groups as Galois groups*
54 J. Le Potier *Lectures on vector bundles*
55 D. Bump *Automorphic forms and representations*
56 G. Laumon *Cohomology of Drinfeld modular varieties II*
57 D.M. Clark & B.A. Davey *Natural dualities for the working algebraist*
58 J. McCleary *A user's guide to spectral sequences II*
59 P. Taylor *Practical foundations of mathematics*
60 M.P. Brodmann & R.Y. Sharp *Local cohomology*
61 J.D. Dixon et al. *Analytic pro-P groups*
62 R. Stanley *Enumerative combinatorics II*
63 R.M. Dudley *Uniform central limit theorems*
64 J. Jost & X. Li-Jost *Calculus of variations*
65 A.J. Berrick & M.E. Keating *An introduction to rings and modules*
66 S. Morosawa *Holomorphic dynamics*
67 A.J. Berrick & M.E. Keating *Categories and modules with K-theory in view*
68 K. Sato *Levy processes and infinitely divisible distributions*
69 H. Hida *Modular forms and Galois cohomology*
70 R. Iorio & V. Iorio *Fourier analysis and partial differential equations*
71 R. Blei *Analysis in integer and fractional dimensions*
72 F. Borceaux & G. Janelidze *Galois theories*
73 B. Bollobas *Random graphs*
74 R.M. Dudley *Real analysis and probability*
75 T. Sheil-Small *Complex polynomials*
76 C. Voisin *Hodge theory and complex algebraic geometry I*
77 C. Voisin *Hodge theory and complex algebraic geometry II*
78 V. Paulsen *Completely bounded maps and operator algebras*
82 G. Tourlakis *Lectures in logic and set theory I*
83 G. Tourlakis *Lectures in logic and set theory II*

ASSOCIATION SCHEMES
DESIGNED EXPERIMENTS, ALGEBRA AND COMBINATORICS

R. A. Bailey
Queen Mary, University of London

CAMBRIDGE
UNIVERSITY PRESS

University Printing House, Cambridge CB2 8BS, United Kingdom

Cambridge University Press is part of the University of Cambridge.

It furthers the University's mission by disseminating knowledge in the pursuit of education, learning and research at the highest international levels of excellence.

www.cambridge.org
Information on this title: www.cambridge.org/9780521824460

First published 2004

A catalogue record for this publication is available from the British Library

Library of Congress Cataloguing in Publication data
Bailey, R. (Rosemary)
Association schemes: designed experiments, algebra, and combinatorics / R.A. Bailey.
p. cm. – (Cambridge studies in advanced mathematics; 84)
Includes bibliographical references and index.
ISBN 0 521 82446 X
1. Association schemes (Combinatorial analysis) 2. Experimental design. I. Title.
II. Series.
QA164.B347 2003
511'.6 – dc21 2003044034

ISBN 978-0-521-82446-0 Hardback
ISBN 978-0-521-18801-2 Paperback

Contents

Preface

Incomplete-block designs for experiments were first developed by Yates at Rothamsted Experimental Station. He produced a remarkable collection of designs for individual experiments. Two of them are shown, with the data from the experiment, in Example 4.3 on page 97 and Exercise 5.9 on page 141. This type of design poses two questions for statisticians: (i) what is the best way of choosing subsets of the treatments to allocate to the blocks, given the resource constraints? (ii) how should the data from the experiment be analysed?

Designs with partial balance help statisticians to answer both of these questions. The designs were formally introduced by Bose and Nair in 1939. The fundamental underlying concept is the association scheme, which was defined in its own right by Bose and Shimamoto in 1952. Theorem 5.2 on page 114 shows the importance of association schemes: the pattern of variances matches the pattern of concurrences.

Many experiments have more than one system of blocks. These can have complicated inter-relationships, like the examples in Section 7.1, which are all taken from real experiments. The general structure is called an orthogonal block structure. Although these were introduced independently of partially balanced incomplete-block designs, they too are association schemes. Thus association schemes play an important role in the design of experiments.

Association schemes also arise naturally in the theory of permutation groups, quite independently of any statistical applications. Much modern literature on association schemes is couched in the language of abstract algebra, with the unfortunate effect that it is virtually inaccessible to statisticians. The result is that many practising statisticians do not know the basic theory of the Bose–Mesner algebra for association schemes, even though it is this algebra that makes everything work. On the other hand, many pure mathematicians working in the subject have no knowledge of the subject's origin in, and continued utility in, designed experiments.

This book is an attempt to bridge the gap, and is intended to be accessible to both audiences. The first half is at a level suitable for final-year undergraduates (on the four-year MSci programme) or MSc students in Mathematics or Statistics. It assumes some linear algebra, modular arithmetic, elementary ideas about graphs, and some probability, but no statistics and very little abstract algebra. The material assumed can be found in almost any standard books on linear algebra, discrete mathematics and probability respectively: for example, [11], [49], [107]. The linear algebra is revised where it is needed (Sections 1.3, 2.1, 3.1), partly to establish the notation used in this book. The same is done for random variables in Section 4.2. Techniques which use finite fields are deliberately avoided, although the reader with some knowledge of them may be able to see where examples can be generalized.

After the basic theory in the first two chapters, the book has three main strands. The first one is the use of association schemes in designed experiments, which is developed in Chapters 4, 5, 7 and parts of 8 and 11. The second is the fruitful interplay between association schemes and partitions: see Chapters 6, 7, 9, 10 and 11. The third gives methods of creating new association schemes from old ones. This starts in Chapter 3 to give us an easy mechanism for developing examples, and continues in Chapters 9 and 10.

Chapters 1–6 form the heart of the book. Chapter 1 introduces association schemes and gives the three different ways of thinking about them: as partitions, as adjacency matrices, as coloured graphs. It also gives many families of association schemes. Chapter 2 moves straight to the Bose–Mesner algebra spanned by the adjacency matrices. The fact that this algebra is commutative implies that these matrices have common eigenspaces, called strata. The relationship between the adjacency matrices and the stratum projectors is called the character table: this is the clue to all calculations in the association scheme. This chapter includes a section on techniques for actually calculating character tables: this should be useful for anyone who needs to make calculations in specific association schemes, for example to calculate the efficiency factors of a block design.

Chapter 3 introduces crossing and nesting, two methods of combining two association schemes to make a new one. Although these are not strictly necessary for an understanding of Chapters 4–6, they provide a wealth of new examples, and a glimpse, in Section 3.5, of the complicated structures that can occur in real experiments.

Chapters 4 and 5 cover incomplete-block designs. Chapter 4 gives the general theory, including enough about data analysis to show what makes a design good. In Chapter 5 this is specialized to partially balanced incomplete-block designs, where the Bose–Mesner algebra gives very pleasing results. Many of Yates's designs are re-examined in the light of the general results.

Chapter 6 introduces the machinery for calculating with partitions of a set. This leads immediately to the definition of orthogonal block structures and derivation of their properties. They yield association schemes which have explicit formulas for the character tables.

The next three chapters build on this core but are almost independent of each other. Chapter 7 covers designs where the experimental units have an orthogonal block structure more complicated than just one system of blocks. There are designs for row-column arrays, for nested blocks, for row-column arrays with split plots, and so on. The idea of partial balance from Chapter 5 is extended to these more complicated designs. Topics covered include efficiency of estimation, combining information from different strata, and randomization. All of this needs the Bose–Mesner algebra.

Group theory is deliberately avoided in Chapters 1–7, because many statisticians do not know any. Cyclic designs are dealt with by using modular arithmetic without any appeal to more general group theory. However, the reader who is familiar with group theory will realise that all such results can be generalized, sometimes to all finite groups, sometimes just to Abelian ones. Chapter 8 revisits Chapters 1–7 and makes this generalization where it can. Later chapters have short sections referring back to this chapter, but the reader without group theory can omit them without losing the main story.

Chapter 9 is devoted to poset block structures, a special class of orthogonal block structures that is very familiar to statisticians. They were developed and understood long before the more general orthogonal block structures but, ironically, have a more complicated theory, which is why they are deferred to this part of the book. Part of the difficulty is that there are two partial orders, which can easily be confused with each other. The idea of poset block structures also gives a new method of combining several association schemes, which generalizes both crossing and nesting.

Chapters 10 and 11 are the most abstract, drawing on all the previous material, and heavily using the methods of calculating with partitions that were developed in Chapter 6. Chapter 10 looks at association schemes with an algebraist's eye, giving some relations between association schemes on different sets. It gives further constructions of new from old, and shows the important role of orthogonal block structures. Some of these results feed into Chapter 11, which looks at association schemes on the same set, trying to answer two statistically motivated questions from earlier chapters. Is there a 'simplest' association scheme with respect to which a given incomplete-block design is partially balanced? This has a positive answer. If there is more than one system of blocks, does it matter if they give partial balance for different association schemes? The answer is messy in general but detailed answers can be given for the three main classes of association schemes discussed in the book: orthogonal block structures, poset block structures, and Abelian group schemes.

The book concludes with two short chapters, looking to the future and summarizing the history.

The reader who simply wants an introduction to association schemes as they occur in designed experiments should read Chapters 1–6. For more emphasis on pure combinatorics, replace parts of Chapters 4 and 5 with parts of Chapters 8 and 10. The reader who designs experiments should also read Chapter 7 and one or both of Chapters 8 and 9. The reader who is more interested in the interplay between partitions and association schemes could read Chapters 1–3, 6 and 8–11.

This book differs from statistics books on incomplete-block designs, such as [70, 134, 137, 140], in that it uses the Bose–Mesner algebra explicitly and centrally. Even books like [91, 201], which cover many partially balanced designs, do not make full use of the Bose–Mesner algebra. Furthermore, orthogonal block structures are developed here in their own right, with a coherent theory, and Chapter 7 gives a more detailed coverage of designs for orthogonal block structures than can be found in any other book on designed experiments.

On the other hand, this book differs from pure mathematical books on association schemes not just because of its attention to statistical matters. The inclusion of explicit sections on calculation (Sections 2.4, 5.2 and 6.4) is unusual. So is the emphasis on imprimitive association schemes, which are deliberately downplayed in books such as [43, 258].

On the other hand, I make no attempt to cover such topics as Krein parameters, polynomial schemes or the links with coding theory.

Most of the material in Chapters 6–7 and 9–11 has not previously appeared in book form, as far as I know.

Some topics in this book already have an established literature, with standard notation, although that may differ between research communities, for example t or v for the number of treatments. I have decided to aim for consistency within the book, as far as is practicable, rather than consistency with the established literatures. Thus I am thinking more of the reader who is new to the subject and who starts at Chapter 1 than I am of the specialist who dips in. If your favourite notation includes an overworked symbol such as n, k, λ or P, be prepared to find something else. I hope that the glossary of notation on page 355 will be helpful.

Web page

Updated information about this book, such as known errors, is available on the following web page.

`http://www.maths.qmul.ac.uk/~rab/Asbook`

R. A. Bailey
School of Mathematical Sciences
Queen Mary, University of London
Mile End Road
London E1 4NS
r.a.bailey@qmul.ac.uk

Acknowledgements

My interest in this subject was lit by two courses which I attended as a DPhil student. G. Higman lectured on graphs, geometries and groups, which included strongly regular graphs and generalized polygons; D. G. Higman gave a course on coherent configurations which became [118]. When I turned to statistics I studied simple orthogonal block structures from [182], and I am grateful to T. P. Speed for pointing out that these could be viewed as association schemes. N. L. Johnson and S. Kotz invited me to write an article [17] on partially balanced designs for the *Encyclopedia of Statistical Science*: I am very glad that they refused to accept my plea of ignorance so that I was forced to see how these designs fitted into the story. Fortunately, I was visiting the University of North Carolina at Chapel Hill at the time, and was able to spend some happy hours in the library reading work by R. C. Bose and his students, one of whom, I. M. Chakravarti, introduced me to Delsarte's thesis. A. W. A. Murray, T. Penttila and P. Diaconis gave early encouragement to the approach taken in [17] and in Chapter 1.

More recently, I am grateful to my colleagues P. J. Cameron and B. A. F. Wehrfritz for encouraging me to develop the material as a lecture course; to those colleagues and students in the Queen Mary Combinatorics Study Group who have commented on draft presentations of the material in Chapters 8–11; and to all those students who have directly or indirectly helped me to develop examples and better ways of explaining things: special thanks here to C. Rutherford, V. Köbberling, E. Postma and E. Gelmi. P. J. Cameron read early drafts of the first six chapters and helped me with the second half of Chapter 13. Of course, the responsibility for the choice of contents, and for the choice of names and notation where there are conflicts between different traditions, is mine, as are all mistakes in the text.

1
Association schemes

1.1 Partitions

Association schemes are about relations between pairs of elements of a set Ω. In this book Ω will always be finite. Recall that $\Omega \times \Omega$ is the set of ordered pairs of elements of Ω; that is,

$$\Omega \times \Omega = \{(\alpha, \beta) : \alpha \in \Omega,\ \beta \in \Omega\}.$$

I shall give three equivalent definitions of association scheme, in terms of partitions, graphs and matrices respectively. Each definition has advantages and disadvantages, depending on the context.

Recall that a *partition* of a set Δ is a set of non-empty subsets of Δ which are mutually disjoint and whose union is Δ.

Let $\mathcal{C}$ be any subset of $\Omega \times \Omega$. Its *dual subset* is $\mathcal{C}'$, where

$$\mathcal{C}' = \{(\beta, \alpha) : (\alpha, \beta) \in \mathcal{C}\}.$$

We say that $\mathcal{C}$ is *symmetric* if $\mathcal{C} = \mathcal{C}'$. One special symmetric subset is the *diagonal* subset $\mathrm{Diag}(\Omega)$ defined by

$$\mathrm{Diag}(\Omega) = \{(\omega, \omega) : \omega \in \Omega\}.$$

First definition of association scheme An *association scheme* with s associate classes on a finite set Ω is a partition of $\Omega \times \Omega$ into sets $\mathcal{C}_0$, $\mathcal{C}_1$, ..., $\mathcal{C}_s$ (called *associate classes*) such that

(i) $\mathcal{C}_0 = \mathrm{Diag}(\Omega)$;
(ii) $\mathcal{C}_i$ is symmetric for $i = 1, \ldots, s$;
(iii) for all i, j, k in $\{0, \ldots, s\}$ there is an integer p_{ij}^k such that, for all (α, β) in $\mathcal{C}_k$,

$$|\{\gamma \in \Omega : (\alpha, \gamma) \in \mathcal{C}_i \text{ and } (\gamma, \beta) \in \mathcal{C}_j\}| = p_{ij}^k.$$

Note that the superscript k in p_{ij}^k does not signify a power.

Elements α and β of Ω are called *i-th associates* if $(\alpha, \beta) \in \mathcal{C}_i$. The number $s+1$, which is the true number of associate classes, is called the *rank* of the association scheme.

We can visualize $\Omega \times \Omega$ as a square array whose rows and columns are indexed by Ω, as in Figure 1.1. If the rows and columns are indexed in the same order, then the diagonal subset consists of the elements on the main diagonal, which are marked 0 in Figure 1.1. Condition (i) says that $\mathcal{C}_0$ is precisely this diagonal subset. Condition (ii) says that every other associate class is symmetric about that diagonal: if the whole picture is reflected about that diagonal then the associate classes remain the same. For example, the set of elements marked $*$ could form an associate class if symmetry were the only requirement. Condition (iii) is much harder to visualize for partitions, but is easier to interpret in the later two definitions.

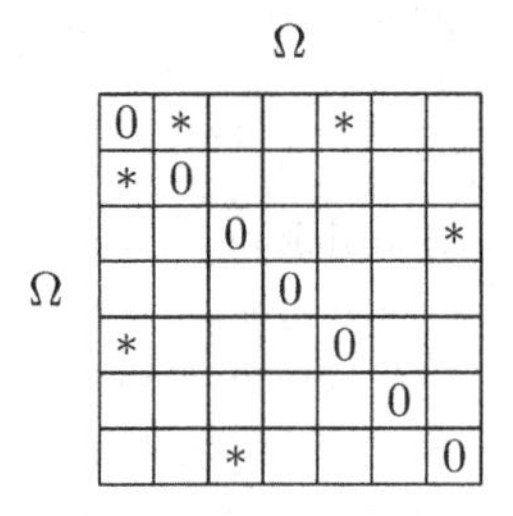

Fig. 1.1. The elements of $\Omega \times \Omega$

Note that condition (ii) implies that $p_{ij}^0 = 0$ if $i \neq j$. Similarly, $p_{0j}^k = 0$ if $j \neq k$ and $p_{i0}^k = 0$ if $i \neq k$, while $p_{0j}^j = 1 = p_{i0}^i$. Condition (iii) implies that every element of Ω has p_{ii}^0 i-th associates, so that in fact the set of elements marked $*$ in Figure 1.1 could not be an associate class. Write $a_i = p_{ii}^0$. This is called the *valency* of the i-th associate class.

(Many authors use k_i or n_i to denote valency, but this conflicts with the very natural use of n_i in Chapters 3 and 6 and k in Chapters 4–7.)

The integers $|\Omega|$, s, a_i for $0 \leqslant i \leqslant s$ and p_{ij}^k for $0 \leqslant i, j, k \leqslant s$ are called the *parameters of the first kind*. Note that $p_{ij}^k = p_{ji}^k$.

Example 1.1 Let $\Delta_1 \cup \cdots \cup \Delta_b$ be a partition of Ω into b subsets of size m, where $b \geqslant 2$ and $m \geqslant 2$. These subsets are traditionally called *groups*, even though they have nothing to do with the algebraic structure called a group. Let α and β be

- first associates if they are in the same group but $\alpha \neq \beta$;
- second associates if they are in different groups.

See Figure 1.2.

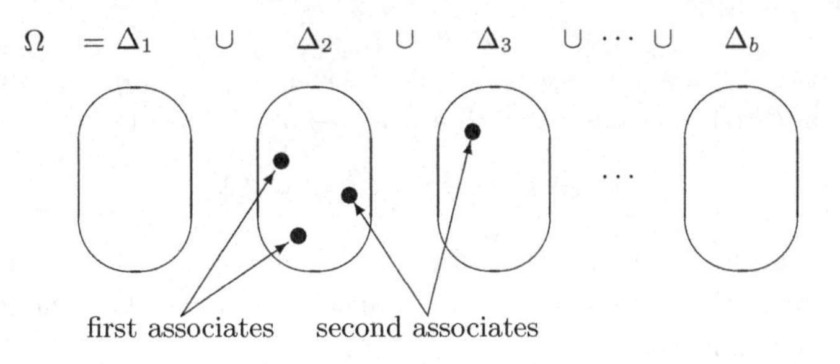

Fig. 1.2. Partition of Ω in Example 1.1

If $\omega \in \Omega$ then ω has $m-1$ first associates and $(b-1)m$ second associates.

If α and β are first associates then the number of γ which are specified associates of α and β are:

	first associate of β	second associate of β
first associate of α	$m-2$	0
second associate of α	0	$(b-1)m$

For example, those elements which are first associates of both α and β are the $m-2$ other elements in the group which contains α and β. If α and β are second associates then the number of γ which are specified associates of α and β are:

	first associate of β	second associate of β
first associate of α	0	$m-1$
second associate of α	$m-1$	$(b-2)m$

So this is an association scheme with $s = 2$, $a_1 = m-1$, $a_2 = (b-1)m$, and

$$p_{11}^1 = m-2 \quad p_{12}^1 = 0 \quad p_{11}^2 = 0 \quad p_{12}^2 = m-1$$
$$p_{21}^1 = 0 \quad p_{22}^1 = (b-1)m \quad p_{21}^2 = m-1 \quad p_{22}^2 = (b-2)m$$

It is called the *group-divisible* association scheme, denoted $\mathrm{GD}(b,m)$ or $\underline{b}/\underline{\underline{m}}$. ■

(The name 'group-divisible' has stuck, because that is the name originally used by Bose and Nair. But anyone who uses the word 'group' in its algebraic sense is upset by this. It seems to me to be quite acceptable to call the scheme just 'divisible'. Some authors tried to compromise by calling it 'groop-divisible', but the nonce word 'groop' has not found wide approval.)

To save writing phrases like 'first associates of α' we introduce the notation $\mathcal{C}_i(\alpha)$ for the set of i-th associates of α. That is,

$$\mathcal{C}_i(\alpha) = \{\beta \in \Omega : (\alpha, \beta) \in \mathcal{C}_i\}.$$

Thus condition (iii) says that $|\mathcal{C}_i(\alpha) \cap \mathcal{C}_j(\beta)| = p_{ij}^k$ if $\beta \in \mathcal{C}_k(\alpha)$. For this reason, the parameters p_{ij}^k are called the *intersection numbers* of the association scheme.

Example 1.2 Let $|\Omega| = n$, let $\mathcal{C}_0$ be the diagonal subset and let

$$\mathcal{C}_1 = \{(\alpha, \beta) \in \Omega \times \Omega : \alpha \neq \beta\} = (\Omega \times \Omega) \setminus \mathcal{C}_0.$$

This is the *trivial* association scheme – the only association scheme on Ω with only one associate class. It has $a_1 = n - 1$ and $p_{11}^1 = n - 2$. I shall denote it $\underline{\underline{n}}$. ■

Example 1.3 Let Ω be an $n \times m$ rectangular array with $n \geqslant 2$ and $m \geqslant 2$, as in Figure 1.3. Note that this is a picture of Ω itself, not of $\Omega \times \Omega$! Put

$$\begin{aligned}
\mathcal{C}_1 &= \{(\alpha, \beta) : \alpha,\ \beta \text{ are in the same row but } \alpha \neq \beta\} \\
\mathcal{C}_2 &= \{(\alpha, \beta) : \alpha,\ \beta \text{ are in the same column but } \alpha \neq \beta\} \\
\mathcal{C}_3 &= \{(\alpha, \beta) : \alpha,\ \beta \text{ are in different rows and columns}\} \\
&= (\Omega \times \Omega) \setminus \mathcal{C}_0 \setminus \mathcal{C}_1 \setminus \mathcal{C}_2.
\end{aligned}$$

Then $a_1 = m - 1$, $a_2 = n - 1$ and $a_3 = (m - 1)(n - 1)$.

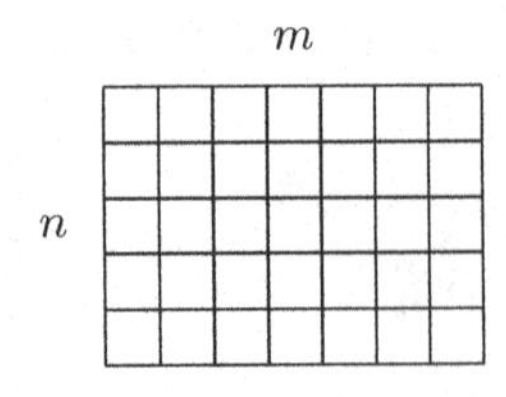

Fig. 1.3. The set Ω in Example 1.3

If $(\alpha, \beta) \in \mathcal{C}_1$ then the number of γ which are specified associates of α and β are:

	$\mathcal{C}_1(\beta)$	$\mathcal{C}_2(\beta)$	$\mathcal{C}_3(\beta)$
$\mathcal{C}_1(\alpha)$	$m-2$	0	0
$\mathcal{C}_2(\alpha)$	0	0	$n-1$
$\mathcal{C}_3(\alpha)$	0	$n-1$	$(n-1)(m-2)$

The entries in the above table are the p_{ij}^1.

If $(\alpha, \beta) \in \mathcal{C}_2$ then the number of γ which are specified associates of α and β are:

	$\mathcal{C}_1(\beta)$	$\mathcal{C}_2(\beta)$	$\mathcal{C}_3(\beta)$
$\mathcal{C}_1(\alpha)$	0	0	$m-1$
$\mathcal{C}_2(\alpha)$	0	$n-2$	0
$\mathcal{C}_3(\alpha)$	$m-1$	0	$(n-2)(m-1)$

The entries in the above table are the p_{ij}^2.

Finally, if $(\alpha, \beta) \in \mathcal{C}_3$ then the number of γ which are specified associates of α and β are:

	$\mathcal{C}_1(\beta)$	$\mathcal{C}_2(\beta)$	$\mathcal{C}_3(\beta)$
$\mathcal{C}_1(\alpha)$	0	1	$m-2$
$\mathcal{C}_2(\alpha)$	1	0	$n-2$
$\mathcal{C}_3(\alpha)$	$m-2$	$n-2$	$(n-2)(m-2)$

and the entries in the above table are the p_{ij}^3.

This is the *rectangular* association scheme $\mathrm{R}(n, m)$ or $\underline{\underline{n}} \times \underline{\underline{m}}$. It has three associate classes. ■

Lemma 1.1 *(i)* $\sum_{i=0}^{s} a_i = |\Omega|$;

(ii) for every i and k, $\sum_j p_{ij}^k = a_i$.

Proof (i) The set Ω is the disjoint union of $\mathcal{C}_0(\alpha), \mathcal{C}_1(\alpha), \ldots, \mathcal{C}_s(\alpha)$.

(ii) Given any (α, β) in $\mathcal{C}_k$, the set $\mathcal{C}_i(\alpha)$ is the disjoint union of the sets $\mathcal{C}_i(\alpha) \cap \mathcal{C}_j(\beta)$ for $j = 0, 1, \ldots, s$. ■

Thus it is sufficient to check constancy of the a_i for all but one value of i, and, for each pair (i, k), to check the constancy of p_{ij}^k for all but one value of j.

Thus construction of tables like those above is easier if we include a row for $\mathcal{C}_0(\alpha)$ and a column for $\mathcal{C}_0(\beta)$. Then the figures in the i-th row and column must sum to a_i, so we begin by putting these totals in the margins of the table, then calculate the easier entries (remembering that the table must be symmetric), then finish off by subtraction.

Example 1.4 Let Ω consist of the vertices of the Petersen graph, which is shown in Figure 1.4. Let $\mathcal{C}_1$ consist of the edges of the graph and

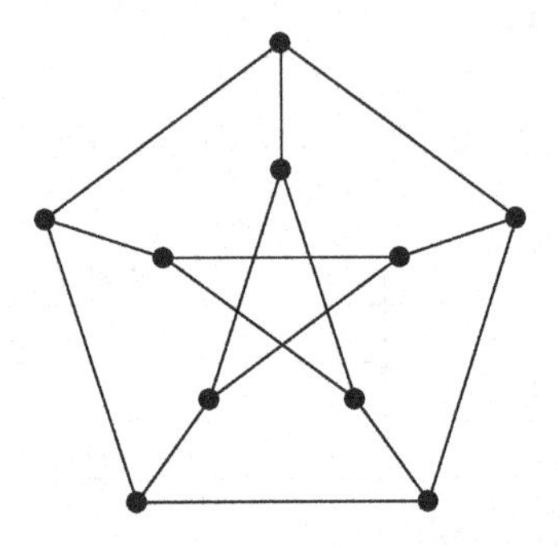

Fig. 1.4. The Petersen graph

$\mathcal{C}_2$ consist of the non-edges (that is, of those pairs of distinct vertices which are not joined by an edge). Inspection of the graph shows that every vertex is joined to three others and so $a_1 = 3$. It follows that $a_2 = 10 - 1 - 3 = 6$.

If $\{\alpha, \beta\}$ is an edge, we readily obtain the partial table

	$\mathcal{C}_0(\beta)$	$\mathcal{C}_1(\beta)$	$\mathcal{C}_2(\beta)$	
$\mathcal{C}_0(\alpha)$	0	1	0	1
$\mathcal{C}_1(\alpha)$	1	0		3
$\mathcal{C}_2(\alpha)$	0			6
	1	3	6	10

because there are no triangles in the graph. To obtain the correct row and column totals, this must be completed as

	$\mathcal{C}_0(\beta)$	$\mathcal{C}_1(\beta)$	$\mathcal{C}_2(\beta)$	
$\mathcal{C}_0(\alpha)$	0	1	0	1
$\mathcal{C}_1(\alpha)$	1	0	2	3
$\mathcal{C}_2(\alpha)$	0	2	4	6
	1	3	6	10

We work similarly for the case that $\{\alpha, \beta\}$ is not an edge. If you are familiar with the Petersen graph you will know that every pair of vertices which are not joined by an edge are both joined to exactly one vertex; if you are not familiar with the graph and its symmetries, you should check that this is true. Thus $p_{11}^2 = 1$. This gives the middle entry of the table, and the three entries in the bottom right-hand corner (p_{12}^2, p_{21}^2 and p_{22}^2) can be calculated by subtraction.

	$\mathcal{C}_0(\beta)$	$\mathcal{C}_1(\beta)$	$\mathcal{C}_2(\beta)$	
$\mathcal{C}_0(\alpha)$	0	0	1	1
$\mathcal{C}_1(\alpha)$	0	1	2	3
$\mathcal{C}_2(\alpha)$	1	2	3	6
	1	3	6	10

So again we have an association scheme with two associate classes. ∎

1.2 Graphs

Now that we have done the example with the Petersen graph, we need to examine graphs a little more formally. Recall that a finite graph is a finite set Γ, whose elements are called *vertices*, together with a set of 2-subsets of Γ called *edges*. (A '2-subset' means a subset of size 2.) Strictly speaking, this is a *simple undirected* graph: in a non-simple graph, there may be two or more edges between a pair of vertices, while in a directed graph each edge is an ordered pair. Vertices γ and δ are said to be *joined by an edge* if $\{\gamma, \delta\}$ is an edge. The graph is *complete* if every 2-subset is an edge.

Example 1.4 suggests a second way of looking at association schemes. Imagine that all the edges in the Petersen graph are blue. For each pair of distinct vertices which are not joined, draw a red edge between them. We obtain a 2-colouring of the complete undirected graph K_{10} on ten vertices. Condition (iii) gives us a special property about the number of triangles of various types through an edge with a given colour.

Second definition of association scheme An *association scheme* with s associate classes on a finite set Ω is a colouring of the edges of the complete undirected graph with vertex-set Ω by s colours such that

(iii)′ for all i, j, k in $\{1, \ldots, s\}$ there is an integer p_{ij}^k such that, whenever $\{\alpha, \beta\}$ is an edge of colour k then

$$|\{\gamma \in \Omega : \{\alpha, \gamma\} \text{ has colour } i \text{ and } \{\gamma, \beta\} \text{ has colour } j\}| = p_{ij}^k;$$

(iv)′ every colour is used at least once;

(v)′ there are integers a_i for i in $\{1,\dots s\}$ such that each vertex is contained in exactly a_i edges of colour i.

The strange numbering of the conditions is to aid comparison with the previous definition. There is no need for an analogue of condition (i), because every edge consists of two distinct vertices, nor for an analogue of condition (ii), because we have specified that the graph be undirected. Condition (iii)′ says that if we fix different vertices α and β, and colours i and j, then the number of triangles which consist of the edge $\{\alpha,\beta\}$ and an i-coloured edge through α and a j-coloured edge through β is exactly p_{ij}^k, where k is the colour of $\{\alpha,\beta\}$, irrespective of the choice of α and β (see Figure 1.5). We did not need a condition (iv) in the partition definition because we specified that the subsets in the partition be non-empty. Finally, because condition (iii)′ does not deal with the analogue of the diagonal subset, we have to put in condition (v)′ explicitly. If we assume that none of the colours used for the edges is white, we can colour all the vertices white (colour 0). Then we recover exactly the parameters of the first kind, with $a_0 = 1$, $p_{ii}^0 = a_i$, $p_{i0}^i = p_{0i}^i = 1$, and $p_{ij}^0 = p_{i0}^j = p_{0j}^i = 0$ if $i \neq j$.

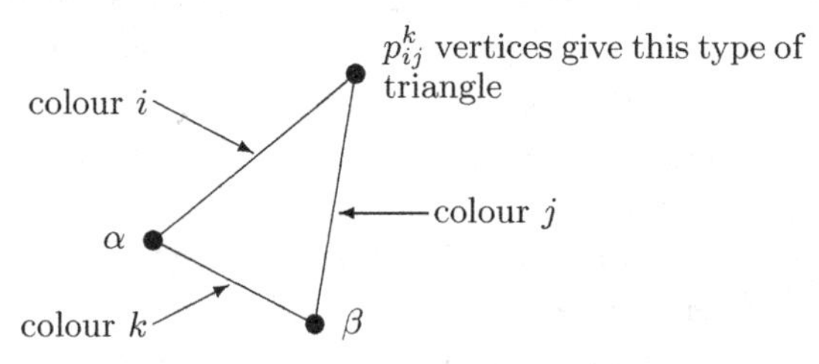

Fig. 1.5. Condition (iii)′

In fact, condition (v)′ is redundant. For suppose that α and β are different vertices and let k be the colour of edge $\{\alpha,\beta\}$. For each i and j, there are exactly p_{ij}^k triangles consisting of $\{\alpha,\beta\}$, an i-coloured edge through α and a j-coloured edge through β. Of course, the edge $\{\alpha,\beta\}$ is an *unordered* pair, so there are also p_{ij}^k triangles consisting of $\{\alpha,\beta\}$, an i-coloured edge through β and a j-coloured edge through α: that is, $p_{ij}^k = p_{ji}^k$. If $k \neq i$ then the number of i-coloured edges through α and the number of i-coloured edges through β are both equal to $\sum_{j=1}^s p_{ij}^k$, as in Lemma 1.1(ii); if $k = i$ then we have to include the edge $\{\alpha,\beta\}$ itself so the number is $\sum_{j=1}^s p_{kj}^k + 1$.

Lemma 1.2 *If $|\Omega|$ is odd then all the non-trivial valencies are even.*

Proof There are a_i edges of colour i through each vertex. If $i \neq 0$ then each of these edges contains two vertices, and so there are $|\Omega|\, a_i/2$ such edges altogether. ■

Example 1.5 Let Ω be the set of the eight vertices of the cube. Colour the edges of the cube yellow, the main diagonals red and the face diagonals black. In Figure 1.6 the yellow edges are shown by solid lines, the red edges by dashed lines, and the black edges are omitted. If you find this picture hard to follow, take any convenient cuboid box, draw black lines along the face diagonals and colour the edges yellow.

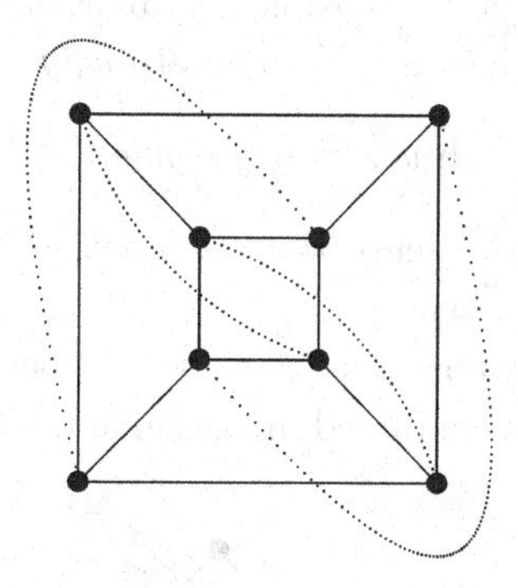

Fig. 1.6. The cube: solid lines are yellow, dashed ones are red

Every vertex is in three yellow edges, one red edge and three black ones, so $a_{\text{yellow}} = 3$, $a_{\text{red}} = 1$ and $a_{\text{black}} = 3$. The values p_{ij}^{yellow} are the entries in

$$\begin{array}{r} \\ \text{white} \\ \text{yellow} \\ \text{red} \\ \text{black} \end{array} \begin{array}{c} \begin{array}{cccc} \text{white} & \text{yellow} & \text{red} & \text{black} \end{array} \\ \left[\begin{array}{cccc} 0 & 1 & 0 & 0 \\ 1 & 0 & 0 & 2 \\ 0 & 0 & 0 & 1 \\ 0 & 2 & 1 & 0 \end{array}\right] \end{array},$$

the values p_{ij}^{red} are the entries in

$$\begin{array}{r} \\ \text{white} \\ \text{yellow} \\ \text{red} \\ \text{black} \end{array} \begin{array}{c} \begin{array}{cccc} \text{white} & \text{yellow} & \text{red} & \text{black} \end{array} \\ \left[\begin{array}{cccc} 0 & 0 & 1 & 0 \\ 0 & 0 & 0 & 3 \\ 1 & 0 & 0 & 0 \\ 0 & 3 & 0 & 0 \end{array}\right] \end{array}$$

and the values p_{ij}^{black} are the entries in

$$\begin{array}{r} \\ \text{white} \\ \text{yellow} \\ \text{red} \\ \text{black} \end{array} \begin{array}{c} \begin{array}{cccc} \text{white} & \text{yellow} & \text{red} & \text{black} \end{array} \\ \left[\begin{array}{cccc} 0 & 0 & 0 & 1 \\ 0 & 2 & 1 & 0 \\ 0 & 1 & 0 & 0 \\ 1 & 0 & 0 & 2 \end{array} \right] \end{array}.$$

Thus we have an association scheme with three associate classes. ■

If an association scheme has two associate classes, we can regard the two colours as 'visible' and 'invisible', as in Example 1.4. The graph formed by the visible edges is said to be *strongly regular*.

Definition A finite graph is *strongly regular* if

(a) it is *regular* in the sense that every vertex is contained in the same number of edges;
(b) every edge is contained in the same number of triangles;
(c) every non-edge is contained in the same number of configurations like

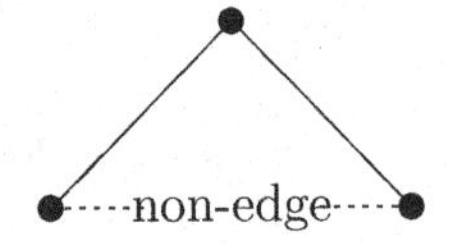

(d) it is neither complete (all pairs are edges) nor null (no pairs are edges).

1.3 Matrices

Given a field F, the set F^Ω of functions from Ω to F forms a vector space. Addition is defined pointwise: for f and g in F^Ω,

$$(f+g)(\omega) = f(\omega) + g(\omega) \qquad \text{for } \omega \text{ in } \Omega.$$

Scalar multiplication is also defined pointwise: for λ in F and f in F^Ω,

$$(\lambda f)(\omega) = \lambda\,(f(\omega)) \qquad \text{for } \omega \text{ in } \Omega.$$

For ω in Ω, let χ_ω be the characteristic function of ω; that is

$$\begin{aligned} \chi_\omega(\omega) &= 1 \\ \chi_\omega(\alpha) &= 0 \qquad \text{for } \alpha \text{ in } \Omega \text{ with } \alpha \neq \omega. \end{aligned}$$

Then $f = \sum_{\omega\in\Omega} f(\omega)\chi_\omega$ for all f in F^Ω, and so $\{\chi_\omega : \omega \in \Omega\}$ forms a natural basis of F^Ω: hence $\dim F^\Omega = |\Omega|$.

For any subset Δ of Ω it is convenient to define χ_Δ in F^Ω by

$$\chi_\Delta(\omega) = \begin{cases} 1 & \text{if } \omega \in \Delta \\ 0 & \text{if } \omega \notin \Delta, \end{cases}$$

so that $\chi_\Delta = \sum_{\omega\in\Delta} \chi_\omega$. Sometimes it is helpful to regard χ_Δ as a column vector with the rows indexed by the elements of Ω. There is a 1 in position ω if $\omega \in \Delta$; otherwise there is a zero. Thus the vector records whether the statement '$\omega \in \Delta$' is true.

This approach is also useful for matrices. Given two finite sets Γ and Δ, we can form their product

$$\Gamma \times \Delta = \{(\gamma, \delta) : \gamma \in \Gamma,\ \delta \in \Delta\},$$

which may be thought of as a rectangle, as in Figure 1.7. Applying the preceding ideas to $\Gamma \times \Delta$, we obtain the vector space $F^{\Gamma\times\Delta}$ of all functions from $\Gamma \times \Delta$ to F. It has dimension $|\Gamma| \times |\Delta|$. If M is such a function, we usually write $M(\gamma, \delta)$ rather than $M((\gamma, \delta))$. In fact, M is just a matrix, with its rows labelled by Γ and its columns labelled by Δ. So long as we retain this labelling, it does not matter in what order we write the rows and columns.

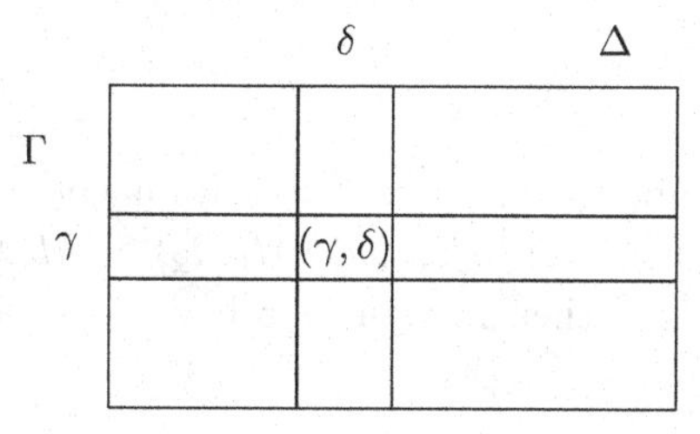

Fig. 1.7. The product set $\Gamma \times \Delta$

In fact, what most people regard as an $m \times n$ matrix over F is just a matrix in $F^{\Gamma\times\Delta}$ with $\Gamma = \{1, \ldots, m\}$ and $\Delta = \{1, \ldots, n\}$. The usual convention is that row 1 appears first, etc., so that order *does* matter but labelling is not needed. Here I use the opposite convention (order does not matter, but labelling is needed) because usually the elements of Γ and Δ are not integers.

Some examples of matrices with such labelled rows and columns were given in Example 1.5.

If $M \in F^{\Gamma\times\Delta}$ and $\Gamma = \Delta$ then we say that M is *square*. This is stronger than merely having the same number of rows as of columns. If we need to emphasize the set, we say that M is square on Δ.

The *transpose* of M is the matrix M' in $F^{\Delta\times\Gamma}$ defined by

$$M'(\delta,\gamma) = M(\gamma,\delta) \qquad \text{for } \delta \text{ in } \Delta \text{ and } \gamma \text{ in } \Gamma.$$

The matrix M in $F^{\Delta\times\Delta}$ is *symmetric* if $M' = M$. There are three special symmetric matrices in $F^{\Delta\times\Delta}$:

$$\begin{aligned} I_\Delta(\delta_1,\delta_2) &= \begin{cases} 1 & \text{if } \delta_1 = \delta_2 \\ 0 & \text{otherwise;} \end{cases} \\ J_\Delta(\delta_1,\delta_2) &= 1 \qquad \text{for all } \delta_1,\ \delta_2 \text{ in } \Delta; \\ O_\Delta(\delta_1,\delta_2) &= 0 \qquad \text{for all } \delta_1,\ \delta_2 \text{ in } \Delta. \end{aligned}$$

Moreover, if $f \in F^\Delta$ we define the *diagonal* symmetric matrix $\mathrm{diag}(f)$ in $F^{\Delta\times\Delta}$ by

$$\mathrm{diag}(f)(\delta_1,\delta_2) = \begin{cases} f(\delta_1) & \text{if } \delta_1 = \delta_2 \\ 0 & \text{otherwise.} \end{cases}$$

Matrix multiplication is possible when the labelling sets are compatible. If $M_1 \in F^{\Gamma\times\Delta}$ and $M_2 \in F^{\Delta\times\Phi}$ then M_1M_2 is the matrix in $F^{\Gamma\times\Phi}$ defined by

$$(M_1M_2)(\gamma,\phi) = \sum_{\delta\in\Delta} M_1(\gamma,\delta)M_2(\delta,\phi).$$

All the usual results about matrix multiplication hold. In particular, matrix multiplication is associative, and $(M_1M_2)' = M_2'M_1'$.

Similarly, if $M \in F^{\Gamma\times\Delta}$ then M defines a linear transformation from F^Δ to F^Γ by

$$f \mapsto Mf$$

where

$$(Mf)(\gamma) = \sum_{\delta\in\Delta} M(\gamma,\delta)f(\delta) \qquad \text{for } \gamma \in \Gamma.$$

If Φ is any subset of $\Gamma\times\Delta$ then its characteristic function χ_Φ satisfies

$$\chi_\Phi(\gamma,\delta) = \begin{cases} 1 & \text{if } (\gamma,\delta) \in \Phi \\ 0 & \text{otherwise.} \end{cases}$$

In the special case that $\Gamma = \Delta = \Omega$ we call χ_Φ the *adjacency matrix* of Φ and write it A_Φ. In particular,

$$A_{\Omega\times\Omega} = J_\Omega$$

$$\begin{aligned} A_{\varnothing} &= O_{\Omega} \\ A_{\mathrm{Diag}(\Omega)} &= I_{\Omega}. \end{aligned}$$

For an association scheme with classes $\mathcal{C}_0$, $\mathcal{C}_1$, ..., $\mathcal{C}_s$, we write A_i for the adjacency matrix $A_{\mathcal{C}_i}$. Thus the (α, β)-entry of A_i is equal to 1 if α and β are i-th associates; otherwise it is equal to 0.

Condition (iii) in the definition of association scheme has a particularly nice consequence for multiplication of the adjacency matrices: it says that

$$A_i A_j = \sum_{k=0}^{s} p_{ij}^k A_k. \tag{1.1}$$

For suppose that $(\alpha, \beta) \in \mathcal{C}_k$. Then the (α, β)-entry of the right-hand side of Equation (1.1) is equal to p_{ij}^k, while the (α, β)-entry of the left-hand side is equal to

$$\begin{aligned} (A_i A_j)(\alpha, \beta) &= \sum_{\gamma \in \Omega} A_i(\alpha, \gamma) A_j(\gamma, \beta) \\ &= |\{\gamma : (\alpha, \gamma) \in \mathcal{C}_i \text{ and } (\gamma, \beta) \in \mathcal{C}_j\}| \\ &= p_{ij}^k \end{aligned}$$

because the product $A_i(\alpha, \gamma) A_j(\gamma, \beta)$ is zero unless $(\alpha, \gamma) \in \mathcal{C}_i$ and $(\gamma, \beta) \in \mathcal{C}_j$, in which case it is 1.

We have just used a useful technique for multiplying matrices all of whose entries are 0 or 1; we shall use it again.

Technique 1.1 Suppose that

$$M_i(\alpha, \beta) = \begin{cases} 1 & \text{if } (\alpha, \beta) \text{ satisfies condition } (i) \\ 0 & \text{otherwise} \end{cases}$$

for $i = 1$, ..., n. Then the (α, β) entry of the product $M_1 M_2 \cdots M_n$ is equal to the number of $(n-1)$-tuples $(\gamma_1, \ldots, \gamma_{n-1})$ such that (α, γ_1) satisfies condition (1), (γ_1, γ_2) satisfies condition (2), ... and (γ_{n-1}, β) satisfies condition (n).

Equation (1.1) leads to a definition of association schemes in terms of the adjacency matrices.

Third definition of association scheme An *association scheme* with s associate classes on a finite set Ω is a set of matrices A_0, A_1, ..., A_s in $\mathbb{R}^{\Omega \times \Omega}$, all of whose entries are equal to 0 or 1, such that

(i)″ $A_0 = I_\Omega$;

(ii)″ A_i is symmetric for $i = 1, \ldots, s$;

(iii)″ for all i, j in $\{1, \ldots, s\}$, the product A_iA_j is a linear combination of A_0, A_1, …, A_s;

(iv)″ none of the A_i is equal to O_Ω, and $\sum_{i=0}^{s} A_i = J_\Omega$.

Notice that we do not need to specify that the coefficients in (iii)″ be integers: *every* entry in A_iA_j is a non-negative integer if all the entries in A_i and A_j are 0 or 1. Condition (iv)″ is the analogue of the sets $\mathcal{C}_0$, …, $\mathcal{C}_s$ forming a partition of $\Omega \times \Omega$.

Since A_i is symmetric with entries in $\{0, 1\}$, the diagonal entries of A_i^2 are the row-sums of A_i. Condition (iii)″ implies that A_i^2 has a constant element, say a_i, on its diagonal. Therefore every row and every column of A_i contains a_i entries equal to 1. Hence $A_iJ_\Omega = J_\Omega A_i = a_iJ_\Omega$. Moreover, $A_0A_i = A_iA_0 = A_i$. Thus condition (iii)″ can be checked, for each i in $\{1, \ldots, s\}$, by verifying that A_i has constant row-sums and that, for all but one value of j in $\{1, \ldots, s\}$, the product A_iA_j is a linear combination of A_0, …, A_s. In fact, the first check is superfluous, because it is a byproduct of checking A_i^2.

Example 1.6 Let Π be a Latin square of size n: an $n \times n$ array filled with n letters in such a way that each letter occurs once in each row and once in each column. A Latin square of size 4 is shown in Figure 1.8(a).

Let Ω be the set of n^2 cells in the array. For α, β in Ω with $\alpha \neq \beta$ let α and β be first associates if α and β are in the same row or are in the same column or have the same letter, and let α and β be second associates otherwise. Figure 1.8(b) shows one element α and marks all its first associates as β.

A	B	C	D
D	A	B	C
C	D	A	B
B	C	D	A

(a) A Latin square

β		β	
β			β
α	β	β	β
β	β		

(b) An element α and all its first associates β

Fig. 1.8. A association scheme of Latin-square type

We need to check that all the nine products A_iA_j are linear combinations of A_0, A_1 and A_2. Five products involve A_0, which is I, so there is nothing to check. (Here and elsewhere we omit the suffix from I, J etc. when the set involved is clear from the context.) I claim that only the product A_1^2 needs to be checked.

	A_0	A_1	A_2
A_0	$\surd$	$\surd$	$\surd$
A_1	$\surd$	?	
A_2	$\surd$		

To check A_1^2, we need a simple expression for A_1. Let R be the adjacency matrix of the subset

$$\{(\alpha,\beta) : \alpha \text{ and } \beta \text{ are in the same row}\},$$

and let C and L be the adjacency matrices of the similarly-defined subsets for columns and letters. Then

$$A_1 = R + C + L - 3I,$$

because the elements of $\mathrm{Diag}(\Omega)$ are counted in each of R, C, and L and need to be removed. Moreover, $A_2 = J - A_1 - I$. These adjacency matrices have constant row-sums, namely $a_1 = 3(n-1)$ and $a_2 = n^2 - 3(n-1) - 1 = (n-2)(n-1)$.

Using Technique 1.1, we see that

$$\begin{aligned} R^2(\alpha,\beta) &= \sum_\gamma R(\alpha,\gamma)R(\gamma,\beta) \\ &= |\{\gamma : \gamma \text{ is the same row as } \alpha \text{ and } \beta\}| \\ &= \begin{cases} n & \text{if } \alpha \text{ and } \beta \text{ are in the same row} \\ 0 & \text{otherwise} \end{cases} \\ &= nR(\alpha,\beta) \end{aligned}$$

so $R^2 = nR$. Similarly $C^2 = nC$ and $L^2 = nL$. Also

$$\begin{aligned} RC(\alpha,\beta) &= \sum_\gamma R(\alpha,\gamma)C(\gamma,\beta) \\ &= \left|\left\{\begin{array}{l} \gamma : \gamma \text{ is in the same row as } \alpha \\ \quad \text{and the same column as } \beta \end{array}\right\}\right| \\ &= 1 \end{aligned}$$

so $RC = J$. Similarly $CR = RL = LR = CL = LC = J$.

Hence $A_1^2 = n(R+C+L) + 6J - 6(R+C+L) + 9I$, which is a linear combination of A_1, J and I, hence of A_1, A_2 and A_0.

Now let us verify my claim that no further products need to be evaluated. We do not need to check A_1A_2, because

$$A_1A_2 = A_1(J - A_1 - I) = a_1J - A_1^2 - A_1.$$

Neither do we need to check A_2A_1, because

$$A_2A_1 = (J - A_1 - I)A_1 = a_1J - A_1^2 - A_1.$$

Finally, we do not need to check A_2^2 either, because

$$A_2^2 = A_2(J - A_1 - I) = a_2J - A_2A_1 - A_2.$$

This association scheme is said to be of *Latin-square type* $\mathrm{L}(3, n)$. ■

It is no accident that $A_1A_2 = A_2A_1$ in Example 1.6.

Lemma 1.3 *If A_0, A_1, …, A_s are the adjacency matrices of an association scheme then $A_iA_j = A_jA_i$ for all i, j in $\{0, 1, \ldots, s\}$.*

Proof

$$\begin{aligned} A_jA_i &= A_j'A_i', && \text{because the adjacency matrices are symmetric,} \\ &= (A_iA_j)' \\ &= \left(\sum_k p_{ij}^k A_k\right)', && \text{by Equation (1.1),} \\ &= \sum_k p_{ij}^k A_k' \\ &= \sum_k p_{ij}^k A_k, && \text{because the adjacency matrices are symmetric,} \\ &= A_iA_j. \quad ■ \end{aligned}$$

As we saw in Example 1.6, there is very little to check when there are only two associate classes.

Lemma 1.4 *Let A be a symmetric matrix in $\mathbb{R}^{\Omega\times\Omega}$ with zeros on the diagonal and all entries in $\{0, 1\}$. Suppose that $A \neq O$ and $A \neq J - I$. Then $\{I, A, J - A - I\}$ is an association scheme on Ω if and only if A^2 is a linear combination of I, A and J.*

Consideration of the α-th row of A_iA_j sheds new light on the graph way of looking at association schemes. Stand at the vertex α. Take a step along an edge coloured i (if $i = 0$ this means stay still). Then take a step along an edge coloured j. Where can you get to? If β is joined to α by a k-coloured edge, then you can get to β in this way if $p_{ij}^k \neq 0$. In fact, there are exactly p_{ij}^k such two-step ways of getting to β from α.

1.4 Some special association schemes

We have already met the trivial, group-divisible and rectangular association schemes and those of Latin-square type L(3, n).

1.4.1 Triangular association schemes

Let Ω consist of all 2-subsets from an n-set Γ, so that $|\Omega| = n(n-1)/2$. For α in Ω, put

$$\begin{aligned} \mathcal{C}_1(\alpha) &= \{\beta \in \Omega : |\alpha \cap \beta| = 1\} \\ \mathcal{C}_2(\alpha) &= \{\beta \in \Omega : \alpha \cap \beta = \varnothing\}. \end{aligned}$$

Then $a_1 = 2(n-2)$ and

$$a_2 = {}^{n-2}\mathrm{C}_2 = (n-2)(n-3)/2.$$

Let ω, η, ζ and θ be distinct elements of Γ. If $\alpha = \{\omega, \eta\}$ and $\beta = \{\omega, \zeta\}$ then

$$\mathcal{C}_1(\alpha) \cap \mathcal{C}_1(\beta) = \{\{\omega, \gamma\} : \gamma \in \Gamma,\ \gamma \neq \omega,\ \eta,\ \zeta\} \cup \{\{\eta, \zeta\}\},$$

which has size $(n-3)+1 = n-2$. On the other hand, if $\alpha = \{\omega, \eta\}$ and $\beta = \{\zeta, \theta\}$ then

$$\mathcal{C}_1(\alpha) \cap \mathcal{C}_1(\beta) = \{\{\omega, \zeta\}, \{\omega, \theta\}, \{\eta, \zeta\}, \{\eta, \theta\}\},$$

which has size 4.

This is called the *triangular* association scheme T(n).

The labelling of the vertices of the Petersen graph in Figure 1.9 shows

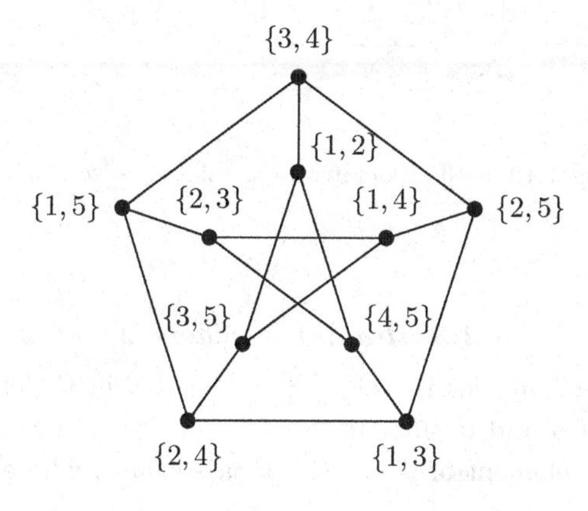

Fig. 1.9. The Petersen graph as T(5).

that the association scheme in Example 1.4 is T(5) (with the names of the classes interchanged – this does not matter).

1.4.2 Johnson schemes

More generally, let Ω consist of all m-subsets of an n-set Γ, where $1 \leqslant m \leqslant n/2$. For $i = 0, 1, \ldots, m$, let α and β be i-th associates if

$$|\alpha \cap \beta| = m - i,$$

so that smaller values of i give as i-th associates subsets which have larger overlap.

It is clear that each element has a_i i-th associates, where

$$a_i = {}^m\mathrm{C}_{m-i} \times {}^{n-m}\mathrm{C}_i$$

(see Figure 1.10).

We shall show later that this is an association scheme. It is called the *Johnson* scheme $\mathrm{J}(n,m)$. In particular, $\mathrm{J}(n,1) = \underline{\underline{n}}$ and $\mathrm{J}(n,2) = \mathrm{T}(n)$. Because it generalizes the triangular scheme, it is also denoted $\mathrm{T}_m(n)$ and called a scheme of *triangular type*.

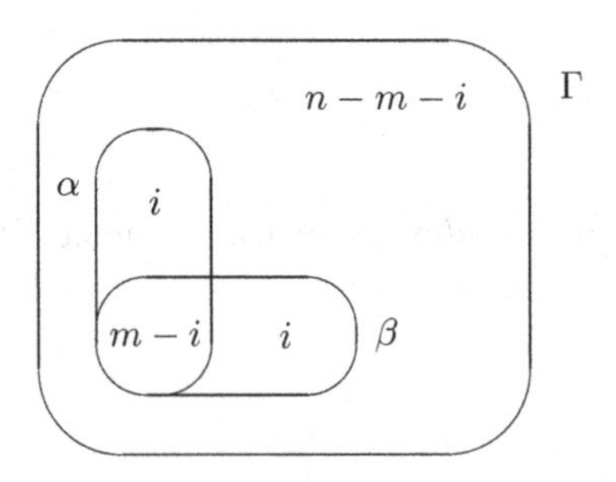

Fig. 1.10. i-th associates in the Johnson scheme

1.4.3 Hamming schemes

Let Γ be an n-set and let $\Omega = \Gamma^m$. For α and β in Ω, let α and β be i-th associates if α and β differ in exactly i positions, where $0 \leqslant i \leqslant m$. Evidently, every element of Ω has a_i i-th associates, where

$$a_i = {}^m\mathrm{C}_i \times (n-1)^i.$$

We shall show later that this is an association scheme. It is called the *Hamming* scheme $\mathrm{H}(m,n)$, because the *Hamming distance* between α and β is defined to be the number of positions in which they differ.

The cube scheme in Example 1.5 is $\mathrm{H}(3,2)$, as the labelling in Figure 1.11 shows. Thus $\mathrm{H}(m,n)$ is also called the *hypercubic* scheme.

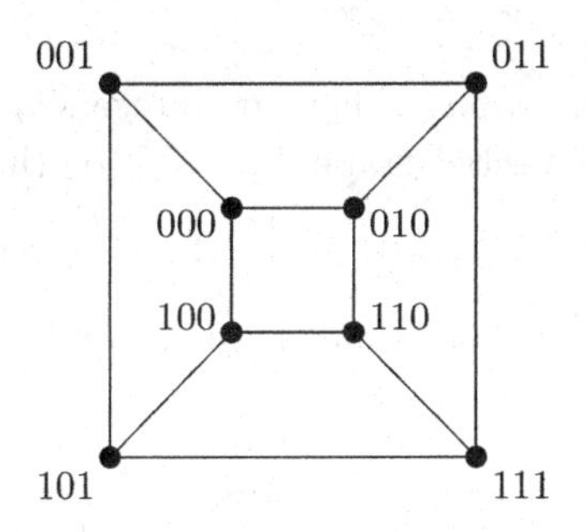

Fig. 1.11. The cube labelled to show it as $\mathrm{H}(3,2)$

1.4.4 Distance-regular graphs

Recall that a *path of length* l between vertices α and β of a graph is a sequence of l edges e_1, e_2, ..., e_l such that $\alpha \in e_1$, $\beta \in e_l$ and $e_i \cap e_{i+1} \neq \varnothing$ for $i = 1, \ldots, l-1$. The graph is *connected* if every pair of vertices is joined by a path.

The only strongly regular graphs which are not connected are disjoint unions of complete graphs of the same size. The graph formed by the red edges of the cube in Example 1.5 is such a graph; so is the graph formed by the black edges of the cube.

In a connected graph, the *distance* between two vertices is the length of a shortest path joining them. In particular, if $\{\alpha,\beta\}$ is an edge then the distance between α and β is 1, while the distance from α to itself is 0. The *diameter* of a connected graph is the maximum distance between any pair of its vertices.

All connected strongly regular graphs have diameter 2. The yellow edges of the cube form a connected graph of diameter 3.

In any connected graph $\mathcal{G}$ it is natural to define subsets $\mathcal{G}_i$ of its vertex set Ω by

$$\mathcal{G}_i = \{(\alpha,\beta) \in \Omega \times \Omega : \text{the distance between } \alpha \text{ and } \beta \text{ is } i\},$$

so that

$$\mathcal{G}_i(\alpha) = \{\beta \in \Omega : \text{the distance between } \alpha \text{ and } \beta \text{ is } i\}.$$

Definition Let $\mathcal{G}$ be a connected graph with diameter s and vertex set Ω. Then $\mathcal{G}$ is *distance-regular* if $\mathcal{G}_0$, $\mathcal{G}_1$, ..., $\mathcal{G}_s$ form an association scheme on Ω.

In a distance-regular graph $\mathcal{G}$, there are integers a_i such that $|\mathcal{G}_i(\alpha)| = a_i$ for all α. In any connected graph, if $\beta \in \mathcal{G}_i(\alpha)$ then

$$\mathcal{G}_1(\beta) \subseteq \mathcal{G}_{i-1}(\alpha) \cup \mathcal{G}_i(\alpha) \cup \mathcal{G}_{i+1}(\alpha);$$

see Figure 1.12. Hence in a distance-regular graph $p^i_{j1} = 0$ if $i \neq 0$ and

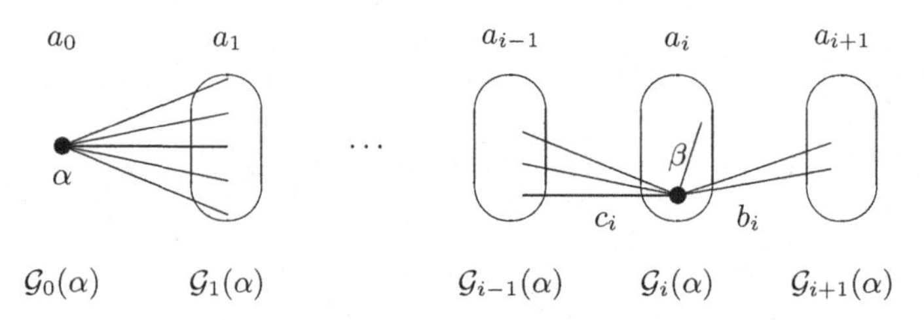

Fig. 1.12. Edges through α and β in a distance-regular graph

$j \notin \{i-1, i, i+1\}$. Put $c_i = p^i_{i-1,1}$ and $b_i = p^i_{i+1,1}$. It is interesting that the constancy of the a_i, b_i and c_i is enough to guarantee that the graph is distance-regular.

Theorem 1.5 *Let Ω be the vertex set of a connected graph $\mathcal{G}$ of diameter s. If there are integers a_i, b_i, c_i for $0 \leqslant i \leqslant s$ such that, for all α in Ω and all i in $\{0, \dots, s\}$,*

(a) $|\mathcal{G}_i(\alpha)| = a_i$ (so a_0 must be 1);
(b) if $\beta \in \mathcal{G}_i(\alpha)$ then $|\mathcal{G}_{i-1}(\alpha) \cap \mathcal{G}_1(\beta)| = c_i$ and $|\mathcal{G}_{i+1}(\alpha) \cap \mathcal{G}_1(\beta)| = b_i$ (so b_s and c_0 must be 0 and $b_0 = a_1$)

then $\mathcal{G}$ is distance-regular.

Proof It is clear that conditions (i) (diagonal subset) and (ii) (symmetry) in the definition of association scheme are satisfied. It remains to prove that A_iA_j is a linear combination of the A_k, where A_i is the adjacency matrix of the subset $\mathcal{G}_i$.

Let $i \in \{1, \dots, s-1\}$. Then

- every element in $\mathcal{G}_{i-1}(\alpha)$ is joined to exactly b_{i-1} elements of $\mathcal{G}_i(\alpha)$,

- every element in $\mathcal{G}_i(\alpha)$ is joined to exactly $a_1 - b_i - c_i$ elements of $\mathcal{G}_i(\alpha)$, and
- every element in $\mathcal{G}_{i+1}(\alpha)$ is joined to exactly c_{i+1} elements of $\mathcal{G}_i(\alpha)$.

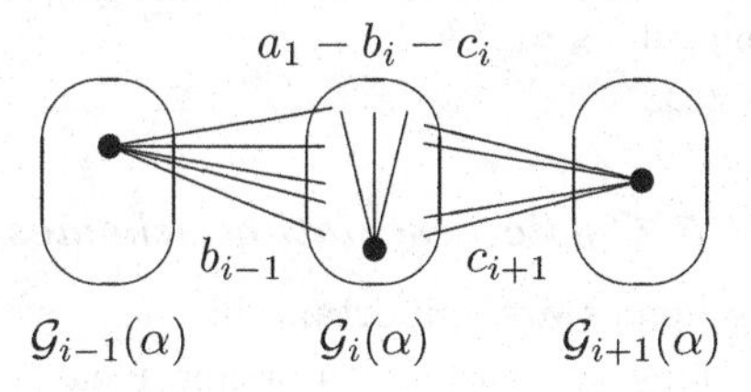

Technique 1.1 shows that

$$A_iA_1 = b_{i-1}A_{i-1} + (a_1 - b_i - c_i)A_i + c_{i+1}A_{i+1}.$$

Using the facts that $A_0 = I$ and all the c_{i+1} are non-zero (because the graph is connected), induction shows that, for $1 \leqslant i \leqslant s$,

- A_i is a polynomial in A_1 of degree i, and
- A_1^i is a linear combination of I, A_1, ..., A_i.

Hence if $i, j \in \{0, \ldots, s\}$ then A_iA_j is also a polynomial in A_1.

Applying a similar argument to $\mathcal{G}_s(\alpha)$ gives

$$A_sA_1 = b_{s-1}A_{s-1} + (a_1 - c_s)A_s$$

so A_1 satisfies a polynomial of degree $s+1$. Thus each product A_iA_j can be expressed as a polynomial in A_1 of degree at most s, so it is a linear combination of A_0, A_1, ..., A_s. Hence condition (iii) is satisfied. ■

Example 1.5 revisited The yellow edges of the cube form a distance-regular graph with diameter 3 and the following parameters:

$$\begin{array}{llll} a_0 = 1 & a_1 = 3 & a_2 = 3 & a_3 = 1 \\ b_0 = 3 & b_1 = 2 & b_2 = 1 & b_3 = 0 \\ c_0 = 0 & c_1 = 1 & c_2 = 2 & c_3 = 3. \end{array}$$ ■

In the Johnson scheme, draw an edge between α and ω if $|\alpha \cap \omega| = m - 1$. Now suppose that $\beta \in \mathcal{C}_i(\alpha)$, so that $|\alpha \cap \beta| = m - i$, as in Figure 1.10. If $\eta \in \mathcal{C}_{i-1}(\alpha) \cap \mathcal{C}_1(\beta)$ then η is obtained from β by replacing one of the i elements in $\beta \setminus \alpha$ by one of the i elements in $\alpha \setminus \beta$, so β is joined to i^2 elements in $\mathcal{C}_{i-1}(\alpha)$. Similarly, if $\theta \in \mathcal{C}_{i+1}(\alpha) \cap \mathcal{C}_1(\beta)$ then θ is obtained from β by replacing one of the $m-i$ elements in $\alpha \cap \beta$ by one of the $n-m-i$ elements in $\Omega \setminus (\alpha \cup \beta)$, so β is joined to $(m-i)(n-m-i)$ elements in $\mathcal{C}_{i+1}(\alpha)$. Hence $\mathcal{C}_1$ defines a distance-regular graph, and so the Johnson scheme is indeed an association scheme.

Similarly, $\mathcal{C}_1$ in the Hamming scheme defines a distance-regular graph, so this is also a genuine association scheme.

(Warning: literature devoted to distance-regular graphs usually has a_i to denote what I am calling $a_1 - b_i - c_i$.)

1.4.5 Cyclic association schemes

Cyclic association schemes are important in their own right, but this subsection also develops two other important ideas. One idea is the expression of the adjacency matrices as linear combinations of other, simpler, matrices whose products we already know, as in Example 1.6. The second is the introduction of algebras, which are central to the theory of association schemes.

Let Ω be $\mathbb{Z}_n$, the integers modulo n, considered as a group under addition.

(The reader who is familiar with group theory will realise that, throughout this subsection, the integers modulo n may be replaced by any finite Abelian group, so long as some of the detailed statements are suitably modified. This extension to general finite Abelian groups will be given in Chapter 8.)

We can define multiplication on F^Ω by

$$\chi_\alpha \chi_\beta = \chi_{\alpha+\beta} \qquad \text{for } \alpha \text{ and } \beta \text{ in } \Omega,$$

extended to the whole of F^Ω by

$$\left(\sum_{\alpha\in\Omega} \lambda_\alpha \chi_\alpha\right)\left(\sum_{\beta\in\Omega} \mu_\beta \chi_\beta\right) = \sum_{\alpha\in\Omega}\sum_{\beta\in\Omega} \lambda_\alpha \mu_\beta \chi_{\alpha+\beta}.$$

Thus, if f and g are in F^Ω then

$$(fg)(\omega) = \sum_{\alpha\in\Omega} f(\alpha) g(\omega - \alpha)$$

so this multiplication is sometimes called *convolution*. It is associative (because addition in $\mathbb{Z}_n$ is), is distributive over vector addition, and commutes with scalar multiplication, so it turns F^Ω into an *algebra*, called the *group algebra* of $\mathbb{Z}_n$, written $F\mathbb{Z}_n$.

Now let

$$\mathcal{D}_\omega = \{(\alpha, \beta) \in \Omega \times \Omega : \beta - \alpha = \omega\}$$

and let $M_\omega = A_{\mathcal{D}_\omega}$, so that the (α, β)-entry of M_ω is equal to 1 if $\alpha = \beta - \omega$ and to 0 otherwise. Technique 1.1 shows that

$$M_\gamma M_\delta = M_{\gamma+\delta}$$

for γ and δ in Ω, and

$$M_\omega \chi_\gamma = \chi_{\gamma-\omega}$$

for ω and γ in Ω. Matrices of the form $\sum_{\omega\in\Omega} \lambda_\omega M_\omega$ for λ_ω in F are called *circulant* matrices.

Example 1.7 For $\mathbb{Z}_5$,

$$M_4 = \begin{array}{c} \\ 0 \\ 1 \\ 2 \\ 3 \\ 4 \end{array} \begin{array}{c} \begin{array}{ccccc} 0 & 1 & 2 & 3 & 4 \end{array} \\ \left[\begin{array}{ccccc} 0 & 0 & 0 & 0 & 1 \\ 1 & 0 & 0 & 0 & 0 \\ 0 & 1 & 0 & 0 & 0 \\ 0 & 0 & 1 & 0 & 0 \\ 0 & 0 & 0 & 1 & 0 \end{array}\right] \end{array}$$

and $M_1 = M_4'$. Circulant matrices are patterned in diagonal stripes. The general circulant matrix for $\mathbb{Z}_5$ is

$$\begin{array}{c} \\ 0 \\ 1 \\ 2 \\ 3 \\ 4 \end{array} \begin{array}{c} \begin{array}{ccccc} 0 & 1 & 2 & 3 & 4 \end{array} \\ \left[\begin{array}{ccccc} \lambda & \mu & \nu & \rho & \sigma \\ \sigma & \lambda & \mu & \nu & \rho \\ \rho & \sigma & \lambda & \mu & \nu \\ \nu & \rho & \sigma & \lambda & \mu \\ \mu & \nu & \rho & \sigma & \lambda \end{array}\right] \end{array}. \quad \blacksquare$$

Let $\mathcal{M}$ be the set of circulant matrices. It forms a vector space over F under matrix addition and scalar multiplication. It is also closed under matrix multiplication, which is associative, is distributive over matrix addition, and commutes with scalar multiplication. So $\mathcal{M}$ is also an algebra. The map

$$\varphi \colon F\mathbb{Z}_n \to \mathcal{M}$$

defined by

$$\varphi\left(\sum_{\omega\in\Omega} \lambda_\omega \chi_\omega\right) = \sum_{\omega\in\Omega} \lambda_\omega M_\omega$$

is a bijection which preserves the three operations (addition, scalar multiplication, multiplication) so it is an *algebra isomorphism*.

Notation If $\Delta \subseteq \mathbb{Z}_n$, write $-\Delta$ for $\{-\delta : \delta \in \Delta\}$ and, if $\gamma \in \mathbb{Z}_n$, write $\gamma + \Delta$ for $\{\gamma + \delta : \delta \in \Delta\}$.

Definition A partition of $\mathbb{Z}_n$ into sets Δ_0, Δ_1, ..., Δ_s is a *blueprint* for $\mathbb{Z}_n$ if

(i) $\Delta_0 = \{0\}$;
(ii) for $i = 1, \ldots, s$, if $\omega \in \Delta_i$ then $-\omega \in \Delta_i$ (that is, $\Delta_i = -\Delta_i$);
(iii) there are integers q_{ij}^k such that if $\beta \in \Delta_k$ then there are precisely q_{ij}^k elements α in Δ_i such that $\beta - \alpha \in \Delta_j$.

Condition (iii) says that there are exactly q_{ij}^k ordered pairs in $\Delta_i \times \Delta_j$ the sum of whose elements is equal to any given element in Δ_k. It implies that

$$\chi_{\Delta_i}\chi_{\Delta_j} = \sum_k q_{ij}^k \chi_{\Delta_k}. \tag{1.2}$$

Note that

$$\chi_{\Delta_0} = \sum_{\omega \in \Delta_0} \chi_\omega = \chi_0;$$

the first '0' is an element of the labelling set $\{0, 1, \ldots, s\}$ and the final '0' is the zero element of $\mathbb{Z}_n$.

Suppose that Δ_0, Δ_1, ..., Δ_s do form a blueprint for $\mathbb{Z}_n$. Put $\mathcal{C}_i = \bigcup_{\omega \in \Delta_i} \mathcal{D}_\omega$, so that

$$\mathcal{C}_i(\alpha) = \{\beta \in \Omega : \beta - \alpha \in \Delta_i\} = \alpha + \Delta_i = \alpha - \Delta_i$$

by condition (ii). In particular, $|\mathcal{C}_i(\alpha)| = |\Delta_i|$. Moreover, each $\mathcal{C}_i$ is symmetric, and $\mathcal{C}_0 = \mathrm{Diag}(\Omega)$. Now the adjacency matrix A_i of $\mathcal{C}_i$ is given by

$$A_i = \sum_{\omega \in \Delta_i} M_\omega = \sum_{\omega \in \Delta_i} \varphi(\chi_\omega) = \varphi\left(\sum_{\omega \in \Delta_i} \chi_\omega\right) = \varphi(\chi_{\Delta_i}),$$

so Equation (1.2) shows that

$$\begin{aligned} A_i A_j &= \varphi(\chi_{\Delta_i})\varphi(\chi_{\Delta_j}) = \varphi(\chi_{\Delta_i}\chi_{\Delta_j}) \\ &= \varphi(\sum_k q_{ij}^k \chi_{\Delta_k}) = \sum_k q_{ij}^k \varphi(\chi_{\Delta_k}) \\ &= \sum_k q_{ij}^k A_k, \end{aligned}$$

and hence the $\mathcal{C}_i$ form an association scheme on Ω. It is called a *cyclic* association scheme.

Thus a partition of the smaller set Ω with the right properties (a blueprint) leads to a partition of the larger set $\Omega \times \Omega$ with the right properties (an association scheme). The former is much easier to check.

Example 1.8 In $\mathbb{Z}_{13}$, put $\Delta_0 = \{0\}$, $\Delta_1 = \{1, 3, 4, -4, -3, -1\}$ and $\Delta_2 = \Omega \setminus \Delta_0 \setminus \Delta_1$. We calculate the sums $\alpha + \beta$ for α, $\beta \in \Delta_1$.

	1	3	4	−4	−3	−1
1	2	4	5	−3	−2	0
3	4	6	−6	−1	0	2
4	5	−6	−5	0	1	3
−4	−3	−1	0	5	6	−5
−3	−2	0	1	6	−6	−4
−1	0	2	3	−5	−4	−2

In the body of the table

$$\begin{array}{rr} 0 & \text{occurs} \quad 6 \text{ times,} \\ 1,\ 3,\ 4,\ -4,\ -3,\ -1 & \text{each occur } 2 \text{ times,} \\ 2,\ 5,\ 6,\ -6,\ -5,\ -2 & \text{each occur } 3 \text{ times,} \end{array}$$

so $\chi_{\Delta_1}\chi_{\Delta_1} = 6\chi_{\Delta_0} + 2\chi_{\Delta_1} + 3\chi_{\Delta_2}$.

As usual for $s = 2$, there is nothing more to check, for

$$\begin{aligned} \chi_{\Delta_1}\chi_{\Delta_2} &= \chi_{\Delta_1}(\chi_\Omega - \chi_0 - \chi_{\Delta_1}) \\ &= 6\chi_\Omega - \chi_{\Delta_1} - \chi_{\Delta_1}\chi_{\Delta_1} \\ &= 6(\chi_0 + \chi_{\Delta_1} + \chi_{\Delta_2}) - \chi_{\Delta_1} - (6\chi_0 + 2\chi_{\Delta_1} + 3\chi_{\Delta_2}) \\ &= 3\chi_{\Delta_1} + 3\chi_{\Delta_2} \end{aligned}$$

and

$$\begin{aligned} \chi_{\Delta_2}\chi_{\Delta_2} &= \chi_{\Delta_2}(\chi_\Omega - \chi_0 - \chi_{\Delta_1}) \\ &= 6\chi_\Omega - \chi_{\Delta_2} - \chi_{\Delta_1}\chi_{\Delta_2} \\ &= 6(\chi_0 + \chi_{\Delta_1} + \chi_{\Delta_2}) - \chi_{\Delta_2} - (3\chi_{\Delta_1} + 3\chi_{\Delta_2}) \\ &= 6\chi_0 + 3\chi_{\Delta_1} + 2\chi_{\Delta_2}. \end{aligned}$$

So Δ_0, Δ_1, Δ_2 form a blueprint for $\mathbb{Z}_{13}$. ■

For any n, the sets $\{0\}$, $\{\pm 1\}$, $\{\pm 2\}$, ... form a blueprint for $\mathbb{Z}_n$. The corresponding cyclic association scheme is the same as the one derived from the circuit C_n with n vertices, which is a distance-regular graph. Call this association scheme $\textcircled{n}$.

Exercises

1.1 Explain why the parameters of an association scheme satisfy

(a) $p_{0j}^{j} = 1$;
(b) $p_{ij}^{k} = p_{ji}^{k}$.

1.2 Let $\mathcal{C}_i$ and $\mathcal{C}_j$ be associate classes in an association scheme on Ω. Suppose that $(\alpha, \beta) \in \mathcal{C}_i$ and $(\beta, \gamma) \in \mathcal{C}_j$. Prove that there is a point δ such that $(\alpha, \delta) \in \mathcal{C}_j$ and $(\delta, \gamma) \in \mathcal{C}_i$.

1.3 Let Ω consist of the integers mod 6, shown round a circle below.

Define

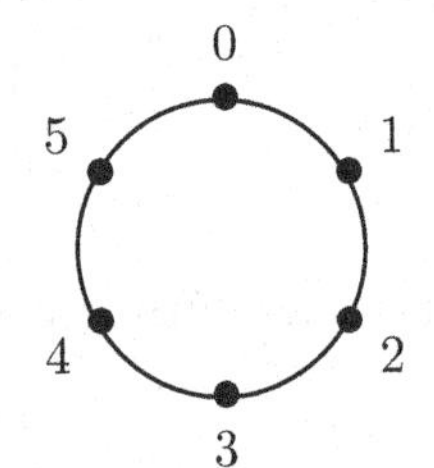

$$\begin{aligned}
\mathcal{C}_0 &= \mathrm{Diag}(\Omega) \\
\mathcal{C}_1 &= \{(\alpha, \beta) \in \Omega \times \Omega : \alpha - \beta = \pm 1\} \\
\mathcal{C}_2 &= \{(\alpha, \beta) \in \Omega \times \Omega : \alpha - \beta = \pm 2\} \\
\mathcal{C}_3 &= \{(\alpha, \beta) \in \Omega \times \Omega : \alpha - \beta = 3\}.
\end{aligned}$$

(a) Write out the sets $\mathcal{C}_0$, $\mathcal{C}_1$, $\mathcal{C}_2$ and $\mathcal{C}_3$ in full. Verify that they are all symmetric. Could you have proved directly from the definition of the classes that they are symmetric?

(b) Write out the sets $\mathcal{C}_1(0)$, $\mathcal{C}_1(4)$, $\mathcal{C}_2(0)$, $\mathcal{C}_2(4)$, $\mathcal{C}_3(0)$ and $\mathcal{C}_3(4)$. Convince yourself that, for $i = 0, \ldots, 3$, each element of Ω has the same number of i-th associates. What are a_0, a_1, a_2 and a_3?

(c) Take α and β to be adjacent points on the circle. By considering the relationship of each of the six points to both α and β, fill in the following table, whose entries are the p_{ij}^{1}. Also fill in the row and column totals.

	$\mathcal{C}_0(\beta)$	$\mathcal{C}_1(\beta)$	$\mathcal{C}_2(\beta)$	$\mathcal{C}_3(\beta)$	
$\mathcal{C}_0(\alpha)$					
$\mathcal{C}_1(\alpha)$					
$\mathcal{C}_2(\alpha)$					
$\mathcal{C}_3(\alpha)$					

(d) Now take α and β to be points on the circle at distance 2 apart. Repeat the above. Then do the same thing for points at distance 3 apart.

(e) Draw coloured lines on the circle as follows. Join adjacent points with a yellow line. Join points at distance 2 with a red line. Join points at distance 3 with a blue line. Then interpret a few of the p_{ij}^k in terms of numbers of triangles of various sorts through an edge of a given colour.

(f) Write down the four adjacency matrices A_0, A_1, A_2 and A_3. Verify that $A_0 + A_1 + A_2 + A_3 = J$.

(g) Calculate all products A_iA_j. Verify that they are all linear combinations of A_0, A_1, A_2 and A_3. Also verify that multiplication is commutative.

(h) For $i = 0, \ldots, 3$, verify that

$$A_i(A_0 + A_1 + A_2 + A_3) = a_i(A_0 + A_1 + A_2 + A_3).$$

Explain why this is so.

(i) Find which association schemes mentioned in this chapter have this one as a special case.

1.4 Verify that a graph is strongly regular if and only if it is neither complete nor null and the sets of edges and non-edges form an association scheme on the set of vertices. Relate the parameters of the association scheme to those of the strongly regular graph.

1.5 Draw a finite graph that is regular but not strongly regular.

1.6 Consider an association scheme on a set Ω of size 6.

(a) Prove that at most one of the classes, in addition to $\mathrm{Diag}(\Omega)$, can have valency 1.

(b) Write down four distinct association schemes on Ω.

(c) Prove that there are exactly four different association schemes on Ω, in the sense that any others are obtained from one of these four by relabelling.

1.7 Let A and B be matrices in $F^{\Gamma\times\Delta}$ and $F^{\Delta\times\Phi}$ respectively, where F is a field and Γ, Δ and Φ are finite sets. Prove that $(AB)' = B'A'$.

1.8 Let A_i and A_j be adjacency matrices of an association scheme. Use Technique 1.1 to interpret the entries of A_iA_j as the number of suitable paths of length 2 in a certain graph. Give a similar interpretation for the entries of A_i^m.

1.9 Write down the adjacency matrices for $\mathrm{GD}(3,2)$ and $\mathrm{R}(2,4)$.

1.10 Write down the four adjacency matrices for the cube association scheme in Example 1.5 explicitly. Calculate all their pairwise products and verify that each one is a linear combination of the four adjacency matrices.

1.11 Suppose that two classes in an association scheme have valencies a_i, a_j and adjacency matrices A_i, A_j respectively. Prove that the sum of the entries in each row of the product A_iA_j is equal to a_ia_j.

1.12 Show that the matrices I, $R-I$, $C-I$, $L-I$ and A_2 in Example 1.6 are the adjacency matrices of an association scheme.

1.13 Use the fact that $p_{ij}^k = p_{ji}^k$ to give an alternative proof of Lemma 1.3.

1.14 Let A satisfy the hypotheses of Lemma 1.4 with $A^2 = xI+yA+zJ$. Find the parameters of the association scheme.

1.15 Two Latin squares of the same size are said to be *orthogonal to each other* if each letter of one square occurs exactly once in the same position as each letter of the second square. Latin squares $\Pi_1, \ldots, \Pi_r$ of the same size are said to be *mutually orthogonal* if Π_i is orthogonal to Π_j for $1 \leqslant i < j \leqslant r$.

Suppose that $\Pi_1, \ldots, \Pi_r$ are mutually orthogonal Latin squares of size n. Let Ω be the set of n^2 cells in the array. For distinct α, β in Ω, let α and β be first associates if α and β are in the same row or are in the same column or have the same letter in any of $\Pi_1, \ldots, \Pi_r$; otherwise α and β are second associates.

(a) Find the size of $\mathcal{C}_1(\alpha)$. Hence find an upper bound on r.
(b) Show that these definitions of $\mathcal{C}_1$ and $\mathcal{C}_2$ make an association scheme on Ω. It is called the *Latin-square type* association scheme $\mathrm{L}(r+2, n)$. When does it have only one associate class?

1.16 Let Ω consist of the 12 edges of the cube. For edges α and β, assign (α, β) to class

$\mathcal{C}_0$ if $\alpha = \beta$;
$\mathcal{C}_1$ if $\alpha \neq \beta$ but α meets β at a vertex;
$\mathcal{C}_2$ if α and β are diagonally opposite;
$\mathcal{C}_3$ if α and β are parallel but not diagonally opposite;
$\mathcal{C}_4$ otherwise.

Calculate the product A_iA_j for $0 \leqslant i, j \leqslant 4$ and hence show that these classes form an association scheme.

1.17 Show that one of the strongly regular graphs defined by T(4) consists of the edges of an octahedron. Which group-divisible association scheme is this the same as?

1.18 Find the parameters of the Johnson scheme J(7, 3).

1.19 Find the parameters of the Hamming scheme H(5, 4).

1.20 Let $\mathcal{G}$ be the graph whose vertices and edges are the vertices and edges of the regular icosahedron. Show that $\mathcal{G}$ is distance-regular. Find the parameters of the corresponding association scheme.

1.21 Repeat the above question for the regular dodecahedron.

1.22 Show that the circuit C_n with n vertices is a distance-regular graph. Find its diameter, and the values of the a_i, b_i and c_i.

1.23 Show that the association scheme ④ defined by the circuit C_4 is the same as a group-divisible association scheme and a Hamming association scheme.

1.24 Prove that, for a distance-regular graph of diameter s,

$$a_i b_i = a_{i+1} c_{i+1}$$

for $i = 0, 1, \ldots, s-1$.

1.25 Verify that $\mathcal{C}_1$ in the Hamming scheme defines a distance-regular graph.

1.26 Draw a connected regular graph that is not distance-regular.

1.27 There is a distance-regular graph with the parameters shown below. Express A_2 and A_3 as polynomials in A_1. Hence find the minimal polynomial of A_1.

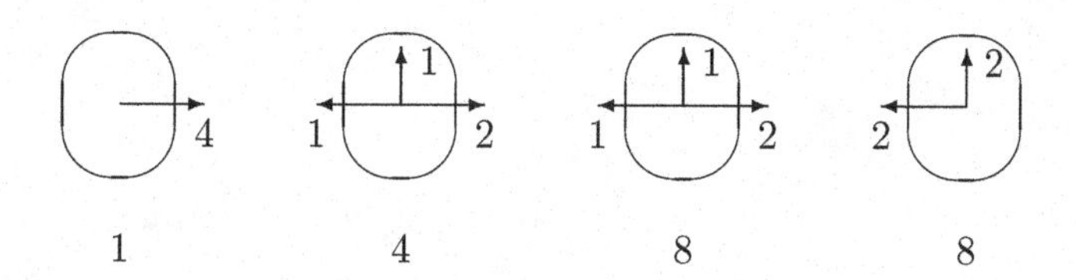

1.28 For $\mathbb{Z}_6$, write down the circulant matrices M_1 and M_3.

1.29 Verify that, for $\mathbb{Z}_n$, $M_\alpha M_\beta = M_{\alpha+\beta}$ and $M_\alpha \chi_\omega = \chi_{\omega-\alpha}$.

1.30 Verify that the sets $\{0\}$, $\{1, 4, -4, -1\}$, $\{2, 7, -7, -2\}$, $\{3, -3\}$, $\{5, -5\}$ and $\{6, -6\}$ form a blueprint for $\mathbb{Z}_{15}$.

1.31 Verify that the sets $\{0\}$, $\{1, 2, 4, 7, -7, -4, -2 - 1\}$, $\{3, 6, -6, -3\}$ and $\{5, -5\}$ form a blueprint for $\mathbb{Z}_{15}$. Show that the corresponding association scheme is the same as a rectangular association scheme.

1.32 Let n and m be coprime integers. In $\mathbb{Z}_{mn}$ let Δ_1 consist of the non-zero multiples of m and Δ_2 consist of the non-zero multiples of n. Put $\Delta_0 = \{0\}$ and $\Delta_3 = \mathbb{Z}_{mn} \setminus \Delta_0 \setminus \Delta_1 \setminus \Delta_2$. Prove that these sets form a blueprint for $\mathbb{Z}_{mn}$ and show that the corresponding association scheme is the same as a rectangular association scheme.

1.33 Write down as many association schemes as you can think of which have 15 elements. Give enough detail about them to show that they are all distinct.

2
The Bose–Mesner algebra

2.1 Orthogonality

This section gives a brief coverage of those aspects of orthogonality that are necessary to appreciate the Bose–Mesner algebra. I hope that most readers have seen most of this material before, even if in a rather different form.

There is a natural inner product $\langle\ ,\ \rangle$ on $\mathbb{R}^\Omega$ defined by

$$\langle f, g\rangle = \sum_{\omega\in\Omega} f(\omega)g(\omega) \qquad \text{for } f \text{ and } g \text{ in } \mathbb{R}^\Omega.$$

Vectors f and g are *orthogonal* to each other (written $f \perp g$) if $\langle f, g\rangle = 0$. If U and W are subspaces of $\mathbb{R}^\Omega$ with $\langle u, w\rangle = 0$ for all u in U and all w in W, we say that U is orthogonal to W (written $U \perp W$).

If W is a subspace of $\mathbb{R}^\Omega$, the *orthogonal complement* $W^\perp$ of W is defined by

$$W^\perp = \left\{v \in \mathbb{R}^\Omega : \langle v, w\rangle = 0 \text{ for all } w \text{ in } W\right\}.$$

Here are some standard facts about orthogonal complements.

- $W^\perp$ is a subspace of $\mathbb{R}^\Omega$;
- $\dim W^\perp + \dim W = \dim \mathbb{R}^\Omega = |\Omega|$;
- $\left(W^\perp\right)^\perp = W$ (recall that we are assuming that $|\Omega|$ is finite);
- $(U + W)^\perp = U^\perp \cap W^\perp$ (recall that $U + W$ is defined to be the set $\{u + w : u \in U \text{ and } w \in W\}$, which is called the *vector space sum* of U and W);
- $(U \cap W)^\perp = U^\perp + W^\perp$;
- $\mathbb{R}^\Omega = W \oplus W^\perp$, which is called the *direct sum* of W and $W^\perp$: this means that if $v \in \mathbb{R}^\Omega$ then there are unique vectors $w \in W$ and $u \in W^\perp$ with $v = w + u$.

The last fact enables us to make the following definition.

Definition The map $P\colon \mathbb{R}^\Omega \to \mathbb{R}^\Omega$ defined by

$$Pv \in W \quad \text{and} \quad v - Pv \in W^\perp$$

is called the *orthogonal projector* onto W.

In the notation at the foot of page 31, $Pv = w$. To show that P is the orthogonal projector onto W it is enough to show that $Pv = v$ for v in W and $Pv = 0$ for v in $W^\perp$.

The orthogonal projector P is a linear transformation. We shall identify it with the matrix that represents it with respect to the basis $\{\chi_\omega : \omega \in \Omega\}$. The phrase 'orthogonal projector' will frequently be abbreviated to 'projector'.

Lemma 2.1 *Let W and U be subspaces of $\mathbb{R}^\Omega$ with orthogonal projectors P and Q respectively. Then*

(i) $P^2 = P$ *(that is, P is* idempotent*);*
(ii) $\langle P\chi_\alpha, \chi_\beta\rangle = \langle \chi_\alpha, P\chi_\beta\rangle$ *and hence P is symmetric;*
(iii) $\dim W = \operatorname{tr}(P)$*;*
(iv) if $U = W^\perp$ then $Q = I_\Omega - P$;
(v) if $U \perp W$ then $PQ = QP = O_\Omega$ and the orthogonal projector onto $U + W$ is $P + Q$;
(vi) if $U \leqslant W$ then $PQ = QP = Q$ and the orthogonal projector onto $W \cap U^\perp$ is $P - Q$.

Commuting projectors have an importance beyond spaces that are actually orthogonal to each other.

Definition Let W and U be subspaces of $\mathbb{R}^\Omega$ with orthogonal projectors P and Q respectively. Then W and U are *geometrically orthogonal* to each other if $PQ = QP$.

Lemma 2.2 *If W and U are geometrically orthogonal then*

(i) PQ is the orthogonal projector onto $W \cap U$;
(ii) $\left(W \cap (W \cap U)^\perp\right) \perp U$.

Proof (i) We need to prove that PQ is the identity on $W \cap U$ and maps everything in $(W \cap U)^\perp$ to the zero vector. If $v \in W \cap U$ then $PQv = Pv = v$. If $v \in (W \cap U)^\perp = W^\perp + U^\perp$ then $v = x + y$ with x in $W^\perp$ and y in $U^\perp$ so $PQv = PQx + PQy =$

$QPx + PQy = 0$ because $Px = Qy = 0$. (This '0' is the zero vector in $\mathbb{R}^\Omega$.)

(ii) If $v \in W \cap (W \cap U)^\perp$ then $Pv = v$ and $PQv = 0$ so $Qv = QPv = PQv = 0$ so $v \in U^\perp$. ■

Property (ii) gives some clue to the strange name 'geometric orthogonality'. The space $W \cap (W \cap U)^\perp$ is the orthogonal complement of $W \cap U$ inside W, and $U \cap (W \cap U)^\perp$ is the orthogonal complement of $W \cap U$ inside U. So geometric orthogonality means that these two complements are orthogonal to each other as well as to $W \cap U$:

$$\left(W \cap (W \cap U)^\perp\right) \perp \left(U \cap (W \cap U)^\perp\right).$$

Thus subspaces W and U are geometrically orthogonal if they are as orthogonal as they can be given that they have a non-zero intersection.

Example 2.1 In $\mathbb{R}^3$, let W be the xy-plane and let U be the xz-plane. Then $W \cap U$ is the x-axis; $W \cap (W \cap U)^\perp$ is the y-axis; and $U \cap (W \cap U)^\perp$ is the z-axis. So the xy-plane is geometrically orthogonal to the xz-plane.

On the other hand, the x-axis is not geometrically orthogonal to the plane $x = y$. ■

Lemma 2.3 *Let $W_1, \ldots, W_r$ be subspaces of $\mathbb{R}^\Omega$ with orthogonal projectors $P_1, \ldots, P_r$ which satisfy*

(i) $\displaystyle\sum_{i=1}^{r} P_i = I_\Omega$;

(ii) $P_iP_j = O_\Omega$ *if* $i \neq j$.

Then $\mathbb{R}^\Omega$ is the direct sum $W_1 \oplus W_2 \oplus \cdots \oplus W_r$.

Proof We need to show that every vector v in $\mathbb{R}^\Omega$ has a unique expression as a sum of vectors in the W_i. Let $v_i = P_iv$. Then

$$v = Iv = \left(\sum P_i\right)v = \sum (P_iv) = \sum v_i$$

with v_i in W_i, so $\mathbb{R}^\Omega = W_1 + W_2 + \cdots + W_r$.

For uniqueness, suppose that $v = \sum w_i$ with w_i in W_i. Then

$$\begin{aligned} v_j = P_jv &= P_j\left(\sum_i w_i\right) = P_j\left(\sum_i P_iw_i\right) \\ &= \left(\sum_i P_jP_iw_i\right) = P_jP_jw_j = w_j. \end{aligned}$$ ■

2.2 The algebra

Given an association scheme on Ω with s associate classes and adjacency matrices A_0, A_1, ..., A_s, let

$$\mathcal{A} = \left\{ \sum_{i=0}^{s} \mu_i A_i : \mu_0, \ldots, \mu_s \in \mathbb{R} \right\}.$$

It is clear that the adjacency matrices are linearly independent, for if $(\alpha, \beta) \in \mathcal{C}_j$ then

$$\left(\sum_i \mu_i A_i \right) (\alpha, \beta) = \mu_j.$$

Therefore $\mathcal{A}$ has dimension $s+1$ as a vector space over $\mathbb{R}$. It is closed under multiplication, because of Equation (1.1), so it is an algebra. It is called the *Bose–Mesner algebra*.

If $M \in \mathcal{A}$ then M is symmetric because every adjacency matrix is symmetric. By a standard result of linear algebra, M is diagonalizable over $\mathbb{R}$. This means that it has distinct real eigenvalues λ_1, ..., λ_r (say, because we do not know the value of r) with eigenspaces W_1, ..., W_r (this means that $W_j = \{v \in \mathbb{R}^\Omega : Mv = \lambda_j v\}$) such that

- $R^\Omega = W_1 \oplus W_2 \oplus \cdots \oplus W_r$;
- the minimal polynomial of M is $\prod_{j=1}^{r} (\mathsf{x} - \lambda_j)$, with no repeated factors.

In fact, not only is M diagonalizable but also $W_i \perp W_j$ if $i \neq j$. In particular, $\ker M = (\operatorname{Im} M)^\perp$.

The orthogonal projector P_i onto W_i is given by

$$P_i = \frac{\prod_{j \neq i} (M - \lambda_j I)}{\prod_{j \neq i} (\lambda_i - \lambda_j)}, \tag{2.1}$$

for if $v \in W_i$ then

$$\left(\prod_{j \neq i} (M - \lambda_j I) \right) v = \left(\prod_{j \neq i} (\lambda_i - \lambda_j) \right) v$$

but if $v \in W_k$ with $k \neq i$ then

$$\left(\prod_{j \neq i} (M - \lambda_j I) \right) v = 0.$$

Because $\mathcal{A}$ is an algebra containing I, $\mathcal{A}$ contains all non-negative powers of M, so $\mathcal{A}$ contains the projectors $P_1, \ldots, P_r$.

Note also that

$$M = \sum_{i=1}^{r} \lambda_i P_i. \tag{2.2}$$

If none of the eigenvalues is zero then M is invertible and

$$M^{-1} = \sum_{i=1}^{r} \frac{1}{\lambda_i} P_i,$$

which is also in $\mathcal{A}$.

We need to consider something more general than an inverse. Suppose that f is any function from any set S to any set T. Then a function $g: T \to S$ is called a *generalized inverse* of f if $fgf = f$ and $gfg = g$. It is evident that set functions always have generalized inverses: if $y \in \mathrm{Im}(f)$ then let $g(y)$ be any element x of S with $f(x) = y$; if $y \notin \mathrm{Im}(f)$ then $g(y)$ may be any element of S. If f is invertible then $g = f^{-1}$; otherwise there are several possibilities for g. However, when there is more structure then generalized inverses may not exist. For example, group homomorphisms do not always have generalized inverses; nor do continuous functions in topology. However, linear transformations (and hence matrices) *do* have generalized inverses.

If M is a diagonalizable matrix then one of its generalized inverses has a special form. It is called the *Drazin* inverse or *Moore–Penrose* generalized inverse and written M^-. This is given by

$$M^- = \sum_{\substack{i=1 \\ \lambda_i \neq 0}}^{r} \frac{1}{\lambda_i} P_i.$$

If $M \in \mathcal{A}$ then $M^- \in \mathcal{A}$.

That concludes the properties of a single element of $\mathcal{A}$. The key fact for dealing with two or more matrices in $\mathcal{A}$ is that $\mathcal{A}$ is *commutative*; that is, if $M \in \mathcal{A}$ and $N \in \mathcal{A}$ then $MN = NM$. This follows from Lemma 1.3. The following rather technical lemma is the building block for the main result.

Lemma 2.4 *Let $U_1, \ldots, U_m$ be non-zero orthogonal subspaces of $\mathbb{R}^\Omega$ such that $\mathbb{R}^\Omega = U_1 \oplus \cdots \oplus U_m$. For $1 \leqslant i \leqslant m$, let Q_i be the orthogonal projector onto U_i. Let M be a symmetric matrix which commutes with Q_i for $1 \leqslant i \leqslant m$. Let the eigenspaces of M be $V_1, \ldots, V_r$. Let the non-zero subspaces among the intersections $U_i \cap V_j$ be $W_1, \ldots, W_t$. Then* *(i)* $\mathbb{R}^\Omega = W_1 \oplus \cdots \oplus W_t$;

(ii) *if* $k \neq l$ *then* $W_k \perp W_l$;

(iii) *for* $1 \leqslant k \leqslant t$, *the space* W_k *is contained in an eigenspace of* M;

(iv) *for* $1 \leqslant k \leqslant t$, *the orthogonal projector onto* W_k *is a polynomial in* $Q_1, \ldots, Q_m$ *and* M.

Proof Part (iii) is immediate from the definition of the W_k. So is part (ii), because $U_{i_1} \perp U_{i_2}$ if $i_1 \neq i_2$ and $V_{j_1} \perp V_{j_2}$ if $j_1 \neq j_2$ so the two spaces $U_{i_1} \cap V_{j_1}$ and $U_{i_2} \cap V_{j_2}$ are either orthogonal or equal.

For $1 \leqslant j \leqslant r$, let P_j be the projector onto V_j. Since P_j is a polynomial in M, it commutes with Q_i, so U_i is geometrically orthogonal to V_j. By Lemma 2.2, the orthogonal projector onto $U_i \cap V_j$ is Q_iP_j, which is a polynomial in Q_i and M. This proves (iv).

Finally, $Q_iP_j = O$ precisely when $U_i \cap V_j = \{0\}$, so the sum of the orthogonal projectors onto the spaces W_k is the sum of the non-zero products Q_iP_j, which is equal to the sum of all the products Q_iP_j. But

$$\sum_i \sum_j Q_iP_j = \left(\sum_i Q_i\right)\left(\sum_j P_j\right) = I^2 = I,$$

so Lemma 2.3 shows that $\mathbb{R}^\Omega = W_1 \oplus \cdots \oplus W_t$, proving (i). ■

Corollary 2.5 *If* M *and* N *are commuting symmetric matrices in* $\mathbb{R}^{\Omega\times\Omega}$ *then* $\mathbb{R}^\Omega$ *is the direct sum of the non-zero intersections of the eigenspaces of* M *and* N. *Moreover, these spaces are mutually orthogonal and their orthogonal projectors are polynomials in* M *and* N.

Proof Apply Lemma 2.4 to M and the eigenspaces of N. ■

If W is contained in an eigenspace of a matrix M, I shall call it a *sub-eigenspace* of M.

Theorem 2.6 *Let* $A_0, A_1, \ldots, A_s$ *be the adjacency matrices of an association scheme on* Ω *and let* $\mathcal{A}$ *be its Bose–Mesner algebra. Then* $\mathbb{R}^\Omega$ *has* $s+1$ *mutually orthogonal subspaces* $W_0, W_1, \ldots, W_s$, *called* strata, *with orthogonal projectors* $S_0, S_1, \ldots, S_s$ *such that*

(i) $\mathbb{R}^\Omega = W_0 \oplus W_1 \oplus \cdots \oplus W_s$;

(ii) *each of* $W_0, W_1, \ldots, W_s$ *is a sub-eigenspace of every matrix in* $\mathcal{A}$;

(iii) *for* $i = 0, 1, \ldots, s$, *the adjacency matrix* A_i *is a linear combination of* $S_0, S_1, \ldots, S_s$;

(iv) *for* $e = 0, 1, \ldots, s$, *the stratum projector* S_e *is a linear combination of* $A_0, A_1, \ldots, A_s$.

Proof The adjacency matrices $A_1, \ldots, A_s$ commute and are symmetric, so $s-1$ applications of Lemma 2.4, starting with the eigenspaces of A_1, give spaces $W_0, \ldots, W_r$ as the non-zero intersections of the eigenspaces of $A_1, \ldots, A_s$, where r is as yet unknown. These spaces W_e are mutually orthogonal and satisfy (i). Since $A_0 = I$ and every matrix in $\mathcal{A}$ is a linear combination of $A_0, A_1, \ldots, A_s$, the spaces W_e clearly satisfy (ii). Each S_e is a polynomial in $A_1, \ldots, A_s$, hence in $\mathcal{A}$, so (iv) is satisfied. Let $C(i,e)$ be the eigenvalue of A_i on W_e. Then Equation (2.2) shows that

$$A_i = \sum_{e=0}^{r} C(i,e) S_e$$

and so (iii) is satisfied.

Finally, (iii) shows that $S_0, \ldots, S_r$ span $\mathcal{A}$. Suppose that there are real scalars $\lambda_0, \ldots, \lambda_r$ such that $\sum_e \lambda_e S_e = O$. Then for $f = 0, \ldots, r$ we have $O = \left(\sum_e \lambda_e S_e\right) S_f = \lambda_f S_f$ so $\lambda_f = 0$. Hence $S_0, \ldots, S_r$ are linearly independent, so they form a basis for $\mathcal{A}$. Thus $r+1 = \dim \mathcal{A} = s+1$ and $r = s$. ∎

We now have two bases for $\mathcal{A}$: the adjacency matrices and the stratum projectors. The former are useful for addition, because $A_j(\alpha,\beta) = 0$ if $(\alpha,\beta) \in \mathcal{C}_i$ and $i \neq j$. The stratum projectors make multiplication easy, because $S_e S_e = S_e$ and $S_e S_f = O$ if $e \neq f$.

Before calculating any eigenvalues, we note that if A is the adjacency matrix of any subset $\mathcal{C}$ of $\Omega \times \Omega$ then $A\chi_\alpha = \sum \{\chi_\beta : (\beta,\alpha) \in \mathcal{C}\}$. In particular, $A_i \chi_\alpha = \chi_{\mathcal{C}_i(\alpha)}$ and $J\chi_\alpha = \chi_\Omega$. Furthermore, if M is any matrix in $\mathbb{R}^{\Omega \times \Omega}$ then $M\chi_\Omega = \sum_{\omega \in \Omega} M\chi_\omega$. If M has constant row-sum r then $M\chi_\Omega = r\chi_\Omega$.

Example 2.2 Consider the group-divisible association scheme $\mathrm{GD}(b,k)$ with b groups of size k and

$$A_1(\alpha,\beta) = \begin{cases} 1 & \text{if } \alpha \text{ and } \beta \text{ are in the same group but } \alpha \neq \beta \\ 0 & \text{otherwise} \end{cases}$$

$$A_2(\alpha,\beta) = \begin{cases} 1 & \text{if } \alpha \text{ and } \beta \text{ are in different groups} \\ 0 & \text{otherwise.} \end{cases}$$

Consider the 1-dimensional space W_0 spanned by χ_Ω. We have

$$\begin{aligned} A_0\chi_\Omega &= I_\Omega\chi_\Omega = & \chi_\Omega \\ A_1\chi_\Omega &= a_1\chi_\Omega = & (k-1)\chi_\Omega \\ A_2\chi_\Omega &= a_2\chi_\Omega = & (b-1)k\chi_\Omega. \end{aligned}$$

Thus W_0 is a sub-eigenspace of every adjacency matrix. This does not prove that W_0 is a stratum, because there might be other vectors which are also eigenvectors of all the adjacency matrices with the same eigenvalues as W_0.

Let α and β be in the same group Δ. Then

$$\begin{aligned} A_1\chi_\alpha &= \chi_\Delta - \chi_\alpha & A_2\chi_\alpha &= \chi_\Omega - \chi_\Delta \\ A_1\chi_\beta &= \chi_\Delta - \chi_\beta & A_2\chi_\beta &= \chi_\Omega - \chi_\Delta \end{aligned}$$

so $A_1(\chi_\alpha - \chi_\beta) = -(\chi_\alpha - \chi_\beta)$ and $A_2(\chi_\alpha - \chi_\beta) = 0$. Let W_{within} be the $b(k-1)$-dimensional *within-groups* subspace spanned by all vectors of the form $\chi_\alpha - \chi_\beta$ with α and β in the same group. Then W_{within} is a sub-eigenspace of A_0, A_1 and A_2 and the eigenvalues for A_1 and A_2 are different from those for W_0.

Since eigenspaces are mutually orthogonal, it is natural to look at the orthogonal complement of $W_0 + W_{\mathrm{within}}$. This is the $(b-1)$-dimensional *between-groups* subspace W_{between}, which is spanned by vectors of the form $\chi_\Delta - \chi_\Gamma$ where Δ and Γ are different groups. Now

$$A_1\chi_\Delta = A_1\sum_{\alpha\in\Delta}\chi_\alpha = k\chi_\Delta - \sum_{\alpha\in\Delta}\chi_\alpha = (k-1)\chi_\Delta$$

and

$$A_2\chi_\Delta = A_2\sum_{\alpha\in\Delta}\chi_\alpha = k(\chi_\Omega - \chi_\Delta),$$

so $A_1(\chi_\Delta - \chi_\Gamma) = (k-1)(\chi_\Delta - \chi_\Gamma)$ and $A_2(\chi_\Delta - \chi_\Gamma) = -k(\chi_\Delta - \chi_\Gamma)$. Thus W_{between} is a sub-eigenspace with eigenvalues different from those for W_{within}. Therefore the strata are W_0, W_{within} and W_{between}. ■

Lemma 2.7 *If $P \in \mathcal{A}$ and P is idempotent then $P = \sum_{e\in\mathcal{F}} S_e$ for some subset $\mathcal{F}$ of $\{0,\ldots,s\}$.*

Proof Let $P = \sum_{e=0}^{s}\lambda_e S_e$. Then $P^2 = \sum_{e=0}^{s}\lambda_e^2 S_e$, which is equal to P if and only if $\lambda_e \in \{0,1\}$ for $e = 0, \ldots, s$. ■

For this reason the stratum projectors are sometimes called *minimal idempotents* or *primitive idempotents*.

Lemma 2.8 *The space W spanned by χ_Ω is always a stratum. Its projector is $|\Omega|^{-1} J_\Omega$.*

Proof The orthogonal projector onto W is $|\Omega|^{-1} J_\Omega$ because

$$J_\Omega \chi_\Omega = \sum_{\omega \in \Omega} J_\Omega \chi_\omega = |\Omega| \chi_\Omega$$

and

$$J_\Omega(\chi_\alpha - \chi_\beta) = 0.$$

This is an idempotent contained in $\mathcal{A}$, so it is equal to $\sum_{e \in \mathcal{F}} S_e$, for some subset $\mathcal{F}$ of $\{0, \ldots, s\}$, by Lemma 2.7. Then

$$1 = \dim W = \operatorname{tr}\left(|\Omega|^{-1} J_\Omega\right) = \sum_{e \in \mathcal{F}} \operatorname{tr} S_e = \sum_{e \in \mathcal{F}} \dim W_e$$

so we must have $|\mathcal{F}| = 1$ and W is itself a stratum. ■

Notation The 1-dimensional stratum spanned by χ_Ω will always be called W_0.

Although there are the same number of strata as associate classes, there is usually no natural bijection between them. When I want to emphasize this, I shall use a set $\mathcal{K}$ to index the associate classes and a set $\mathcal{E}$ to index the strata. However, there are some association schemes for which $\mathcal{E}$ and $\mathcal{K}$ are naturally the same but for which W_0 does not correspond to A_0. So the reader should interpret these two subscripts '0' as different sorts of zero.

I shall always write d_e for $\dim W_e$.

2.3 The character table

For i in $\mathcal{K}$ and e in $\mathcal{E}$ let $C(i, e)$ be the eigenvalue of A_i on W_e and let $D(e, i)$ be the coefficient of A_i in the expansion of S_e as a linear combination of the adjacency matrices. That is:

$$A_i = \sum_{e \in \mathcal{E}} C(i, e) S_e \tag{2.3}$$

and

$$S_e = \sum_{i \in \mathcal{K}} D(e, i) A_i. \tag{2.4}$$

Lemma 2.9 *The matrices C in $\mathbb{R}^{\mathcal{K}\times\mathcal{E}}$ and D in $\mathbb{R}^{\mathcal{E}\times\mathcal{K}}$ are mutual inverses.*

We note some special values of $C(i,e)$ and $D(e,i)$:

$$C(0,e) = 1 \qquad \text{because } A_0 = I = \sum_{e\in\mathcal{E}} S_e;$$

$$C(i,0) = a_i \qquad \text{because } A_i\chi_\Omega = a_i\chi_\Omega;$$

$$D(0,i) = \frac{1}{|\Omega|} \qquad \text{because } S_0 = \frac{1}{|\Omega|}J = \frac{1}{|\Omega|}\sum_{i\in\mathcal{K}} A_i;$$

$$D(e,0) = \frac{d_e}{|\Omega|} \qquad \text{because } d_e = \operatorname{tr}(S_e) = \sum_{i\in\mathcal{K}} D(e,i)\operatorname{tr}(A_i) = |\Omega|\, D(e,0).$$

Lemma 2.10 *The map $\varphi_e\colon \mathcal{A} \to \mathcal{A}$ defined by*

$$\varphi_e\colon A_i \mapsto C(i,e)S_e$$

and extended linearly is an algebra homomorphism.

Corollary 2.11 *The maps ϑ_0, ..., $\vartheta_s\colon \mathcal{A} \to \mathbb{R}$ defined by*

$$\vartheta_e\colon \sum_{i\in\mathcal{K}} \lambda_i A_i \mapsto \sum_{i\in\mathcal{K}} \lambda_i C(i,e)$$

are algebra homomorphisms.

Definition The maps ϑ_0, ..., ϑ_s are *characters* of the association scheme. The matrix C, whose columns are the characters, is the *character table* of the association scheme.

Example 2.2 revisited The character table is

	0	within	between
0	1	1	1
1	$k-1$	-1	$k-1$
2	$(b-1)k$	0	$-k$

The entries in the 0-th row are equal to 1; those in the 0-th column are the valencies. ■

Theorem 2.12 (Orthogonality relations for the associate classes)

$$\sum_{e\in\mathcal{E}} C(i,e)C(j,e)d_e = \begin{cases} a_i\,|\Omega| & \text{if } i=j \\ 0 & \text{otherwise.} \end{cases}$$

Proof We calculate the trace of A_iA_j in two different ways. First

$$\begin{aligned} A_iA_j &= \left(\sum_e C(i,e)S_e\right)\left(\sum_f C(j,f)S_f\right) \\ &= \sum_e C(i,e)C(j,e)S_e \end{aligned}$$

so

$$\begin{aligned} \operatorname{tr}(A_iA_j) &= \sum_e C(i,e)C(j,e)\operatorname{tr}(S_e) \\ &= \sum_e C(i,e)C(j,e)d_e. \end{aligned}$$

But $A_iA_j = \sum_k p_{ij}^k A_k$ so $\operatorname{tr}(A_iA_j) = p_{ij}^0\,|\Omega|$; and $p_{ij}^0 = 0$ if $i \neq j$, while $p_{ii}^0 = a_i$. ■

Corollary 2.13 *If* $|\Omega| = n$ *then*

$$D = \frac{1}{n}\operatorname{diag}(d)C'\operatorname{diag}(a)^{-1}.$$

Proof The equation in Theorem 2.12 can be written as

$$C\operatorname{diag}(d)C' = n\operatorname{diag}(a)$$

so

$$C\operatorname{diag}(d)C'\operatorname{diag}(a)^{-1} = nI.$$

But $D = C^{-1}$ so $D = n^{-1}\operatorname{diag}(d)C'\operatorname{diag}(a)^{-1}$. ■

Thus C is inverted by transposing it, multiplying the rows by the dimensions, dividing the columns by the valencies, and finally dividing all the entries by the size of Ω.

Example 2.2 revisited Here $n = bk$,

$$\operatorname{diag}(a) = \begin{bmatrix} 1 & 0 & 0 \\ 0 & k-1 & 0 \\ 0 & 0 & (b-1)k \end{bmatrix}$$

and

$$\operatorname{diag}(d) = \begin{bmatrix} 1 & 0 & 0 \\ 0 & b(k-1) & 0 \\ 0 & 0 & b-1 \end{bmatrix}$$

so D is equal to

$$\frac{1}{bk}\begin{bmatrix} 1 & 0 & 0 \\ 0 & b(k-1) & 0 \\ 0 & 0 & b-1 \end{bmatrix}\begin{bmatrix} 1 & k-1 & (b-1)k \\ 1 & -1 & 0 \\ 1 & k-1 & -k \end{bmatrix}\begin{bmatrix} 1 & 0 & 0 \\ 0 & \frac{1}{k-1} & 0 \\ 0 & 0 & \frac{1}{(b-1)k} \end{bmatrix}$$

which is

$$\frac{1}{bk}\begin{bmatrix} 1 & 1 & 1 \\ b(k-1) & -b & 0 \\ b-1 & b-1 & -1 \end{bmatrix}.$$

Note that the entries in the top row are all equal to $1/bk$, while those in the first column are the dimensions divided by bk.

From D we can read off the stratum projectors as

$$S_0 = \frac{1}{bk}(A_0 + A_1 + A_2) = \frac{1}{bk}J,$$

$$S_{\text{within}} = \frac{1}{bk}\left(b(k-1)A_0 - bA_1\right) = I - \frac{1}{k}(A_0 + A_1) = I - \frac{1}{k}G,$$

where $G = A_0 + A_1$, which is the adjacency matrix for the relation 'is in the same group as', and

$$\begin{aligned} S_{\text{between}} &= \frac{1}{bk}\left((b-1)(A_0 + A_1) - A_2\right) \\ &= \frac{1}{bk}\left((b-1)G - (J-G)\right) = \frac{1}{k}G - \frac{1}{bk}J. \quad \blacksquare \end{aligned}$$

Corollary 2.14 (Orthogonality relations for the characters)

$$\sum_{i \in \mathcal{K}} \frac{C(i,e)C(i,f)}{a_i} = \begin{cases} \dfrac{|\Omega|}{d_e} & \textit{if } e = f \\ 0 & \textit{otherwise.} \end{cases}$$

Proof Let $n = |\Omega|$. Now $DC = I$ so

$$\frac{1}{n}\operatorname{diag}(d)C' \operatorname{diag}(a)^{-1}C = I$$

so

$$C' \operatorname{diag}(a)^{-1}C = n \operatorname{diag}(d)^{-1},$$

as required. ■

Corollary 2.15 (Orthogonality relations for D)

$$\text{(i)} \sum_{i\in\mathcal{K}} D(e,i)D(f,i)a_i = \begin{cases} \dfrac{d_e}{|\Omega|} & \textit{if } e=f \\ 0 & \textit{otherwise;} \end{cases}$$

$$\text{(ii)} \sum_{e\in\mathcal{E}} \frac{D(e,i)D(e,j)}{d_e} = \begin{cases} \dfrac{1}{|\Omega|\, a_i} & \textit{if } i=j \\ 0 & \textit{otherwise.} \end{cases}$$

The entries in the matrices C and D are called *parameters of the second kind.*

2.4 Techniques

Given an association scheme in terms of its parameters of the first kind, we want to find

- its strata;
- the dimensions of the strata;
- the matrix D expressing the stratum projectors as linear combinations of the adjacency matrices;
- the minimal polynomial of each adjacency matrix;
- the eigenvalues of each adjacency matrix;
- the character table (the matrix C).

There are several techniques for doing this.

Definition A subspace W of $\mathbb{R}^\Omega$ is *invariant* under a matrix M in $\mathbb{R}^{\Omega\times\Omega}$ if $Mw \in W$ for all w in W. It is invariant under $\mathcal{A}$ if it is invariant under every matrix in $\mathcal{A}$.

Of course, the strata are invariant under every matrix in the Bose–Mesner algebra, but they are not the only such subspaces, so our first technique needs a little luck.

Technique 2.1 Use knowledge of or experience of or intuition about the symmetry of the association scheme to find "natural" invariant subspaces. By taking intersections and complements, refine these to a set of $s+1$ mutually orthogonal subspaces, including W_0, whose sum is $\mathbb{R}^\Omega$. Then verify that each of these subspaces is a sub-eigenspace of each adjacency matrix.

This is the technique we adopted in Example 2.2. If it works it is a marvellous technique, because it gives the most insight into the strata and their dimensions, and it gives the character table (hence all of the eigenvalues) as part of the verification. But this is cold comfort to someone without the insight to guess the strata. The next technique is completely systematic and always works.

Technique 2.2 Choose one of the adjacency matrices A_i and express its powers in terms of $A_0, \ldots, A_s$, using the equations

$$A_iA_j = \sum_k p_{ij}^k A_k.$$

Hence find the minimal polynomial of A_i, which has degree at most $s+1$. Factorize this minimal polynomial to obtain the eigenvalues of A_i. If you are lucky, it has $s+1$ distinct eigenvalues $\lambda_0, \ldots, \lambda_s$. Then the eigenspaces of A_i are the strata and the projectors onto them are $S_0, \ldots, S_s$ where

$$S_e = \frac{\prod_{f\neq e}(A_i - \lambda_f I)}{\prod_{f\neq e}(\lambda_e - \lambda_f)}.$$

Expressing these in terms of $A_0, \ldots, A_s$ (we already have the powers of A_i in this form) gives the entries in D.

Example 2.3 (Example 1.5 continued) In the association scheme defined by the cube, write $Y = A_{\mathrm{yellow}}$, $B = A_{\mathrm{black}}$ and $R = A_{\mathrm{red}}$. Using the values of $p^i_{\mathrm{yellow,black}}$ from Example 1.5, we obtain

$$YB = \sum_i p^i_{\mathrm{yellow,black}} A_i = 2Y + 3R.$$

Equivalently, start at a point on the cube, take one yellow step and then one black one: where can you get to? The point at the end of the red edge from the starting point can be reached in three ways along a yellow-black path, and each point at the end of a yellow edge from the starting point can be reach in two ways along such a path. Similarly,

$$Y^2 = 3I + 2B \tag{2.5}$$

and

$$YR = B.$$

So

$$Y^3 = 3Y + 2YB = 3Y + 4Y + 6R = 7Y + 6R$$

and

$$Y^4 = 7Y^2 + 6YR = 7Y^2 + 6B = 7Y^2 + 3(Y^2 - 3I) = 10Y^2 - 9I.$$

Thus

$$O = Y^4 - 10Y^2 + 9I = (Y^2 - 9I)(Y^2 - I),$$

so the minimal polynomial of Y is

$$(\mathsf{x} - 3)(\mathsf{x} + 3)(\mathsf{x} - 1)(\mathsf{x} + 1)$$

and the eigenvalues of Y are ± 3 and ± 1. (This verifies one eigenvalue we already know: the valency of yellow, which is 3.)

The eigenspace W_{+1} of Y with eigenvalue $+1$ has projector S_{+1} given by

$$\begin{aligned} S_{+1} &= \frac{(Y^2 - 9I)(Y + I)}{(1^2 - 9)(1 + 1)} = \frac{(3I + 2B - 9I)(Y + I)}{-16} \\ &= \frac{(3I - B)(Y + I)}{8} = \frac{3I + 3Y - B - YB}{8} = \frac{3I + Y - B - 3R}{8}. \end{aligned}$$

The other three eigenprojectors are found similarly. The dimensions follow immediately: for example

$$d_{+1} = \dim W_{+1} = \operatorname{tr}(S_{+1}) = 3. \quad \blacksquare$$

There are two small variations on Technique 2.2. The first is necessary if your chosen A_i has fewer than $s + 1$ distinct eigenvalues. Then you need to find the eigenprojectors for at least one more adjacency matrix, and take products of these with those for A_i. Continue until you have $s + 1$ different non-zero projectors. A good strategy is to choose a complicated adjacency matrix to start with, so that it is likely to have many eigenvalues. That is why we used Y instead of R in Example 2.3.

The other variation is always possible, and saves some work. We know that χ_Ω is always an eigenvector of A_i with eigenvalue a_i. So we can find the other eigenvalues of A_i by working on the orthogonal complement $W_0^\perp$ of χ_Ω. We need to find the polynomial $p(A_i)$ of lowest degree such that $p(A_i)v = 0$ for all v in $W_0^\perp$. Every power of A_i has the form

$\sum_j \mu_j A_j$. If $v \in W_0^\perp$ then $Jv = 0$ so $(\sum_j A_j)v = 0$, so

$$\sum_{j=0}^{s} \mu_j A_j v = \sum_{j=0}^{s-1} (\mu_j - \mu_s) A_j v.$$

Thus for our calculations we can pretend that $J = O$ and work with one fewer of the A_j.

Example 2.3 revisited If we put $J = O$ we get $R = -(I + Y + B)$, so

$$Y^3 = 7Y - 6(I + Y + B) = Y - 6I - 3(Y^2 - 3I)$$

so

$$Y^3 + 3Y^2 - Y - 3I = O$$

so

$$(Y^2 - I)(Y + 3I) = O.$$

Now, $a_{\text{yellow}} = 3$ and $Y - 3I$ is not already a factor of this polynomial, so the minimal polynomial is

$$(\mathsf{x} - 3)(\mathsf{x}^2 - 1)(\mathsf{x} + 3),$$

as before. ■

Technique 2.3 If the eigenvalues $C(i, e)$ of A_i are known, and if A_j is a known polynomial function $F(A_i)$ of A_i, then calculate the corresponding eigenvalues of A_j as $F(C(i, e))$.

Example 2.3 revisited The eigenvalues of Y are 3, -3, 1 and -1. From Equation (2.5), $B = (Y^2 - 3I)/2$. Hence the corresponding eigenvalues of B are 3, 3, -1 and -1 respectively. Thus part of the character table is

$$\begin{array}{cc} 0 & (I) \\ 1 & (Y) \\ 2 & (B) \\ 3 & (R) \end{array} \begin{bmatrix} 1 & 1 & 1 & 1 \\ 3 & -3 & 1 & -1 \\ 3 & 3 & -1 & -1 \\ 1 & & & \end{bmatrix}. \quad ■$$

Technique 2.4 If C and the dimensions are known, find D by using

$$D = \frac{1}{n} \operatorname{diag}(d) C' \operatorname{diag}(a)^{-1}. \tag{2.6}$$

If D is known, find the dimensions from the first column and then find C by using

$$C = n\,\mathrm{diag}(a)D'\,\mathrm{diag}(d)^{-1}. \tag{2.7}$$

Technique 2.5 Use the orthogonality relations and/or the fact that the dimensions must be integers to complete C or D from partial information. In particular

$$\sum_{i\in\mathcal{K}} C(i,e) = 0 \qquad \text{if } e \neq 0; \tag{2.8}$$

$$\sum_{e\in\mathcal{E}} C(i,e)d_e = 0 \qquad \text{if } i \neq 0; \tag{2.9}$$

$$\sum_{e\in\mathcal{E}} D(e,i) = 0 \qquad \text{if } i \neq 0 \tag{2.10}$$

and

$$\sum_{i\in\mathcal{K}} D(e,i)a_i = 0 \qquad \text{if } e \neq 0. \tag{2.11}$$

Example 2.4 In the association scheme ⑤ defined by the 5-circuit, let A_1 be the adjacency matrix for edges. Then

$$A_1^2 = 2I + (J - A_1 - I).$$

Ignoring J, we have $A_1^2 + A_1 - I = O$, so we find that the eigenvalues of A_1 on $W_0^\perp$ are

$$\frac{-1 \pm \sqrt{5}}{2}.$$

Let the other two strata be W_1 and W_2, with dimensions d_1 and d_2. Then the incomplete character table is

$$\begin{array}{cc} & \\ & \\ 0 & (1) \\ 1 & (2) \\ 2 & (2) \end{array}
\begin{array}{ccc} W_0 & W_1 & W_2 \\ (1) & (d_1) & (d_2) \\ \end{array}$$

$$\begin{array}{cc|ccc} & & W_0 & W_1 & W_2 \\ & & (1) & (d_1) & (d_2) \\ \hline 0 & (1) & 1 & 1 & 1 \\ 1 & (2) & 2 & \frac{-1+\sqrt{5}}{2} & \frac{-1-\sqrt{5}}{2} \\ 2 & (2) & 2 & & \end{array}$$

(it is helpful to show the valencies and dimensions in parentheses like this). Then

$$0 = \sum_e C(1,e)d_e = 2 + \left(\frac{-1+\sqrt{5}}{2}\right) d_1 + \left(\frac{-1-\sqrt{5}}{2}\right) d_2.$$

To get rid of the $\sqrt{5}$, we must have $d_1 = d_2$. But $\sum_e d_e = 5$, so $d_1 = d_2 = 2$.

The sums of the entries in the middle column and the final column must both be 0, so the complete character table is as follows.

$$\begin{array}{cc} & \\ & \\ 0 & (1) \\ 1 & (2) \\ 2 & (2) \end{array} \begin{array}{c} \begin{array}{ccc} W_0 & W_1 & W_2 \\ (1) & (2) & (2) \end{array} \\ \left[\begin{array}{ccc} 1 & 1 & 1 \\ 2 & \dfrac{-1+\sqrt{5}}{2} & \dfrac{-1-\sqrt{5}}{2} \\ 2 & \dfrac{-1-\sqrt{5}}{2} & \dfrac{-1+\sqrt{5}}{2} \end{array}\right] \end{array} \quad \blacksquare$$

There is an alternative to Technique 2.2, which uses a remarkable algebra isomorphism.

Theorem 2.16 *For k in $\mathcal{K}$ define the matrix P_k in $\mathbb{R}^{\mathcal{K}\times\mathcal{K}}$ by*

$$P_k(i,j) = p^i_{jk} = p^i_{kj}$$

(so that $P_k(i,j)$ is the number of k-coloured edges from β to $\mathcal{C}_j(\alpha)$ if $\beta \in \mathcal{C}_i(\alpha)$: see Figure 2.1). Let $\mathcal{P} = \left\{\sum_{i\in\mathcal{K}} \lambda_i P_i : \lambda_0, \ldots, \lambda_s \in \mathbb{R}\right\}$. Define $\varphi\colon \mathcal{A} \to \mathcal{P}$ by $\varphi(A_i) = P_i$, extended linearly. Then φ is an algebra isomorphism, called the Bose–Mesner isomorphism.

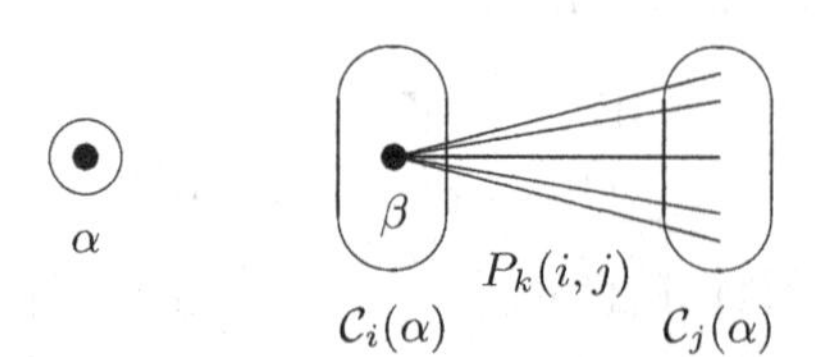

Fig. 2.1. Definition of the matrix P_k

Proof The set of matrices $\{P_i : i \in \mathcal{K}\}$ is linearly independent, because $P_k(0,j) = 0$ unless $j = k$. Hence it is sufficient to prove that $\varphi(A_iA_j) = \varphi(A_i)\varphi(A_j)$.

Matrix multiplication is associative, so

$$(A_iA_j)A_z = A_i(A_jA_z)$$

for i, j and z in $\mathcal{K}$. But

$$(A_iA_j)A_z = \sum_{k\in\mathcal{K}} p_{ij}^k A_k A_z = \sum_{k\in\mathcal{K}}\sum_{x\in\mathcal{K}} p_{ij}^k p_{kz}^x A_x$$

and

$$A_i(A_jA_z) = A_i\left(\sum_{y\in\mathcal{K}} p_{jz}^y A_y\right) = \sum_{x\in\mathcal{K}}\sum_{y\in\mathcal{K}} p_{iy}^x p_{jz}^y A_x.$$

The adjacency matrices are linearly independent, so

$$\sum_k p_{ij}^k p_{kz}^x = \sum_y p_{iy}^x p_{jz}^y \tag{2.12}$$

for i, j, x and z in $\mathcal{K}$.

Now

$$\begin{aligned}(P_iP_j)(x,z) &= \sum_y P_i(x,y)P_j(y,z)\\ &= \sum_y p_{iy}^x p_{jz}^y\\ &= \sum_k p_{ij}^k p_{kz}^x\\ &= \sum_k p_{ij}^k P_k(x,z),\end{aligned}$$

and so $P_iP_j = \sum_k p_{ij}^k P_k$. Therefore $\varphi(A_i)\varphi(A_j) = \sum_k p_{ij}^k \varphi(A_k) = \varphi(\sum_k p_{ij}^k A_k) = \varphi(A_iA_j)$. ■

Corollary 2.17 *The matrices A_i and P_i have the same minimal polynomial and hence the same eigenvalues.*

Technique 2.6 Working in $\mathbb{R}^{\mathcal{K}\times\mathcal{K}}$, find the eigenvalues and minimal polynomial of P_i. These are the eigenvalues and minimal polynomial of A_i.

Example 2.5 (Example 1.4 continued) In the Petersen graph,

$$P_1 = \begin{bmatrix} 0 & 3 & 0 \\ 1 & 0 & 2 \\ 0 & 1 & 2 \end{bmatrix}.$$

By inspection,

$$P_1 + 2I = \begin{bmatrix} 2 & 3 & 0 \\ 1 & 2 & 2 \\ 0 & 1 & 4 \end{bmatrix},$$

which is singular, so -2 is an eigenvalue. We know that 3 is an eigenvalue, because $a_1 = 3$. The sum of the eigenvalues is equal to $\mathrm{tr}(P_1)$, which is 2, so the third eigenvalue is 1.

Let d_1 and d_2 be the dimensions of the strata corresponding to eigenvalues -2, 1 respectively. Then

$$10 = \sum_e d_e = 1 + d_1 + d_2$$

and

$$0 = \sum_e C(1,e)d_e = 3 - 2d_1 + d_2,$$

so $d_1 = 4$ and $d_2 = 5$. Now the incomplete character table is

$$\begin{array}{cc} & \\ & \\ 0 & (1) \\ 1 & (3) \\ 2 & (6) \end{array} \begin{array}{c} \begin{array}{ccc} 0 & 1 & 2 \\ (1) & (4) & (5) \end{array} \\ \begin{bmatrix} 1 & 1 & 1 \\ 3 & -2 & 1 \\ 6 & & \end{bmatrix} \end{array}.$$

Apart from the 0-th column, the column sums must be zero, so

$$C = \begin{array}{cc} & \\ & \\ 0 & (1) \\ 1 & (3) \\ 2 & (6) \end{array} \begin{array}{c} \begin{array}{ccc} 0 & 1 & 2 \\ (1) & (4) & (5) \end{array} \\ \begin{bmatrix} 1 & 1 & 1 \\ 3 & -2 & 1 \\ 6 & 1 & -2 \end{bmatrix} \end{array}.$$

Then

$$D = \frac{1}{10} \begin{bmatrix} 1 & 0 & 0 \\ 0 & 4 & 0 \\ 0 & 0 & 5 \end{bmatrix} \begin{bmatrix} 1 & 3 & 6 \\ 1 & -2 & 1 \\ 1 & 1 & -2 \end{bmatrix} \begin{bmatrix} 1 & 0 & 0 \\ 0 & \frac{1}{3} & 0 \\ 0 & 0 & \frac{1}{6} \end{bmatrix}$$

$$= \frac{1}{10} \begin{bmatrix} 1 & 1 & 1 \\ 4 & -\frac{8}{3} & \frac{2}{3} \\ 5 & \frac{5}{3} & -\frac{5}{3} \end{bmatrix}. \quad \blacksquare$$

Theorem 2.18 *In the cyclic association scheme defined by the blueprint $\{\Delta_i : i \in \mathcal{K}\}$ of $\mathbb{Z}_n$, the eigenvalues of A_i are*

$$\sum_{\alpha \in \Delta_i} \epsilon^\alpha$$

as ϵ ranges over the complex n-th roots of unity.

Proof Consider the complex vector $v = \sum_{\beta \in \Omega} \epsilon^\beta \chi_\beta$, where $\epsilon^n = 1$. Here $A_i = \sum_{\alpha \in \Delta_i} M_\alpha$, so

$$\begin{aligned} A_i v &= \sum_{\alpha \in \Delta_i} M_\alpha \sum_{\beta \in \Omega} \epsilon^\beta \chi_\beta \\ &= \sum_{\alpha \in \Delta_i} \sum_{\beta \in \Omega} \epsilon^\beta M_\alpha \chi_\beta \\ &= \sum_{\alpha \in \Delta_i} \epsilon^\alpha \sum_{\beta \in \Omega} \epsilon^{\beta - \alpha} \chi_{\beta - \alpha} \\ &= \sum_{\alpha \in \Delta_i} \epsilon^\alpha v. \end{aligned}$$

Note that subtraction is modulo n in both $\epsilon^{\beta-\alpha}$ and $\chi_{\beta-\alpha}$, so there is no problem about inconsistency. Now, $\Delta_i = -\Delta_i$, so not only is

$$\sum_{\alpha \in \Delta_i} \epsilon^{-\alpha} = \sum_{\alpha \in \Delta_i} \epsilon^\alpha$$

but also this value is real. If $\epsilon \in \{1, -1\}$ then v is real; otherwise v and its complex conjugate $\bar{v}$ have the same real eigenvalue so $v + \bar{v}$ and $\mathrm{i}(v - \bar{v})$ are distinct real vectors with eigenvalue $\sum_{\alpha \in \Delta_i} \epsilon^\alpha$. ■

Technique 2.7 For a cyclic association scheme on $\mathbb{Z}_n$, calculate the eigenvalues $\sum_{\alpha \in \Delta_i} \epsilon^\alpha$, where $\epsilon^n = 1$, and amalgamate those spaces which have the same eigenvalue on every adjacency matrix.

Example 2.6 The 6-circuit gives the blueprint $\{0\}$, $\{\pm 1\}$, $\{\pm 2\}$, $\{3\}$

of $\mathbb{Z}_6$. Then $A_1 = M_1 + M_5$, so the eigenvalues of A_1 are

$$\begin{array}{ll} \epsilon_6 + \epsilon_6^5 & \text{(twice)} \\ \epsilon_6^2 + \epsilon_6^4 & \text{(twice)} \\ \epsilon_6^3 + \epsilon_6^3 & \text{(once)} \\ \epsilon_6^6 + \epsilon_6^6 & \text{(once)}, \end{array}$$

where ϵ_6 is a primitive sixth root of unity in $\mathbb{C}$. But $\epsilon_6^6 = 1$, $\epsilon_6^3 = -1$, $\epsilon_6^2 + \epsilon_6^4 = -1$ (because the cube roots of unity sum to zero) and $\epsilon_6 + \epsilon_6^5 = 1$ (because the sixth roots of unity sum to zero). So a portion of the character table is

$$\begin{array}{cc} & \\ & \\ 0 & (1) \\ \pm 1 & (2) \\ \pm 2 & (2) \\ 3 & (1) \end{array} \begin{array}{cccc} \epsilon_6^6 & \epsilon_6^{\pm 1} & \epsilon_6^{\pm 2} & \epsilon_6^3 \\ (1) & (2) & (2) & (1) \\ \left[\begin{array}{c} 1 \\ 2 \\ 2 \\ 1 \end{array} \right. & \begin{array}{c} 1 \\ 1 \\ \\ \\ \end{array} & \begin{array}{c} 1 \\ -1 \\ \\ \\ \end{array} & \left. \begin{array}{c} 1 \\ -2 \\ \\ \\ \end{array} \right] \end{array}.$$

In this case the matrix is already square, so there is no amalgamation of columns. If there are fewer than $\lceil (n+1)/2 \rceil$ rows, complete the table and then amalgamate identical columns. ■

The techniques are summarized in Figure 2.2.

Note that many authors use P and Q for the matrices that I call C and nD. Delsarte established this notation in his important work [89] on the connection between association schemes and error-correcting codes, and it was popularized by MacWilliams and Sloane's book [167]. However, I think that P and Q are already over-used in this subject. Quite apart from the use of P for probability in the context of random responses in designed experiments (see Chapter 4), P and Q are well established notation for projectors, and P is also the obvious letter for the matrices in Theorem 2.16. Moreover, I find C and D more memorable, because C contains the **c**haracters while D contains the **d**imensions.

2.5 Parameters of strongly regular graphs

Let $\mathcal{G}$ be a strongly regular graph on n vertices. Put $a_1 = a$, $p_{11}^1 = p$ and $p_{11}^2 = q$, so that every vertex has valency a, every edge is contained in p triangles, and every non-edge is contained in q paths of length 2. See Figure 2.3.

Counting the edges between $\mathcal{G}_1(\alpha)$ and $\mathcal{G}_2(\alpha)$ in two different ways

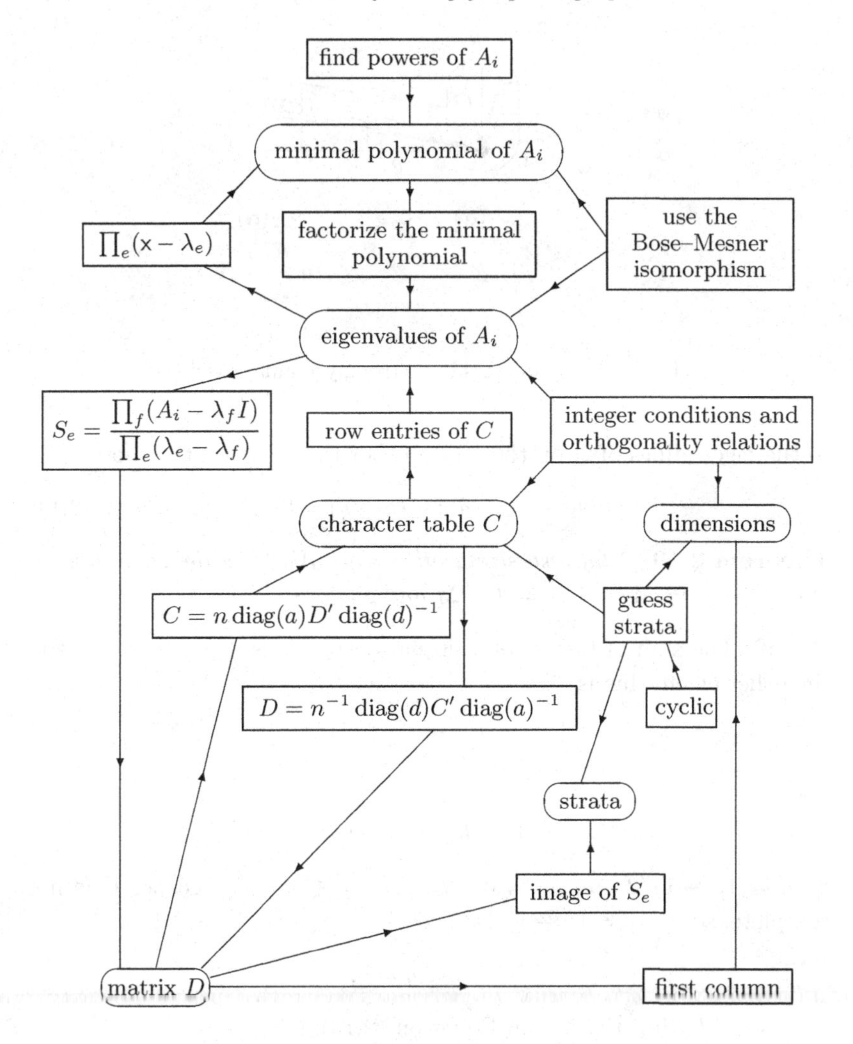

Fig. 2.2. Techniques for finding parameters of the second kind

shows that

$$a(a-p-1) = (n-a-1)q. \tag{2.13}$$

Moreover,

$$A^2 = aI + pA + q(J - A - I),$$

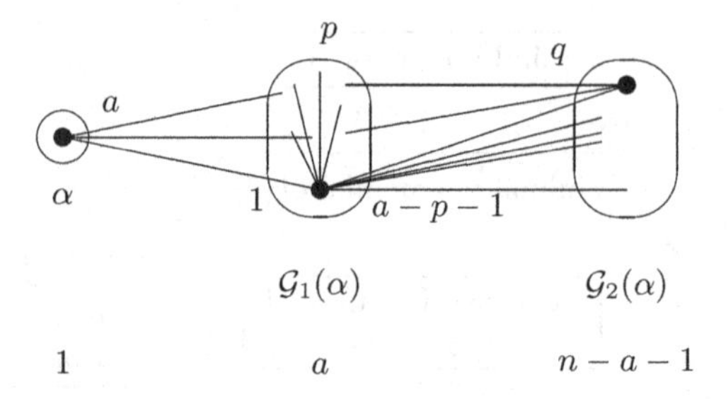

Fig. 2.3. Parameters of a strongly regular graph

so the eigenvalues of A on the strata other than W_0 are the roots of

$$\mathsf{x}^2 + (q-p)\mathsf{x} - (a-q) = 0. \tag{2.14}$$

Theorem 2.19 *If the two strata other than W_0 have the same dimension d then $n = 4q+1$, $a = d = 2q$ and $p = q-1$.*

Proof The sum of the roots of Equation (2.14) is equal to $p-q$, and the other eigenvalue is a, so

$$1 + 2d = n$$

and

$$a + d(p-q) = 0.$$

So $d = (n-1)/2$ and d divides a. But $a \leqslant n-2$, because $\mathcal{G}$ is not complete, so $a = d = (n-1)/2$. Then

$$1 + p - q = 0,$$

so $p = q-1$. Substitution in Equation (2.13) gives

$$a(a-q) = (2a+1-a-1)q$$

and so $a = 2q$. ■

Theorem 2.20 *If the two strata other than W_0 have different dimensions then $(q-p)^2 + 4(a-q)$ is a perfect square t^2, and t divides $(n-1)(p-q)+2a$, and the quotient has the same parity as $n-1$.*

Proof Put $u = (q-p)^2 + 4(a-q)$. The roots of Equation (2.14) are

$$\frac{p-q \pm \sqrt{u}}{2}.$$

Let d_1 and d_2 be the corresponding dimensions. Then

$$a + d_1\left(\frac{p-q+\sqrt{u}}{2}\right) + d_2\left(\frac{p-q-\sqrt{u}}{2}\right) = 0.$$

Since d_1 and d_2 are integers, if they are different then $\sqrt{u}$ is rational. But u is an integer, so it must be a perfect square t^2. Then

$$1 + d_1 + d_2 = n$$

and

$$2a + (d_1 + d_2)(p - q) + t(d_1 - d_2) = 0,$$

whose solutions are

$$\begin{aligned} d_1 &= \frac{1}{2}\left[(n-1) - \frac{(n-1)(p-q)+2a}{t}\right] \\ d_2 &= \frac{1}{2}\left[(n-1) + \frac{(n-1)(p-q)+2a}{t}\right]. \end{aligned}$$ ■

Example 2.7 Let $n = 11$ and $a = 4$. Then Equation (2.13) gives

$$4(3 - p) = 6q,$$

whose only solutions in non-negative integers are $p = 0$, $q = 2$ and $p = 3$, $q = 0$. Because 11 is not congruent to 1 modulo 4, Theorem 2.20 applies. If $p = 0$ and $q = 2$ then

$$(q - p)^2 + 4(a - q) = 2^2 + 4(4 - 2) = 12,$$

which is not a perfect square. If $p = 3$ and $q = 0$ then the graph is the disjoint union of complete graphs on four vertices, which is impossible because 4 does not divide 11. Hence there is no strongly regular graph with valency 4 on 11 vertices.

Since interchanging edges and non-edges (taking the complement) of a strongly regular graph yields another strongly regular graph, we see immediately that there is no strongly regular graph with valency 6 on 11 vertices. ■

Note that authors who specialize in strongly regular graphs use n, k, λ and μ where I use n, a, p and q. My choice of notation here is not only for consistency. I can see no need to use two different alphabets for these four parameters. In this subject, k is already over-used, both as a member of the index set $\mathcal{K}$ and as the block size (see Chapter 4). So is λ, which is not only a general scalar and a general eigenvalue, but which is firmly established as the notation for concurrence in Chapter 4. And μ is an occasional mate for λ, and more specifically the Möbius function in Chapter 6.

Exercises

2.1 Give an alternative proof that the space W spanned by χ_Ω is always a stratum by showing directly that

(a) χ_Ω is an eigenvector of every adjacency matrix;
(b) there is at least one matrix M in $\mathcal{A}$ which has W as a whole eigenspace.

2.2 Let $\mathcal{A}$ be the Bose–Mesner algebra of the group-divisible association scheme $\mathrm{GD}(b,k)$ on a set Ω of size bk. Let G be the adjacency matrix for the relation 'is in the same group as' on Ω.

(a) Show that $\{I, G, J\}$ is a basis for $\mathcal{A}$.
(b) Write down the multiplication table for $\{I, G, J\}$.
(c) Find all the idempotents in $\mathcal{A}$, expressing them in terms of this basis.

2.3 Prove Lemma 2.10.

2.4 Prove Corollary 2.15.

2.5 Verify the four sets of orthogonality relations for the group-divisible association scheme $\mathrm{GD}(b,k)$.

2.6 (a) Finish Example 2.3 (the cube association scheme) by finding the remaining three stratum projectors, the dimensions of the strata and the character table.

(b) Verify that the four vectors below are eigenvectors, with different eigenvalues, of the adjacency matrix Y for the cube association scheme, and that the stratum projectors found in part (a) do indeed have the right effect on these vectors.

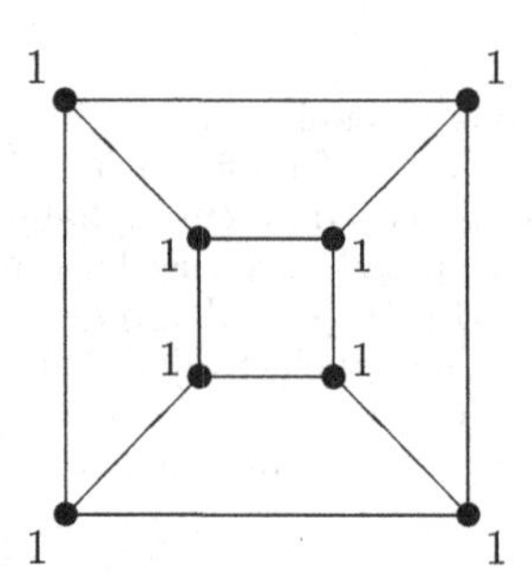

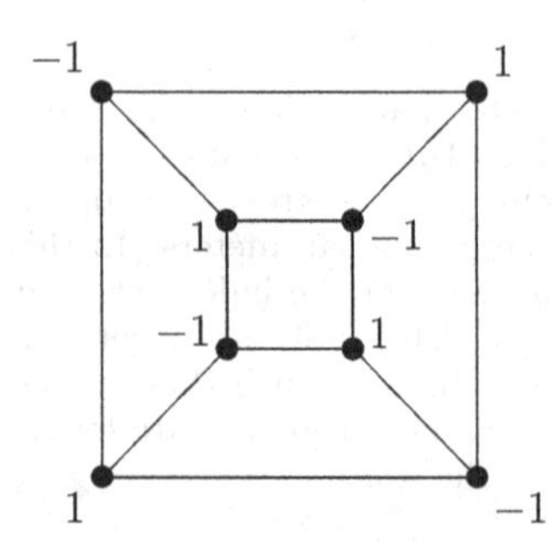

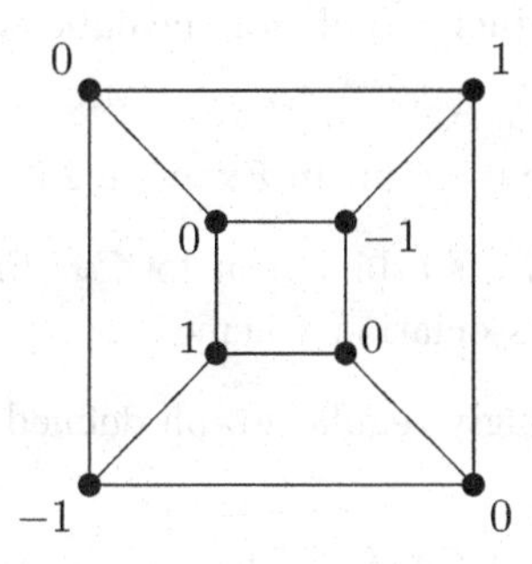

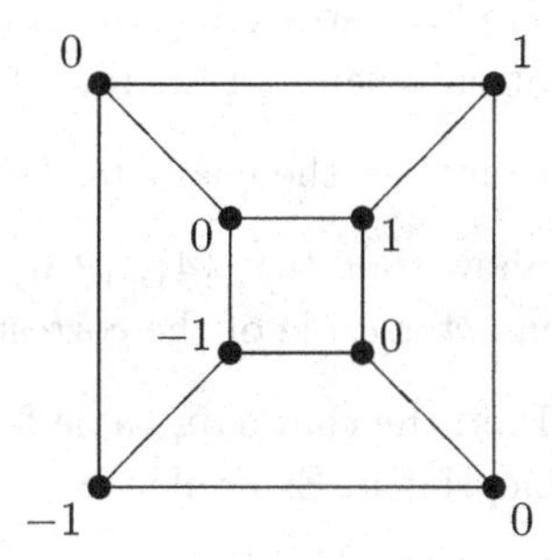

2.7 Find the minimal polynomial for the adjacency matrix A_1 in the Hamming association scheme $\mathrm{H}(4,2)$. Hence find the character table for $\mathrm{H}(4,2)$.

2.8 Let Ω be the set of 2-subsets of an n-set Γ. Consider the triangular association scheme on Ω.

For γ in Γ, define the subset $F(\gamma)$ of Ω by $F(\gamma) = \{\omega \in \Omega : \gamma \in \omega\}$. Let W_{parents} be the subspace of $\mathbb{R}^\Omega$ spanned by vectors of the form $\chi_{F(\gamma)} - \chi_{F(\delta)}$ for γ and δ in Γ. Prove that W_{parents} is a sub-eigenspace of A_1.

(The terminology comes from genetics, where $\{\gamma, \delta\}$ represents the genotype obtained by crossing two pure lines γ and δ.)

Put $W_{\text{offspring}} = (W_0 + W_{\text{parents}})^\perp$. Then $W_{\text{offspring}}$ is spanned by vectors such as $\chi_{\{\gamma,\delta\}} + \chi_{\{\eta,\zeta\}} - \chi_{\{\gamma,\eta\}} - \chi_{\{\delta,\zeta\}}$ where γ, δ, η and ζ are distinct elements of Γ. Show that $W_{\text{offspring}}$ is also a sub-eigenspace of A_1.

Hence find all the parameters of the second kind for the triangular scheme $\mathrm{T}(n)$.

The preceding approach uses Technique 2.1. Now find the parameters using Technique 2.2. (Hint: you should obtain Equation (5.1) on the way.)

2.9 Derive the four equations (2.8)–(2.11).

2.10 Find the matrices C and D for the icosahedral association scheme from Exercise 1.20.

2.11 Find the stratum projectors and the character table for the distance-regular graph in Exercise 1.27.

2.12 Consider the association scheme of Latin-square type $\mathrm{L}(3,n)$. Find its character table and the dimensions of its strata. Find its stratum projectors in terms of the matrices I, J, R, C and L used in Example 1.6.

2.13 Find the stratum projectors and character table for the cube-edge association scheme in Exercise 1.16.

2.14 Complete the character table for the 6-circuit in Example 2.6.

2.15 Show that $\{0\}$, $\{4\}$, $\{2,6\}$, $\{1,3,5,7\}$ is a blueprint for $\mathbb{Z}_8$. Find the character table of the corresponding association scheme.

2.16 Find the character table for the strongly regular graph defined by the blueprint in Example 1.8.

2.17 For each of the values of n, a, p and q below, either describe a strongly regular graph on n vertices with valency a such that every edge is contained in p triangles and every non-edge is contained in q paths of length 2, or prove that no such strongly regular graph exists.

(a) $n = 10,\quad a = 6,\quad p = 3,\quad q = 4.$
(b) $n = 13,\quad a = 6,\quad p = 2,\quad q = 3.$
(c) $n = 13,\quad a = 6,\quad p = 5,\quad q = 0.$
(d) $n = 16,\quad a = 9,\quad p = 4,\quad q = 6.$
(e) $n = 16,\quad a = 9,\quad p = 6,\quad q = 3.$
(f) $n = 20,\quad a = 4,\quad p = 3,\quad q = 0.$
(g) $n = 20,\quad a = 6,\quad p = 1,\quad q = 2.$
(h) $n = 170,\quad a = 13,\quad p = 0,\quad q = 1.$

2.18 A connected regular graph with diameter s is a *Moore graph* if it contains no circuits of length less than $2s + 1$.

(a) Prove that a Moore graph with diameter 2 is a strongly regular graph in which $n = a^2 + 1$, $p = 0$ and $q = 1$.
(b) Prove that, for such a Moore graph, $a \in \{2, 3, 7, 57\}$.
(c) Draw a Moore graph with diameter 2.

3

Combining association schemes

3.1 Tensor products

In this chapter we take an association scheme on Ω_1 and an association scheme on Ω_2, and combine them to obtain an association scheme on $\Omega_1 \times \Omega_2$. We need some preliminary notions about the space $\mathbb{R}^{\Omega_1 \times \Omega_2}$ of all functions from $\Omega_1 \times \Omega_2$ to $\mathbb{R}$.

For f in $\mathbb{R}^{\Omega_1}$ and g in $\mathbb{R}^{\Omega_2}$, define the *tensor product* $f \otimes g$ in $\mathbb{R}^{\Omega_1 \times \Omega_2}$ by

$$f \otimes g \colon (\omega_1, \omega_2) \mapsto f(\omega_1)g(\omega_2) \qquad \text{for } (\omega_1, \omega_2) \text{ in } \Omega_1 \times \Omega_2.$$

For example,

$$(f \otimes \chi_{\Omega_2})(\omega_1, \omega_2) = f(\omega_1).$$

For subspaces U of $\mathbb{R}^{\Omega_1}$ and V of $\mathbb{R}^{\Omega_2}$ define the *tensor product* $U \otimes V$ to be the subspace of $\mathbb{R}^{\Omega_1 \times \Omega_2}$ spanned by $\{f \otimes g : f \in U \text{ and } g \in V\}$. If $\{u_1, \ldots, u_m\}$ is a basis for U and $\{v_1, \ldots, v_r\}$ is a basis for V, then

$$\{u_i \otimes v_j : 1 \leqslant i \leqslant m,\ 1 \leqslant j \leqslant r\}$$

is a basis for $U \otimes V$, so

$$\dim(U \otimes V) = \dim(U)\dim(V).$$

The natural bases for $\mathbb{R}^{\Omega_1}$ and $\mathbb{R}^{\Omega_2}$ respectively are $\{\chi_\alpha : \alpha \in \Omega_1\}$ and $\{\chi_\beta : \beta \in \Omega_2\}$. But $\chi_\alpha \otimes \chi_\beta = \chi_{(\alpha,\beta)}$ and $\left\{\chi_{(\alpha,\beta)} : (\alpha, \beta) \in \Omega_1 \times \Omega_2\right\}$ is a basis for $\mathbb{R}^{\Omega_1 \times \Omega_2}$, so $\mathbb{R}^{\Omega_1} \otimes \mathbb{R}^{\Omega_2} = \mathbb{R}^{\Omega_1 \times \Omega_2}$.

Tensor products of matrices are also defined. Let M be a matrix in $\mathbb{R}^{\Omega_1 \times \Omega_1}$ and N be a matrix in $\mathbb{R}^{\Omega_2 \times \Omega_2}$. Since matrices are just functions on special sets, the preceding discussion shows that $M \otimes N$ should be a function in $\mathbb{R}^{(\Omega_1 \times \Omega_1) \times (\Omega_2 \times \Omega_2)}$. But it is convenient if $M \otimes N$ is square when M and N are square, so we use the natural bijection between

$\mathbb{R}^{(\Omega_1\times\Omega_1)\times(\Omega_2\times\Omega_2)}$ and $\mathbb{R}^{(\Omega_1\times\Omega_2)\times(\Omega_1\times\Omega_2)}$ and define $M\otimes N$ to be the matrix in $\mathbb{R}^{(\Omega_1\times\Omega_2)\times(\Omega_1\times\Omega_2)}$ given by

$$M\otimes N\left((\alpha_1,\alpha_2),(\beta_1,\beta_2)\right) = M(\alpha_1,\beta_1)N(\alpha_2,\beta_2) \qquad \text{for } \alpha_1,\ \beta_1 \text{ in } \Omega_1 \text{ and } \alpha_2,\ \beta_2 \text{ in } \Omega_2.$$

For example,

$$\begin{aligned} I_{\Omega_1}\otimes I_{\Omega_2} &= I_{\Omega_1\times\Omega_2};\\ J_{\Omega_1}\otimes J_{\Omega_2} &= J_{\Omega_1\times\Omega_2};\\ M\otimes O_{\Omega_2} = O_{\Omega_1}\otimes N &= O_{\Omega_1\times\Omega_2}. \end{aligned}$$

A concrete way of constructing $M\otimes N$ is to take each entry $M(\alpha,\beta)$ of M and replace it by the whole matrix N, all multiplied by $M(\alpha,\beta)$. Alternatively, one can start with the matrix N and replace each entry $N(\alpha,\beta)$ by the whole matrix M multiplied by $N(\alpha,\beta)$. So long as the rows and columns of $M\otimes N$ are *labelled*, these two constructions give the same thing.

Some standard facts about tensor products are gathered into the following proposition, whose proof is straightforward.

Proposition 3.1 *Let M, M_1 and M_2 be matrices in $\mathbb{R}^{\Omega_1\times\Omega_1}$ and N, N_1 and N_2 be matrices in $\mathbb{R}^{\Omega_2\times\Omega_2}$. Then*

(i) if $f\in\mathbb{R}^{\Omega_1}$ and $g\in\mathbb{R}^{\Omega_2}$ then

$$(M\otimes N)\,(f\otimes g) = (Mf)\otimes(Ng)$$

in the sense that

$$((M\otimes N)(f\otimes g))\,(\omega_1,\omega_2) = (Mf)\,(\omega_1)\times(Ng)\,(\omega_2);$$

(ii) if u is an eigenvector of M with eigenvalue λ and v is an eigenvector of N with eigenvalue μ then $u\otimes v$ is an eigenvector of $M\otimes N$ with eigenvalue $\lambda\mu$;

(iii) $(M\otimes N)' = M'\otimes N'$;

(iv) if M and N are symmetric then so is $M\otimes N$;

(v) for scalars λ and μ,

$$M\otimes(\lambda N_1+\mu N_2) = \lambda M\otimes N_1 + \mu M\otimes N_2$$

and

$$(\lambda M_1+\mu M_2)\otimes N = \lambda M_1\otimes N + \mu M_2\otimes N;$$

(vi) $(M_1\otimes N_1)(M_2\otimes N_2) = M_1M_2\otimes N_1N_2$.

Corollary 3.2 *If U is a subspace of $\mathbb{R}^{\Omega_1}$ with orthogonal projector P and V is a subspace of $\mathbb{R}^{\Omega_2}$ with orthogonal projector Q then the orthogonal projector onto $U \otimes V$ is $P \otimes Q$.*

Proof We need to prove that $P \otimes Q$ is the identity on $U \otimes V$ and is zero on $(U \otimes V)^\perp$.

Proposition 3.1(i) shows that if $u \in U$ and $v \in V$ then $(P \otimes Q)(u \otimes v) = (Pu) \otimes (Qv) = u \otimes v$.

Clearly, $\mathrm{Im}(P \otimes Q) = U \otimes V$. Proposition 3.1(iv) shows that $P \otimes Q$ is symmetric, so $\ker(P \otimes Q) = (\mathrm{Im}(P \otimes Q))^\perp = (U \otimes V)^\perp$. Therefore $P \otimes Q$ is zero on $(U \otimes V)^\perp$. ■

Note: tensor products are sometimes called *Kronecker products*.

3.2 Crossing

For the rest of this chapter we need notation for two different association schemes which is as close as possible to that used heretofore without being cumbersome. For symbols which are not already suffixed the obvious solution is to affix suffices 1 and 2. Otherwise the only clean solution is to use two different but related symbols, as shown in Table 3.1.

	first scheme	second scheme
set	Ω_1	Ω_2
size of set	n_1	n_2
associate classes ...	$\mathcal{C}$	$\mathcal{D}$
... indexed by	$\mathcal{K}_1$	$\mathcal{K}_2$
number of classes	s_1	s_2
parameters	p_{ij}^k	q_{xy}^z
valencies	a_i	b_x
adjacency matrices	A_i	B_x
strata ...	U_e	V_f
... indexed by	$\mathcal{E}_1$	$\mathcal{E}_2$
stratum projectors	S_e	T_f
dimensions	d_e	not used
character table ...	C_1	C_2
... and inverse	D_1	D_2

Table 3.1. *Notation for two association schemes*

We also need an abuse of notation similar to the one we made when defining the tensor product of matrices. If $\mathcal{C} \subseteq \Omega_1 \times \Omega_1$ and $\mathcal{D} \subseteq \Omega_2 \times \Omega_2$ then strictly speaking $\mathcal{C} \times \mathcal{D}$ is a subset of $(\Omega_1 \times \Omega_1) \times (\Omega_2 \times \Omega_2)$. We need to regard it as a subset of $(\Omega_1 \times \Omega_2) \times (\Omega_1 \times \Omega_2)$. So we make the convention that, for such sets $\mathcal{C}$ and $\mathcal{D}$,

$$\mathcal{C} \times \mathcal{D} = \{((\alpha_1, \alpha_2), (\beta_1, \beta_2)) : (\alpha_1, \beta_1) \in \mathcal{C} \text{ and } (\alpha_2, \beta_2) \in \mathcal{D}\}.$$

This convention has the happy consequence that

$$A_{\mathcal{C} \times \mathcal{D}} = A_{\mathcal{C}} \otimes A_{\mathcal{D}}. \tag{3.1}$$

Definition For $t = 1, 2$, let $\mathcal{Q}_t$ be a set of subsets of $\Omega_t \times \Omega_t$. The *direct product* $\mathcal{Q}_1 \times \mathcal{Q}_2$ of $\mathcal{Q}_1$ and $\mathcal{Q}_2$ is the set $\{\mathcal{C} \times \mathcal{D} : \mathcal{C} \in \mathcal{Q}_1 \text{ and } \mathcal{D} \in \mathcal{Q}_2\}$ of subsets of $(\Omega_1 \times \Omega_2) \times (\Omega_1 \times \Omega_2)$.

The operation of forming this direct product is called *crossing* $\mathcal{Q}_1$ and $\mathcal{Q}_2$.

Any sets of subsets of $\Omega_1 \times \Omega_1$ and $\Omega_2 \times \Omega_2$ can be crossed. It is clear that if $\mathcal{Q}_t$ is a partition of $\Omega_t \times \Omega_t$ for $t = 1, 2$ then $\mathcal{Q}_1 \times \mathcal{Q}_2$ is a partition of $(\Omega_1 \times \Omega_2) \times (\Omega_1 \times \Omega_2)$. The result of crossing is most interesting when the two components have some nice structure, such as being association schemes.

Theorem 3.3 *Let $\mathcal{Q}_1$ be an association scheme on Ω_1 with s_1 associate classes, valencies a_i for i in $\mathcal{K}_1$ and adjacency matrices A_i for i in $\mathcal{K}_1$. Let $\mathcal{Q}_2$ be an association scheme on Ω_2 with s_2 associate classes, valencies b_x for x in $\mathcal{K}_2$ and adjacency matrices B_x for x in $\mathcal{K}_2$. Then $\mathcal{Q}_1 \times \mathcal{Q}_2$ is an association scheme on $\Omega_1 \times \Omega_2$ with $s_1 s_2 + s_1 + s_2$ associate classes, valencies $a_i b_x$ for (i, x) in $\mathcal{K}_1 \times \mathcal{K}_2$ and adjacency matrices $A_i \otimes B_x$ for (i, x) in $\mathcal{K}_1 \times \mathcal{K}_2$.*

Proof Equation (3.1) shows that all the adjacency matrices have the required form. For (i, x) in $\mathcal{K}_1 \times \mathcal{K}_2$, the set of (i, x)-th associates of (α_1, α_2) is

$$\{(\beta_1, \beta_2) \in \Omega_1 \times \Omega_2 : \beta_1 \in \mathcal{C}_i(\alpha_1) \text{ and } \beta_2 \in \mathcal{D}_x(\alpha_2)\},$$

which has size $a_i b_x$.

The first two conditions for an association scheme are easy to check, for $\mathrm{Diag}(\Omega_1 \times \Omega_2) = \mathrm{Diag}(\Omega_1) \times \mathrm{Diag}(\Omega_2)$ and symmetry follows from Proposition 3.1(iv). For the third condition we check the product of adjacency matrices:

$$(A_i \otimes B_x)(A_j \otimes B_y) \quad = \quad A_i A_j \otimes B_x B_y$$

$$= \left(\sum_{k\in\mathcal{K}_1} p_{ij}^k A_k\right) \otimes \left(\sum_{z\in\mathcal{K}_2} q_{xy}^z B_z\right)$$

$$= \sum_{(k,z)\in\mathcal{K}_1\times\mathcal{K}_2} p_{ij}^k q_{xy}^z A_k \otimes B_z,$$

which is a linear combination of the adjacency matrices. So $\mathcal{Q}_1 \times \mathcal{Q}_2$ is indeed an association scheme.

The number of associate classes is one less than the number of sets in the partition $\mathcal{Q}_1 \times \mathcal{Q}_2$, which is equal to $(s_1 + 1)(s_2 + 1)$. So there are $s_1s_2 + s_1 + s_2$ associate classes. ■

Example 3.1 The direct product of the trivial association schemes $\underline{\underline{n}}$ and $\underline{\underline{m}}$ is indeed the rectangular association scheme $\mathrm{R}(n, m)$, with adjacency matrices I_{mn}, $I_n \otimes (J_m - I_m)$, $(J_n - I_n) \otimes I_m$ and $(J_n - I_n) \otimes (J_m - I_m)$. ■

Theorem 3.4 *Let the character tables of the association schemes $\mathcal{Q}_1$, $\mathcal{Q}_2$ and $\mathcal{Q}_1 \times \mathcal{Q}_2$ be C_1, C_2 and C with inverses D_1, D_2 and D. Let the strata for $\mathcal{Q}_1$ be U_e for e in $\mathcal{E}_1$ and the strata for $\mathcal{Q}_2$ be V_f for f in $\mathcal{E}_2$. Then*

(i) the strata for $\mathcal{Q}_1 \times \mathcal{Q}_2$ are $U_e \otimes V_f$ for (e, f) in $\mathcal{E}_1 \times \mathcal{E}_2$;
(ii) $C = C_1 \otimes C_2$;
(iii) $D = D_1 \otimes D_2$.

Proof (i) The spaces $U_e \otimes V_f$ for (e, f) in $\mathcal{E}_1 \times \mathcal{E}_2$ are pairwise orthogonal and sum to $\mathbb{R}^{\Omega_1\times\Omega_2}$. Proposition 3.1(ii) shows that every subspace of $\mathbb{R}^{\Omega_1\times\Omega_2}$ of the form $U_e \otimes V_f$ is a sub-eigenspace of every adjacency matrix of $\mathcal{Q}_1 \times \mathcal{Q}_2$, so every stratum is a sum of one or more of these spaces. But the number of strata is equal to the number of adjacency matrices, by Theorem 2.6, which is $(s_1 + 1)(s_2 + 1)$. The same theorem shows that $|\mathcal{E}_1| = s_1 + 1$ and $|\mathcal{E}_2| = s_2 + 1$. Hence the number of spaces of the form $U_e \otimes V_f$ is the same as the number of strata, and so these spaces are exactly the strata.

(ii) The eigenvalue of A_i on U_e is $C_1(i, e)$ and the eigenvalue of B_x on V_f is $C_2(x, f)$, so the eigenvalue of $A_i \otimes B_x$ on $U_e \otimes V_f$ is $C_1(i, e)C_2(x, f)$, by Proposition 3.1(ii). Hence $C((i, x)(e, f)) = C_1(i, e)C_2(x, f)$ and so $C = C_1 \otimes C_2$.

(iii) Lemma 2.9 shows that $D_1C_1 = I_{\Omega_1}$ and $D_2C_2 = I_{\Omega_2}$. Hence $(D_1 \otimes D_2)(C_1 \otimes C_2) = (D_1C_1) \otimes (D_2C_2) = I_{\Omega_1} \otimes I_{\Omega_2} = I_{\Omega_1\times\Omega_2}$. Thus $D_1 \otimes D_2 = (C_1 \otimes C_2)^{-1} = C^{-1} = D$. ■

Example 3.1 revisited The trivial association scheme $\underline{\underline{n}}$ has strata U_0 and $U_0^\perp$. The character table and inverse are

$$C_{\underline{\underline{n}}} = \begin{array}{ll} & \\ & \\ \text{same} & (1) \\ \text{different} & (n-1) \end{array} \begin{array}{c} \begin{array}{cc} U_0 & U_0^\perp \\ (1) & (n-1) \end{array} \\ \left[\begin{array}{cc} 1 & 1 \\ n-1 & -1 \end{array}\right] \end{array}$$

$$D_{\underline{\underline{n}}} = \begin{array}{ll} & \\ & \\ U_0 & (1) \\ U_0^\perp & (n-1) \end{array} \begin{array}{c} \begin{array}{cc} \text{same} & \text{different} \\ (1) & (n-1) \end{array} \\ \left[\begin{array}{cc} \dfrac{1}{n} & \dfrac{1}{n} \\ \dfrac{n-1}{n} & \dfrac{-1}{n} \end{array}\right] \end{array}.$$

(Why is it not correct to write '$nD_{\underline{\underline{n}}} = C_{\underline{\underline{n}}}$'?)

So the strata for $\underline{\underline{n}} \times \underline{\underline{m}}$ are $U_0 \otimes V_0$, $U_0 \otimes V_0^\perp$, $U_0^\perp \otimes V_0$ and $U_0^\perp \otimes V_0^\perp$. Showing the associate classes in the order 'same', 'same row', 'same column, 'other', the character table $C_{\underline{\underline{n}} \times \underline{\underline{m}}}$ is

$$\begin{array}{c} \begin{array}{cccc} U_0 \otimes V_0 & U_0 \otimes V_0^\perp & U_0^\perp \otimes V_0 & U_0^\perp \otimes V_0^\perp \\ (1) & (m-1) & (n-1) & ((m-1)(n-1)) \end{array} \\ \left[\begin{array}{cccc} 1 & 1 & 1 & 1 \\ m-1 & -1 & m-1 & -1 \\ n-1 & n-1 & -1 & -1 \\ (m-1)(n-1) & -(n-1) & -(m-1) & 1 \end{array}\right] \end{array}$$

and the inverse $D_{\underline{\underline{n}} \times \underline{\underline{m}}}$ is (showing the rows in the same order as the columns of $C_{\underline{\underline{n}} \times \underline{\underline{m}}}$)

$$\begin{array}{c} \begin{array}{cccc} \text{same} & \text{same row} & \text{same column} & \text{other} \\ (1) & (m-1) & (n-1) & ((m-1)(n-1)) \end{array} \\ \left[\begin{array}{cccc} \dfrac{1}{mn} & \dfrac{1}{mn} & \dfrac{1}{mn} & \dfrac{1}{mn} \\ \dfrac{m-1}{mn} & \dfrac{-1}{mn} & \dfrac{m-1}{mn} & \dfrac{-1}{mn} \\ \dfrac{n-1}{mn} & \dfrac{n-1}{mn} & \dfrac{-1}{mn} & \dfrac{-1}{mn} \\ \dfrac{(m-1)(n-1)}{mn} & \dfrac{-(n-1)}{mn} & \dfrac{-(m-1)}{mn} & \dfrac{1}{mn} \end{array}\right] \end{array}. \quad \blacksquare$$

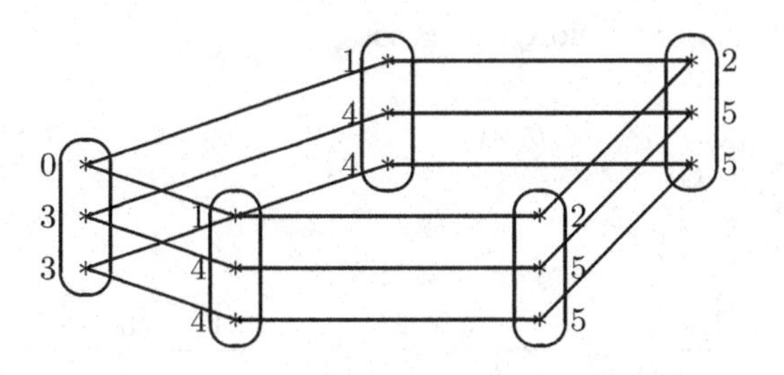

Fig. 3.1. The association scheme $\underline{\underline{3}} \times \textcircled{5}$

Example 3.2 The underlying set for the association scheme $\underline{\underline{3}} \times \textcircled{5}$ consists of five columns of three points each, with the columns arranged in a pentagon as shown in Figure 3.1. The (names of the) associate classes in $\underline{\underline{3}}$ are 'same' and 'different', while those in $\textcircled{5}$ are 'same', 'edge' and 'non-edge'. Each ordered pair of these gives an associate class in $\underline{\underline{3}} \times \textcircled{5}$.

One point in Figure 3.1 is labelled 0. The remaining points are labelled i if they are i-th associates of the point 0, according to the following table.

(same, same)	0	(different, same)	3
(same, edge)	1	(different, edge)	4
(same, non-edge)	2	(different, non-edge)	5

The character tables of the component association schemes are

$$C_{\underline{\underline{3}}} \quad = \quad \begin{array}{lc} & \\ & \\ \text{same} & (1) \\ \text{different} & (2) \end{array} \begin{array}{c} \begin{array}{cc} U_0 & U_0^\perp \\ (1) & (2) \end{array} \\ \left[\begin{array}{cc} 1 & 1 \\ 2 & -1 \end{array} \right] \end{array}$$

and

$$C_{\textcircled{5}} \quad = \quad \begin{array}{lc} & \\ & \\ \text{same} & (1) \\ \text{edge} & (2) \\ \text{non-edge} & (2) \end{array} \begin{array}{c} \begin{array}{ccc} V_0 & V_1 & V_2 \\ (1) & (2) & (2) \end{array} \\ \left[\begin{array}{ccc} 1 & 1 & 1 \\ 2 & \dfrac{-1+\sqrt{5}}{2} & \dfrac{-1-\sqrt{5}}{2} \\ 2 & \dfrac{-1-\sqrt{5}}{2} & \dfrac{-1+\sqrt{5}}{2} \end{array} \right] \end{array}$$

from Example 2.4. Therefore $C_{\underline{\underline{3}}\times ⑤}$ is

$$\begin{array}{cccccc} U_0\times V_0 & U_0\times V_1 & U_0\times V_2 & U_0^\perp\times V_0 & U_0^\perp\times V_1 & U_0^\perp\times V_2 \\ (1) & (2) & (2) & (2) & (4) & (4) \end{array}$$

$$\left[\begin{array}{cccccc} 1 & 1 & 1 & 1 & 1 & 1 \\ 2 & \frac{-1+\sqrt{5}}{2} & \frac{-1-\sqrt{5}}{2} & 2 & \frac{-1+\sqrt{5}}{2} & \frac{-1-\sqrt{5}}{2} \\ 2 & \frac{-1-\sqrt{5}}{2} & \frac{-1+\sqrt{5}}{2} & 2 & \frac{-1-\sqrt{5}}{2} & \frac{-1+\sqrt{5}}{2} \\ 2 & 2 & 2 & -1 & -1 & -1 \\ 4 & -1+\sqrt{5} & -1-\sqrt{5} & -2 & \frac{1-\sqrt{5}}{2} & \frac{1+\sqrt{5}}{2} \\ 4 & -1-\sqrt{5} & -1+\sqrt{5} & -2 & \frac{1+\sqrt{5}}{2} & \frac{1-\sqrt{5}}{2} \end{array}\right],$$

where the associate classes are shown in the order 0, 1, ..., 5. ■

3.3 Isomorphism

Definition Let $\mathcal{Q}_1$ be an association scheme on Ω_1 with classes $\mathcal{C}_i$ for i in $\mathcal{K}_1$, and let $\mathcal{Q}_2$ be an association scheme on Ω_2 with classes $\mathcal{D}_j$ for j in $\mathcal{K}_2$. Then $\mathcal{Q}_1$ is *isomorphic* to $\mathcal{Q}_2$ if there are bijections $\phi\colon \Omega_1 \to \Omega_2$ and $\pi\colon \mathcal{K}_1 \to \mathcal{K}_2$ such that

$$(\alpha, \beta) \in \mathcal{C}_i \iff (\phi(\alpha), \phi(\beta)) \in \mathcal{D}_{\pi(i)}.$$

The pair (ϕ, π) is an *isomorphism* between association schemes. We write $\mathcal{Q}_1 \cong \mathcal{Q}_2$.

If $\mathcal{K}_1 = \mathcal{K}_2$ then an isomorphism (ϕ, π) is a *strong isomorphism* if π is the identity. In this situation an isomorphism in which π is not necessarily so constrained is a *weak isomorphism*.

If $\mathcal{Q}_1 = \mathcal{Q}_2$ then an isomorphism (ϕ, π) is an *automorphism* of the association scheme.

Example 3.3 Let $\mathcal{Q}_1$ be GD(2, 2), which gives the strongly regular graph on the left of Figure 3.2, with classes $\mathcal{C}_0$, $\mathcal{C}_1$ and $\mathcal{C}_2$, where

$$\mathcal{C}_1 = \{(a, b), (b, a), (c, d), (d, c)\}.$$

Let $\mathcal{Q}_2$ be ④, which gives the strongly regular graph on the right of

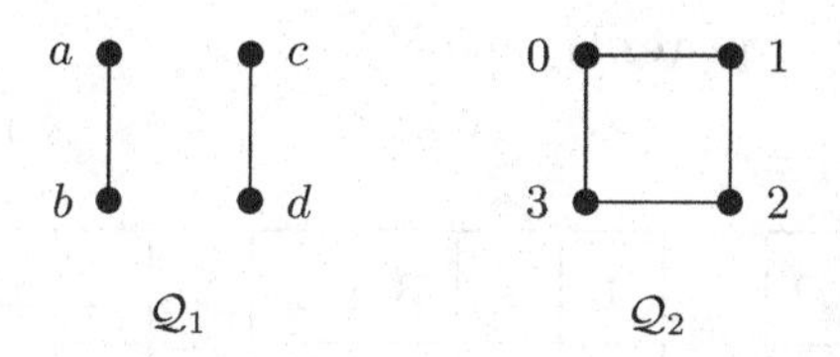

Fig. 3.2. Two association schemes in Example 3.3

Figure 3.2, with classes $\mathcal{D}_0$, $\mathcal{D}_1$ and $\mathcal{D}_2$, where

$$\mathcal{D}_1 = \{(0,1),(1,0),(1,2),(2,1),(2,3),(3,2),(3,0),(0,3)\}.$$

Let

$$\phi(a) = 0 \qquad \phi(b) = 2 \qquad \phi(c) = 1 \qquad \phi(d) = 3$$

and

$$\pi(0) = 0 \qquad \pi(1) = 2 \qquad \pi(2) = 1.$$

Then (ϕ, π) is an isomorphism from $\mathcal{Q}_1$ to $\mathcal{Q}_2$, so that GD(2, 2) is isomorphic to ④. If we consider that the labels of the classes $\mathcal{C}_i$ are the same as the labels of the classes $\mathcal{D}_i$ then (ϕ, π) is a weak isomorphism but not a strong one, because it carries edges to non-edges. The two strongly regular graphs in Figure 3.2 are not isomorphic *as graphs* even though they define isomorphic association schemes. ■

Proposition 3.5 *Let Δ_0, Δ_1, ..., Δ_s be a blueprint for $\mathbb{Z}_n$. Suppose that m is coprime to n. Let ϕ be the permutation of $\mathbb{Z}_n$ defined by $\phi(\omega) = m\omega$ for ω in $\mathbb{Z}_n$. If there is a permutation π of $\{0,\dots,s\}$ such that $\{\phi(\omega) : \omega \in \Delta_i\} = \Delta_{\pi(i)}$ for $i = 0$, ..., s then (ϕ, π) is a weak automorphism of the cyclic association scheme defined by this blueprint.*

Example 3.4 (Example 1.8 continued) We know that Δ_0, Δ_1, Δ_2 is a blueprint for $\mathbb{Z}_{13}$, where $\Delta_0 = \{0\}$, $\Delta_1 = \{1,3,4,-4,-3,-1\}$ and $\Delta_2 = \{2,5,6,-6,-5,-2\}$. Let $\phi(\omega) = 2\omega$ for ω in $\mathbb{Z}_{13}$. Then $\phi(\Delta_0) = \Delta_0$, $\phi(\Delta_1) = \Delta_2$ and $\phi(\Delta_2) = \Delta_1$ so ϕ induces a weak automorphism of the cyclic association scheme defined by Δ_0, Δ_1, Δ_2. ■

It is evident that isomorphic association schemes have the same parameters, but the converse is not true.

Example 3.5 Let $\mathcal{Q}_1$ and $\mathcal{Q}_2$ be the association schemes of L(3, 4) type defined by the Latin squares Π_1 and Π_2.

$\Pi_1 =$

A	B	C	D
D	A	B	C
C	D	A	B
B	C	D	A

$\Omega =$

1	2	3	4
5	6	7	8
9	10	11	12
13	14	15	16

$\Pi_2 =$

A	B	C	D
B	A	D	C
C	D	A	B
D	C	B	A

Draw edges between first associates. In both schemes $p_{11}^1 = 4$ so every edge is contained in four triangles. In Π_2 every pair of points in the same row or column or letter is contained in a 2×2 Latin subsquare, so the edge is contained in two complete graphs of size 4:

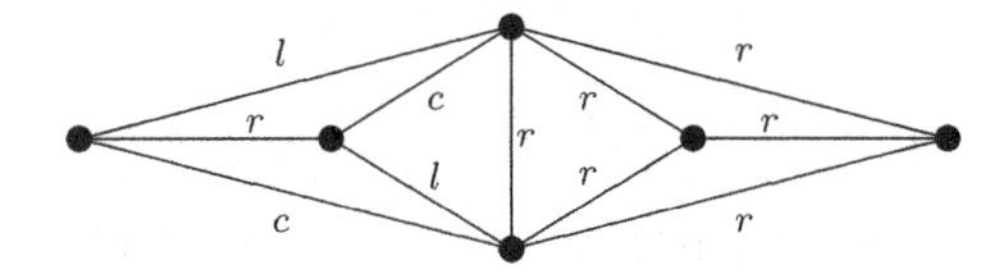

(the letters r, c and l denote 'same row', 'same column' and 'same letter' respectively). However, there are edges in $\mathcal{Q}_1$ which are not contained in two complete graphs of size 4. For example, the triangles through $\{1, 2\}$ are as follows.

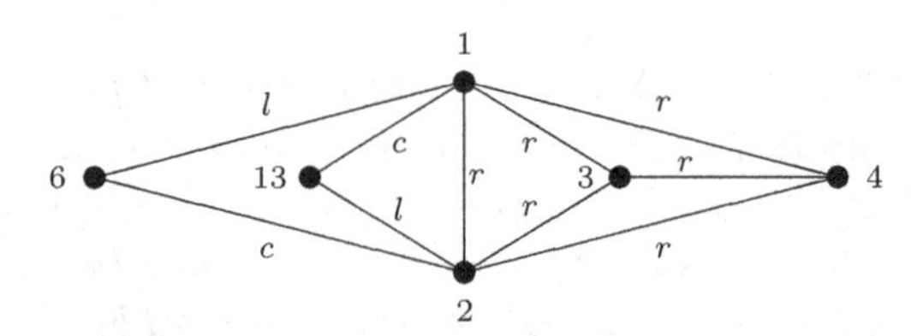

So $\mathcal{Q}_1$ cannot be isomorphic to $\mathcal{Q}_2$. ■

Theorem 3.6 *Crossing is commutative in the sense that $\mathcal{Q}_1 \times \mathcal{Q}_2$ is isomorphic to $\mathcal{Q}_2 \times \mathcal{Q}_1$.*

Proof Take $\phi((\omega_1, \omega_2)) = (\omega_2, \omega_1)$ and $\pi((i, x)) = (x, i)$. ■

Theorem 3.7 *Crossing is associative in the sense that $\mathcal{Q}_1 \times (\mathcal{Q}_2 \times \mathcal{Q}_3) \cong (\mathcal{Q}_1 \times \mathcal{Q}_2) \times \mathcal{Q}_3$.*

3.4 Nesting

When we cross $\mathcal{Q}_1$ with $\mathcal{Q}_2$, the new underlying set can be thought of as the rectangle $\Omega_1 \times \Omega_2$. When we nest $\mathcal{Q}_2$ within $\mathcal{Q}_1$ we replace each element of Ω_1 by a copy of Ω_2. Then the old adjacencies within Ω_1 apply to whole copies of Ω_2, while the old adjacencies within Ω_2 apply only within each separate copy.

Definition For $t = 1, 2$, let $\mathcal{Q}_t$ be a set of subsets of $\Omega_t \times \Omega_t$. Suppose that $\mathcal{Q}_1$ contains $\mathrm{Diag}(\Omega_1)$. The *wreath product* $\mathcal{Q}_1/\mathcal{Q}_2$ of $\mathcal{Q}_1$ and $\mathcal{Q}_2$ is the set of subsets

$$\{\mathcal{C} \times (\Omega_2 \times \Omega_2) : \mathcal{C} \in \mathcal{Q}_1,\ \mathcal{C} \neq \mathrm{Diag}(\Omega_1)\} \cup \{\mathrm{Diag}(\Omega_1) \times \mathcal{D} : \mathcal{D} \in \mathcal{Q}_2\}$$

of $(\Omega_1 \times \Omega_2) \times (\Omega_1 \times \Omega_2)$. The operation of forming this wreath product is called *nesting* $\mathcal{Q}_2$ within $\mathcal{Q}_1$.

It is clear that $\mathcal{Q}_1/\mathcal{Q}_2$ is a partition of $(\Omega_1 \times \Omega_2) \times (\Omega_1 \times \Omega_2)$ if $\mathcal{Q}_1$ is a partition of $\Omega_1 \times \Omega_1$ containing $\mathrm{Diag}(\Omega_1)$ and $\mathcal{Q}_2$ is a partition of $\Omega_2 \times \Omega_2$.

Theorem 3.8 *Let $\mathcal{Q}_1$ be an association scheme on a set Ω_1 of size n_1 with s_1 associate classes, valencies a_i for i in $\mathcal{K}_1$ and adjacency matrices A_i for i in $\mathcal{K}_1$. Let $\mathcal{Q}_2$ be an association scheme on a set Ω_2 of size n_2 with s_2 associate classes, valencies b_x for x in $\mathcal{K}_2$ and adjacency matrices B_x for x in $\mathcal{K}_2$. Then $\mathcal{Q}_1/\mathcal{Q}_2$ is an association scheme on $\Omega_1 \times \Omega_2$ with $s_1 + s_2$ associate classes, valencies $a_i n_2$ for i in $\mathcal{K}_1 \setminus \{0\}$ and b_x for x in $\mathcal{K}_2$, and adjacency matrices $A_i \otimes J_{\Omega_2}$ for i in $\mathcal{K}_1 \setminus \{0\}$ and $I_{\Omega_1} \otimes B_x$ for x in $\mathcal{K}_2$.*

Proof Everything follows immediately from the definition except the fact that the product of two adjacency matrices is a linear combination of adjacency matrices. However,

$$\begin{aligned} & (A_i \otimes J_{\Omega_2})(A_j \otimes J_{\Omega_2}) \\ = {} & A_iA_j \otimes n_2 J_{\Omega_2} = n_2 \left(\sum_{k\in\mathcal{K}_1} p_{ij}^k A_k \right) \otimes J_{\Omega_2} \\ = {} & n_2 \sum_{k\in\mathcal{K}_1\setminus\{0\}} p_{ij}^k A_k \otimes J_{\Omega_2} + n_2 p_{ij}^0 \sum_{x\in\mathcal{K}_2} I_{\Omega_1} \otimes B_x. \end{aligned}$$

Moreover,

$$(A_i \otimes J_{\Omega_2})(I_{\Omega_1} \otimes B_x) = A_i \otimes b_x J_{\Omega_2},$$

and

$$(I_{\Omega_1} \otimes B_x)(I_{\Omega_1} \otimes B_y) = I_{\Omega_1} \otimes B_x B_y = \sum_{k \in \mathcal{K}_2} q_{xy}^z I_{\Omega_1} \otimes B_z. \quad \blacksquare$$

Example 3.6 The wreath product of the trivial association schemes $\underline{\underline{b}}$ and $\underline{\underline{k}}$ is indeed the group-divisible association scheme GD(b,k), with adjacency matrices $(J_b - I_b) \otimes J_k$, $I_b \otimes I_k$ and $I_b \otimes (J_k - I_k)$. ■

This example shows immediately that nesting is not commutative. For example, $\underline{\underline{10}}/\underline{\underline{3}}$ has valencies 1, 2 and 27 while $\underline{\underline{3}}/\underline{\underline{10}}$ has valencies 1, 9 and 20.

Example 3.7 The underlying set for the association scheme $\underline{\underline{3}}$/⑤ consists of three separate pentagons, as shown in Figure 3.3. One point is labelled 0. Each other point is labelled i if it is an i-th associate of the point 0, according to the following table.

(same, same)	0	(same, non-edge)	2
(same, edge)	1	(different, any)	3

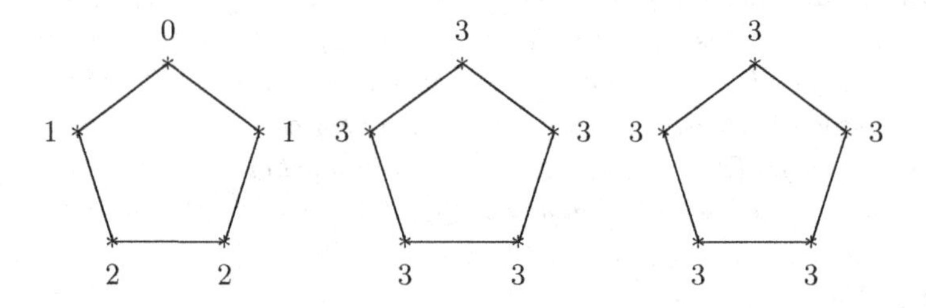

Fig. 3.3. The association scheme $\underline{\underline{3}}$/⑤

Contrast this with Example 3.2. Also contrast it with the association scheme ⑤/$\underline{\underline{3}}$, where the underlying set is a pentagon with three points at each vertex. This is shown in Figure 3.4, labelled as follows.

(same, same)	0	(edge, any)	2
(same, different)	1	(non-edge, any)	3

■

Theorem 3.9 *Let the strata for $\mathcal{Q}_1$ be U_e, for e in $\mathcal{E}_1$, and the strata for $\mathcal{Q}_2$ be V_f, for f in $\mathcal{E}_2$. Then the strata for $\mathcal{Q}_1/\mathcal{Q}_2$ are $U_e \otimes V_0$, for e in $\mathcal{E}_1$, and $\mathbb{R}^{\Omega_1} \otimes V_f$, for f in $\mathcal{E}_2 \setminus \{0\}$.*

Moreover, let C be the character table of $\mathcal{Q}_1/\mathcal{Q}_2$, and D its inverse,

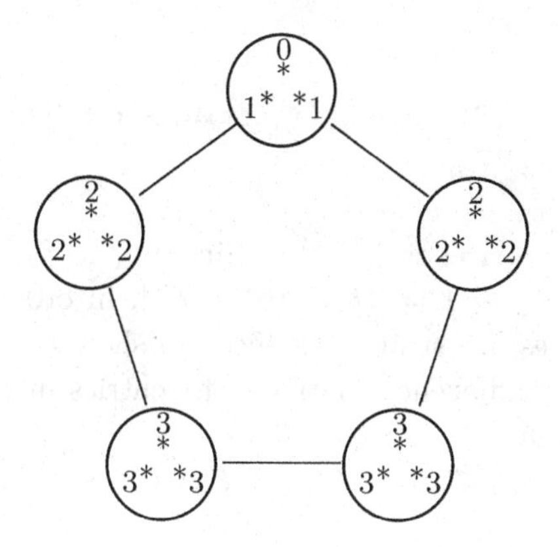

Fig. 3.4. The association scheme ⑤/$\underline{\underline{3}}$

so that the rows of C and columns of D are indexed by $(\mathcal{K}_1 \setminus \{0\}) \cup \mathcal{K}_2$, *while the columns of C and the rows of D are indexed by* $\mathcal{E}_1 \cup (\mathcal{E}_2 \setminus \{0\})$. *If* $\mathcal{Q}_1$ *and* $\mathcal{Q}_2$ *have character tables* C_1 *and* C_2 *with inverses* D_1 *and* D_2 *then*

$$\begin{aligned}
C(i,e) &= n_2 C_1(i,e) && \text{for } i \in \mathcal{K}_1 \setminus \{0\} \text{ and } e \in \mathcal{E}_1,\\
C(i,f) &= 0 && \text{for } i \in \mathcal{K}_1 \setminus \{0\} \text{ and } f \in \mathcal{E}_2 \setminus \{0\},\\
C(x,e) &= b_x && \text{for } x \in \mathcal{K}_2 \text{ and } e \in \mathcal{E}_1,\\
C(x,f) &= C_2(x,f) && \text{for } x \in \mathcal{K}_2 \text{ and } f \in \mathcal{E}_2 \setminus \{0\},
\end{aligned}$$

and

$$\begin{aligned}
D(e,i) &= \frac{1}{n_2} D_1(e,i) && \text{for } e \in \mathcal{E}_1 \text{ and } i \in \mathcal{K}_1 \setminus \{0\},\\
D(e,x) &= \frac{d_e}{n_1 n_2} && \text{for } e \in \mathcal{E}_1 \text{ and } x \in \mathcal{K}_2,\\
D(f,i) &= 0 && \text{for } f \in \mathcal{E}_2 \setminus \{0\} \text{ and } i \in \mathcal{K}_1 \setminus \{0\},\\
D(f,x) &= D_2(f,x) && \text{for } f \in \mathcal{E}_2 \setminus \{0\} \text{ and } x \in \mathcal{K}_2.
\end{aligned}$$

Proof The spaces $U_e \otimes V_0$, for e in $\mathcal{E}_1$, and $\mathbb{R}^{\Omega_1} \otimes V_f$, for f in $\mathcal{E}_2 \setminus \{0\}$, are pairwise orthogonal. Also,

$$\sum_{e \in \mathcal{E}_1} (U_e \otimes V_0) = \mathbb{R}^{\Omega_1} \otimes V_0$$

and

$$\sum_{f\in\mathcal{E}_2\setminus\{0\}} (\mathbb{R}^{\Omega_1}\otimes V_f) = \mathbb{R}^{\Omega_1}\otimes V_0^{\perp},$$

so the sum of these spaces is $\mathbb{R}^{\Omega_1\times\Omega_2}$. Since $|\mathcal{E}_1\cup(\mathcal{E}_2\setminus\{0\})| = |\mathcal{E}_1| + |\mathcal{E}_2| - 1 = |\mathcal{K}_1| + |\mathcal{K}_2| - 1 = |(\mathcal{K}_1\setminus\{0\})\cup\mathcal{K}_2|$, in order to prove that the named subspaces are strata it suffices to show that they are sub-eigenspaces of every adjacency matrix. The entries in C come as part of this demonstration.

Let $i\in\mathcal{K}_1\setminus\{0\}$, $x\in\mathcal{K}_2$, $e\in\ \mathcal{E}_1$, $f\in\mathcal{E}_2\setminus\{0\}$, $u\in U_e$, $w\in\mathbb{R}^{\Omega_1}$ and $v\in V_f$. Then

$$(A_i\otimes J_{\Omega_2})(u\otimes\chi_{\Omega_2}) = A_iu\otimes J_{\Omega_2}\chi_{\Omega_2} = C_1(i,e)u\otimes n_2\chi_{\Omega_2};$$

$$(A_i\otimes J_{\Omega_2})(w\otimes v) = A_iw\otimes J_{\Omega_2}v = A_iw\otimes 0_{\Omega_2} = 0_{\Omega_1\times\Omega_2};$$

$$(I_{\Omega_1}\otimes B_x)(u\otimes\chi_{\Omega_2}) = I_{\Omega_1}u\otimes B_x\chi_{\Omega_2} = u\otimes b_x\chi_{\Omega_2};$$

$$(I_{\Omega_1}\otimes B_x)(w\otimes v) = I_{\Omega_1}w\otimes B_xv = w\otimes C_2(x,f)v.$$

Finally, we find the entries in D by expressing the stratum projectors as linear combinations of the adjacency matrices. If S_e is the projector onto U_e and T_f is the projector onto V_f then the projectors onto $U_e\otimes V_0$ and $\mathbb{R}^{\Omega_1}\otimes V_f$ are $S_e\otimes T_0$ and $I_{\Omega_1}\otimes T_f$, by Corollary 3.2. But

$$\begin{aligned} S_e\otimes T_0 &= \left(\sum_{i\in\mathcal{K}_1} D_1(e,i)A_i\right)\otimes\frac{1}{n_2}J_{\Omega_2} \\ &= \frac{1}{n_2}\sum_{i\in\mathcal{K}_1\setminus\{0\}} D_1(e,i)A_i\otimes J_{\Omega_2} + \frac{D_1(e,0)}{n_2}\sum_{x\in\mathcal{K}_2} I_{\Omega_1}\otimes B_x \\ &= \frac{1}{n_2}\sum_{i\in\mathcal{K}_1\setminus\{0\}} D_1(e,i)A_i\otimes J_{\Omega_2} + \frac{d_e}{n_1n_2}\sum_{x\in\mathcal{K}_2} I_{\Omega_1}\otimes B_x, \end{aligned}$$

while

$$I_{\Omega_1}\otimes T_f = I_{\Omega_1}\otimes\left(\sum_{x\in\mathcal{K}_2} D_2(f,x)B_x\right). \quad \blacksquare$$

Using an obvious extension to the notation J and O, we have shown that C is

	$\mathcal{E}_1$	$\mathcal{E}_2 \setminus \{0\}$
$\mathcal{K}_1 \setminus \{0\}$	$n_2 \times$ [C_1 omitting 0-th row]	$O_{\mathcal{K}_1\setminus\{0\},\mathcal{E}_2\setminus\{0\}}$
$\mathcal{K}_2$	$\operatorname{diag}(b) J_{\mathcal{K}_2,\mathcal{E}_1}$	[C_2 omitting 0-th column]

while D is

	$\mathcal{K}_1 \setminus \{0\}$	$\mathcal{K}_2$
$\mathcal{E}_1$	$\frac{1}{n_2}$[D_1 omitting 0-th column]	$\frac{1}{n_1 n_2} \operatorname{diag}(d) J_{\mathcal{E}_1,\mathcal{K}_2}$
$\mathcal{E}_2 \setminus \{0\}$	$O_{\mathcal{E}_2\setminus\{0\},\mathcal{K}_1\setminus\{0\}}$	[D_2 omitting 0-th row].

Example 3.6 revisited For $\underline{\underline{b}}/\underline{\underline{k}}$ we have $n_1 = b$, $n_2 = k$,

$$C_1 = \begin{bmatrix} 1 & 1 \\ b-1 & -1 \end{bmatrix}, \qquad D_1 = \frac{1}{b}\begin{bmatrix} 1 & 1 \\ b-1 & -1 \end{bmatrix},$$

$$C_2 = \begin{bmatrix} 1 & 1 \\ k-1 & -1 \end{bmatrix}, \quad \text{and} \quad D_2 = \frac{1}{k}\begin{bmatrix} 1 & 1 \\ k-1 & -1 \end{bmatrix}.$$

So

$$C = \left[\begin{array}{cc|c} (b-1)k & -k & 0 \\ \hline 1 & 1 & 1 \\ k-1 & k-1 & -1 \end{array}\right]$$

and

$$D = \frac{1}{bk}\left[\begin{array}{c|cc} 1 & 1 & 1 \\ -1 & b-1 & b-1 \\ \hline 0 & b(k-1) & -b \end{array}\right].$$

This agrees with Example 2.2, except for the order in which the rows and the columns are written. ■

We see that it is not possible to simultaneously (i) label both the rows and the columns of C and D by (some of) the indices for $\mathcal{Q}_1$ followed by (some of) the indices for $\mathcal{Q}_2$ and (ii) show both the diagonal associate class and the stratum W_0 as the first row and column of these matrices. This is because the diagonal class of $\mathcal{Q}_1/\mathcal{Q}_2$ is labelled by the diagonal class of $\mathcal{Q}_2$ while the 0-th stratum of $\mathcal{Q}_1/\mathcal{Q}_2$ is labelled by the 0-th stratum of $\mathcal{Q}_1$. As explained before, where there is a natural bijection, or even partial bijection, between $\mathcal{K}$ and $\mathcal{E}$, it may not carry the 0-th associate class to the 0-th stratum.

Theorem 3.10 *Nesting is associative in the sense that* $\mathcal{Q}_1/(\mathcal{Q}_2/\mathcal{Q}_3)$ *is isomorphic to* $(\mathcal{Q}_1/\mathcal{Q}_2)/\mathcal{Q}_3$.

Note that algebraists write wreath products with the operands in the reverse order. Both orders are in use by combinatorialists.

3.5 Iterated crossing and nesting

Very many association schemes can now be formed from known ones by iterated crossing and nesting: for example,

$$\left(\mathrm{T}(5)/\underline{\underline{2}}\right) \times \underline{\underline{6}},$$

$$\underline{\underline{5}}/\left(\mathrm{H}(3,2) \times \textcircled{7}\right)/\underline{\underline{4}}.$$

Definition An association scheme $\mathcal{Q}$ is said to be *atomic* if whenever $\mathcal{Q} \cong \mathcal{Q}_1 \times \mathcal{Q}_2$ or $\mathcal{Q} \cong \mathcal{Q}_1/\mathcal{Q}_2$ then either $\mathcal{Q}_1 = \underline{\underline{1}}$ or $\mathcal{Q}_2 = \underline{\underline{1}}$.

The structures which occur most often on the experimental units for a designed experiment are those formed from trivial association schemes by crossing and nesting. An experiment to compare drugs which should relieve the symptoms of a chronic disease might use 200 patients for 6 years, changing each patient's drug each year: the structure (ignoring the drug treatments) is $\underline{\underline{200}} \times \underline{\underline{6}}$. An experiment on feeding piglets will use piglets from several litters. Typically the litters are not exactly the same size, but the experimenter uses the same number from each litter, say the nine biggest piglets. If there are 12 sows, this gives the structure $\underline{\underline{12}}/\underline{\underline{9}}$.

More interesting are association schemes derived from three trivial association schemes.

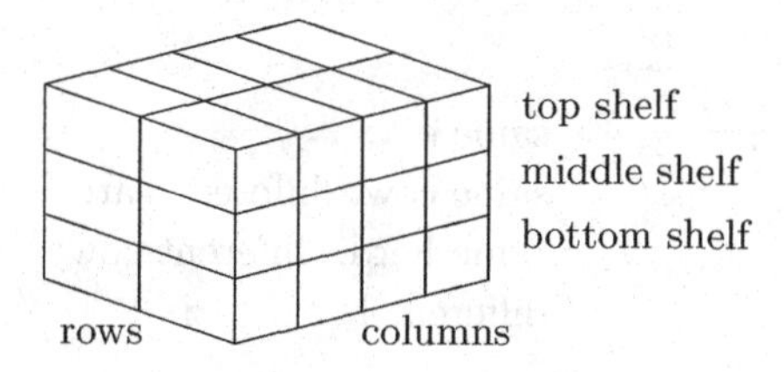

Fig. 3.5. The association scheme $\underline{\underline{2}} \times \underline{\underline{4}} \times \underline{\underline{3}}$ in Example 3.8

Example 3.8 Repeated crossing gives $\underline{\underline{n}} \times \underline{\underline{m}} \times \underline{\underline{r}}$, which is appropriate for experimental units in a 3-dimensional array, such as boxes of mushrooms stacked on shelves in a shed. See Figure 3.5. The associate classes are:

same box;
same row, same column, different shelf;
same row, different column, same shelf;
different row, same column, same shelf;
same row, different column, different shelf;
different row, same column, different shelf;
different row, different column, same shelf;
different row, different column, different shelf. ■

Example 3.9 Repeated nesting gives a structure such as $\underline{\underline{3}}/\underline{\underline{15}}/\underline{\underline{2}}$, which occurs in an experiment on cows' ears, using 15 cows from each of three herds. See Figure 3.6. The associate classes are

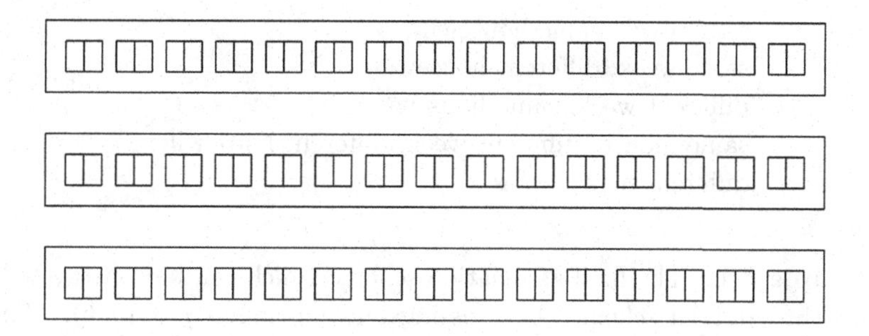

Fig. 3.6. The association scheme $\underline{\underline{3}}/\underline{\underline{15}}/\underline{\underline{2}}$ in Example 3.9

same ear;
same cow, different ear;
same herd, different cow;
different herd. ■

Example 3.10 Consumer experiments give some interesting examples. If each of twelve housewives tests two new vacuum cleaners per week for four weeks we have $(\underline{\underline{4}} \times \underline{\underline{12}})/\underline{\underline{2}}$. See Figure 3.7. The classes are

same week, same housewife, same test;
same week, same housewife, different test;
same week, different housewife;
different week, same housewife;
different week, different housewife. ■

	hw 1	hw 2	hw 3	hw 4	hw 5	hw 6	hw 7	hw 8	hw 9	hw 10	hw 11	hw 12
week 1	* *	* *	* *	* *	* *	* *	* *	* *	* *	* *	* *	* *
week 2	* *	* *	* *	* *	* *	* *	* *	* *	* *	* *	* *	* *
week 3	* *	* *	* *	* *	* *	* *	* *	* *	* *	* *	* *	* *
week 4	* *	* *	* *	* *	* *	* *	* *	* *	* *	* *	* *	* *

Fig. 3.7. The association scheme $(\underline{\underline{4}} \times \underline{\underline{12}})/\underline{\underline{2}}$ in Example 3.10

Example 3.11 On the other hand, if four housewives test one cleaner each week for the first four weeks, then four more housewives test cleaners for the next four weeks, then four more for a third "month" of four weeks, we have $\underline{\underline{3}}/(\underline{\underline{4}} \times \underline{\underline{4}})$. See Figure 3.8. Now the classes are

same week, same housewife;
same week, different housewife;
different week, same housewife;
same month, different week, different housewife;
different month. ■

Example 3.12 The final association scheme which can be constructed from three trivial schemes by crossing and nesting is $\underline{\underline{n}} \times (\underline{\underline{b}}/\underline{\underline{k}})$. An example is given by a trial of winter feed for sheep, shown in Figure 3.9. There were 12 sheep, kept in three houses of four sheep each. For a

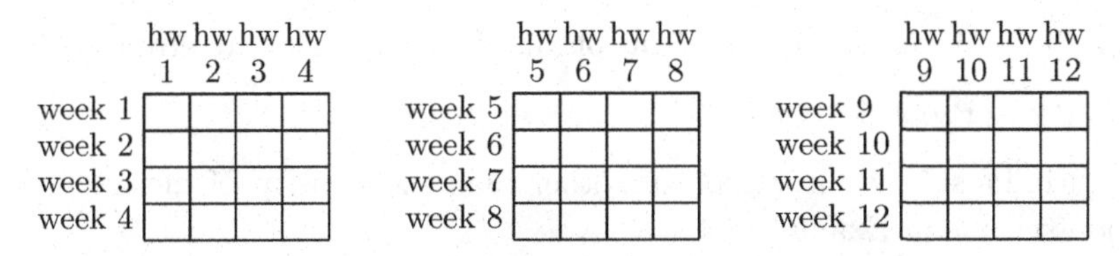

Fig. 3.8. The association scheme $\underline{\underline{3}}/(\underline{\underline{4}} \times \underline{\underline{4}})$ in Example 3.11

four-month period, each sheep was given a new feed each month. The association scheme on the sheep-months (ignoring the feeds) was thus $\underline{\underline{4}} \times (\underline{\underline{3}}/\underline{\underline{4}})$. The classes are

same sheep, same month;
same sheep, different month;
same house, different sheep, same month;
same house, different sheep, different month;
different house, same month;
different house, different month. ■

	house 1				house 2				house 3			
	sh 1	sh 2	sh 3	sh 4	sh 5	sh 6	sh 7	sh 8	sh 9	sh 10	sh 11	sh 12
month 1												
month 2												
month 3												
month 4												

Fig. 3.9. The association scheme $\underline{\underline{4}} \times (\underline{\underline{3}}/\underline{\underline{4}})$ in Example 3.12

There are other ways than crossing and nesting of making new association schemes from old. Some are described in Chapters 9 and 10.

Exercises

3.1 Prove Proposition 3.1.

3.2 Find the character tables of $\mathrm{T}(5) \times \underline{\underline{2}}$ and of $\underline{\underline{n}} \times \underline{\underline{m}} \times \underline{\underline{r}}$.

3.3 Find two strongly regular graphs with $n = 16$, $a = 6$, $p = 2$ and $q = 2$ which are not isomorphic to each other.

3.4 Find all weak automorphisms of ⑤. Which of them are strong?

3.5 Prove Proposition 3.5.

3.6 Find a strong automorphism, other than the identity, of the association scheme in Example 3.4.

3.7 For each of the following association schemes, find the numbers of weak and strong automorphisms. Describe the automorphisms in words.

(a) $\mathrm{GD}(2,3)$
(b) $\mathrm{H}(2,3)$
(c) $\mathrm{T}(5)$
(d) $\underline{\underline{2}} \times \underline{\underline{2}} \times \underline{\underline{2}}$

3.8 Let $\mathcal{L}$ be the association scheme of $\mathrm{L}(3,4)$ type on 16 points defined by the 4×4 Latin square Π_2 in Example 3.5. Describe the following association schemes.

(a) $\mathcal{L} \times \underline{\underline{3}}$
(b) $\underline{\underline{3}} \times \mathcal{L}$
(c) $\underline{\underline{3}}/\mathcal{L}$
(d) $\mathcal{L}/\underline{\underline{3}}$

3.9 Describe the association schemes $\underline{\underline{4}}$/⑥ and ⑥/$\underline{\underline{4}}$. Find their character tables and D matrices.

3.10 Show that $\mathrm{H}(3,3)$ is atomic.

3.11 In Example 3.9, what is the appropriate association scheme if the difference between right ears and left ears is deemed important?

3.12 Find the character table of each of the following association schemes.

(a) $\underline{\underline{3}}/\underline{\underline{15}}/\underline{\underline{2}}$
(b) $(\underline{\underline{4}} \times \underline{\underline{12}})/\underline{\underline{2}}$
(c) $\underline{\underline{3}}/(\underline{\underline{4}} \times \underline{\underline{4}})$
(d) $\underline{\underline{4}} \times (\underline{\underline{3}}/\underline{\underline{4}})$

3.13 Describe the association schemes obtainable from four trivial association schemes by crossing and nesting.

3.14 Write down as many association schemes as you can think of which have eight elements. Give enough detail about them to show that they are all distinct. (There are ten, and you know them all.)

4

Incomplete-block designs

4.1 Block designs

Let Ω be a set of experimental units, called *plots*, to which we can apply treatments in an experiment. Suppose that Ω is partitioned into b blocks of k plots each. A *block design* on Ω for a given treatment set Θ is this partition combined with a function $\psi\colon \Omega \to \Theta$ allocating treatments to plots. We say that 'treatment θ occurs on plot ω' if $\psi(\omega) = \theta$. Define the *design matrix* X in $\mathbb{R}^{\Omega\times\Theta}$ by

$$X(\omega, \theta) = \begin{cases} 1 & \text{if } \psi(\omega) = \theta \\ 0 & \text{otherwise.} \end{cases}$$

The partition into blocks defines a group-divisible association scheme $\mathrm{GD}(b, k)$ on Ω: its adjacency matrices are I, $B - I$ and $J - B$, where B is the matrix in $\mathbb{R}^{\Omega\times\Omega}$ defined by

$$B(\alpha, \beta) = \begin{cases} 1 & \text{if } \alpha \text{ and } \beta \text{ are in the same block} \\ 0 & \text{otherwise.} \end{cases}$$

Strictly speaking, the design is the quadruple $(\Omega, \mathcal{G}, \Theta, \psi)$, where Ω is the set of plots, $\mathcal{G}$ is the group-divisible association scheme on Ω, Θ is the set of treatments and ψ is the design function from Ω to Θ.

Definition The *replication* r_θ of treatment θ is

$$|\{\omega \in \Omega : \psi(\omega) = \theta\}|.$$

The design is *equi-replicate* if all treatments have the same replication.

If there are t treatments with replication r then

$$tr = bk. \tag{4.1}$$

In such a design, $X'X = rI$.

The trick for remembering the parameters is that b and k are to do with **b**loc**k**s while t and r concern **tr**eatments.

(In his first paper on incomplete-block designs, Yates was considering agricultural field trials to compare different varieties. So he called the treatments *varieties*, and let their number be v. This gives the mnemonic **v**a**r**ieties. In his second paper on the subject he switched to the less specific word *treatment*, and established the notation t, r, b, k used here. Statisticians have tended to follow this second usage, but pure mathematicians have retained v for the number of treatments, even though they often call them *points*.)

Example 4.1 A small design with 12 experimental units is shown in Figure 4.1. The set Ω consists of three blocks δ_1, δ_2, δ_3 of four plots each, so $b = 3$ and $k = 4$. The treatment set Θ is $\{1,2,3,4,5,6\}$, so $t = 6$. The design is equi-replicate with $r = 2$.

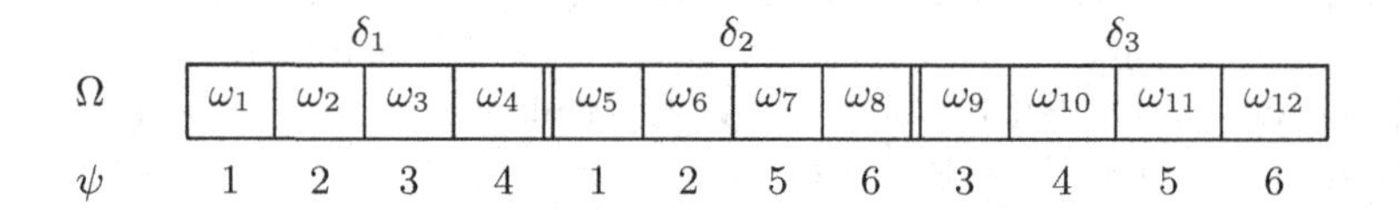

Fig. 4.1. Design in Example 4.1

With the plots in the obvious order, the blocks matrix B is given by

$$B = \begin{bmatrix} 1 & 1 & 1 & 1 & 0 & 0 & 0 & 0 & 0 & 0 & 0 & 0 \\ 1 & 1 & 1 & 1 & 0 & 0 & 0 & 0 & 0 & 0 & 0 & 0 \\ 1 & 1 & 1 & 1 & 0 & 0 & 0 & 0 & 0 & 0 & 0 & 0 \\ 1 & 1 & 1 & 1 & 0 & 0 & 0 & 0 & 0 & 0 & 0 & 0 \\ 0 & 0 & 0 & 0 & 1 & 1 & 1 & 1 & 0 & 0 & 0 & 0 \\ 0 & 0 & 0 & 0 & 1 & 1 & 1 & 1 & 0 & 0 & 0 & 0 \\ 0 & 0 & 0 & 0 & 1 & 1 & 1 & 1 & 0 & 0 & 0 & 0 \\ 0 & 0 & 0 & 0 & 1 & 1 & 1 & 1 & 0 & 0 & 0 & 0 \\ 0 & 0 & 0 & 0 & 0 & 0 & 0 & 0 & 1 & 1 & 1 & 1 \\ 0 & 0 & 0 & 0 & 0 & 0 & 0 & 0 & 1 & 1 & 1 & 1 \\ 0 & 0 & 0 & 0 & 0 & 0 & 0 & 0 & 1 & 1 & 1 & 1 \\ 0 & 0 & 0 & 0 & 0 & 0 & 0 & 0 & 1 & 1 & 1 & 1 \end{bmatrix}.$$

The stratum projectors are $\frac{1}{12}J$, $\frac{1}{4}B - \frac{1}{12}J$ and $I - \frac{1}{4}B$.

With the plots in the same order as above and treatments also in the obvious order, the design matrix X is given by

$$X = \begin{bmatrix} 1 & 0 & 0 & 0 & 0 & 0 \\ 0 & 1 & 0 & 0 & 0 & 0 \\ 0 & 0 & 1 & 0 & 0 & 0 \\ 0 & 0 & 0 & 1 & 0 & 0 \\ 1 & 0 & 0 & 0 & 0 & 0 \\ 0 & 1 & 0 & 0 & 0 & 0 \\ 0 & 0 & 0 & 0 & 1 & 0 \\ 0 & 0 & 0 & 0 & 0 & 1 \\ 0 & 0 & 1 & 0 & 0 & 0 \\ 0 & 0 & 0 & 1 & 0 & 0 \\ 0 & 0 & 0 & 0 & 1 & 0 \\ 0 & 0 & 0 & 0 & 0 & 1 \end{bmatrix}.$$

Then $X'X = 2I$. ■

Let Δ be the set of blocks. The *incidence matrix* is the matrix N in $\mathbb{R}^{\Delta\times\Theta}$ given by

$$N(\delta,\theta) = |\{\omega \in \delta : \psi(\omega) = \theta\}|.$$

The design is *binary* if $N(\delta,\theta) \in \{0,1\}$ for all δ in Δ and all θ in Θ. In this case we often identify block δ with $\{\theta \in \Theta : N(\delta,\theta) = 1\}$ and identify the design with the family of blocks.

The design is a *complete-block design* if it is binary and $k = |\Theta|$, so that $N = J_{\Delta,\Theta}$. It is an *incomplete-block design* if it is binary and $k < |\Theta|$, so that each block is incomplete in the sense of not containing all of the treatments. For the remainder of Chapters 4–5, only binary designs are considered.

(When Yates introduced incomplete-block designs he called them 'incomplete block designs' to contrast them with complete block designs. However, when pure mathematicians took up his term, they assumed that 'incomplete' referred to the *design* and meant that not every k-subset of the treatments was a block. That is why I insist on the hyphen in 'incomplete-block design'.)

Definition The *concurrence* $\Lambda(\theta,\eta)$ of treatments θ and η is

$$|\{(\alpha,\beta) \in \Omega\times\Omega : \alpha,\ \beta \text{ in the same block, } \psi(\alpha) = \theta,\ \psi(\beta) = \eta\}|.$$

The matrix Λ is called the *concurrence matrix*. Equi-replicate block designs in which every non-diagonal concurrence is in $\{0,1\}$ are called *configurations*.

Lemma 4.1 *(i) If the design is binary then $\Lambda(\theta,\theta)$ is equal to the replication of θ, and, for $\eta \neq \theta$, $\Lambda(\theta,\eta)$ is equal to the number of blocks in which θ and η both occur.*

(ii) $\sum_{\eta\in\Theta}\Lambda(\theta,\eta) = kr_\theta$; in particular, in an equi-replicate binary design

$$\sum_{\substack{\eta\in\Theta\\ \eta\neq\theta}} \Lambda(\theta,\eta) = r(k-1).$$

(iii) $\Lambda = X'BX = N'N$.

Proof The first two parts are simple counting, while the third uses Technique 1.1. ■

Example 4.1 revisited The incidence matrix N and concurrence matrix Λ are shown with their rows and columns labelled for clarity.

$$N = \begin{array}{c} \\ \delta_1 \\ \delta_2 \\ \delta_3 \end{array}\begin{array}{c} \begin{array}{cccccc} 1 & 2 & 3 & 4 & 5 & 6 \end{array} \\ \left[\begin{array}{cccccc} 1 & 1 & 1 & 1 & 0 & 0 \\ 1 & 1 & 0 & 0 & 1 & 1 \\ 0 & 0 & 1 & 1 & 1 & 1 \end{array}\right] \end{array} \qquad \Lambda = \begin{array}{c} \\ 1 \\ 2 \\ 3 \\ 4 \\ 5 \\ 6 \end{array}\begin{array}{c} \begin{array}{cccccc} 1 & 2 & 3 & 4 & 5 & 6 \end{array} \\ \left[\begin{array}{cccccc} 2 & 2 & 1 & 1 & 1 & 1 \\ 2 & 2 & 1 & 1 & 1 & 1 \\ 1 & 1 & 2 & 2 & 1 & 1 \\ 1 & 1 & 2 & 2 & 1 & 1 \\ 1 & 1 & 1 & 1 & 2 & 2 \\ 1 & 1 & 1 & 1 & 2 & 2 \end{array}\right] \end{array}.$$

It is clear that $\Lambda = N'N = X'BX$. ■

The *dual* block design to $(\Omega, \mathcal{G}, \Theta, \psi)$ is obtained by interchanging the roles of Θ and Δ. So the experimental units of the dual design are still the elements of Ω, but the blocks are the sets $\psi^{-1}(\theta)$ for θ in Θ. These new blocks define a group-divisible association scheme $\mathcal{G}^*$ on Ω of type $\mathrm{GD}(t,r)$. The dual design is the quadruple $(\Omega, \mathcal{G}^*, \Delta, \psi^*)$ where the dual design function $\psi^*\colon \Omega \to \Delta$ is given by

$$\psi^*(\omega) = \text{old block containing } \omega.$$

The incidence matrix for the dual design is N'.

Example 4.2 The dual of the design in Example 4.1 has treatment set $\{\delta_1, \delta_2, \delta_3\}$. Its blocks are $\{\delta_1, \delta_2\}$, $\{\delta_1, \delta_2\}$, $\{\delta_1, \delta_3\}$, $\{\delta_1, \delta_3\}$, $\{\delta_2, \delta_3\}$ and $\{\delta_2, \delta_3\}$.

Definition The *treatment-concurrence* graph of an incomplete-block design is the (not necessarily simple) graph whose vertices are the treatments and in which the number of edges from θ to η is $\Lambda(\theta,\eta)$ if $\theta \neq \eta$.

Definition An incomplete-block design is *connected* if its treatment-concurrence graph is connected.

4.2 Random variables

This section is a very brief recall of the facts we need about random variables. I will not attempt to explain what they are.

Given a set Γ, suppose that there are random variables $Y(\gamma)$, for γ in Γ, with a joint distribution. Then functions of two or more of these, such as $Y(\alpha)Y(\beta)$, are also random variables. The components $Y(\gamma)$ can be assembled into a random vector Y.

The random variable $Y(\gamma)$ has an *expectation* $\mathbb{E}(Y(\gamma))$ in $\mathbb{R}$. (I assume that the sum or integral which defines the expectation does converge for all the random variables that we consider.) Then we define $\mathbb{E}(Y)$ in $\mathbb{R}^\Gamma$ by

$$\mathbb{E}(Y)(\gamma) = \mathbb{E}(Y(\gamma)).$$

The main result we need about expectation is the following.

Proposition 4.2 *Expectation is* affine *in the sense that if* $M \in \mathbb{R}^{\Delta\times\Gamma}$ *and* $f \in \mathbb{R}^\Delta$ *then*

$$\mathbb{E}(MY + f) = M\,\mathbb{E}(Y) + f.$$

The *covariance* of random variables $Y(\alpha)$ and $Y(\beta)$ is defined by

$$\operatorname{cov}\left(Y(\alpha), Y(\beta)\right) = \mathbb{E}\left[\left(Y(\alpha) - \mathbb{E}(Y(\alpha))\right)\left(Y(\beta) - \mathbb{E}(Y(\beta))\right)\right].$$

The covariance of $Y(\alpha)$ with itself is called the *variance* of $Y(\alpha)$, written $\operatorname{Var}(Y(\alpha))$. The *covariance matrix* $\operatorname{Cov}(Y)$ of the random vector Y is defined by

$$\operatorname{Cov}(Y)(\alpha, \beta) = \operatorname{cov}\left(Y(\alpha), Y(\beta)\right).$$

Lemma 4.3 *Covariance is* bi-affine *in the sense that*

(i) *if* $g \in \mathbb{R}^\Gamma$ *then* $\operatorname{Cov}(Y + g) = \operatorname{Cov}(Y)$;
(ii) *if* $M \in \mathbb{R}^{\Delta\times\Gamma}$ *then* $\operatorname{Cov}(MY) = M\operatorname{Cov}(Y)M'$.

Proof (i) Put $Z = Y + g$. Proposition 4.2 shows that $\mathbb{E}(Z) = \mathbb{E}(Y) + g$, and therefore $Z - \mathbb{E}(Z) = Y - \mathbb{E}(Y)$: in particular $Z(\alpha) - \mathbb{E}(Z(\alpha)) = Y(\alpha) - \mathbb{E}(Y(\alpha))$.

(ii) By (i), we can assume that $\mathbb{E}(Y) = 0$. Then

$$\begin{aligned}
\operatorname{Cov}(MY)(\alpha,\beta) &= \operatorname{cov}((MY)(\alpha),(MY)(\beta)) \\
&= \mathbb{E}[((MY)(\alpha))((MY)(\beta))] \\
&= \mathbb{E}\left[\left(\sum_{\gamma\in\Gamma} M(\alpha,\gamma)Y(\gamma)\right)\left(\sum_{\delta\in\Gamma} M(\beta,\delta)Y(\delta)\right)\right] \\
&= \sum_{\gamma}\sum_{\delta} M(\alpha,\gamma)\,[\mathbb{E}(Y(\gamma)Y(\delta))]\,M'(\delta,\beta) \\
&= \sum_{\gamma}\sum_{\delta} M(\alpha,\gamma)\operatorname{Cov}(Y)(\gamma,\delta)M'(\delta,\beta) \\
&= (M\operatorname{Cov}(Y)M')\,(\alpha,\beta). \quad \blacksquare
\end{aligned}$$

Corollary 4.4 *(i) If λ is a real number then*

$$\operatorname{Var}(\lambda Y(\alpha)) = \lambda^2 \operatorname{Var}(Y(\alpha));$$

(ii) if $\omega_1, \ldots, \omega_n$ are distinct elements of Ω and $Z = \sum_{i=1}^n Y(\omega_n)$ then

$$\begin{aligned}
\operatorname{Var}(Z) &= \sum_{i=1}^{n}\sum_{j=1}^{n} \operatorname{cov}(Y(\omega_i), Y(\omega_j)) \\
&= \sum_{i=1}^{n} \operatorname{Var}(Y(\omega_i)) + 2\sum_{i=1}^{n}\sum_{j=i+1}^{n} \operatorname{cov}(Y(\omega_i), Y(\omega_j)).
\end{aligned}$$

Lemma 4.5 *The expectation of the sum of the squares of the entries in the random vector Y is equal to $\mathbb{E}(Y'Y)$ and to $\operatorname{tr}(\operatorname{Cov}(Y)) + \mathbb{E}(Y)'\,\mathbb{E}(Y)$.*

Proof The sum of the squares in Y is $Y'Y$, whose expectation is $\mathbb{E}(Y'Y)$. Now,

$$\begin{aligned}
\mathbb{E}(Y'Y) &= \sum_{\omega} \mathbb{E}\left((Y(\omega)^2)\right) \\
&= \sum_{\omega} \left(\operatorname{Var}(Y(\omega)) + (\mathbb{E}(Y(\omega)))^2\right) \\
&= \operatorname{tr}(\operatorname{Cov}(Y)) + \mathbb{E}(Y)'\,\mathbb{E}(Y). \quad \blacksquare
\end{aligned}$$

4.3 Estimating treatment effects

Let $Y(\omega)$ be the response on plot ω when our incomplete-block design is used for an experiment. For example, Table 4.1 on page 98 shows the responses on one particular experiment on potatoes. Given these responses, we want to know how much better treatment 1 is than treatment 2 (as well as other treatment differences) given that they do not always occur in the same block. To do this, we need to assume a model for the response.

First, we assume that there is a vector τ in $\mathbb{R}^\Theta$ whose entries are the treatment parameters that we want to estimate. If there were no blocks and there were no random variation then the response on plot ω would be $\tau(\psi(\omega))$ and the overall response vector would be $X\tau$.

To allow for the effect of blocks, put

$$V_B = \left\{v \in \mathbb{R}^\Omega : v(\alpha) = v(\beta) \text{ if } \alpha \text{ and } \beta \text{ are in the same block}\right\}.$$

Then the characteristic functions of the blocks form an orthogonal basis for V_B, so $\dim V_B = b$. Also

$$w \in V_B^\perp \iff \sum_{\alpha \in \delta} w(\alpha) = 0 \quad \text{for each block } \delta.$$

Let P and Q be the orthogonal projectors onto V_B and $V_B^\perp$. It can be easily checked that $P = k^{-1}B$ and $Q = I - P$. Now, a response vector that depends only on the block is simply a vector h in V_B.

Our first assumption is therefore that

$$\mathbb{E}(Y) = X\tau + h, \tag{4.2}$$

where τ is an unknown vector in $\mathbb{R}^\Theta$ and h is an unknown vector in V_B. That is, the expectation of $Y(\omega)$ is the sum of two parts, one depending on the treatment applied to ω and the other depending on the block containing ω.

To allow for random variation, the simplest assumption we can make is that the responses on different plots are uncorrelated and all have the same variance σ^2, where σ^2 is an unknown positive constant. That is, $\mathrm{Var}(Y(\omega)) = \sigma^2$ for every plot ω, and $\mathrm{cov}(Y(\alpha), Y(\beta)) = 0$ whenever α and β are different plots. This assumption can be written in matrix form as follows:

$$\mathrm{Cov}(Y) = I\sigma^2. \tag{4.3}$$

We want to use the observed values of the $Y(\omega)$ from the experiment to estimate τ. However, let U_0 be the subspace of $\mathbb{R}^\Theta$ spanned by χ_Θ. Then

$$X\chi_\Theta = \chi_\Omega \in V_B,$$

so we cannot estimate τ: the best we can hope to do is to estimate τ

up to a multiple of χ_Θ. Then we could estimate differences such as $\tau(\theta) - \tau(\eta)$.

Definition A vector x in $\mathbb{R}^\Theta$ is a *contrast* if $x \in U_0^\perp$. It is a *simple* contrast if there are θ, η in Θ such that $x(\theta) = 1$, $x(\eta) = -1$ and $x(\zeta) = 0$ for ζ in $\Theta \setminus \{\theta, \eta\}$.

We want to estimate linear combinations such as $\sum_\theta x(\theta)\tau(\theta)$ for x in $U_0^\perp$. In order to use the results of the previous section in a straightforward way, it is convenient to make a slight shift of perspective on our vectors. I have defined x to be a function from Θ to $\mathbb{R}$. However, the definitions of the action of a matrix on a vector, and of matrix multiplication, are consistent with the idea that x is a column vector, that is, an element of $\mathbb{R}^{\Theta\times\{1\}}$. So we can define the transpose of x as a matrix x' in $\mathbb{R}^{\{1\}\times\Theta}$. Then

$$\sum_{\theta\in\Theta} x(\theta)\tau(\theta) = \langle x, \tau\rangle = x'\tau.$$

Such a linear combination is called a *treatment effect*.

Definition An *unbiased estimator* for $x'\tau$ is a function of Y and of the design (but not of τ, h or σ^2) whose expectation is equal to $x'\tau$.

Theorem 4.6 *If there is a vector z in $\mathbb{R}^\Theta$ with $X'QXz = x$ then $z'X'QY$ is an unbiased estimator for $x'\tau$ and its variance is $z'X'QXz\sigma^2$.*

Proof

$$\begin{aligned}
\mathbb{E}(z'X'QY) &= z'X'Q\,\mathbb{E}(Y), \qquad \text{by Proposition 4.2,}\\
&= z'X'Q(X\tau + h)\\
&= z'X'QX\tau, \qquad \text{because } Qh = 0,\\
&= x'\tau
\end{aligned}$$

because $Q' = Q$. Then, by Lemma 4.3,

$$\begin{aligned}
\mathrm{Var}(z'X'QY) &= z'X'Q(I\sigma^2)Q'Xz\\
&= (z'X'Q^2Xz)\sigma^2\\
&= z'X'QXz\sigma^2
\end{aligned}$$

because Q is idempotent. ■

We can show that Theorem 4.6 is, in some sense, best possible. A little linear algebra is needed first.

Lemma 4.7 *(i) The kernel of $X'QX$ is equal to the kernel of QX.*

(ii) The image of $X'QX$ is equal to the image of $X'Q$.

(iii) If the design is equi-replicate with replication r then the matrix of orthogonal projection onto the image of X is $r^{-1}XX'$.

Proof (i) Let z be in $\mathbb{R}^\Theta$. If $z \in \ker QX$ then $QXz = 0$ so $X'QXz = 0$ so $z \in \ker X'QX$. Conversely, if $z \in \ker X'QX$ then $\langle QXz, QXz\rangle = z'X'Q'QXz = z'X'QXz = 0$; but $\langle\ ,\ \rangle$ is an inner product, so $QXz = 0$ and $z \in \ker QX$.

(ii) Clearly, $\operatorname{Im} X'QX \leqslant \operatorname{Im} X'Q$. However, part (i) shows that $\dim(\ker X'QX) = \dim(\ker QX)$, and therefore $\dim(\operatorname{Im} X'QX) = \dim \mathbb{R}^\Theta - \dim(\ker X'QX) = \dim \mathbb{R}^\Theta - \dim(\ker QX) = \operatorname{rank} QX = \operatorname{rank} X'Q = \dim(\operatorname{Im} X'Q)$. Hence $\operatorname{Im} X'QX = \operatorname{Im} X'Q$.

(iii) Let w be in $\mathbb{R}^\Omega$. If $w = Xz$ then $(r^{-1}XX')\,w = (r^{-1}XX')\,Xz = r^{-1}X(X'X)z = r^{-1}X(rI)z = Xz = w$. If $w \in (\operatorname{Im} X)^\perp$ then $w'Xz = 0$ for all z in $\mathbb{R}^\Theta$, so $X'w = 0$ and $r^{-1}XX'w = 0$. ■

Theorem 4.8 *Let x be in $\mathbb{R}^\Theta$.*

(i) If $x \notin \operatorname{Im} X'QX$ then there is no linear unbiased estimator for $x'\tau$.

(ii) If there is a vector z in $\mathbb{R}^\Theta$ with $X'QXz = x$ then $z'X'QY$ has the smallest variance among linear unbiased estimators for $x'\tau$.

Proof Since h is unknown, any linear unbiased estimator for $x'\tau$ must be a linear function of QY, so it has the form $w'QY$ for some w in $\mathbb{R}^{\Omega\times\{1\}}$. Then $\mathbb{E}(w'QY) = w'QX\tau$; this is equal to $x'\tau$ for all τ in $\mathbb{R}^\Theta$ if and only if $w'QX = x'$, that is $X'Qw = x$.

(i) From Lemma 4.7(ii), x has the form $X'Qw$ only if $x \in \operatorname{Im} X'QX$.

(ii) Suppose that $X'QXz = x$ and $X'Qw = x$. Put $u = w - Xz$. Then $X'Qu = 0$. Hence

$$
\begin{aligned}
\operatorname{Var}(w'QY) &= w'Qw\sigma^2 \\
&= (Xz+u)'Q(Xz+u)\sigma^2 \\
&= (z'X'QXz + z'X'Qu + u'QXz + u'Qu)\,\sigma^2 \\
&= (z'X'QXz + u'Qu)\,\sigma^2
\end{aligned}
$$

because $X'Qu = 0$. Now, $u'Qu = u'Q'Qu = \langle Qu, Qu\rangle \geqslant 0$ and so $\operatorname{Var}(w'QY) \geqslant z'X'QXz\sigma^2 = \operatorname{Var}(z'X'QY)$. ■

Theorem 4.9 *The kernel of $X'QX$ is spanned by the characteristic functions of the connected components of the treatment-concurrence graph.*

Proof Let z be in $\mathbb{R}^\Theta$. By Lemma 4.7(i),

$$\begin{aligned} z \in \ker X'QX &\iff QXz = 0 \\ &\iff PXz = Xz \\ &\iff Xz \in V_B \\ &\iff z(\theta) = z(\eta) \text{ whenever } \Lambda(\theta,\eta) > 0 \\ &\iff z \text{ is constant on each component of the treatment-concurrence graph.} \end{aligned}$$

■

Corollary 4.10 *For a connected incomplete-block design,* $\mathrm{Im}(X'QX) = U_0^\perp$.

Proof If the design is connected then $\ker X'QX = U_0$. But $X'QX$ is symmetric, so $\mathrm{Im}(X'QX) = (\ker X'QX)^\perp$. ■

Definition The matrix $X'QX$ is the *information matrix* of the design. Write $L = X'QX$.

The matrix kL is sometimes called the *Laplacian*, particularly when $k = 2$.

Theorem 4.6 says that if $Lz = x$ then $z'XQY$ is an unbiased estimator of $x'\tau$ with variance $z'Lz\sigma^2$. Recall from Section 2.2 that L has a generalized inverse L^- such that $LL^-L = L$. Thus $z'Lz\sigma^2 = z'LL^-Lz\sigma^2 = x'L^-x\sigma^2$ because L is symmetric, so we obtain an expression for the variance of the estimator of $x'\tau$ in terms of x rather than z. In particular, if the design is connected then we can estimate every difference $\tau(\theta) - \tau(\eta)$ and the variance of the estimator is

$$\mathrm{Var}(\widehat{\tau(\theta) - \tau(\eta)}) = \left(L^-(\theta,\theta) - L^-(\theta,\eta) - L^-(\eta,\theta) + L^-(\eta,\eta)\right)\sigma^2, \tag{4.4}$$

where we have used the statisticians' notation $\widehat{}$ for an estimator.

In an equi-replicate block design,

$$\begin{aligned} L = X'QX &= X'(I-P)X \\ &= X'X - k^{-1}X'BX \\ &= rI_\Theta - k^{-1}\Lambda. \end{aligned} \tag{4.5}$$

Example 4.1 revisited Let $x = \chi_1 - \chi_2$. Since treatments 1 and 2 always occur together in a block, $Nx = 0$. We say that x is 'orthogonal to blocks' to indicate that $\sum_{\alpha\in\delta}(Xx)(\alpha) = 0$ for every block δ; that is, $Xx \in V_B^\perp$ and $PXx = 0$. Equivalently, the 1- and 2-rows of Λ are identical, so $\Lambda x = 0$.

Now Equation (4.5) gives $Lx = rx = 2x$ so we may take $z = \frac{1}{2}x$ in

Theorem 4.6. Then $z'X'QY = \frac{1}{2}x'X'QY = \frac{1}{2}x'X'(I-P)Y = \frac{1}{2}x'X'Y$ so we estimate $\tau(1) - \tau(2)$ by

$$\frac{Y(\omega_1) - Y(\omega_2) + Y(\omega_5) - Y(\omega_6)}{2}.$$

In other words, we take the difference between the response on treatments 1 and 2 in each block where they occur, and average these differences. Corollary 4.4 shows that the variance of this estimator is $(\sigma^2 + \sigma^2 + \sigma^2 + \sigma^2)/2^2 = \sigma^2$.

Now put $x = \chi_1 - \chi_3$. This contrast is not orthogonal to blocks, so we will have to do some explicit calculations. We have

$$\begin{aligned} L &= 2I - \frac{1}{4}\begin{bmatrix} 2 & 2 & 1 & 1 & 1 & 1 \\ 2 & 2 & 1 & 1 & 1 & 1 \\ 1 & 1 & 2 & 2 & 1 & 1 \\ 1 & 1 & 2 & 2 & 1 & 1 \\ 1 & 1 & 1 & 1 & 2 & 2 \\ 1 & 1 & 1 & 1 & 2 & 2 \end{bmatrix} \\ &= \frac{1}{4}\begin{bmatrix} 6 & -2 & -1 & -1 & -1 & -1 \\ -2 & 6 & -1 & -1 & -1 & -1 \\ -1 & -1 & 6 & -2 & -1 & -1 \\ -1 & -1 & -2 & 6 & -1 & -1 \\ -1 & -1 & -1 & -1 & 6 & -2 \\ -1 & -1 & -1 & -1 & -2 & 6 \end{bmatrix}. \end{aligned}$$

Put $z = \frac{1}{12}(7\chi_1 + \chi_2 - 7\chi_3 - \chi_4)$. Direct calculation shows that $Lz = x$, so we estimate $\tau(1) - \tau(3)$ by $z'X'QY$. Now, the effect of Q is to subtract the block average from every entry in a block, so

$$\begin{aligned} 12z'X'QY &= 7Y(\omega_1) + Y(\omega_2) - 7Y(\omega_3) - Y(\omega_4) \\ &\quad + 5Y(\omega_5) - Y(\omega_6) - 2Y(\omega_7) - 2Y(\omega_8) \\ &\quad - 5Y(\omega_9) + Y(\omega_{10}) + 2Y(\omega_{11}) + 2Y(\omega_{12}). \end{aligned}$$

This time the response on *every* plot contributes to the estimator of $\tau(1) - \tau(3)$, whose variance is

$$\frac{\sigma^2}{12^2}(7^2 + 1^2 + 7^2 + 1^2 + 5^2 + 1^2 + 2^2 + 2^2 + 5^2 + 1^2 + 2^2 + 2^2) = \frac{7\sigma^2}{6}.$$

Since so many more responses are involved, it is, perhaps, not surprising that this variance is greater than the variance of the estimator of the difference $\tau(1) - \tau(2)$. (This issue will be discussed in Section 5.3.)

Finally, we look at a non-simple contrast. Put $x = \chi_1 + \chi_2 - \chi_3 - \chi_4$.

Direct calculation shows that $Lx = \frac{3}{2}x$, so the estimator of $\tau(1)+\tau(2)-\tau(3)-\tau(4)$ is $(2/3)x'X'QY$, which is

$$\frac{1}{3}\left(\begin{array}{c} 2Y(\omega_1)+2Y(\omega_2)-2Y(\omega_3)-2Y(\omega_4) \\ +Y(\omega_5)+Y(\omega_6)-Y(\omega_7)-Y(\omega_8) \\ -Y(\omega_9)-Y(\omega_{10})+Y(\omega_{11})+Y(\omega_{12}) \end{array}\right).$$

This is the sum of the estimators of $\tau(1)-\tau(3)$ and $\tau(2)-\tau(4)$, which is no surprise, because estimation is linear in x. Its variance is $(8/3)\sigma^2$. ■

4.4 Efficiency factors

From now until Chapter 12, all designs are assumed to be equi-replicate. For comparison we consider a complete-block design where the variance of each response is σ^2_{CBD}. In such a design, $\Lambda = rJ_\Theta$ and $k = t$, so Equation (4.5) gives

$$L = r(I_\Theta - t^{-1}J_\Theta).$$

Now, $(I_\Theta - t^{-1}J_\Theta)$ is the projector onto $U_0^\perp$, so

$$L^- = \frac{1}{r}(I_\Theta - t^{-1}J_\Theta)$$

and the variance of the estimator of $x'\tau$ is $(x'L^-x)\sigma^2_{\mathrm{CBD}}$, which is equal to $r^{-1}x'x\sigma^2_{\mathrm{CBD}}$.

Definition The *efficiency* for a contrast x in a given equi-replicate incomplete-block design with variance σ^2 and replication r relative to a complete-block design with variance σ^2_{CBD} and the same replication is

$$\frac{x'x}{rx'L^-x}\,\frac{\sigma^2_{\mathrm{CBD}}}{\sigma^2}$$

and the *efficiency factor* for x is

$$\frac{x'x}{rx'L^-x}.$$

If x is the simple contrast for the difference between treatments θ and η then $x'x = 2$. Thus

$$\mathrm{Var}(\widehat{\tau(\theta)-\tau(\eta)}) = \frac{2}{r}\,\frac{\sigma^2}{\text{efficiency factor for } x}. \tag{4.6}$$

Theorem 4.11 *Let ε be the efficiency factor for a contrast x. Then $0 \leqslant \varepsilon \leqslant 1$.*

Proof Suppose that $Lz = x$. Then $x'L^-x = z'Lz = z'X'QXz = \langle QXz, QXz\rangle \geqslant 0$. Also, Lemma 4.7(iii) shows that $r^{-1}XX'$ is a projector, so the length of $r^{-1}XX'QXz$ is at most the length of QXz. Thus

$$0 \leqslant r^{-2}z'X'QXX'XX'QXz \leqslant z'X'QXz$$

so

$$0 \leqslant r^{-1}z'X'QXX'QXz \leqslant z'Lz$$

so

$$0 \leqslant r^{-1}x'x \leqslant x'L^-x. \quad \blacksquare$$

Example 4.1 revisited Here $r = 2$, so the efficiency factor for a simple contrast is σ^2/variance. Thus the efficiency factor for the simple contrast $\chi_1 - \chi_2$ is 1 while the efficiency factor for the simple contrast $\chi_1 - \chi_3$ is 6/7.

When $x = \chi_1 + \chi_2 - \chi_3 - \chi_4$ then $x'x = 4$ and $\mathrm{Var}(\widehat{x'\tau}) = 8\sigma^2/3$, so the efficiency factor for x is

$$\frac{4}{2} \times \frac{3}{8} = \frac{3}{4}. \quad \blacksquare$$

We want estimators with low variance. Efficiency is defined by comparing the reciprocals of the variances, so that low variance corresponds to high efficiency. In practice, neither σ^2 nor σ^2_{CBD} is known before the experiment is done. Indeed, the usual reason for doing an experiment in incomplete blocks is that large enough blocks are not available. Even if they are available, there may be some prior knowledge about the likely relative sizes of σ^2 and σ^2_{CBD}. Part of the statistician's job in deciding what blocks to use is to assess whether the ratio $\sigma^2/\sigma^2_{\mathrm{CBD}}$ is likely to be less than the ratio $x'x/rx'L^-x$. The latter ratio, the efficiency factor, is a function of the design and the contrast, so it can be used for comparing different designs of the same size, at least for the single contrast x.

The efficiency factor for x has a particularly simple form if x is an eigenvector of L, for if $Lx = \mu x$ then $x'L^-x = \mu^{-1}x'x$ and so the efficiency factor is μ/r.

Definition A *basic contrast* of a given equi-replicate incomplete-block design is a contrast which is an eigenvector of the information matrix L.

Definition The *canonical efficiency factors* of a given equi-replicate incomplete-block design with replication r are $\mu_1/r, \ldots, \mu_{t-1}/r$, where $\mu_1, \ldots, \mu_{t-1}$ are the eigenvalues of L on $U_0^\perp$, with multiplicities.

Technique 4.1 To find the canonical efficiency factors, find the eigenvalues of Λ, divide by rk, subtract from 1, and ignore one of the zero values. There will always be a zero value corresponding to the eigenvector χ_Θ; if the design is connected then that will be the only zero value.

Once the eigenvectors and eigenvalues of L are known, they can be used to find the efficiency factors of all contrasts if the design is connected.

Technique 4.2 If the eigenvalues of L are known, use them to write down the minimal polynomial of L on $(\ker L)^\perp$. Hence find the Moore–Penrose generalized inverse L^- of L. If the design is connected and x is any contrast then $x'L^-x$ is not zero, so calculate the efficiency factor for x as $x'x/rx'L^-x$. Ignore the contribution of J to L^-, because $Jx = 0$ for all contrasts x.

Example 4.1 revisited We have seen that $Lx = 2x$ if x is $\chi_1 - \chi_2$, $\chi_3 - \chi_4$ or $\chi_5 - \chi_6$ and $Lx = (3/2)x$ if x is $\chi_1 + \chi_2 - \chi_3 - \chi_4$ or $\chi_1 + \chi_2 - \chi_5 - \chi_6$. These five contrasts span $U_0^\perp$, so the eigenvalues of L on $U_0^\perp$ are 2 and 3/2. Thus, on $U_0^\perp$,

$$(L - 2I)(L - \frac{3}{2}I) = 0,$$

whence

$$L^2 - \frac{7}{2}L + 3I = 0$$

so

$$L(L - \frac{7}{2}I) = -3I$$

and the inverse of L on $U_0^\perp$ is $(1/6)(7I - 2L)$. Thus

$$L^- = \frac{1}{6}(7I - 2L) + cJ$$

for some constant c. If x is a contrast then $Jx = 0$ and so

$$x'L^-x = \frac{1}{6}(7x'x - 2x'Lx).$$

In particular, if $x = \chi_1 - \chi_3$ then $x'x = 2$ and

$$x'L^-x = \frac{1}{6}\left(14 - 2\left(L(1,1) - L(1,3) - L(3,1) + L(3,3)\right)\right) = \frac{7}{6},$$

so the efficiency factor for $\chi_1 - \chi_3$ is 6/7, as we found on page 91. ■

Technique 4.3 If the basic contrasts and their canonical efficiency factors are known, express an arbitrary contrast x as a sum $x_1 + \cdots + x_m$ of basic contrasts with different canonical efficiency factors $\varepsilon_1, \ldots, \varepsilon_m$. If the design is connected then none of $\varepsilon_1, \ldots, \varepsilon_m$ is zero. Since the distinct eigenspaces of L are orthogonal to each other, $x_i' L^- x_j = 0$ if $i \neq j$, so

$$rx'L^-x = \sum_{i=1}^{m} \frac{x_i'x_i}{\varepsilon_i}.$$

Example 4.1 revisited Put $x = \chi_1 - \chi_3$. Then $x = x_1 + x_2$ where $x_1 = (\chi_1 - \chi_2 - \chi_3 + \chi_4)/2$ and $x_2 = (\chi_1 + \chi_2 - \chi_3 - \chi_4)/2$. But x_1 is a basic contrast with canonical efficiency factor $\varepsilon_1 = 1$ and x_2 is a basic contrast with canonical efficiency factor $\varepsilon_2 = 3/4$, so

$$rx'L^-x = x_1'x_1 + \frac{4}{3}x_2'x_2 = 1 + \frac{4}{3} = \frac{7}{3}$$

and $x'x/rx'L^-x = 6/7$, as before. ∎

We need an overall measure of the efficiency of a design. It is tempting to take the arithmetic mean of the canonical efficiency factors. But

$$\begin{aligned}
\sum \text{canonical efficiency factors} &= \sum \text{canonical efficiency factors} + 0 \\
&= \frac{1}{r}\left(\sum \text{eigenvalues of } L\right) \\
&= \frac{1}{r}\operatorname{tr} L = \frac{1}{r}\left(rt - \frac{rt}{k}\right) \\
&= \frac{t(k-1)}{k},
\end{aligned} \tag{4.7}$$

which is independent of the design. Instead we measure the overall efficiency of the design by the harmonic mean of the canonical efficiency factors. (The harmonic mean of a collection of positive numbers is the reciprocal of the arithmetic mean of their reciprocals.) This overall efficiency factor is called A (not to be confused with an adjacency matrix!). That is, if the canonical efficiency factors are $\varepsilon_1, \ldots, \varepsilon_{t-1}$ then

$$A = \left(\frac{\displaystyle\sum_{i=1}^{t-1} \frac{1}{\varepsilon_i}}{t-1} \right)^{-1}.$$

The next theorem shows that the choice of harmonic mean is not entirely arbitrary.

Theorem 4.12 *In a connected equi-replicate incomplete-block design with replication r, the average variance of simple contrasts is equal to $2\sigma^2/(rA)$.*

Proof The information matrix is zero in its action on U_0, so the same is true of its generalized inverse L^-. That is, the row and column sums of L^- are all zero. Equation (4.4) shows that the average variance of simple contrasts

$$
\begin{aligned}
&= \frac{\sigma^2}{t(t-1)}\sum_\eta\sum_{\theta\neq\eta}\left(L^-(\theta,\theta)-L^-(\theta,\eta)-L^-(\eta,\theta)+L^-(\eta,\eta)\right)\\
&= \frac{\sigma^2}{t(t-1)}\sum_\eta\sum_{\theta}\left(L^-(\theta,\theta)-L^-(\theta,\eta)-L^-(\eta,\theta)+L^-(\eta,\eta)\right)\\
&= \frac{\sigma^2}{t(t-1)}\sum_\eta\sum_{\theta}\left(L^-(\theta,\theta)+L^-(\eta,\eta)\right), \qquad \text{because the row and column sums of } L^- \text{ are zero,}\\
&= \frac{\sigma^2}{t(t-1)}2t\operatorname{tr}L^-\\
&= \frac{2\sigma^2}{t-1}\left(\frac{1}{\mu_1}+\cdots+\frac{1}{\mu_{t-1}}\right), \qquad \text{where } \mu_1,\ldots,\mu_{t-1} \text{ are the eigenvalues of } L \text{ on } U_0^\perp,\\
&= \frac{2\sigma^2}{r(t-1)}\left(\frac{r}{\mu_1}+\cdots+\frac{r}{\mu_{t-1}}\right)\\
&= \frac{2\sigma^2}{r}\times\frac{1}{\text{harmonic mean of } \frac{\mu_1}{r},\ldots,\frac{\mu_{t-1}}{r}}\\
&= \frac{2\sigma^2}{rA}. \quad \blacksquare
\end{aligned}
$$

Theorem 4.13 *The canonical efficiency factors, including multiplicities, of an equi-replicate incomplete-block design and its dual are the same, apart from $|b-t|$ values equal to* 1.

Proof Let ε be a canonical efficiency factor different from 1 for the original design and let x be a corresponding eigenvector of L. Then $Lx = r\varepsilon x$. Lemma 4.1 and Equation (4.5) show that $N'Nx = \Lambda x = k(rI - L)x = kr(1-\varepsilon)x$. Therefore $NN'Nx = kr(1-\varepsilon)Nx$. The dual design has replication k and information matrix $kI_\Delta - r^{-1}NN'$. Now,

$$\left(kI_\Delta - \frac{1}{r}NN'\right)Nx = k\varepsilon Nx.$$

Thus x has canonical efficiency factor equal to ε in the original design and Nx has canonical efficiency factor equal to ε in the dual.

The maps

$$x \mapsto Nx \qquad \text{and} \qquad y \mapsto \frac{1}{rk(1-\varepsilon)} N'y$$

are mutual inverses on the spaces of contrasts in the two designs which have canonical efficiency factor ε, so the dimensions of these spaces are equal.

All remaining canonical efficiency factors of both designs must be equal to 1. ■

Note that if the canonical efficiency factors are $\varepsilon_1, \ldots, \varepsilon_{t-1}$ then the $rk\varepsilon_i$ are the zeros of the monic integer polynomial

$$\det(\mathrm{x}I - kL) \tag{4.8}$$

in $\mathbb{Z}[\mathrm{x}]$. So each $rk\varepsilon_i$ is an algebraic integer, so is either an integer or irrational. This fact helps to identify the canonical efficiency factors exactly if a computer program finds them numerically. Moreover,

$$\frac{1}{rk}\sum \frac{1}{\varepsilon_i} = \frac{1}{rk}\frac{\sum_i \prod_{j\neq i} \varepsilon_j}{\prod_i \varepsilon_i} = \frac{\sum_i \prod_{j\neq i} rk\varepsilon_j}{\prod_i rk\varepsilon_i}:$$

both numerator and denominator are elementary symmetric functions in the zeros of the polynomial (4.8), so they are integers. Hence A is always rational.

Definition An equi-replicate incomplete-block design is *A-optimal* if it has the highest value of A among all incomplete-block designs with the same values of t, r, b and k.

4.5 Estimating variance

Given a contrast x and information matrix L, we know that the variance of the estimator for $x'\tau$ is $x'L^{-}x\sigma^2$. Efficiency factors enable us to compare

- variances of different contrasts in the same design (change x, keep L constant)
- variances of the same contrast in different designs (keep x constant but change L).

However, to estimate, rather than compare, the variance once the experiment has been done, we need to estimate σ^2 from the data. This requires the projector onto $V_B^\perp \cap (\operatorname{Im} X)^\perp$.

Lemma 4.14 *Let $\varepsilon_1, \ldots, \varepsilon_m$ be the distinct non-zero canonical efficiency factors. For $i = 1, \ldots, m$, let U_i be the eigenspace of L with eigenvalue $r\varepsilon_i$ and let P_i be the matrix of orthogonal projection onto U_i. Then*

(i) *the dimension of the image of QXP_i is equal to* $\dim U_i$;
(ii) *if $i \neq j$ then* $\operatorname{Im} QXP_i$ *is orthogonal to* $\operatorname{Im} QXP_j$;
(iii) *the matrix of orthogonal projection onto* $\operatorname{Im} QXP_i$ *is equal to* $(r\varepsilon_i)^{-1}QXP_iX'Q$;
(iv) *the matrix of orthogonal projection onto $V_B^\perp \cap (\operatorname{Im} X)^\perp$ is*

$$Q - \sum_{i=1}^{m} \frac{1}{r\varepsilon_i} QXP_iX'Q$$

and the dimension of this space is equal to

$$\dim V_B^\perp - \sum_{i=1}^{m} \dim U_i.$$

Proof (i) If $x \in U_i$ then $X'QXx = Lx = r\varepsilon_i x$ so the maps

$$x \mapsto QXx \qquad \text{for } x \text{ in } U_i$$

and

$$y \mapsto \frac{1}{r\varepsilon_i} X'y \qquad \text{for } y \text{ in } \operatorname{Im} QXP_i$$

are mutual inverses.

(ii) Let $x \in U_i$ and $z \in U_j$. Then

$$\langle QXx, QXz \rangle = x'X'QXz = x'Lz = r\varepsilon_i x'z.$$

If $i \neq j$ then $U_i \perp U_j$ and so $x'z = 0$.

(iii) Let w be in $\mathbb{R}^\Omega$. If $w = QXz$ for some z in U_i then $QXP_iX'Qw = QXP_iX'QXz = QXP_iLz = r\varepsilon_i QXP_iz = r\varepsilon_i QXz = r\varepsilon_i w$. If $w \in (\operatorname{Im} QXP_i)^\perp$ then $w'QXP_iz = 0$ for all z in $\mathbb{R}^\Theta$, and so $P_iX'Qw = 0$ and $QXP_iX'Qw = 0$.

(iv) Part (ii) shows that the projection of $\operatorname{Im} X$ onto $V_B^\perp$ is the orthogonal direct sum

$$\bigoplus_{i+1}^{m} \operatorname{Im} QXP_i,$$

whose dimension is $\sum_{i=1}^{m} \dim U_i$, by part (i), and whose orthogonal projector is

$$\sum_{i=1}^{m} \frac{1}{r\varepsilon_i} QXP_iX'Q,$$

by part (iii) and Lemma 2.1(v). Then Lemma 2.1(vi) gives the result. ■

Theorem 4.15 *Let R be the matrix of orthogonal projection onto the space $V_B^\perp \cap (\operatorname{Im} X)^\perp$ and let d^* be its dimension. Then $\mathbb{E}(Y'RY) = d^*\sigma^2$.*

Proof Equation (4.2) shows that

$$\mathbb{E}(RY) = RX\tau + Rh = 0$$

because $X\tau \in \operatorname{Im} X$ and $h \in V_B$. The matrix R is symmetric and idempotent, so Lemma 4.5 gives

$$\begin{aligned}
\mathbb{E}(Y'RY) &= \mathbb{E}((RY)'(RY)) \\
&= \operatorname{tr}(\operatorname{Cov}(RY)) \\
&= \operatorname{tr}(R(I\sigma^2)R'), \qquad \text{by Lemma 4.3 and Equation (4.3),} \\
&= (\operatorname{tr} R)\sigma^2 \\
&= \dim\left(V_B^\perp \cap (\operatorname{Im} X)^\perp\right)\sigma^2 \\
&= d^*\sigma^2. \quad \blacksquare
\end{aligned}$$

Corollary 4.16 *An unbiased estimator for σ^2 is*

$$\left(Y'QY - \sum_{i=1}^{m} \frac{1}{r\varepsilon_i} Y'QXP_iX'QY\right) \Bigg/ \left(\dim V_B^\perp - \sum_{i=1}^{m} \dim U_i\right).$$

Note that, for a connected block design for t treatments in b blocks of size k,

$$\dim V_B^\perp = b(k-1) \quad \text{and} \quad \sum_{i=1}^{m} \dim U_i = t-1$$

so $d^* = bk - b - t + 1$.

Example 4.3 Here is a worked example on an experiment on potatoes reported by Yates [255, Section 4d]. The first three columns of Table 4.1 show the allocation of the eight fertilizer treatments to the 32 plots, which were arranged in eight blocks of four.

Ω	Block	Θ	Y	PY	QY
1	1	1	101	290.75	−189.75
2	1	2	291	290.75	0.25
3	1	4	373	290.75	82.25
4	1	3	398	290.75	107.25
5	2	8	265	283.25	−18.25
6	2	7	106	283.25	−177.25
7	2	6	312	283.25	28.75
8	2	5	450	283.25	166.75
9	3	7	89	276.50	−187.50
10	3	3	407	276.50	130.50
11	3	4	338	276.50	61.50
12	3	8	272	276.50	−4.50
13	4	5	449	296.25	152.75
14	4	6	324	296.25	27.75
15	4	1	106	296.25	−190.25
16	4	2	306	296.25	9.75
17	5	2	334	302.00	32.00
18	5	6	323	302.00	21.00
19	5	7	128	302.00	−174.00
20	5	3	423	302.00	121.00
21	6	8	279	290.25	−11.25
22	6	1	87	290.25	−203.25
23	6	4	324	290.25	33.75
24	6	5	471	290.25	180.75
25	7	8	302	314.75	−12.75
26	7	4	361	314.75	46.25
27	7	2	272	314.75	−42.75
28	7	6	324	314.75	9.25
29	8	1	131	279.00	−148.00
30	8	5	437	279.00	158.00
31	8	7	103	279.00	−176.00
32	8	3	445	279.00	166.00
layout and design			vectors in $\mathbb{R}^\Omega$		

Table 4.1. *Experiment on potatoes: see Example 4.3*

With treatments written in the order 1, 2, 3, 4, 5, 6, 7, 8, we have

$$\Lambda = \begin{bmatrix} 4 & 2 & 2 & 2 & 3 & 1 & 1 & 1 \\ 2 & 4 & 2 & 2 & 1 & 3 & 1 & 1 \\ 2 & 2 & 4 & 2 & 1 & 1 & 3 & 1 \\ 2 & 2 & 2 & 4 & 1 & 1 & 1 & 3 \\ 3 & 1 & 1 & 1 & 4 & 2 & 2 & 2 \\ 1 & 3 & 1 & 1 & 2 & 4 & 2 & 2 \\ 1 & 1 & 3 & 1 & 2 & 2 & 4 & 2 \\ 1 & 1 & 1 & 3 & 2 & 2 & 2 & 4 \end{bmatrix}$$

so

$$L = 4I - \frac{1}{4}\Lambda = \frac{1}{4}\begin{bmatrix} 12 & -2 & -2 & -2 & -3 & -1 & -1 & -1 \\ -2 & 12 & -2 & -2 & -1 & -3 & -1 & -1 \\ -2 & -2 & 12 & -2 & -1 & -1 & -3 & -1 \\ -2 & -2 & -2 & 12 & -1 & -1 & -1 & -3 \\ -3 & -1 & -1 & -1 & 12 & -2 & -2 & -2 \\ -1 & -3 & -1 & -1 & -2 & 12 & -2 & -2 \\ -1 & -1 & -3 & -1 & -2 & -2 & 12 & -2 \\ -1 & -1 & -1 & -3 & -2 & -2 & -2 & 12 \end{bmatrix}.$$

As we shall show in Chapter 5, or as can be verified by direct calculation, the non-zero eigenvalues of L are 3 and 4, with multiplicities 4 and 3 respectively and eigenprojectors

$$P_3 = \frac{1}{8}\begin{bmatrix} 4 & 0 & 0 & 0 & 2 & -2 & -2 & -2 \\ 0 & 4 & 0 & 0 & -2 & 2 & -2 & -2 \\ 0 & 0 & 4 & 0 & -2 & -2 & 2 & -2 \\ 0 & 0 & 0 & 4 & -2 & -2 & -2 & 2 \\ 2 & -2 & -2 & -2 & 4 & 0 & 0 & 0 \\ -2 & 2 & -2 & -2 & 0 & 4 & 0 & 0 \\ -2 & -2 & 2 & -2 & 0 & 0 & 4 & 0 \\ -2 & -2 & -2 & 2 & 0 & 0 & 0 & 4 \end{bmatrix}$$

$$P_4 = \frac{1}{8}\begin{bmatrix} 3 & -1 & -1 & -1 & -3 & 1 & 1 & 1 \\ -1 & 3 & -1 & -1 & 1 & -3 & 1 & 1 \\ -1 & -1 & 3 & -1 & 1 & 1 & -3 & 1 \\ -1 & -1 & -1 & 3 & 1 & 1 & 1 & -3 \\ -3 & 1 & 1 & 1 & 3 & -1 & -1 & -1 \\ 1 & -3 & 1 & 1 & -1 & 3 & -1 & -1 \\ 1 & 1 & -3 & 1 & -1 & -1 & 3 & -1 \\ 1 & 1 & 1 & -3 & -1 & -1 & -1 & 3 \end{bmatrix}.$$

The succeeding three columns of Table 4.1 show Y, which is the response in pounds of potatoes; PY, which is calculated by replacing each value of Y by the average value of Y over that block; and QY, which is calculated as $Y - PY$.

The columns in Table 4.2 are vectors in $\mathbb{R}^{\Theta}$. Thus $X'QY$ is calculated from QY by summing over each treatment. For $i = 3, 4$, $P_iX'QY/(r\varepsilon_i)$ is calculated from $X'QY$ by using the matrix P_i. If $x \in U_i$ then $x'\tau$ is estimated as $x'P_iX'QY/(r\varepsilon_i)$. For example, let $x_3 = \chi_1 - \chi_2 + \chi_5 - \chi_6$.

Θ	$X'QY$	$P_3X'QY/3$	$P_4X'QY/4$
1	−731.25	−10.792	−174.719
2	−0.75	15.708	−11.969
3	524.75	−30.292	153.906
4	223.75	30.875	32.781
5	658.25	−13.542	174.719
6	86.75	12.958	11.969
7	−714.75	−33.042	−153.906
8	−46.75	28.125	−32.781

vectors in $\mathbb{R}^\Theta$

Table 4.2. *Calculations on the responses in Table 4.1*

Then $x_3 \in U_3$ and so

$$\begin{aligned}\widehat{\tau(1)} - \widehat{\tau(2)} + \widehat{\tau(5)} - \widehat{\tau(6)} &= -10.792 - 15.708 - 13.542 - 12.958 \\ &= -53.000\end{aligned}$$

with variance $4\sigma^2/3$. Let $x_4 = \chi_1 - \chi_2 - \chi_5 + \chi_6$. Then $x_4 \in U_4$ and so

$$\begin{aligned}\widehat{\tau(1)} - \widehat{\tau(2)} - \widehat{\tau(5)} + \widehat{\tau(6)} &= -174.719 + 11.969 - 174.719 + 11.969 \\ &= -325.500\end{aligned}$$

with variance $4\sigma^2/4$. Thus $\widehat{\tau(1)} - \widehat{\tau(2)} = -189.25$ with variance given by Technique 4.3 as

$$\frac{1}{4}\left(\frac{4}{3} + \frac{4}{4}\right)\sigma^2 = \frac{7}{12}\sigma^2.$$

The sum of the squares of the entries in QY is $Y'QY$. For $i = 3$ and $i = 4$, the sum of the squares of the entries in $P_iX'QY/(r\varepsilon_i)$ is equal to $Y'QXP_iX'QY/(r\varepsilon_i)^2$; multiplying by $r\varepsilon_i$ gives the quantity to subtract from $Y'QY$ in the calculation of $Y'RY$. Thus

vector	sum of squares	multiply by	
QY	462 280.7500	1	462 280.750
$P_3X'QY/3$	4 468.1250	−3	−13 404.375
$P_4X'QY/4$	110 863.2735	−4	−443 453.094
			5 423.281

Since $d^* = 32 - 8 - 7 = 17$, we estimate σ^2 as $5423.281/17 = 319$. ■

The business of drawing inferences about contrasts $x'\tau$ in the light of the estimated magnitude of σ^2 is covered in textbooks on statistical inference and is beyond the scope of this book.

```
***** Analysis of variance *****

Variate: yield

Source of variation     d.f.        s.s.        m.s.     v.r.

block stratum
fertiliz                   4      3724.9       931.2     3.61
Residual                   3       774.1       258.0     0.81

block.plot stratum
fertiliz                   7    456857.5     65265.4   204.58
Residual                  17      5423.3       319.0

Total                     31    466779.7


***** Information summary *****

Model term                  e.f.  non-orthogonal terms

block stratum
  frow                     0.250
  fcol                     0.250

block.plot stratum
  frow                     0.750  block
  fcol                     0.750  block

Aliased model terms
fertiliz
```

Fig. 4.2. Part of Genstat output for Example 4.3

Of course, in practice nobody does hand calculations like those in Table 4.2. Every reasonable statistical computing package does this with a single command once the blocks, treatments, responses (and possibly some indication of the eigenspaces U_i) have been entered. However, it is important for the statistician to understand efficiency factors so that an efficient design can be chosen for the experiment.

Figure 4.2 shows part of the output when the statistical computing package Genstat is used to analyse the data in Table 4.1. The part labelled *Analysis of variance* includes the calculation of 319 as the estimate of σ^2. The part labelled *Information summary* shows that there is a canonical efficiency factor equal to 0.75; canonical efficiency factors

```
***** Tables of means *****

Variate: yield

Grand mean   291.6

 fertiliz          1         2         3         4         5         6
     frow          1         1         1         1         2         2
     fcol          1         2         3         4         1         2
               106.1     295.3     415.2     355.3     452.8     316.5

 fertiliz          7         8
     frow          2         2
     fcol          3         4
               104.6     286.9


*** Standard errors of differences of means ***

Table               fertiliz
rep.                       4
d.f.                      17
s.e.d.                 14.12
Except when comparing means with the same level(s) of
 frow                  13.64
 fcol                  13.15
```

Fig. 4.3. Further Genstat output for Example 4.3

equal to 0 or 1 are not shown. The items called frow and fcol will be explained in Section 5.2.

Further output is shown in Figure 4.3. The part labelled *Tables of means* gives estimates of $\widehat{\tau(1)}$ to $\widehat{\tau(8)}$. We have already noted that you cannot estimate these values. However, if the figures given in the output are substituted in contrasts then they do give the correct estimates. For example, $\widehat{\tau(1)} - \widehat{\tau(2)} = 106.1 - 295.3 = -189.2$, which is correct to one decimal place.

The final section of output gives *Standard errors of differences of means*, which are the square roots of the estimated variances of the estimators of differences $\tau(\theta) - \tau(\eta)$. Treatments 1 and 2 have the same value (called *level*) of frow, so the standard error of their difference is 13.64, which is indeed equal to

$$\sqrt{\frac{7}{12} \times 319}.$$

4.6 Building block designs

If $bk = tr$ it is always possible to construct a binary incomplete-block design for t treatments replicated r times within b blocks of size k. Identify the treatment set Θ with $\mathbb{Z}_t$. The blocks are $\{1, 2, \ldots, k\}$, $\{k+1, k+2, \ldots, 2k\}$, ..., $\{(b-1)k+1, (b-1)k+2, \ldots, bk\}$, all addition being done in $\mathbb{Z}_t$. This design may not be very efficient: in fact, it is not even connected if k divides t.

At the other extreme, if $b = {}^t\mathrm{C}_k$ then we can take all the k-subsets of Θ as blocks. Such a design is called *unreduced*. We shall show in Section 5.7 that unreduced designs have as high a value of the overall efficiency factor as possible.

Other specific constructions will be left to Sections 5.4–5.6. However, it is worth mentioning some ways of building new block designs out of smaller ones, because the techniques apply quite generally.

4.6.1 Juxtaposition

Suppose that $(\Omega_1, \mathcal{G}_1, \Theta, \psi_1)$ and $(\Omega_2, \mathcal{G}_2, \Theta, \psi_2)$ are block designs for the same treatment set Θ, with b_1, b_2 blocks respectively, all of size k. If $\Omega_1 \cap \Omega_2 = \varnothing$ then $\mathcal{G}_1$ and $\mathcal{G}_2$ define a partition of $\Omega_1 \cup \Omega_2$ into $b_1 + b_2$ blocks of size k, with a corresponding group-divisible association scheme $\mathcal{G}$ on $\Omega_1 \cup \Omega_2$. Define $\psi\colon \Omega_1 \cup \Omega_2 \to \Theta$ by $\psi(\omega) = \psi_1(\omega)$ if $\omega \in \Omega_1$ and $\psi(\omega) = \psi_2(\omega)$ if $\omega \in \Omega_2$. Then $(\Omega_1 \cup \Omega_2, \mathcal{G}, \Theta, \psi)$ is a block design with $b_1 + b_2$ blocks of size k.

Let Λ be the concurrence matrix of the new design, and let Λ_1 and Λ_2 be the concurrence matrices of $(\Omega_1, \mathcal{G}_1, \Theta, \psi_1)$ and $(\Omega_2, \mathcal{G}_2, \Theta, \psi_2)$. Then $\Lambda = \Lambda_1 + \Lambda_2$.

Suppose that x in $\mathbb{R}^\Theta$ is a basic contrast of both component designs, with canonical efficiency factors ε_1 and ε_2 respectively. Then $\Lambda_i x = kr_i(1 - \varepsilon_i)x$ for $i = 1$, 2, where r_i is the replication in $(\Omega_i, \mathcal{G}_i, \Theta, \psi_i)$. Thus

$$\begin{aligned} \Lambda x &= k(r_1 - r_1\varepsilon_1 + r_2 - r_2\varepsilon_2)x \\ &= k(r_1 + r_2)(1 - \varepsilon)x, \end{aligned}$$

where

$$\varepsilon = \frac{r_1\varepsilon_1 + r_2\varepsilon_2}{r_1 + r_2}. \tag{4.9}$$

It follows that x is also a basic contrast of the new design and its canonical efficiency factor is ε, the weighted average of ε_1 and ε_2, with weights equal to the replication in the component designs.

This technique can be used to make a good design out of two poor ones, by matching up the high canonical efficiency factors of one with the low canonical efficiency factors of the other. The two component designs do not even need to be connected.

Example 4.4 Take $\Theta = \{1,2,3,4\}$ and $b_1 = b_2 = k = 2$, so that $r_1 = r_2 = 1$. If the first design has blocks $\{1,2\}$ and $\{3,4\}$ and the second has blocks $\{1,3\}$ and $\{2,4\}$ then we obtain the canonical efficiency factors in Table 4.3. ■

x	ε_1	ε_2	ε
$\chi_1 + \chi_2 - \chi_3 - \chi_4$	0	1	$\frac{1}{2}$
$\chi_1 - \chi_2 + \chi_3 - \chi_4$	1	0	$\frac{1}{2}$
$\chi_1 - \chi_2 - \chi_3 + \chi_4$	1	1	1

Table 4.3. *Canonical efficiency factors in Example 4.4*

The technique of juxtaposition can obviously be iterated, as the block size stays the same. If all the components have the same basic contrasts then these contrasts are also the basic contrasts of the final design, and the canonical efficiency factors for the final design are the weighted averages of those for the component designs, with weights equal to the replication in the component designs. In particular, if x has all its canonical efficiency factors in the component designs equal to 0 or 1 then its canonical efficiency factor in the final design is equal to the proportion of the experimental units lying in those component designs where it has canonical efficiency factor equal to 1.

Example 4.4 revisited Now take the first design to be the 2-replicate design constructed above, and the second to have blocks $\{1,4\}$ and $\{2,3\}$. The union is the unreduced design for four treatments in blocks of size 2. Now the canonical efficiency factors are shown in Table 4.4. We shall show in Section 5.2 that the canonical efficiency factors for an unreduced design all have the same value, namely $t(k-1)/(t-1)k$. ■

x	ε_1	ε_2	$\varepsilon = (2\varepsilon_1 + \varepsilon_2)/3$
$\chi_1 + \chi_2 - \chi_3 - \chi_4$	$\frac{1}{2}$	1	$\frac{2}{3}$
$\chi_1 - \chi_2 + \chi_3 - \chi_4$	$\frac{1}{2}$	1	$\frac{2}{3}$
$\chi_1 - \chi_2 - \chi_3 + \chi_4$	1	0	$\frac{2}{3}$

Table 4.4. *Canonical efficiency factors in the three-replicate version of Example 4.4*

4.6.2 Inflation

If we have a block design for t treatments in blocks of size k, its *m-fold inflation* is a design for mt treatments in blocks of size mk, obtained by replacing every treatment in the original design by m new treatments.

Formally, let the original design be $(\Omega, \mathcal{G}, \Theta, \psi)$ and let $\Gamma = \{1, \ldots, m\}$. In the new design the set of experimental units is $\Omega \times \Gamma$ and the set of treatments is $\Theta \times \Gamma$. The group-divisible association scheme $\mathcal{G}^*$ is defined on $\Omega \times \Gamma$ by putting (ω_1, γ_1) and (ω_2, γ_2) in the same (new) block whenever ω_1 and ω_2 are in the same block in $(\Omega, \mathcal{G})$. The new design function ψ^* is defined by $\psi^*(\omega, \gamma) = (\psi(\omega), \gamma)$. The concurrence matrix Λ^* of $(\Omega \times \Gamma, \mathcal{G}^*, \Theta \times \Gamma, \psi^*)$ is related to the concurrence matrix Λ of $(\Omega, \mathcal{G}, \Theta, \psi)$ by $\Lambda^* = \Lambda \otimes J_\Gamma$.

Example 4.5 The unreduced design for treatment set $\{\theta, \eta, \zeta\}$ in blocks of size two has blocks $\{\theta, \eta\}$, $\{\theta, \zeta\}$ and $\{\eta, \zeta\}$. For the 2-fold inflation, replace θ by new treatments 1 and 2, replace η by new treatments 3 and 4, and replace ζ by new treatments 5 and 6. This gives the block design in Example 4.1. ■

Let x be a basic contrast of the original design with canonical efficiency factor ε, so that $\Lambda x = rk(1 - \varepsilon)x$. Consider the vector $x \otimes \chi_\Gamma$. Now

$$\begin{aligned}
\Lambda^*(x \otimes \chi_\Gamma) &= (\Lambda \otimes J_\Gamma)(x \otimes \chi_\Gamma) \\
&= \Lambda x \otimes J_\Gamma \chi_\Gamma \\
&= rk(1 - \varepsilon)x \otimes m\chi_\Gamma \\
&= r(mk)(1 - \varepsilon)x \otimes \chi_\Gamma
\end{aligned}$$

so $x \otimes \chi_\Gamma$ is a basic contrast of the new design and its canonical efficiency factor is ε.

Orthogonal to all these contrasts is the $t(m-1)$-dimensional space $\mathbb{R}^\Theta \otimes W_0^\perp$, where W_0 is the subspace of $\mathbb{R}^\Gamma$ spanned by χ_Γ. If $z \in \mathbb{R}^\Theta$ and $y \in W_0^\perp$ then $\Lambda^*(z \otimes y) = 0$ because $J_\Gamma y = 0$, so $z \otimes y$ is a basic contrast of the new design with canonical efficiency factor equal to 1. Thus the canonical efficiency factors of the new design are the canonical efficiency factors of the old design plus $t(m-1)$ canonical efficiency factors equal to 1.

4.6.3 Orthogonal superposition

Let $(\Omega_1, \mathcal{G}_1, \Theta_1, \psi_1)$ be a block design for t_1 treatments replicated r times in b blocks of size k_1, and let $(\Omega_2, \mathcal{G}_2, \Theta_2, \psi_2)$ be a block design for t_2 treatments replicated r times in b blocks of size k_2. Suppose that $\Omega_1 \cap \Omega_2 = \varnothing$ and $\Theta_1 \cap \Theta_2 = \varnothing$ and that there is a bijection π from the block set Δ_1 of $(\Omega_1, \mathcal{G}_1)$ to the block set Δ_2 of $(\Omega_2, \mathcal{G}_2)$ which satisfies the following *orthogonality condition*:

for all θ in Θ_1 and all η in Θ_2,

$$\sum_{\delta \in \Delta_1} |N_1(\delta, \theta)|\,|N_2(\pi(\delta), \eta)| = \frac{r^2}{b}, \qquad (4.10)$$

where N_1 and N_2 are the incidence matrices for the two designs.

Define a group-divisible association scheme $\mathcal{G}$ on $\Omega_1 \cup \Omega_2$ by letting its blocks be the sets $\delta \cup \pi(\delta)$ for blocks δ in Δ_1. Define the function $\psi\colon \Omega_1 \cup \Omega_2 \to \Theta_1 \cup \Theta_2$ by $\psi(\omega) = \psi_1(\omega)$ if $\omega \in \Omega_1$ and $\psi(\omega) = \psi_2(\omega)$ if $\omega \in \Omega_2$. Then $(\Omega_1 \cup \Omega_2, \mathcal{G}, \Theta_1 \cup \Theta_2, \psi)$ is a block design with b blocks of size $k_1 + k_2$.

Like juxtaposition, orthogonal superposition can be used to create a good design from two poor components. It can also be iterated.

Example 4.6 Take $b = 16$, $t_1 = t_2 = 4$ and $k_1 = k_2 = 1$, so that the component designs are not even connected. If $\Theta_1 = \{1, 2, 3, 4\}$ and $\Theta_2 = \{5, 6, 7, 8\}$ then the first design has four blocks $\{1\}$, and so on, while the second design has four blocks $\{5\}$, and so on. We can choose π to satisfy the orthogonality condition by letting π map the four blocks $\{1\}$ to one each of the blocks $\{5\}$, $\{6\}$, $\{7\}$, $\{8\}$ and so on. The superposed design has the following 16 blocks:

$$\begin{array}{cccc} \{1,5\} & \{1,6\} & \{1,7\} & \{1,8\} \\ \{2,5\} & \{2,6\} & \{2,7\} & \{2,8\} \\ \{3,5\} & \{3,6\} & \{3,7\} & \{3,8\} \\ \{4,5\} & \{4,6\} & \{4,7\} & \{4,8\}. \end{array}$$

Now we can iterate the construction. Let $\Theta_3 = \{9, 10, 11, 12\}$. Consider the design for Θ_3 whose 16 singleton blocks are laid out in a Latin square as follows:

$$\begin{array}{cccc} 9 & 10 & 11 & 12 \\ 12 & 9 & 10 & 11 \\ 11 & 12 & 9 & 10 \\ 10 & 11 & 12 & 9. \end{array}$$

This Latin square gives us a bijection to the previous set of 16 blocks, and the bijection satisfies the orthogonality condition. The new superposed design has the following 16 blocks:

$$\begin{array}{cccc} \{1,5,\ 9\} & \{1,6,10\} & \{1,7,11\} & \{1,8,12\} \\ \{2,5,12\} & \{2,6,\ 9\} & \{2,7,10\} & \{2,8,11\} \\ \{3,5,11\} & \{3,6,12\} & \{3,7,\ 9\} & \{3,8,10\} \\ \{4,5,10\} & \{4,6,11\} & \{4,7,12\} & \{4,8,\ 9\}. \end{array} \blacksquare$$

Let Λ_1, Λ_2 and Λ be the concurrence matrices of the two component designs and the superposed design. Then

$$\Lambda = \begin{array}{c} \\ \Theta_1 \\ \\ \Theta_2 \end{array} \begin{array}{c} \begin{array}{cc} \Theta_1 & \Theta_2 \end{array} \\ \left[\begin{array}{cc} \Lambda_1 & \frac{r^2}{b}J \\ \frac{r^2}{b}J & \Lambda_2 \end{array} \right] \end{array}.$$

Let x be a basic contrast of the first component design with canonical efficiency factor ε. Define $\tilde{x}$ in $\mathbb{R}^{\Theta_1 \cup \Theta_2}$ by $\tilde{x}(\theta) = x(\theta)$ for θ in Θ_1 and $\tilde{x}(\theta) = 0$ for θ in Θ_2. Then $\tilde{x}$ is also a contrast, and $\Lambda\tilde{x} = \widetilde{\Lambda_1 x} = rk_1(1-\varepsilon)\tilde{x}$. Therefore $\tilde{x}$ is a basic contrast of the superposed design and its canonical efficiency factor is equal to

$$\frac{k_1\varepsilon + k_2}{k_1 + k_2}. \tag{4.11}$$

Similarly, canonical efficiency factors of the second component design give rise to canonical efficiency factors of the superposed design. Finally, $\Lambda(k_2\chi_{\Theta_1} - k_1\chi_{\Theta_2}) = 0$, because $b/r = t_1/k_1 = t_2/k_2$, so $k_2\chi_{\Theta_1} - k_1\chi_{\Theta_2}$ has canonical efficiency factor equal to 1.

Example 4.6 revisited In each component design with singleton blocks, all canonical efficiency factors are zero. So the superimposed design for 8 treatments in 16 blocks of size two has six canonical efficiency factors equal to $\frac{1}{2}$ and one canonical efficiency factor equal to 1.

Superposing this on the third component design gives nine canonical efficiency factors equal to 2/3 and two canonical efficiency factors equal to 1. ■

4.6.4 Products

For $i = 1, 2$, let $(\Omega_i, \mathcal{G}_i, \Theta_i, \psi_i)$ be a block design for t_i treatments replicated r_i times in b_i blocks of size k_i, with block set Δ_i and concurrence matrix Λ_i. Put $\Omega = \Omega_1 \times \Omega_2$ and $\Theta = \Theta_1 \times \Theta_2$. Define a group-divisible association scheme $\mathcal{G}$ on Ω by letting its blocks be $\delta_1 \times \delta_2$ for δ_1 in Δ_1 and δ_2 in Δ_2. Define $\psi: \Omega \to \Theta$ by $\psi(\omega_1, \omega_2) = (\psi_1(\omega_1), \psi_2(\omega_2))$. Then $(\Omega, \mathcal{G}, \Theta, \psi)$ is a block design for t_1t_2 treatments replicated r_1r_2 times in b_1b_2 blocks of size k_1k_2: it is the *product* of the two component designs.

The concurrence matrix Λ of the product design is $\Lambda_1 \otimes \Lambda_2$. Let x and y be basic contrasts of the component designs with canonical efficiency factors ε_1 and ε_2 respectively. Then

$$
\begin{aligned}
\Lambda(x \otimes y) &= (\Lambda_1 \otimes \Lambda_2)(x \otimes y) \\
&= \Lambda_1 x \otimes \Lambda_2 y \\
&= r_1k_1(1-\varepsilon_1)x \otimes r_2k_2(1-\varepsilon_2)y \\
&= r_1r_2k_1k_2(1-\varepsilon_1-\varepsilon_2+\varepsilon_1\varepsilon_2)x \otimes y
\end{aligned}
$$

so $x \otimes y$ is a basic contrast of the product design with canonical efficiency factor $\varepsilon_1 + \varepsilon_2 - \varepsilon_1\varepsilon_2$. Similarly, $x \otimes \chi_{\Omega_2}$ and $\chi_{\Omega_1} \otimes y$ are basic contrast of the product design with canonical efficiency factors ε_1 and ε_2 respectively.

Example 4.7 The product of the first design in Example 4.4 with the unreduced design in Example 4.5 has the following twelve blocks:

$$
\begin{array}{ll}
\{(1,\theta),(1,\eta),(2,\theta),(2,\eta)\} & \{(3,\theta),(3,\eta),(4,\theta),(4,\eta)\} \\
\{(1,\theta),(1,\eta),(3,\theta),(3,\eta)\} & \{(2,\theta),(2,\eta),(4,\theta),(4,\eta)\} \\
\{(1,\theta),(1,\zeta),(2,\theta),(2,\zeta)\} & \{(3,\theta),(3,\zeta),(4,\theta),(4,\zeta)\} \\
\{(1,\theta),(1,\zeta),(3,\theta),(3,\zeta)\} & \{(2,\eta),(2,\zeta),(4,\eta),(4,\zeta)\} \\
\{(1,\eta),(1,\zeta),(2,\eta),(2,\zeta)\} & \{(3,\eta),(3,\zeta),(4,\eta),(4,\zeta)\} \\
\{(1,\eta),(1,\zeta),(3,\eta),(3,\zeta)\} & \{(2,\eta),(2,\zeta),(4,\eta),(4,\zeta)\}.
\end{array}
$$ ■

Inflation is the special case of the product construction when $b_2 = r_2 = 1$ and $k_2 = t_2$.

Exercises

4.1 Write down the concurrence matrix for each of the following incomplete-block designs.

(a) The treatment set is $\{a, b, \ldots, n\}$ and the blocks are $\{a, b, d, h, i, k\}$, $\{b, c, e, i, j, l\}$, $\{c, d, f, j, k, m\}$, $\{d, e, g, k, l, n\}$, $\{e, f, a, l, m, h\}$, $\{f, g, b, m, n, i\}$ and $\{g, a, c, n, h, j\}$.

(b) The treatment set is $\{1, 2, 3, 4, 5, 6, 7\}$ and the blocks are $\{1, 2, 3\}$, $\{1, 4, 5\}$, $\{1, 6, 7\}$, $\{2, 3, 7\}$, $\{2, 4, 6\}$, $\{3, 5, 6\}$ and $\{4, 5, 7\}$.

(c) The treatment set is $\{1, 2, 3, 4, 5, 6\}$ and the blocks are $\{1, 2, 4\}$, $\{2, 3, 5\}$, $\{3, 4, 1\}$, $\{4, 5, 2\}$, $\{5, 1, 3\}$, $\{1, 2, 6\}$, $\{2, 3, 6\}$, $\{3, 4, 6\}$, $\{4, 5, 6\}$ and $\{5, 1, 6\}$.

(d) The treatment set is $\{0, 1, 2, 3, 4, 5, 6, 7, 8, 9\}$ and the blocks are $\{0, 1\}$, $\{1, 2\}$, $\{2, 3\}$, $\{3, 4\}$, $\{4, 5\}$, $\{0, 5\}$, $\{0, 6\}$, $\{0, 7\}$, $\{0, 8\}$, $\{0, 9\}$, $\{1, 6\}$, $\{1, 7\}$, $\{1, 8\}$, $\{1, 9\}$, $\{2, 6\}$, $\{2, 7\}$, $\{2, 8\}$, $\{2, 9\}$, $\{3, 6\}$, $\{3, 7\}$, $\{3, 8\}$, $\{3, 9\}$, $\{4, 6\}$, $\{4, 7\}$, $\{4, 8\}$, $\{4, 9\}$, $\{5, 6\}$, $\{5, 7\}$, $\{5, 8\}$ and $\{5, 9\}$.

(e) The treatment set is $\{a, b, \ldots, l\}$ and the blocks are $\{a, b, c\}$, $\{d, e, f\}$, $\{g, h, i\}$, $\{j, k, l\}$, $\{d, g, j\}$, $\{a, h, k\}$, $\{b, e, l\}$, $\{c, f, i\}$, $\{e, i, k\}$, $\{a, f, j\}$, $\{b, d, h\}$ and $\{c, g, l\}$.

(f) The treatment set is $\{1, 2, 3, 4, 5, 6, 7, 8\}$ and the blocks are $\{1, 5, 8\}$, $\{2, 6, 1\}$, $\{3, 7, 2\}$, $\{4, 8, 3\}$, $\{5, 1, 4\}$, $\{6, 2, 5\}$, $\{7, 3, 6\}$ and $\{8, 4, 7\}$.

4.2 Construct the dual of each of the incomplete-block designs in the preceding question.

4.3 Prove that $k^{-1}B$ is the orthogonal projector onto V_B.

4.4 Verify Theorem 4.12 for the design in Example 4.1.

4.5 Construct the dual to the design in Example 4.4 revisited. Find its canonical efficiency factors.

4.6 (a) Construct a design for six treatments in nine blocks of size two using the method of orthogonal superposition. Find its canonical efficiency factors.

(b) Form the 2-fold inflation of the design in part (a) and find its canonical efficiency factors.

(c) Construct two designs like the one in part (a) and find a bijection π between their blocks which satisfies the orthogonality condition. Hence construct a design for 12 treatments in nine blocks of size four and find its canonical efficiency factors.

(d) Which of the designs in parts (b) and (c) has the higher value of A?

4.7 (a) Find the canonical efficiency factors of the design for four treatments in 12 blocks of size two which consists of three copies of $\{1,3\}$, $\{1,4\}$, $\{2,3\}$ and $\{2,4\}$.

(b) Hence find the canonical efficiency factors of the following design for eight treatments in 12 blocks of size four.

$$\left\{\begin{array}{cc} 1, & 3, \\ 5, & 7 \end{array}\right\}, \left\{\begin{array}{cc} 1, & 3, \\ 5, & 8 \end{array}\right\}, \left\{\begin{array}{cc} 1, & 3, \\ 6, & 7 \end{array}\right\}, \left\{\begin{array}{cc} 1, & 4, \\ 5, & 7 \end{array}\right\},$$

$$\left\{\begin{array}{cc} 1, & 4, \\ 6, & 8 \end{array}\right\}, \left\{\begin{array}{cc} 1, & 4, \\ 6, & 8 \end{array}\right\}, \left\{\begin{array}{cc} 2, & 3, \\ 5, & 8 \end{array}\right\}, \left\{\begin{array}{cc} 2, & 3, \\ 6, & 8 \end{array}\right\},$$

$$\left\{\begin{array}{cc} 2, & 3, \\ 6, & 7 \end{array}\right\}, \left\{\begin{array}{cc} 2, & 4, \\ 5, & 7 \end{array}\right\}, \left\{\begin{array}{cc} 2, & 4, \\ 5, & 8 \end{array}\right\}, \left\{\begin{array}{cc} 2, & 4, \\ 6, & 7 \end{array}\right\}.$$

(c) Find the canonical efficiency factors of the design for four treatments in 12 blocks of size two which consists of two copies of $\{1,2\}$, $\{1,3\}$, $\{1,4\}$, $\{2,3\}$, $\{2,4\}$ and $\{3,4\}$.

(d) Hence find the canonical efficiency factors of the following design for eight treatments in 12 blocks of size four.

$$\left\{\begin{array}{cc} 1, & 2, \\ 5, & 6 \end{array}\right\}, \left\{\begin{array}{cc} 1, & 2, \\ 6, & 7 \end{array}\right\}, \left\{\begin{array}{cc} 1, & 3, \\ 5, & 7 \end{array}\right\}, \left\{\begin{array}{cc} 1, & 3, \\ 6, & 8 \end{array}\right\},$$

$$\left\{\begin{array}{cc} 1, & 4, \\ 5, & 8 \end{array}\right\}, \left\{\begin{array}{cc} 1, & 4, \\ 7, & 8 \end{array}\right\}, \left\{\begin{array}{cc} 2, & 3, \\ 5, & 8 \end{array}\right\}, \left\{\begin{array}{cc} 2, & 3, \\ 7, & 8 \end{array}\right\},$$

$$\left\{\begin{array}{cc} 2, & 4, \\ 5, & 7 \end{array}\right\}, \left\{\begin{array}{cc} 2, & 4, \\ 6, & 8 \end{array}\right\}, \left\{\begin{array}{cc} 3, & 4, \\ 5, & 6 \end{array}\right\}, \left\{\begin{array}{cc} 3, & 4, \\ 6, & 7 \end{array}\right\}.$$

(e) Which of the designs in parts (b) and (d) has the higher value of A?

5
Partial balance

5.1 Partially balanced incomplete-block designs

Definition Let $\mathcal{Q}$ be an association scheme on a set Θ with non-diagonal associate classes $\mathcal{C}_1$, $\mathcal{C}_2$, ..., $\mathcal{C}_s$ and adjacency matrices A_0, A_1, ..., A_s. An incomplete-block design whose treatment set is Θ is *partially balanced* with respect to $\mathcal{Q}$ if there are integers λ_0, λ_1, ..., λ_s such that

$$(\theta, \eta) \in \mathcal{C}_i \Longrightarrow \Lambda(\theta, \eta) = \lambda_i;$$

in other words, the concurrence matrix Λ satisfies

$$\Lambda = \sum_{i=0}^{s} \lambda_i A_i.$$

More generally, an incomplete-block design is partially balanced if there exists some association scheme on the treatment set with respect to which it is partially balanced.

Partial balance implies equal replication. We always write r for λ_0.

Definition An incomplete-block design is *balanced* if it is partially balanced with respect to the trivial association scheme.

Statisticians often use the abbreviation PBIBD for a partially balanced incomplete-block design and BIBD for a balanced incomplete-block design. Pure mathematicians often know BIBDs as '2-designs'; some people call them simply 'designs'. We always write λ for λ_1 in a BIBD.

Of course, a given incomplete-block design may be partially balanced with respect to more than one association scheme on Θ: it is not necessary for λ_i to differ from λ_j if $i \neq j$. For example, a BIBD is partially balanced with respect to every association scheme on Θ. We usually give the association scheme with the smallest number of associate classes among those with respect to which the design is partially balanced. In Chapter 11 we shall show that such a "simplest" association scheme is well defined.

Definition An incomplete-block design is a *group-divisible*, *triangular*, ... design if it is partially balanced with respect to any group-divisible, triangular, ... association scheme on Θ. It is a *transversal design* if it is group-divisible with $\lambda_1 = 0$ and $\lambda_2 = 1$ and every block contains one treatment from each group.

Example 5.1 (Example 4.1 continued) The concurrence matrix on page 82 shows that this block design is group-divisible with groups $1,2 \parallel 3,4 \parallel 5,6$ and concurrences $\lambda_1 = 2$, $\lambda_2 = 1$. ■

Lemma 5.1 *In a partially balanced incomplete-block design,*

$$\sum_{i=1}^{s} a_i \lambda_i = r(k-1).$$

In particular, in a balanced incomplete-block design,

$$(t-1)\lambda = r(k-1).$$

Proof Use Lemma 4.1(ii). ■

Example 5.2 (Example 4.4 continued) The design with $t = 4$, $b = 6$, $k = 2$, $r = 3$ and blocks

$$\begin{array}{lll} a = \{1,2\}, & b = \{1,3\}, & c = \{1,4\}, \\ d = \{2,3\}, & e = \{2,4\}, & f = \{3,4\} \end{array}$$

is balanced. Its dual has $t = 6$, $b = 4$, $k = 3$, $r = 2$ and blocks

$$1 = \{a,b,c\}, \quad 2 = \{a,d,e\}, \quad 3 = \{b,d,f\}, \quad 4 = \{c,e,f\}.$$

This is group-divisible for the groups $a, f \parallel b, e \parallel c, d$. In fact, it is a transversal design. ■

Warning: duals of partially balanced incomplete-block designs are not always partially balanced. I have introduced duals to enlarge the class of designs we can study.

Example 5.3 Let $\Theta = \{0,1,2,3,4,5,6\}$. The blocks $\{1,2,4\}$, $\{2,3,5\}$, $\{3,4,6\}$, $\{4,5,0\}$, $\{5,6,1\}$, $\{6,0,2\}$, $\{0,1,3\}$ form a balanced incomplete-block design with $t = b = 7$, $r = k = 3$ and $\lambda = 1$. ■

A balanced incomplete-block design for which $k = 3$ and $\lambda = 1$ is called a *Steiner triple system*.

Example 5.4 By taking the complement of each block in Example 5.3, we obtain the following balanced incomplete-block design with $t = b = 7$, $r = k = 4$ and $\lambda = 2$. The blocks are $\{0,3,5,6\}$, $\{1,4,6,0\}$, $\{2,5,0,1\}$, $\{3,6,1,2\}$, $\{4,0,2,3\}$, $\{5,1,3,4\}$ and $\{6,2,4,5\}$. ∎

Example 5.5 Let Θ consist of all 2-subsets of $\{1,2,3,4,5\}$. Consider the ten blocks of the form

$$\{\{i,j\},\{i,k\},\{j,k\}\}.$$

They form a triangular design. It has $t = b = 10$, $r = k = 3$, $\lambda_1 = 1$ and $\lambda_2 = 0$. ∎

Example 5.6 More generally, choose positive integers l, m n such that $m < l < n$ and $n \geqslant 2m$. Let Θ consist of all m-subsets of an n-set Γ. For each l-subset Φ of Γ, form the block $\{\theta \in \Theta : \theta \subset \Phi\}$. The set of such blocks is an incomplete-block design with $t = {}^n\mathrm{C}_m$, $b = {}^n\mathrm{C}_l$, $k = {}^l\mathrm{C}_m$ and $r = {}^{n-m}\mathrm{C}_{n-l}$. It is partially balanced with respect to the Johnson association scheme $\mathrm{J}(n,m)$.

Example 5.7 Let Θ be the set of 16 cells in a 4×4 square array and let Π be any 4×4 Latin square. Construct an incomplete-block design with 12 blocks of size 4 as follows. Each row is a block. Each column is a block. For each letter of Π, the cells which have that letter form a block.

For example, if

$\Theta =$

1	2	3	4
5	6	7	8
9	10	11	12
13	14	15	16

and $\Pi =$

A	*B*	*C*	*D*
B	*A*	*D*	*C*
C	*D*	*A*	*B*
D	*C*	*B*	*A*

then the blocks are

$$\begin{array}{cccc}
\{1,2,3,4\}, & \{5,6,7,8\}, & \{9,10,11,12\}, & \{13,14,15,16\}, \\
\{1,5,9,13\}, & \{2,6,10,14\}, & \{3,7,11,15\}, & \{4,8,12,16\}, \\
\{1,6,11,16\}, & \{2,5,12,15\}, & \{3,8,9,14\} & \{4,7,10,13\}.
\end{array}$$

This design is partially balanced with respect to the association scheme of $\mathrm{L}(3,4)$-type defined by Π. ∎

5.2 Variance and efficiency

Theorem 5.2 *In a partially balanced incomplete-block design with non-diagonal associate classes $\mathcal{C}_1, \ldots, \mathcal{C}_s$ there are constants $\kappa_1, \ldots, \kappa_s$ such that the variance of the estimator of $\tau(\theta) - \tau(\eta)$ is equal to $\kappa_i \sigma^2$ if $(\theta, \eta) \in \mathcal{C}_i$.*

Proof By definition, $\Lambda = \sum_{i=0}^{s} \lambda_i A_i \in \mathcal{A}$ so Equation (4.5) shows that $L \in \mathcal{A}$. From Section 2.2, $L^- \in \mathcal{A}$. Thus there are constants $\nu_0, \ldots, \nu_s$ such that $L^- = \sum_{i=0}^{s} \nu_i A_i$. Now Equation (4.4) shows that

$$\begin{aligned}
\mathrm{Var}(\widehat{\tau(\theta) - \tau(\eta)}) &= \left(L^-(\theta,\theta) - L^-(\theta,\eta) - L^-(\eta,\theta) + L^-(\eta,\eta)\right)\sigma^2, \\
&= (\nu_0 - \nu_i - \nu_i + \nu_0)\sigma^2, \qquad \text{if } (\theta,\eta) \in \mathcal{C}_i, \\
&= \kappa_i \sigma^2
\end{aligned}$$

with $\kappa_i = 2(\nu_0 - \nu_i)$. ■

Theorem 5.2 shows why partially balanced incomplete-block designs were invented. To a combinatorialist the pattern of concurrences which defines partial balance is interesting. To a statistician, the pattern of variances demonstrated by Theorem 5.2 is important, far more important than combinatorial patterns. Some statisticians are puzzled that in general incomplete-block designs the pattern of variances of simple contrasts does not match the pattern of concurrences. The technical condition about p_{ij}^k in the definition of association scheme is there precisely to give Theorem 5.2. The irony is that some statisticians who are interested in the pattern of variances reject the p_{ij}^k condition as "of no importance"; see [192, 229].

Example 5.8 In a balanced incomplete-block design, $\Lambda = rI + \lambda(J - I)$, so

$$L = \left(\frac{r(k-1)+\lambda}{k}\right) I - \frac{\lambda}{k} J = \frac{\lambda t}{k}\left(I - \frac{1}{t}J\right),$$

because $\lambda(t-1) = r(k-1)$. Thus

$$L^- = \frac{k}{\lambda t}\left(I - \frac{1}{t}J\right),$$

so

$$\mathrm{Var}(\widehat{\tau(\theta) - \tau(\eta)}) = \frac{k}{\lambda t}(1+1)\sigma^2 = \frac{2k}{\lambda t}\sigma^2 = \frac{2}{r}\frac{k}{(k-1)}\frac{(t-1)}{t}\sigma^2.$$

Equation (4.6) shows that the efficiency factor for every simple contrast is equal to

$$\frac{t}{t-1}\frac{k-1}{k},$$

which is indeed the eigenvalue of $r^{-1}L$ on the whole of $U_0^\perp$. ■

Theorem 5.3 (Fisher's Inequality) *If a balanced incomplete-block design has t treatments and b blocks then $b \geqslant t$.*

Proof If $t > b$ then Theorem 4.13 shows that the design has at least $t - b$ canonical efficiency factors equal to 1. But Example 5.8 shows that no canonical efficiency factor is equal to 1 in a BIBD. ■

Technique 5.1 Pretending that $J = O$, find the inverse of L on $U_0^\perp$. If L is a polynomial in a single adjacency matrix A, use the minimal polynomial of A on $U_0^\perp$ to calculate this inverse.

Example 5.5 revisited Let A be the adjacency matrix for first associates in the triangular association scheme T(n). The argument at the end of Section 1.4.4, or the parameters given in Section 1.4.1, show that

$$A^2 = (2n-4)I + (n-2)A + 4(J - A - I). \tag{5.1}$$

The incomplete-block design in Example 5.5 is partially balanced with respect to the triangular association scheme T(5), for which

$$A^2 = 2I - A + 4J.$$

We have $\Lambda = 3I + A$ and thus $L = \frac{1}{3}(6I - A)$. If we pretend that $J = O$ then

$$LA = \frac{1}{3}(6A - A^2) = \frac{1}{3}(6A - 2I + A) = \frac{1}{3}(7A - 2I)$$

and so

$$L(A + 7I) = \frac{40}{3}I.$$

Therefore

$$L^- = \frac{3}{40}(A + 7I) + cJ$$

for some c. If θ and η are first associates then the variance of the estimator of $\tau(\theta) - \tau(\eta)$ is

$$2\sigma^2 \times \frac{3}{40} \times (7-1) = \frac{9\sigma^2}{10};$$

otherwise it is $21\sigma^2/20$.

Each treatment has six first associates and three second associates, so the average variance of simple contrasts is

$$\frac{1}{3}\left[2 \times \frac{18\sigma^2}{20} + \frac{21\sigma^2}{20}\right] = \frac{19\sigma^2}{20}.$$

By Theorem 4.12, the harmonic mean efficiency factor A is equal to 40/57. ■

Theorem 5.4 *In a partially balanced incomplete-block design, the strata (in $\mathbb{R}^\Theta$) are sub-eigenspaces of the information matrix, and the canonical efficiency factors are*

$$1 - \frac{1}{rk}\sum_{i=0}^{s} \lambda_i C(i,e)$$

with multiplicity d_e, for e in $\mathcal{E} \setminus \{0\}$, where C is the character table of the association scheme.

Proof We have

$$L = rI - \frac{1}{k}\Lambda = rI - \frac{1}{k}\sum_{i=0}^{s} \lambda_i A_i.$$

If x is in the stratum U_e then $A_i x = C(i,e)x$ and so

$$Lx = rx - \frac{1}{k}\sum_{i=0}^{s} \lambda_i C(i,e)x. \quad ■$$

Example 5.9 Consider nine treatments in a 3×3 square, forming the Hamming association scheme H(2, 3). The nine blocks of shape

	*	*
*		
*		

give a partially balanced incomplete-block design with $t = b = 9$, $k = 4$, $\lambda_0 = 4$, $\lambda_1 = 1$ and $\lambda_2 = 2$.

Now $A_1^2 = 4I + A_1 + 2A_2 = 4I + A_1 + 2(J - I - A_1)$. Ignoring J, we get $A_1^2 + A_1 - 2I = O$ so $(A_1 + 2I)(A_1 - I) = O$. This leads to the character table

$$\begin{array}{lll} & & & (1) & (4) & (4) \\ \lambda_0 = 4 & \text{0th associates} & (1) & 1 & 1 & 1 \\ \lambda_1 = 1 & \text{1st associates} & (4) & 4 & -2 & 1 \\ \lambda_2 = 2 & \text{2nd associates} & (4) & 4 & 1 & -2 \end{array}.$$

We use Theorem 5.4 on the three columns of the character table:

$$\begin{aligned} 1-\frac{1}{16}(4+4+2\times 4) &= 0, \quad \text{as it must do,} \\ 1-\frac{1}{16}(4-2+2\times 1) &= \frac{3}{4}, \\ 1-\frac{1}{16}(4+1+2\times(-2)) &= \frac{15}{16}. \end{aligned}$$

Thus the canonical efficiency factors are 3/4 and 15/16, with multiplicity 4 each, and

$$A = \left(\frac{\frac{4}{3}+\frac{16}{15}}{2}\right)^{-1} = \frac{5}{6}. \quad \blacksquare$$

It is possible for the canonical efficiency factors for two different strata to be the same even if the λ_i are all different.

Example 5.10 (Example 4.3 continued) If we write the treatments in the rectangular array

1	2	3	4
5	6	7	8

we see that

$$\Lambda = 4I + 2A_1 + 3A_2 + A_3,$$

where the A_i are the adjacency matrices of the rectangular association scheme $\underline{\underline{2}}\times\underline{\underline{4}}$ in Example 1.3. The character table is given in Example 3.1 on page 64:

$$\begin{aligned} I &= S_0 + S_1 + S_2 + S_3 \\ A_1 &= 3S_0 - S_1 + 3S_2 - S_3 \\ A_2 &= S_0 + S_1 - S_2 - S_3 \\ A_3 &= 3S_0 - S_1 - 3S_2 + S_3. \end{aligned}$$

Thus

$$\Lambda = 16S_0 + 4S_1 + 4S_2$$

so

$$L = 4I - \frac{1}{4}\Lambda = 3(S_1 + S_2) + 4S_3.$$

The canonical efficiency factors are 3/4 and 1, with multiplicities 4 and 3 respectively. In the notation of Example 4.3, $P_3 = S_1 + S_2$ and $P_4 = S_3$.

The Genstat analysis shown in Figures 4.2–4.3 used this rectangular association scheme. Treatments in the top row of the rectangular array have frow = 1, while those on the bottom row have frow = 2. Similarly, the four columns of the array are distinguished by the values of fcol. ■

Technique 5.2 Even if you do not remember the whole character table for an association scheme, do remember its strata. Find the canonical efficiency factor for each stratum by applying Λ to any vector in that stratum.

Example 5.11 The group-divisible design in Example 5.2 has $k = 3$, $r = 2$ and groups $a, f \parallel b, e \parallel c, d \parallel$. The concurrence matrix is

$$\Lambda = \begin{array}{c} \\ a \\ f \\ b \\ e \\ c \\ d \end{array} \begin{array}{c} \begin{array}{cccccc} a & f & b & e & c & d \end{array} \\ \left[\begin{array}{cccccc} 2 & 0 & 1 & 1 & 1 & 1 \\ 0 & 2 & 1 & 1 & 1 & 1 \\ 1 & 1 & 2 & 0 & 1 & 1 \\ 1 & 1 & 0 & 2 & 1 & 1 \\ 1 & 1 & 1 & 1 & 2 & 0 \\ 1 & 1 & 1 & 1 & 0 & 2 \end{array}\right] \end{array}$$

One within-groups vector is $\chi_a - \chi_f$, which is an eigenvector of Λ with eigenvalue 2. So the within-groups canonical efficiency factor is equal to $1 - 2/6 = 2/3$: it has multiplicity 3. One between-groups vector is $\chi_a + \chi_f - \chi_b - \chi_e$. This is an eigenvector of Λ with eigenvalue 0, so the between-groups canonical efficiency factor is equal to 1, with multiplicity 2.

The dual design is balanced, with all canonical efficiency factors equal to $(4/3) \times (1/2) = 2/3$, as shown in Example 5.8. This agrees with Theorem 4.13. ■

Technique 5.3 If you remember the stratum projectors for the association scheme, express L in terms of them. The coefficients of the projectors are the eigenvalues of L. Moreover, L^- is obtained by replacing each non-zero coefficient by its reciprocal.

Example 5.11 revisited As we saw in Section 2.3, it is useful to let G be the adjacency matrix for the relation 'is in the same group as'. Then the stratum projectors are

$$\frac{1}{6}J, \quad \frac{1}{2}G - \frac{1}{6}J \quad \text{and} \quad I - \frac{1}{2}G,$$

with corresponding dimensions 1, 2 and 3 respectively. Now,

$$\Lambda = 2I + (J - G)$$

so

$$\begin{aligned} L &= 2I - \frac{1}{3}(2I + J - G) = \frac{4I - J + G}{3} \\ &= \frac{4}{3}\left(I - \frac{1}{2}G\right) + 2\left(\frac{1}{2}G - \frac{1}{6}J\right). \end{aligned}$$

(Note that the coefficient of $t^{-1}J$ *must* be zero, so here is a check on the arithmetic.) From this we read off the eigenvalues of L as 4/3 and 2, with multiplicities 3 and 2 respectively. Dividing these by 2 gives the canonical efficiency factors 2/3 and 1 that we found before.

Moreover,

$$L^{-} = \frac{3}{4}\left(I - \frac{1}{2}G\right) + \frac{1}{2}\left(\frac{1}{2}G - \frac{1}{6}J\right).$$

For treatments that are first associates, the relevant 2×2 submatrices of G and J are both equal to

$$\left[\begin{array}{cc} 1 & 1 \\ 1 & 1 \end{array}\right],$$

which make no contribution to the variance of the difference, which is therefore equal to $2 \times (3/4)\sigma^2 = (3/2)\sigma^2$. This agrees with what we already know, because we have already found that the efficiency factor for the simple contrast of first associates is equal to 2/3. From Equation (4.6), the variance is equal to

$$\frac{2}{r}\frac{\sigma^2}{\text{efficiency factor}},$$

which is equal to

$$\frac{2}{2}\frac{3\sigma^2}{2}$$

in this case.

For treatments that are second associates we can still ignore J, but the relevant 2×2 submatrices of G and I are both equal to

$$\left[\begin{array}{cc} 1 & 0 \\ 0 & 1 \end{array}\right]$$

so the variance of the difference is equal to

$$2\left[\frac{3}{4}\left(1 - \frac{1}{2}\right) + \frac{1}{2}\left(\frac{1}{2}\right)\right]\sigma^2 = \frac{5}{4}\sigma^2.$$

Each treatment has one first associate and four second associates, so the average variance of simple contrasts is

$$\frac{\frac{3}{2}+4\times\frac{5}{4}}{5}\sigma^2=\frac{13}{10}\sigma^2.$$

We can also calculate A directly as

$$A=\left(\frac{3\times\frac{3}{2}+2\times 1}{5}\right)^{-1}=\frac{10}{13}.$$

The values of A and of the average variance agree with Theorem 4.12. ■

Technique 5.4 Calculate the canonical efficiency factors and then check the arithmetic by verifying that Equation (4.7) holds. Alternatively, if there is one stratum whose vectors are more difficult for calculations, then find the other canonical efficiency factors and deduce the missing canonical efficiency factor from Equation (4.7); that is

$$\sum_{e=1}^{s} d_e\varepsilon_e=\frac{t(k-1)}{k},$$

where ε_e is the canonical efficiency factor for stratum U_e.

Example 5.12 Let Θ be the set of 2-subsets of $\{1,2,3,4,5,6\}$. Consider the incomplete-block design whose blocks are all fifteen subsets of Θ like $\{\{1,2\},\{3,4\},\{5,6\}\}$ in which no parent is repeated. This is a triangular design, partially balanced for the association scheme T(6). We know from Exercise 2.8 that the strata for this association scheme are U_0, U_{parents} and $U_{\mathrm{offspring}}$ with dimensions 1, 5 and 9 respectively. Let the corresponding canonical efficiency factors be $\varepsilon_{\mathrm{parents}}$ and $\varepsilon_{\mathrm{offspring}}$ respectively. Then

$$5\varepsilon_{\mathrm{parents}}+9\varepsilon_{\mathrm{offspring}}=15\times\frac{2}{3}=10.$$

Every parent occurs once in each block, so if $x\in U_{\mathrm{parents}}$ then $BXx=0$ so $\Lambda x=0$. Thus $\varepsilon_{\mathrm{parents}}=1$. Now

$$5+9\varepsilon_{\mathrm{offspring}}=10,$$

so $\varepsilon_{\mathrm{offspring}}=5/9$. ■

5.3 Concurrence and variance

To many statisticians, it is intuitively reasonable that the higher is the concurrence $\Lambda(\theta,\eta)$ the lower should be the variance of the estimator of

$\tau(\theta) - \tau(\eta)$, as we have seen in all of the examples so far. Why is this? Suppose that treatments θ and η but not ζ occur in one block, while treatments η and ζ but not θ occur in the next, so that a fragment of the design is

		Block 1		Block 2		
Ω	...	ω_1	ω_2	ω_3	ω_4	...
ψ		θ	η	η	ζ	

An unbiased estimator of $\tau(\theta) - \tau(\eta)$ from Block 1 is $Y(\omega_1) - Y(\omega_2)$, which has variance $2\sigma^2$; an unbiased estimator of $\tau(\theta) - \tau(\zeta)$ from these two blocks is $Y(\omega_1) - Y(\omega_2) + Y(\omega_3) - Y(\omega_4)$, which has variance $4\sigma^2$. Thus every path in the treatment-concurrence graph from one treatment to another gives an unbiased estimator of the difference between those treatments, and the variance of the estimator is proportional to the length of the path. Hence the conventional wisdom that the more closely that θ and η are connected in the graph then the lower is the variance of the estimator of $\tau(\theta) - \tau(\eta)$.

Unfortunately, this does *not* imply that higher concurrence always leads to lower variance, as the following example shows.

Example 5.13 Let Θ consist of 36 treatments with the association scheme $\underline{\underline{2}} \times (\underline{\underline{9}}/\underline{\underline{2}})$ shown in Figure 5.1.

	box 1		box 2		box 3		box 4		box 5		box 6		box 7		box 8		box 9	
	col 1	col 2	col 3	col 4	col 5	col 6	col 7	col 8	col 9	col 10	col 11	col 12	col 13	col 14	col 15	col 16	col 17	col 18
row 1	θ	ζ																
row 2	η																	

Fig. 5.1. The association scheme $\underline{\underline{2}} \times (\underline{\underline{9}}/\underline{\underline{2}})$ in Example 5.13

Define a group of treatments to be a set of two treatments which occur in the same row and the same box. Let R, C, B and G be the adjacency matrices for the relations 'is in the same row as', 'is in the same column as', 'is in the same box as', and 'is in the same group as' respectively. Then the adjacency matrices of the association scheme are I, $G - I$, $R - G$, $C - I$, $B - G - C + I$ and $J - R - B + G$. The methods of Chapter 3 (see also Example 6.4) show that the stratum projectors are

$$\frac{1}{36}J,\quad \frac{1}{4}B - \frac{1}{36}J,\quad \frac{1}{2}C - \frac{1}{4}B,\quad \frac{1}{18}R - \frac{1}{36}J,$$

$$\frac{1}{2}G - \frac{1}{18}R - \frac{1}{4}B + \frac{1}{36}J \quad \text{and} \quad I - \frac{1}{2}G - \frac{1}{2}C + \frac{1}{4}B.$$

Consider the incomplete-block design in 324 blocks of size two in which each column occurs as two blocks and there is also one block for each pair of treatments which are in the same row but different groups. Then

$$\Lambda = 18I + 2(C - I) + (R - G) = 16I + 2C + R - G$$

so

$$\begin{aligned} L &= \frac{1}{2}(20I - 2C - R + G) \\ &= \frac{1}{2}\left[20\left(I - \frac{1}{2}G - \frac{1}{2}C + \frac{1}{4}B\right) + 22\left(\frac{1}{2}G - \frac{1}{18}R - \frac{1}{4}B + \frac{1}{36}J\right)\right. \\ &\qquad \left. + 4\left(\frac{1}{18}R - \frac{1}{36}J\right) + 16\left(\frac{1}{2}C - \frac{1}{4}B\right) + 18\left(\frac{1}{4}B - \frac{1}{36}J\right)\right] \end{aligned}$$

and

$$\begin{aligned} L^- &= \frac{1}{10}\left(I - \frac{1}{2}G - \frac{1}{2}C + \frac{1}{4}B\right) + \frac{1}{11}\left(\frac{1}{2}G - \frac{1}{18}R - \frac{1}{4}B + \frac{1}{36}J\right) \\ &\qquad + \frac{1}{2}\left(\frac{1}{18}R - \frac{1}{36}J\right) + \frac{1}{8}\left(\frac{1}{2}C - \frac{1}{4}B\right) + \frac{1}{9}\left(\frac{1}{4}B - \frac{1}{36}J\right) \\ &= \frac{1}{15\,840}(1584I - 72G + 360R + 198C - 19B) - \frac{103}{198}\frac{1}{36}J. \end{aligned}$$

For treatments that are in the same column, the relevant 2×2 submatrices of C, B and J are all equal to

$$\begin{bmatrix} 1 & 1 \\ 1 & 1 \end{bmatrix},$$

which makes no contribution to the variance, while the relevant submatrices of I, G and R are all equal to

$$\begin{bmatrix} 1 & 0 \\ 0 & 1 \end{bmatrix}.$$

Thus the variance of the estimator of the difference is equal to

$$2 \times \frac{1}{15\,840} \times (1584 - 72 + 360) = \frac{1872}{7920}.$$

Similarly, for treatments that are in the same group we can ignore G, R, B and J: we find that the variance of the estimator of the difference is

$$2 \times \frac{1}{15\,840} \times (1584 + 198) = \frac{1782}{7920}.$$

For treatments in the same row but different groups, the variance is

$$2 \times \frac{1}{15\,840} \times (1584 - 72 + 198 - 19) = \frac{1691}{7920};$$

for treatments in the same box but different rows and columns, it is

$$2 \times \frac{1}{15\,840} \times (1584 - 72 + 360 + 198) = \frac{2070}{7920};$$

for treatments in different rows and boxes, it is

$$2 \times \frac{1}{15\,840} \times (1584 - 72 + 360 + 198 - 19) = \frac{2051}{7920}.$$

Ordering the associate classes by ascending size of variance, we have Table 5.1.

Associate class	7920 × variance	concurrence
same row, different group	1691	1
same group	1782	0
same column	1872	2
different row and box	2051	0
same box, different row and column	2070	0

Table 5.1. *Concurrences and variances in Example 5.13*

Consider the treatments θ, η and ζ in Figure 5.1. Table 5.1 shows that $\Lambda(\theta, \eta) = 2$ and $\Lambda(\theta, \zeta) = 0$ but the variance of the estimator of $\tau(\theta) - \tau(\eta)$ is higher than the variance of the estimator of $\tau(\theta) - \tau(\zeta)$. In some sense, this is because the 16 paths of length 2 between θ and ζ connect them more closely than the two paths of length 1 which connect θ and η directly. ■

There are many paths connecting a given pair of treatments in the graph, and their corresponding estimators may not be independent. It is not simple to work out the variance directly from the graph. But the number of paths of length l between treatments θ and η is equal to the (θ, η)-entry of $(\Lambda - rI)^l$. If the design has m distinct canonical efficiency factors then L^- is a polynomial in L of degree $m-1$ (ignoring J), and so is also a polynomial in $\Lambda - rI$ of degree $m-1$ (ignoring J). So variance is a linear combination of the numbers of paths of lengths 0, 1, 2, ..., $m-1$ between the two treatments. This is especially straightforward when m is equal to 1 (corresponding to balanced incomplete-block designs) or 2.

Theorem 5.5 *If a connected equi-replicate incomplete-block design with* *replication r and block size k has exactly two canonical efficiency factors ε_1 and ε_2 then the variance of the estimator of $\tau(\theta) - \tau(\eta)$ is equal to*

$$\frac{2\sigma^2}{r^2 k \varepsilon_1 \varepsilon_2}\left[rk(\varepsilon_1 + \varepsilon_2) - r(k-1) - \Lambda(\theta,\eta)\right].$$

Proof For $i = 1, 2$, let P_i be the matrix of orthogonal projection onto the eigenspace of $r^{-1}L$ with eigenvalue ε_i. Then

$$\varepsilon_1 P_1 + \varepsilon_2 P_2 = r^{-1}L = I - (rk)^{-1}\Lambda. \tag{5.2}$$

Moreover,

$$I = P_1 + P_2 + t^{-1}J,$$

because the design is connected. From these two equations we obtain

$$(\varepsilon_1 - \varepsilon_2)P_1 = (1-\varepsilon_2)I - (rk)^{-1}\Lambda + \varepsilon_2 t^{-1}J;$$

$$(\varepsilon_2 - \varepsilon_1)P_2 = (1-\varepsilon_1)I - (rk)^{-1}\Lambda + \varepsilon_1 t^{-1}J.$$

The spectral decomposition of $r^{-1}L$ in Equation (5.2) gives

$$\begin{aligned} L^- &= r^{-1}\varepsilon_1^{-1}P_1 + r^{-1}\varepsilon_2^{-1}P_2 \\ &= \frac{1}{r\varepsilon_1\varepsilon_2}\left[(\varepsilon_1+\varepsilon_2-1)I + (rk)^{-1}\Lambda - (\varepsilon_1+\varepsilon_2)t^{-1}J\right]. \end{aligned}$$

From Equation (4.4),

$$\mathrm{Var}\left(\widehat{\tau(\theta) - \tau(\eta)}\right) = \frac{2}{r\varepsilon_1\varepsilon_2}\left(\varepsilon_1 + \varepsilon_2 - 1 + \frac{1}{k} - \frac{\Lambda(\theta,\eta)}{rk}\right)\sigma^2. \quad \blacksquare$$

Corollary 5.6 *In a partially balanced incomplete-block design with two associate classes, those pairs of treatments with the higher concurrence have the lower variance.*

This corollary agrees with our findings in Examples 4.1, 5.5 and 5.11.

5.4 Cyclic designs

Let $\Theta = \mathbb{Z}_t$. If $\Upsilon \subseteq \Theta$, a *translate* of Υ is a set of the form

$$\Upsilon + \theta = \{\upsilon + \theta : \upsilon \in \Upsilon\}$$

for some θ in Θ. Of course, Υ is a translate of itself.

It is possible to have $\Upsilon + \theta_1 = \Upsilon + \theta_2$ even when $\theta_1 \neq \theta_2$. Then $\Upsilon + (\theta_1 - \theta_2) = \Upsilon$. Let l be the number of distinct translates of Υ: we shall abuse group-theoretic terminology slightly and refer to l as the *index* of Υ. Then $\Upsilon + (l \bmod t) = \Upsilon$. Moreover, l is the smallest positive integer with this property, and l divides t (for if not, the remainder on dividing t by l is a smaller positive number l' with $\Upsilon + (l' \bmod t) = \Upsilon$).

Definition An incomplete-block design with treatment set $\mathbb{Z}_t$ is a *thin cyclic design* if there is some subset Υ of $\mathbb{Z}_t$ such that the blocks are all the distinct translates of Υ: the design is said to be *generated* by Υ. An incomplete-block design is a *cyclic* design if its blocks can be partitioned into sets of blocks such that each set is a thin cyclic design.

Example 5.14 Let $\Upsilon = \{0, 1, 3\} \subset \mathbb{Z}_8$. This has index 8, so it generates the following thin cyclic design:

$\{0, 1, 3\}$, $\{1, 2, 4\}$, $\{2, 3, 5\}$, $\{3, 4, 6\}$, $\{4, 5, 7\}$, $\{5, 6, 0\}$, $\{6, 7, 1\}$, $\{7, 0, 2\}$. ■

Example 5.15 Here is a cyclic design for $\mathbb{Z}_6$ which is not thin:

$\{0, 1, 4\}$, $\{1, 2, 5\}$, $\{2, 3, 0\}$, $\{3, 4, 1\}$, $\{4, 5, 2\}$, $\{5, 0, 3\}$, $\{0, 2, 4\}$, $\{1, 3, 5\}$.

The index of $\{0, 1, 4\}$ is 6 and the index of $\{0, 2, 4\}$ is 2. ■

Theorem 5.7 *Let $\Upsilon \subset \mathbb{Z}_t$ and let l be the index of Υ. For θ in $\mathbb{Z}_t$, let*

$$m_\theta(\Upsilon) = \left|\{(v_1, v_2) \in \Upsilon \times \Upsilon : v_1 - v_2 = \theta\}\right|,$$

so that

$$\chi_\Upsilon \chi_{-\Upsilon} = \sum_{\theta \in \Theta} m_\theta(\Upsilon) \chi_\theta.$$

Then, in the thin cyclic design generated by Υ,

$$\Lambda(0, \theta) = m_\theta(\Upsilon) \times \frac{l}{t}$$

and

$$\Lambda(\eta, \zeta) = \Lambda(0, \zeta - \eta). \tag{5.3}$$

Proof Treatments 0 and θ concur in the translate $\Upsilon + \rho$ if and only if there are v_1, v_2 in Υ such that $v_1 + \rho = \theta$ and $v_2 + \rho = 0$, that is $\rho = -v_2$ and $\theta = v_1 - v_2$. If $l = t$ then $\Lambda(0, \theta) = m_\theta(\Upsilon)$. In general, the family of sets Υ, $\Upsilon + 1$, ..., $\Upsilon + t - 1$ consists of t/l copies of the l distinct translates Υ, $\Upsilon + 1$, ..., $\Upsilon + l - 1$, so the concurrence in the thin design is $(l/t) m_\theta(\Upsilon)$.

Moreover, treatments 0 and θ concur in $\Upsilon + \rho$ if and only if treatments η and $\eta + \theta$ concur in $\Upsilon + \rho + \eta$, so $\Lambda(0, \theta) = \Lambda(\eta, \eta + \theta)$. ■

Corollary 5.8 *Every cyclic design is partially balanced with respect to the cyclic association scheme on $\mathbb{Z}_t$ defined by the blueprint* $\{0\}$, $\{\pm 1\}$, $\{\pm 2\}$, ... *(It* may *be partially balanced with respect to a cyclic association scheme with fewer associate classes.)*

Proof Since Equation (5.3) holds in each thin component of the design, it holds overall, and

$$\Lambda = \sum_{\theta \in \Theta} \Lambda(0, \theta) M_\theta,$$

where

$$M_\theta(\eta, \zeta) = \begin{cases} 1 & \text{if } \zeta - \eta = \theta \\ 0 & \text{otherwise,} \end{cases}$$

as in Section 1.4.5. However, Λ is symmetric, so Equation (5.3) gives $\Lambda(0, \theta) = \Lambda(-\theta, 0) = \Lambda(0, -\theta)$. The adjacency matrices for the cyclic association scheme defined by the blueprint $\{0\}$, $\{\pm 1\}$, $\{\pm 2\}$ ... are $(M_\theta + M_{-\theta})$ if $2\theta \neq 0$ and M_θ if $2\theta = 0$, so Λ is a linear combination of the adjacency matrices, and so the design is partially balanced with respect to this association scheme.

Suppose that Δ_0, Δ_1, ..., Δ_s is a blueprint for $\mathbb{Z}_t$ such that $\Lambda(0, \theta)$ is constant λ_i for θ in Δ_i. Putting $A_i = \sum_{\theta \in \Delta_i} M_\theta$ gives $\Lambda = \sum_i \lambda_i A_i$, and so the design is partially balanced with respect to the cyclic association scheme defined by the blueprint. ■

Now write λ_θ for $\Lambda(0, \theta)$.

Technique 5.5 To calculate the concurrences in the thin design which is generated by Υ, form the *table of differences* for Υ. Try to find the coarsest blueprint such that λ_θ is constant on each set in the partition.

Example 5.14 revisited In $\mathbb{Z}_8$, the block $\{0, 1, 3\}$ gives the following table of differences.

	0	1	3
0	0	1	3
1	7	0	2
3	5	6	0

Therefore $\lambda_0 = 3$, $\lambda_1 = \lambda_2 = \lambda_3 = \lambda_5 = \lambda_6 = \lambda_7 = 1$ and $\lambda_4 = 0$. Hence the design is partially balanced for the association scheme defined by the blueprint $\{0\}$, $\{4\}$, $\{1, 2, 3, 5, 6, 7\}$ (so this design is group-divisible with groups $0, 4 \parallel 1, 5 \parallel 2, 6 \parallel 3, 7$). ■

Definition A subset Υ of $\mathbb{Z}_t$ is a *perfect difference set* for $\mathbb{Z}_t$ if there are integers r, λ such that

$$\chi_\Upsilon \chi_{-\Upsilon} = r\chi_0 + \lambda(\chi_{\mathbb{Z}_t} - \chi_0);$$

in other words, $m_\theta(\Upsilon) = \lambda$ for all θ with $\theta \neq 0$.

Proposition 5.9 *The thin cyclic design generated by Υ is balanced if and only if Υ is a perfect difference set.*

Example 5.3 revisited The subset $\{1, 2, 4\}$ is a perfect difference set for $\mathbb{Z}_7$.

	1	2	4
1	0	1	3
2	6	0	2
4	4	5	0

Its table of differences contains every non-zero element of $\mathbb{Z}_7$ exactly once. ■

Theorem 5.10 *The canonical efficiency factors of a cyclic design are*

$$1 - \frac{1}{rk} \sum_{\theta \in \mathbb{Z}_t} \lambda_\theta \epsilon^\theta$$

for complex t-th roots of unity ϵ with $\epsilon \neq 1$.

Proof Use Theorems 5.4 and 2.18. ■

Technique 5.6 Let $\epsilon_t = \exp\left(\frac{2\pi \mathrm{i}}{t}\right)$. Then ϵ is a complex t-th root of unity if there is an integer m such that $\epsilon = \epsilon_t^m$. To calculate canonical efficiency factors of cyclic designs numerically, replace $\epsilon^\theta + \epsilon^{-\theta}$ by $2\cos\left(\frac{2\pi\theta m}{t}\right)$. To calculate the harmonic mean efficiency factor A as an exact rational number, leave everything in powers of ϵ_t.

Example 5.16 Consider the thin cyclic design generated by $\{0, 1, 3, 7\}$ in $\mathbb{Z}_9$.

	0	1	3	7
0	0	1	3	7
1	8	0	2	6
3	6	7	0	4
7	2	3	5	0

Thus the eigenvalues of Λ are

$$4 + (\epsilon + \epsilon^{-1}) + 2(\epsilon^2 + \epsilon^{-2}) + 2(\epsilon^3 + \epsilon^{-3}) + (\epsilon^4 + \epsilon^{-4})$$

where $\epsilon^9 = 1$. If $\epsilon^3 = 1$ and $\epsilon \neq 1$ then $\epsilon + \epsilon^{-1} = -1$ (the cube roots of unity sum to zero) so the eigenvalue is

$$4 - 1 - 2 + 4 - 1 = 4;$$

otherwise it is

$$4 + \epsilon^2 + \epsilon^{-2} - 2 = 2 + \epsilon^2 + \epsilon^{-2},$$

because the primitive ninth roots of unity sum to zero (because all the ninth roots do). Let ϵ_9 be a fixed primitive ninth root of unity, and put $x = \epsilon_9 + \epsilon_9^{-1}$, $y = \epsilon_9^2 + \epsilon_9^{-2}$ and $z = \epsilon_9^4 + \epsilon_9^{-4}$. Then the canonical efficiency factors are

$$\frac{3}{4}, \quad \frac{14 - x}{16}, \quad \frac{14 - y}{16}, \quad \frac{14 - z}{16},$$

all with multiplicity 2.

Substituting $x = 2\cos 40^o$, $y = 2\cos 80^o$, $z = 2\cos 160^o$ gives

$$0.7500, \quad 0.7792, \quad 0.8533 \quad \text{and} \quad 0.9925$$

to four decimal places, and $A = 0.8340$.

To do the exact calculation, we note first that $x + y + z = 0$. Then

$$\begin{aligned} xy &= (\epsilon_9 + \epsilon_9^{-1})(\epsilon_9^2 + \epsilon_9^{-2}) \\ &= \epsilon_9 + \epsilon_9^3 + \epsilon_9^{-3} + \epsilon_9^{-1} \\ &= x - 1, \end{aligned}$$

and similarly $yz = y - 1$ and $zx = z - 1$. Therefore $xy + yz + zx = x + y + z - 3 = -3$ and $xyz = (x - 1)z = xz - z = z - 1 - z = -1$.

Now

$$\begin{aligned} &\frac{1}{14 - x} + \frac{1}{14 - y} + \frac{1}{14 - x} \\ &\quad = \frac{(14 - x)(14 - y) + (14 - x)(14 - z) + (14 - y)(14 - z)}{(14 - x)(14 - y)(14 - z)} \\ &\quad = \frac{3 \cdot 14^2 - 28(x + y + z) + (xy + yz + zx)}{14^3 - 14^2(x + y + z) + 14(xy + yz + zx) - xyz} \\ &\quad = \frac{3 \cdot 14^2 - 3}{14^3 - 3 \cdot 14 + 1} = \frac{195}{901}, \end{aligned}$$

so

$$4A^{-1} = \frac{4}{3} + \frac{16 \times 195}{901}$$

so

$$A^{-1} = \frac{1}{3} + \frac{4 \times 195}{901} = \frac{3241}{2703}$$

and

$$A = \frac{2703}{3241}. \quad \blacksquare$$

5.5 Lattice designs

Several useful families of incomplete-block design come under the collective heading of *lattice design.*

5.5.1 Simple lattice designs

Let Θ consist of the cells of an $n \times n$ square array. Construct two disconnected designs, each with n blocks of size n. In the first design the rows of Θ are the blocks; in the second the blocks are the columns of Θ. The juxtaposition of these two component designs is a *simple lattice* design for n^2 treatments. It is partially balanced with respect to the Hamming association scheme $\mathrm{H}(2, n)$, with $\Lambda = 2I + A_1$.

A *contrast between rows* of Θ is a contrast of the form $\chi_\Phi - \chi_\Psi$ where Φ and Ψ are distinct rows of Θ. Any such contrast is a basic contrast of both component designs, with canonical efficiency factors 0 and 1 respectively. Contrasts between columns are similar. Every contrast which is orthogonal to both rows and columns has canonical efficiency factor 1 in both component designs. From Section 4.6.1, the canonical efficiency factors of the simple lattice design are 1/2, with multiplicity $2(n-1)$, and 1, with multiplicity $(n-1)^2$. Now Theorem 5.5 with $r = 2$ and $k = n$ shows that

$$\begin{aligned}
\mathrm{Var}\left(\widehat{\tau(\theta) - \tau(\eta)}\right) &= \frac{2\sigma^2}{2n}\left(3n - 2(n-1) - \Lambda(\theta, \eta)\right) \\
&= \frac{\sigma^2}{n}\left(n + 2 - \Lambda(\theta, \eta)\right),
\end{aligned}$$

which is equal to $(n+1)\sigma^2/n$ for first associates and to $(n+2)\sigma^2/n$ for second associates.

The first design in Example 4.4 is a simple lattice design with $n = 2$.

5.5.2 Square lattice designs

Square lattice designs are a generalization of simple lattice designs. Like them, they are formed by juxtaposing r disconnected component designs each with n blocks of size n. Let $\Pi_1, \ldots, \Pi_{r-2}$ be mutually orthogonal $n \times n$ Latin squares. From Exercise 1.15 we know that $r \leqslant n+1$. Let Θ be the set of n^2 cells. The first two components are the same as those of the simple lattice design. For $i = 3, \ldots, r$, each block of the i-th component design consists of the cells which have a given letter of the square Π_{i-2}. The juxtaposition of the components forms a *square lattice* design for n^2 treatments in nr blocks of size n.

Example 5.7 is a square lattice design with $n = 4$ and $r = 3$.

Square lattice designs are partially balanced with respect to the association scheme of L(r,n)-type defined by $\Pi_1, \ldots, \Pi_{r-2}$. They have $\Lambda = rI + A_1$. An alternative name for square lattice designs is *nets*. If $r = n+1$ then L$(r,n) = \underline{\underline{n^2}}$ and the design is balanced: in this case it is called an *affine plane*.

Arguing as for simple lattice designs, we see that a square lattice design has canonical efficiency factors $(r-1)/r$, with multiplicity $r(n-1)$, and 1, with multiplicity $(n+1-r)(n-1)$. Thus the harmonic mean efficiency factor A is equal to

$$\left(\frac{r \times \frac{r}{r-1} + (n+1-r)}{n+1}\right)^{-1} = \frac{(r-1)(n+1)}{rn+2r-n-1}.$$

Theorem 5.5 shows that

$$\begin{aligned}
\operatorname{Var}\left(\widehat{\tau(\theta) - \tau(\eta)}\right) &= \frac{2\sigma^2}{rn(r-1)}\left[rn\left(\frac{2r-1}{r}\right) - r(n-1) - \Lambda(\theta,\eta)\right] \\
&= \frac{2\sigma^2}{rn(r-1)}\left(rn - n + r - \Lambda(\theta,\eta)\right),
\end{aligned}$$

which is equal to

$$\frac{2\sigma^2(n+1)}{rn} \quad \text{and} \quad \frac{2\sigma^2(rn-n+r)}{rn(r-1)}$$

for first and second associates respectively.

Each treatment has $r(n-1)$ first associates and $(n+1-r)(n-1)$ second associates, so the average variance of simple contrasts is

$$\frac{2\sigma^2}{rn(r-1)}\left[\frac{r(n-1)(rn-n+r-1) + (n+1-r)(n-1)(rn-n+r)}{n^2-1}\right],$$

which is equal to

$$\frac{2\sigma^2}{r}\frac{(rn+2r-n-1)}{(r-1)(n+1)}.$$

This is equal to $2\sigma^2/rA$, in agreement with Theorem 4.12.

The dual of a square lattice design is a transversal design for r groups of size n. Such a design may be constructed directly by orthogonal superposition, as in Example 4.6.

5.5.3 Cubic and higher-dimensional lattices

These are generalizations of simple lattice designs to higher dimensions. For the *cubic lattice* design for n^3 treatments we take $\Theta = \Gamma^3$, where Γ is an n-set. For $i = 1, 2, 3$, construct a disconnected incomplete-block design in n^2 blocks of size n: each block consists of all the treatments having a specified value of the j-th coordinate for all $j \neq i$. Juxtapose these three components. Then $\Lambda = 3I + A_1$, where A_1 is the adjacency matrix for first associates in the Hamming association scheme $\mathrm{H}(3,n)$. Thus $L = \frac{1}{n}[3(n-1)I - A_1]$.

We can find L^- by a technique similar to Technique 5.1. From Section 1.4.4 and Exercise 1.25 we have

$$A_1^2 = 3(n-1)I + (n-2)A_1 + 2A_2$$

and

$$\begin{aligned} A_1A_2 &= 2(n-1)A_1 + 2(n-2)A_2 + 3A_3 \\ &= -3I + (2n-5)A_1 + (2n-7)A_2 + 3J. \end{aligned}$$

Therefore

$$\begin{aligned} LA_1 &= \frac{1}{n}\left[3(n-1)A_1 - A_1^2\right] \\ &= \frac{1}{n}\left[-3(n-1)I + (2n-1)A_1 - 2A_2\right]. \end{aligned}$$

If we pretend that $J = O$ then

$$\begin{aligned} LA_2 &= \frac{1}{n}\left[3(n-1)A_2 - A_1A_2\right] \\ &= \frac{1}{n}\left[3I - (2n-5)A_1 + (n+4)A_2\right] \end{aligned}$$

and so

$$L\left[(2n^2+3n+6)I + (n+4)A_1 + 2A_2\right] = 6n^2I.$$

Therefore

$$L^- = \frac{1}{6n^2}\left[(2n^2+3n+6)I + (n+4)A_1 + 2A_2\right] + cJ$$

for some c. If θ and η are first associates then the variance of the estimator of $\tau(\theta) - \tau(\eta)$ is

$$2\sigma^2 \times \frac{1}{6n^2}\left[(2n^2+3n+6) - (n+4)\right] = \frac{2(n^2+n+1)}{3n^2}\sigma^2;$$

if they are second associates then it is

$$\frac{(2n^2+3n+4)}{3n^2}\sigma^2;$$

and if they are third associates it is

$$\frac{(2n^2+3n+6)}{3n^2}\sigma^2.$$

Higher-dimensional lattice designs are constructed similarly. There are n^m treatments, with the Hamming association scheme $\mathrm{H}(m,n)$. There are mn^{m-1} blocks of size n, and the replication is m. The concurrence matrix is $mI + A_1$.

Although square lattices are the most efficient designs possible for their size (see Section 5.7), higher-dimensional lattice designs become less efficient (for their size) as m increases.

There is an obvious generalization of Latin square called a *Latin cube*. This suggests a generalization of the association scheme of Latin-square type in Example 1.6 and hence cubic lattice designs with replication bigger than three. Unfortunately, the obvious generalization of Example 1.6 does not give an association scheme unless the Latin cube is obtained from an Abelian group (see Theorem 8.21), in which case the association scheme is covered by Chapter 8.

5.5.4 Rectangular lattice designs

Rectangular lattices are designs for $n(n-1)$ treatments in rn blocks of size $n-1$, where $r \leqslant n$. The construction is a modification of the construction of square lattice designs.

Let $\Pi_1, \ldots, \Pi_{r-2}$ be mutually orthogonal $n \times n$ Latin squares. A *common transversal* of such squares is a set of n cells which contains one cell from each row, one cell from each column, and, for $i = 1, \ldots, r-2$, one cell with each letter of Π_i. If there is a Latin square Π orthogonal to each of $\Pi_1, \ldots, \Pi_{r-2}$ then every letter of Π is a common transversal; but some common transversals are not of this form. If a common transversal exists, then $r \leqslant n$ and the rows of the squares may be simultaneously permuted so that the transversal occupies the main diagonal.

Now let Θ be the set of $n(n-1)$ cells in the $n \times n$ square array, omitting those in the common transversal. The construction proceeds exactly as for square lattice designs.

Example 5.17 Take $n = r = 4$. The following Latin squares Π_1 and Π_2 are mutually orthogonal, and have the main diagonal as a common transversal.

$\Pi_1 =$

A	B	C	D
C	D	A	B
D	C	B	A
B	A	D	C

$\Pi_2 =$

A	B	C	D
D	C	B	A
B	A	D	C
C	D	A	B

$\Theta =$

	1	2	3
4		5	6
7	8		9
10	11	12	

The blocks of the design are $\{1,2,3\}$, $\{4,5,6\}$, $\{7,8,9\}$, $\{10,11,12\}$, $\{4,7,10\}$, $\{1,8,11\}$, $\{2,5,12\}$, $\{3,6,9\}$, $\{5,9,11\}$ $\{1,6,10\}$, $\{2,4,8\}$, $\{3,7,12\}$, $\{6,8,12\}$, $\{1,5,7\}$, $\{2,9,10\}$ and $\{3,4,11\}$. ■

In general, rectangular lattice designs are not partially balanced, but their duals are. For $i = 1, \ldots, n$ and $j = 1, \ldots, r$ let δ_{ij} be the block in the j-th component design which would contain the i-th cell of the common transversal if that transversal were not missing. In the dual design

$$\Lambda(\delta_{ij}, \delta_{i'j'}) = \begin{cases} n-1 & \text{if } i = i' \text{ and } j = j' \\ 0 & \text{if } i = i' \text{ but } j \neq j' \\ 0 & \text{if } i \neq i' \text{ but } j = j' \\ 1 & \text{if } i \neq i' \text{ and } j \neq j' \end{cases}$$

so that the dual is partially balanced with respect to $\underline{\underline{n}} \times \underline{\underline{r}}$, with

$$\Lambda = (n-1)I + A_{\text{other}}.$$

In Example 3.1 we found that the stratum projectors S_0, S_1, S_2, S_3 for $\underline{\underline{n}} \times \underline{\underline{r}}$ have ranks 1, $r-1$, $n-1$ and $(n-1)(r-1)$, and that

$$A_{\text{other}} = (n-1)(r-1)S_0 - (n-1)S_1 - (r-1)S_2 + S_3.$$

Therefore,

$$\Lambda = (n-1)rS_0 + (n-r)S_2 + nS_3,$$

which has eigenvalues $(n-1)r$, 0, $n-r$ and n, with multiplicities 1, $r-1$, $n-1$ and $(n-1)(r-1)$ respectively. Technique 4.1 shows that the canonical efficiency factors of this dual design are

$$1, \quad \frac{n(r-1)}{(n-1)r} \quad \text{and} \quad \frac{(nr-r-n)}{(n-1)r}$$

with multiplicities $r-1$, $n-1$ and $(n-1)(r-1)$ respectively. Applying Theorem 4.13 with $t = n(n-1)$ and $b = rn$, we see that the canonical efficiency factors of the rectangular lattice are the same as those of its dual except that the multiplicity of 1 is increased by $n(n-1-r)$. Thus if $r = n$ the canonical efficiency factors are 1 and $(n-2)/(n-1)$ with multiplicities $n-2$ and $(n-1)^2$; if $r \leqslant n-1$ they are

$$1, \quad \frac{n(r-1)}{(n-1)r} \quad \text{and} \quad \frac{nr-r-n}{(n-1)r}$$

with multiplicities $n^2 - n - rn + r - 1$, $n-1$ and $(n-1)(r-1)$.

Although rectangular lattice designs are not partially balanced in general, they *are* partially balanced for three particular values of r: $r = n$, $r = n-1$ and $r = 2$. When $r = n$ the number of mutually orthogonal Latin squares is only one short of the maximum possible. In this case it can be shown that there is a further $n \times n$ Latin square Π orthogonal to each of them: the common transversal must correspond to one of the letters in Π. Partition Θ into $n-1$ groups of size n according to the remaining letters in Π. The design is partially balanced with respect to this association scheme $\mathrm{GD}(n-1, n)$: in fact, it is a transversal design.

When $r = n-1$, the existence of $r-2$ mutually orthogonal Latin squares with a common transversal also implies the existence of two more Latin squares Π and Π' orthogonal to each of them and to each other, with one letter of Π giving the common transversal. Now each element of Θ is uniquely defined by its letter in Π ($n-1$ possibilities) and its letter in Π' (n possibilities), so Π and Π' define the rectangular association scheme $\mathrm{R}(n, n-1)$ on Θ. The design is partially balanced with respect to this association scheme.

When $r = 2$ the design is partially balanced with respect to an association scheme $\mathrm{Pair}(n)$ which we have not met before. The underlying set is $\Gamma^2 \setminus \mathrm{Diag}(\Gamma)$, where Γ is an n-set. One associate class consists of pairs with one coordinate in common. Its adjacency matrix A has

$A((\gamma_1,\gamma_2),(\gamma_3,\gamma_4))$ equal to 1 if $\gamma_1 = \gamma_3$ and $\gamma_2 \neq \gamma_4$, or if $\gamma_1 \neq \gamma_3$ and $\gamma_2 = \gamma_4$, and to 0 otherwise. The second class consists of mirror-image pairs $((\gamma_1,\gamma_2),(\gamma_2,\gamma_1))$ for $\gamma_1 \neq \gamma_2$. Let M be its adjacency matrix. The third class consists of other pairs with opposite coordinates in common: its adjacency matrix B has $B((\gamma_1,\gamma_2),(\gamma_3,\gamma_4))$ equal to 1 if $\gamma_1 = \gamma_4$ and $\gamma_2 \neq \gamma_3$, or if $\gamma_1 \neq \gamma_4$ and $\gamma_2 = \gamma_3$, and to 0 otherwise. The final class, with adjacency matrix G, consists of all other non-diagonal pairs. The valencies of the associate classes are $2(n-2)$, 1, $2(n-2)$ and $(n-2)(n-3)$ respectively, so there are four associate classes so long as $n \geqslant 4$.

Figure 5.2 shows the underlying set of Pair(5), which contains 20 elements. One is labelled $*$. Every other element θ is labelled by the unique adjacency matrix whose $(*,\theta)$-entry is equal to 1.

	$*$	A	A	A
M		B	B	B
B	A		G	G
B	A	G		
B	A	G	G	

Fig. 5.2. The association scheme Pair(5): one element $*$ and its associates

It is clear that $M^2 = I$, $AM = B$ and

$$A^2 = 2(n-2)I + (n-3)A + B + 2G.$$

Therefore these classes do form an association scheme.

Because rectangular lattices are not, in general, partially balanced, their pattern of variances of simple contrasts can be quite complicated. There are three canonical efficiency factors if $r < n$, so L^- is effectively a quadratic in $\Lambda - rI$. Hence the variance of the estimator of $\tau(\theta) - \tau(\eta)$ is a linear function of the numbers of paths from θ to η of lengths 1 and 2 in the treatment-concurrence graph. The number of paths of length 2 can take several different values.

5.6 General constructions

The preceding two sections gave constructions for specific association schemes. Here we give a few constructions that work for all association schemes.

5.6.1 Elementary designs

Let $\mathcal{Q}$ be an association scheme on Θ with associate classes $\mathcal{C}_i$ for i in $\mathcal{K}$. Choose a fixed i in $\mathcal{K}$ and form the incomplete-block design whose blocks are $\mathcal{C}_i(\theta)$ for θ in Θ. If $(\eta,\zeta) \in \mathcal{C}_j$ then $\Lambda(\eta,\zeta) = p^j_{ii}$, so the design is partially balanced with respect to $\mathcal{Q}$. Examples 5.5 and 5.9 are constructed in this way from T(5) and H(2, 3) respectively.

More generally, let $\mathcal{J}$ be any non-empty subset of $\mathcal{K}$ and form blocks $\bigcup_{i\in\mathcal{J}} \mathcal{C}_i(\theta)$ for θ in Θ. If $(\eta,\zeta) \in \mathcal{C}_j$ then $\Lambda(\eta,\zeta) = \sum_{i\in\mathcal{J}}\sum_{l\in\mathcal{J}} p^j_{il}$, so once again the design is partially balanced.

The designs in these two paragraphs are both called *elementary*.

5.6.2 New designs from old

There are several ways of creating new partially balanced incomplete-block designs from old ones. Unfortunately, many of the resulting designs are too large or too inefficient for practical use.

Proposition 5.11 *If two incomplete-block designs with the same block size are partially balanced with respect to the same association scheme then so is their juxtaposition.*

Proposition 5.12 *If an incomplete-block design is partially balanced with respect to the association scheme $\mathcal{Q}$ then its m-fold inflation is partially balanced with respect to $\mathcal{Q}/\underline{\underline{m}}$.*

In particular, every inflation of a balanced incomplete-block design is group-divisible. Such group-divisible designs are sometimes called *singular*. The within-group canonical efficiency factor is equal to 1.

There is also a form of inflation where each block is replaced by a set of blocks.

Proposition 5.13 *Consider an incomplete-block design for t treatments in b blocks of size k. Suppose that there is a balanced incomplete-block design for k treatments in b' blocks of size k'. If each block of the first design is replaced by the b' blocks of such a BIBD on the set of treatments in that block, we obtain a new incomplete-block design for t treatments in bb' blocks of size k'. If the first design is partially balanced with respect to an association scheme $\mathcal{Q}$ then so is the final design.*

We can use both forms of inflation together to combine two partially balanced incomplete-block designs while nesting their association schemes. For $i = 1, 2, 3$, let $\mathcal{D}_i$ be an incomplete-block design for t_i treatments replicated r_i times in b_i blocks of size k_i, partially balanced with respect to an association scheme $\mathcal{Q}_i$ on treatment set Θ_i. Assume that $\mathcal{Q}_3$ is $\mathrm{GD}(k_1, b_2)$ (so that $t_3 = k_1 b_2$).

Let δ be a block in $\mathcal{D}_1$. Consider δ as a subset of Θ_1 and form the Cartesian product $\delta \times \Delta_2$, where Δ_2 is the block set of $\mathcal{D}_2$. This Cartesian product has k_1 rows of size b_2, so we may identify it with Θ_3. Thus each block γ of $\mathcal{D}_3$ consists of some pairs of the form (θ_1, δ_2) for $\theta_1 \in \delta \subseteq \Theta_1$ and $\delta_2 \in \Delta_2$. Form the subset $\Phi_{\delta\gamma}$ of $\Theta_1 \times \Theta_2$ which consists of all (θ_1, θ_2) for which $\theta_1 \in \delta$ and there exists a block δ_2 in Δ_2 such that $(\theta_1, \delta_2) \in \gamma$ and $\theta_2 \in \delta_2$. These subsets $\Phi_{\delta\gamma}$ are the blocks of the final design, which has $b_1 b_3$ blocks of size $k_2 k_3$ for treatment set $\Theta_1 \times \Theta_2$ replicated $r_1 r_2 r_3$ times. It is partially balanced with respect to $\mathcal{Q}_1/\mathcal{Q}_2$.

Superposition can also give nested association schemes. First we need a definition.

Definition Block designs $(\Omega_1, \mathcal{G}_1, \Theta_1, \psi_1)$ and $(\Omega_2, \mathcal{G}_2, \Theta_2, \psi_2)$ are *isomorphic* to each other if there are bijections $\phi\colon \Omega_1 \to \Omega_2$ and $\pi\colon \Theta_1 \to \Theta_2$ such that

(i) $(\phi, \text{identity})$ is an isomorphism between the group-divisible association schemes $\mathcal{G}_1$ and $\mathcal{G}_2$, that is, $\phi(\alpha)$ is in the same block as $\phi(\beta)$ if and only if α is in the same block as β, for α, β in Ω_1; and

(ii) $\psi_2(\phi(\omega)) = \pi(\psi_1(\omega))$ for all ω in Ω_1.

The pair (ϕ, π) is an *isomorphism* between block designs.

Proposition 5.14 *Suppose that there are m incomplete-block designs on disjoint treatment sets, each isomorphic to a design which is partially balanced with respect to an association scheme $\mathcal{Q}$. If these designs can be orthogonally superposed then the resulting design is partially balanced with respect to $\underline{\underline{m}}/\mathcal{Q}$.*

In particular, if the component designs are balanced then the resulting design is group-divisible. Such group-divisible designs are sometimes called *semi-regular*. The between-group canonical efficiency factor is equal to 1. A group-divisible design which is neither singular nor semi-regular is called *regular*.

We can weaken the condition of orthogonal superposition. Suppose that Θ is the underlying set of $\mathcal{Q}$, and that (ϕ_i, π_i) is the isomorphism from the basic design to the i-th component design $\mathcal{D}_i$. Let the incidence matrix of $\mathcal{D}_i$ be N_i. Let ρ_i be a bijection from the block set Δ of the basic design to the block set of $\mathcal{D}_i$. Replace Condition (4.10) by

there are constants μ_1, μ_2 such that, for all $1 \leqslant i < j \leqslant m$ and all θ, η in Θ,

$$\sum_{\delta \in \Delta} |N_i(\rho_i(\delta), \pi_i(\theta))| \, |N_j(\rho_j(\delta), \pi_j(\eta))| = \begin{cases} \mu_1 & \text{if } \theta = \eta \\ \mu_2 & \text{otherwise.} \end{cases} \tag{5.4}$$

If this new condition is satisfied then the superposed design is partially balanced with respect to $\underline{\underline{m}} \times \mathcal{Q}$.

Example 5.18 Two copies of the unreduced design for five treatments in ten blocks of size two can be superposed to satisfy Condition (5.4) with $\mu_1 = 0$ and $\mu_2 = 2$. The resulting design has ten blocks of size four, and is partially balanced with respect to $\underline{\underline{2}} \times \underline{\underline{5}}$. The association scheme is

A	B	C	D	E
a	b	c	d	e

and the superposed design has blocks $\{A, B, c, d\}$, $\{A, C, d, e\}$, $\{A, D, b, e\}$, $\{A, E, b, c\}$, $\{B, C, a, e\}$, $\{B, D, c, e\}$, $\{B, E, a, d\}$, $\{C, D, a, b\}$, $\{C, E, b, d\}$ and $\{D, E, a, c\}$. ■

Proposition 5.15 *Given two incomplete-block designs partially balanced with respect to association schemes $\mathcal{Q}_1$ and $\mathcal{Q}_2$ respectively, their product is partially balanced with respect to $\mathcal{Q}_1 \times \mathcal{Q}_2$.*

The design in Example 4.7 is partially balanced with respect to the association scheme $\mathrm{H}(2, 2) \times \underline{\underline{3}}$.

5.7 Optimality

Our intuition that "balance is good" is justified by the following results, all of which assume that we are considering only binary equi-replicate designs.

Theorem 5.16 *For an equi-replicate incomplete-block design for t treatments in blocks of size k, the harmonic mean efficiency factor A is less than or equal to*

$$\frac{t}{t-1}\frac{k-1}{k}.$$

This value is achieved if and only if the design is balanced.

Proof From Equation (4.7), the sum of the canonical efficiency factors is equal to $t(k-1)/k$, so their arithmetic mean is $t(k-1)/[(t-1)k]$. The harmonic mean of a set of positive numbers is always less than or equal to their arithmetic mean, with equality when, and only when, the numbers are all equal. ■

Corollary 5.17 *Balanced incomplete-block designs are A-optimal. Furthermore, if there exists a BIBD for given values of t, r, b and k then only the BIBDs are A-optimal.*

Theorem 5.18 *If an equi-replicate incomplete-block design is A-optimal then so is its dual.*

Proof Without loss of generality, suppose that $b \geqslant t$. Let A be the harmonic mean efficiency factor of the design. Theorem 4.13 shows that the harmonic mean efficiency factor of the dual is equal to

$$\frac{b-1}{b-t+\frac{t-1}{A}}. \quad ■$$

Sometimes other partially balanced incomplete-block designs are A-optimal. I quote two results without proof.

Theorem 5.19 *If a group-divisible design with two groups has $\lambda_2 = \lambda_1 + 1$ then it is A-optimal.*

Corollary 5.20 *Transversal designs with two groups are A-optimal.*

Theorem 5.21 *If a partially balanced incomplete-block design has two associate classes, if its non-diagonal concurrences differ by 1, and if one of its canonical efficiency factors is equal to 1, then it is A-optimal.*

Corollary 5.22 *Square lattice designs are A-optimal.*

Corollary 5.23 *Transversal designs with any number of groups are A-optimal.*

The triangular design in Example 5.12 is therefore A-optimal.

Sometimes A-optimal designs are found by exhaustive search. That is how the design in Example 5.16 was found to be A-optimal.

Unfortunately, the class of A-optimal designs does not always include any partially balanced designs, even when partially balanced designs exist for those values of t, r, b and k. For $t = 10$, $r = 6$, $b = 30$ and $k = 2$, the design in Exercise 4.1(d) is A-optimal: it is better than the elementary design constructed from T(5).

Exercises

5.1 For each of the incomplete-block designs in Exercise 4.1, decide whether it is partially balanced. If it is partially balanced, give the association scheme with the smallest number of classes with respect to which it is partially balanced. If it is not partially balanced, explain why it is not.

5.2 For each of the following designs, decide whether it is (i) partially balanced (ii) balanced (iii) group-divisible (iv) a transversal design.

(a) The design in Example 4.4 with blocks $\{1,2\}$, $\{3,4\}$, $\{1,3\}$ and $\{2,4\}$.
(b) The design in Example 4.6 for eight treatments in 16 blocks of size two.
(c) The design in Example 4.6 for 12 treatments in 16 blocks of size three.
(d) The design in Exercise 4.6 part (b).
(e) The design in Exercise 4.6 part (c).
(f) The design in Exercise 4.7 part (b).
(g) The design in Exercise 4.7 part (d).
(h) The duals of the last five designs.

5.3 Consider the unreduced incomplete-block design whose blocks are all 2-subsets of an n-set. Show that its dual is a triangular design.

5.4 Let N be the incidence matrix for an incomplete-block design with block set Δ and treatment set Θ. The *complement* of this design has incidence matrix N^*, where $N^*(\delta,\theta) = 1 - N(\delta,\theta)$. Prove that the complement of a partially balanced incomplete-block design is partially balanced with respect to the same association scheme.

5.5 Suppose that there is a balanced incomplete-block design with $\lambda = 1$. Choose one treatment, and remove all the blocks containing it. Show that the remaining blocks form a group-divisible design. When is it semi-regular?

5.6 Find the canonical efficiency factors of the design in Example 5.5 and verify Theorem 4.12.

5.7 Describe the design in Example 5.6 with $m = 2$ and $l = n - 1$ and then describe its complement. Find the canonical efficiency factors of both designs.

5.8 For each of the following designs, find the variances of all simple contrasts and verify that Theorem 4.12 is satisfied.

(a) The design in Example 5.9.
(b) The design in Example 5.12.

5.9 Yates [255, Section 13] describes an experiment on potatoes in which 12 treatments $a, \ldots, l$ were compared in six blocks of six plots. Table 5.2 shows the design and the responses in pounds of potatoes.

Ω	Block	Θ	Y	Ω	Block	Θ	Y	Ω	Block	Θ	Y
1	1	e	172	13	3	b	208	25	5	c	176
2	1	g	161	14	3	f	144	26	5	l	186
3	1	h	192	15	3	j	190	27	5	a	132
4	1	c	145	16	3	e	104	28	5	d	242
5	1	f	227	17	3	a	113	29	5	b	196
6	1	d	232	18	3	i	131	30	5	k	178
7	2	i	231	19	4	c	158	31	6	j	238
8	2	a	166	20	4	d	171	32	6	h	198
9	2	k	204	21	4	h	171	33	6	g	180
10	2	b	253	22	4	k	135	34	6	e	175
11	2	j	231	23	4	l	146	35	6	f	230
12	2	l	214	24	4	g	103	36	6	i	216

Table 5.2. *Data for Exercise 5.9*

(a) Show that the design is partially balanced with respect to the association scheme $(\underline{\underline{2}} \times \underline{\underline{3}})/\underline{\underline{2}}$ indicated below.

a	b	c	d	c	f
g	h	i	j	k	l

(b) Calculate its canonical efficiency factors.
(c) By mimicking the calculations in Example 4.3, estimate $x'\tau$ for x in a basis of $W_0^\perp$; also estimate the variance of each estimator.
(d) Construct a partially balanced design for this situation which is more efficient than the one in Table 5.2.

5.10 Compare the incomplete-block design in Example 5.13 with the group-divisible design which has one block for each pair of treatments in different rows (the association scheme is GD(2, 18)).

5.11 Construct a further partially balanced incomplete-block design which, like the design in Example 5.13, has at least one pair of treatments with both higher concurrence and higher variance than another pair of treatments.

5.12 Prove an analogue of Theorem 5.5 for designs with three distinct canonical efficiency factors.

5.13 Find the concurrence matrix and canonical efficiency factors for the cyclic design in Example 5.15.

5.14 Consider the following incomplete-block designs for eight treatments in six blocks of size 4.

(a) The treatments are the vertices of the cube. The blocks are the faces of the cube.
(b) The cyclic design generated by $\{0,2,4,6\}$ and $\{0,1,4,5\}$ modulo 8.

Show that each design is partially balanced. Find its canonical efficiency factors, and the variances of all simple contrasts. Which design do you think is better?

5.15 Consider the following incomplete-block designs for 15 treatments in 60 blocks of size 3.

(a) The treatments are all 2-subsets of a 6-set. The blocks are all triples like $\{\{1,2\},\{1,3\},\{2,3\}\}$, each occurring three times. (Compare Example 5.6.)
(b) The treatments are all 2-subsets of a 6-set. The blocks are all triples like $\{\{1,2\},\{3,4\},\{5,6\}\}$, each occurring four times. (Compare Example 5.12.)
(c) The cyclic design generated by $\{0,5,13\}$, $\{0,1,5\}$, $\{0,3,14\}$ and $\{0,6,8\}$ modulo 15.

Show that each design is partially balanced. Find its canonical efficiency factors, and the variances of all simple contrasts. Which design do you think is best?

5.16 Construct a square lattice design for 25 treatments in 15 blocks of size 5.

5.17 Construct an affine plane for 49 treatments.

5.18 Construct a transversal design for four groups of size 5.

5.19 Find the canonical efficiency factors of the cubic lattice design.

5.20 Compare the following two incomplete-block designs for eight treatments in 12 blocks of size 2:

(a) the cubic lattice;

(b) the cyclic design generated by $\{0,1\}$ and $\{0,4\}$ modulo 8.

5.21 Construct a rectangular lattice design for 20 treatments in 15 blocks of size 4. Show that it is not partially balanced.

5.22 Identify Pair(3) with a familiar association scheme.

5.23 For a rectangular lattice design with $r < n$, express L^- on $U_0^\perp$ as a quadratic polynomial in A, where $A = \Lambda - rI$.

Find A^2 for the design constructed in Exercise 5.21 and hence find the variances of all simple contrasts in that design.

5.24 Find the character table for the association scheme Pair(n), where $n \geqslant 4$. Identify the strata in terms of the characteristic vectors of rows and columns, and symmetric and antisymmetric vectors. Here a vector v is defined to be *symmetric* if $v(i,j) = v(j,i)$ and *antisymmetric* if $v(i,j) = -v(j,i)$.

5.25 (a) The first construction in Section 5.6.1 can be applied to the 'edge' class of a strongly regular graph. If the graph has n vertices and valency a, the resulting elementary block design has n treatments in n blocks of size a. Show that this block design is balanced if and only if $p = q$, in the notation of Section 2.5.

(b) Hence construct a balanced incomplete-block design for 16 treatments in 16 blocks of size 6.

(c) In Example 3.5, we showed a pair of non-isomorphic strongly regular graphs defined by two different 4×4 Latin squares. Use the non-edges of these graphs to construct two balanced incomplete-block designs for 16 treatments in 16 blocks of size 6. Are these block designs isomorphic to each other? Or to the block design constructed in part (b)?

(d) Apply the second construction in Section 5.6.1 to a strongly regular graph, using $\mathcal{I} = \{\text{diagonal, edge}\}$. Show that the resulting block design is balanced if and only if $q = p + 2$, in the notation of Section 2.5.

(e) Hence construct another balanced incomplete-block design.

5.26 Prove that all transversal designs are semi-regular. Is the converse true?

5.27 Consider the affine plane constructed from mutually orthogonal $n \times n$ Latin squares $\Pi_1, \ldots, \Pi_{n-1}$. Let $\eta_1, \ldots, \eta_{n+1}$ be new treatments. Make a new block design with n^2+n+1 blocks of size $n+1$ by adjoining η_i to each block of the i-th component design, for $1 \leqslant i \leqslant n+1$, and adjoining the new block $\{\eta_1, \ldots, \eta_{n+1}\}$. Prove that the new design is balanced. (Such a design is called a *projective plane*.)

5.28 Consider a balanced incomplete-block design with $b > t$ and $\lambda = 1$.

(a) Prove that if δ_1 and δ_2 are two distinct blocks then the number of treatments that occur in both δ_1 and δ_2 is equal to either 0 or 1. Prove that both possibilities do arise.

(b) Construct a graph as follows. The vertices are the blocks. There is an edge between two blocks if there is a treatment which occurs in both blocks. Prove that the graph is strongly regular.

5.29 Consider a balanced incomplete-block design with $b = t$ and $\lambda = 1$. Let

$$\Gamma = \{(\theta, \delta) : \text{there is a plot } \omega \text{ in block } \delta \text{ with } \psi(\omega) = \theta\}.$$

Let $\mathcal{C}$ be the subset of $\Gamma \times \Gamma$ consisting of pairs $((\theta_1, \delta_1), (\theta_2, \delta_2))$ for which either $\theta_1 = \theta_2$ or $\delta_1 = \delta_2$ but not both. Show that the pairs in $\mathcal{C}$ form the edges of a distance-regular graph. (Applying this construction to the balanced incomplete-block design in Example 5.3 gives the distance-regular graph in Exercise 1.27.)

5.30 Find an A-optimal incomplete-block design for each set of values of t, r, b and k below.

(a) $t = 6$, $r = 6$, $b = 9$, $k = 4$.
(b) $t = 6$, $r = 9$, $b = 18$, $k = 3$.
(c) $t = 7$, $r = 8$, $b = 14$, $k = 4$.
(d) $t = 8$, $r = 6$, $b = 12$, $k = 4$.
(e) $t = 9$, $r = 8$, $b = 18$, $k = 4$.
(f) $t = 10$, $r = 2$, $b = 5$, $k = 4$.
(g) $t = 14$, $r = 4$, $b = 7$, $k = 8$.
(h) $t = 16$, $r = 3$, $b = 12$, $k = 4$.
(i) $t = 20$, $r = 5$, $b = 25$, $k = 4$.

6
Families of partitions

6.1 A partial order on partitions

We have seen that the group-divisible association scheme is defined by a partition of Ω into blocks of equal size. Exercise 2.2 shows that a particularly tidy basis for the Bose–Mesner algebra $\mathcal{A}$ in this case is $\{I, B, J\}$, where B is the adjacency matrix for the relation 'is in the same block as'. Similarly, in the examples of iterated crossing and nesting, starting from only the trivial association schemes, in Section 3.5, all of the association relations can be described in terms such as 'is in the same such-and-such as ... but in different other such-and-such'. The main object of this chapter is to give a single general construction for association schemes defined by families of partitions of Ω.

(I use the letter F for a typical partition because statisticians call partitions *factors* if they have some physical basis. In Example 3.12, most statisticians would regard months, houses and sheep as factors, but there would be some uncertainty as to whether sheep had four or twelve classes: this is covered in Chapter 9. Not everyone would agree that the partition into the twelve house-months is a factor. Thus it seems clearer to use the abstract notion of partition, formalizing the idea behind house-months as the *infimum*.)

Let F be a partition of Ω into n_F subsets, which I shall call F-classes. Define the relation matrix R_F in $\mathbb{R}^{\Omega\times\Omega}$ by

$$R_F(\alpha,\beta) = \begin{cases} 1 & \text{if } \alpha \text{ and } \beta \text{ are in the same } F\text{-class} \\ 0 & \text{otherwise,} \end{cases} \tag{6.1}$$

just as B was defined in Section 4.1. Then, as in Section 4.3, define the subspace V_F of $\mathbb{R}^\Omega$ to be

$$\left\{v \in \mathbb{R}^\Omega : v(\alpha) = v(\beta) \text{ whenever } \alpha \text{ and } \beta \text{ are in the same } F\text{-class}\right\}.$$

Then $\dim V_F = n_F$. Let P_F be the orthogonal projector onto V_F. Then $(P_F v)(\alpha)$ is equal to the average of the values $v(\beta)$ for β in the same F-class as α.

Definition The partition F is *uniform* if all classes of F have the same size.

(Unfortunately, there is no consensus about what to call such a partition. The words *uniform*, *regular*, *balanced* and *proper* are all in use.)

Proposition 6.1 *If the partition F is uniform with classes of size k_F then $P_F = k_F^{-1} R_F$.*

So much for a single partition. Now we consider what happens when we have two partitions.

If F and G are both partitions of Ω, write $F \preccurlyeq G$ if every F-class is contained in a G-class; write $F \prec G$ if $F \preccurlyeq G$ and $F \neq G$. We may pronounce $F \preccurlyeq G$ as 'F is finer than G' or 'G is coarser than F'.

(Statisticians often say 'F is nested in G', but this meaning of *nesting* merges into the one in Chapter 3 without being identical. They also often use the phrase *is nested in* for a related, but different, concept that we shall meet in Chapter 9.)

Example 6.1 In the association scheme $\underline{\underline{3}}/(\underline{\underline{2}} \times \underline{\underline{4}})$ the set Ω consists of 24 elements, divided into three rectangles as follows.

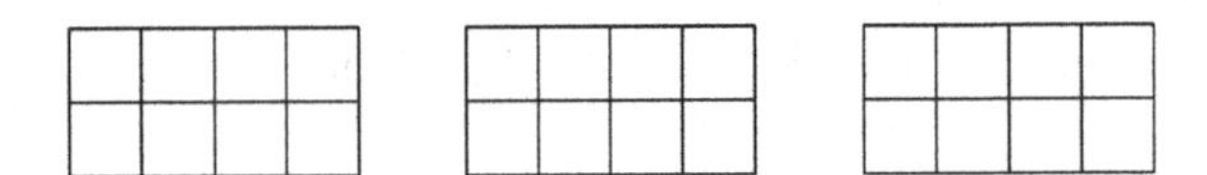

The set is partitioned in three ways—into rows, columns and rectangles. Identifying the names of the partitions with the names of their classes, we have rows $\prec$ rectangles and columns $\prec$ rectangles. ■

Lemma 6.2 *If $F \preccurlyeq G$ then $V_G \leqslant V_F$.*

Proof Any function which is constant on G-classes must be constant on F-classes too if each F-class is contained in a G-class. ■

There are two special, but trivial, partitions on every set with more than one element. The *universal* partition U consists of a single class containing the whole of Ω (the *universe*). At the other extreme, the *equality* partition E has as its classes all the singleton subsets of Ω; in other words, α and β are in the same E-class if and only if $\alpha = \beta$. Note that $E \preccurlyeq F \preccurlyeq U$ for all partitions F.

The relation $\preccurlyeq$ satisfies:

(i) (**reflexivity**) for every partition F, $F \preccurlyeq F$;
(ii) (**antisymmetry**) if $F \preccurlyeq G$ and $G \preccurlyeq F$ then $F = G$;
(iii) (**transitivity**) if $F \preccurlyeq G$ and $G \preccurlyeq H$ then $F \preccurlyeq H$.

This means that $\preccurlyeq$ is a *partial order*. Partial orders are often shown on *Hasse diagrams*: there is a dot for each element (partition in this case); if $F \prec G$ then F is drawn below G and is joined to G by a line or sequence of lines, all going generally upwards.

Example 6.1 revisited The Hasse diagram for the example consisting of rectangular arrays is in Figure 6.1. ■

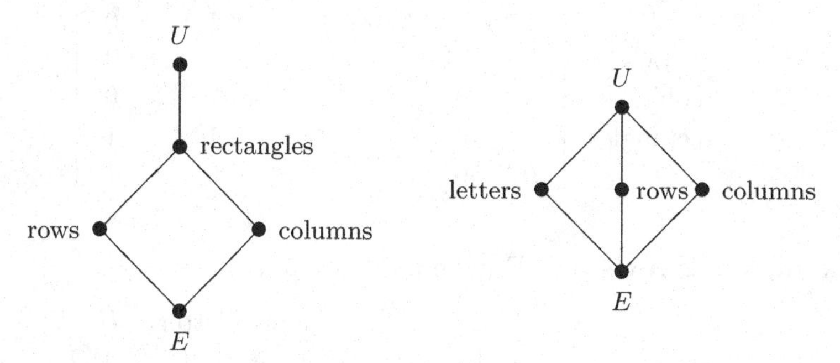

Fig. 6.1. Hasse diagram for Example 6.1

Fig. 6.2. Hasse diagram for Example 6.2

Example 6.2 The set of n^2 cells in a $n \times n$ Latin square has three non-trivial partitions – into rows, columns and letters. The Hasse diagram in shown in Figure 6.2. ■

A set with a partial order on it is often called a *poset*. Every poset has a zeta function and a Möbius function, which I shall now define for our particular partial order.

Let $\mathcal{F}$ be any set of partitions of Ω. Define ζ in $\mathbb{R}^{\mathcal{F}\times\mathcal{F}}$ by

$$\zeta(F,G) = \begin{cases} 1 & \text{if } F \preccurlyeq G \\ 0 & \text{otherwise.} \end{cases}$$

The partial order $\preccurlyeq$ is often considered to be the subset $\{(F,G) : F \preccurlyeq G\}$ of $\mathcal{F} \times \mathcal{F}$. From this viewpoint, ζ is just the adjacency matrix of $\preccurlyeq$.

The elements of $\mathcal{F}$ can be written in an order such that F comes before G if $F \prec G$: see Exercise 6.2. Then ζ is an upper triangular matrix with 1s on the diagonal, so it has an inverse matrix μ which is also upper triangular with integer entries and 1s on the diagonal. The matrix μ is called the *Möbius* function of the poset $(\mathcal{F}, \preccurlyeq)$.

Example 6.1 revisited Here

$$\zeta = \begin{array}{r} \\ E \\ \text{rows} \\ \text{columns} \\ \text{rectangles} \\ U \end{array} \begin{array}{c} \begin{array}{ccccc} E & \text{rows} & \text{columns} & \text{rectangles} & U \end{array} \\ \left[\begin{array}{ccccc} 1 & 1 & 1 & 1 & 1 \\ 0 & 1 & 0 & 1 & 1 \\ 0 & 0 & 1 & 1 & 1 \\ 0 & 0 & 0 & 1 & 1 \\ 0 & 0 & 0 & 0 & 1 \end{array}\right] \end{array}$$

so

$$\mu = \begin{array}{r} \\ E \\ \text{rows} \\ \text{columns} \\ \text{rectangles} \\ U \end{array} \begin{array}{c} \begin{array}{ccccc} E & \text{rows} & \text{columns} & \text{rectangles} & U \end{array} \\ \left[\begin{array}{ccccc} 1 & -1 & -1 & 1 & 0 \\ 0 & 1 & 0 & -1 & 0 \\ 0 & 0 & 1 & -1 & 0 \\ 0 & 0 & 0 & 1 & -1 \\ 0 & 0 & 0 & 0 & 1 \end{array}\right] \end{array}. \quad \blacksquare$$

Example 6.2 revisited For the Latin square,

$$\zeta = \begin{array}{r} \\ E \\ \text{rows} \\ \text{columns} \\ \text{letters} \\ U \end{array} \begin{array}{c} \begin{array}{ccccc} E & \text{rows} & \text{columns} & \text{letters} & U \end{array} \\ \left[\begin{array}{ccccc} 1 & 1 & 1 & 1 & 1 \\ 0 & 1 & 0 & 0 & 1 \\ 0 & 0 & 1 & 0 & 1 \\ 0 & 0 & 0 & 1 & 1 \\ 0 & 0 & 0 & 0 & 1 \end{array}\right] \end{array}$$

and so

$$\mu = \begin{array}{r} \\ E \\ \text{rows} \\ \text{columns} \\ \text{letters} \\ U \end{array} \begin{array}{c} \begin{array}{ccccc} E & \text{rows} & \text{columns} & \text{letters} & U \end{array} \\ \left[\begin{array}{ccccc} 1 & -1 & -1 & -1 & 2 \\ 0 & 1 & 0 & 0 & -1 \\ 0 & 0 & 1 & 0 & -1 \\ 0 & 0 & 0 & 1 & -1 \\ 0 & 0 & 0 & 0 & 1 \end{array}\right] \end{array}. \quad \blacksquare$$

If F and G are partitions, let $F \wedge G$ be the partition whose classes are the non-empty intersections of F-classes with G-classes. Then

(i) $F \wedge G \preccurlyeq F$ and $F \wedge G \preccurlyeq G$, and
(ii) if $H \preccurlyeq F$ and $H \preccurlyeq G$ then $H \preccurlyeq F \wedge G$,

so $F \wedge G$ is the *infimum* (or *greatest lower bound*) of F and G. Dually, the *supremum* (or *least upper bound*) $F \vee G$ of F and G is defined by

(i) $F \preccurlyeq F \vee G$ and $G \preccurlyeq F \vee G$, and
(ii) if $F \preccurlyeq H$ and $G \preccurlyeq H$ then $F \vee G \preccurlyeq H$.

To construct $F \vee G$, draw a coloured graph whose vertex set is Ω. There is a red edge between α and β if α and β are in the same F-class, and a blue edge if α and β are in the same G-class. Then the classes of $F \vee G$ are the connected components of the red-and-blue graph.

Example 6.1 revisited In the example with rectangular arrays, we have rows $\wedge$ columns $= E$ and rows $\vee$ columns $=$ rectangles.

Example 6.2 revisited In the Latin square, the supremum of any two of the non-trivial partitions is U and the infimum of any two is E.

Of course, if $F \preccurlyeq G$ then $F \wedge G = F$ and $F \vee G = G$. Moreover, just as the empty sum is 0 and the empty product is 1, we make the convention that the empty infimum is U and the empty supremum is E.

Lemma 6.3 *If F and G are partitions of Ω, then $V_F \cap V_G = V_{F \vee G}$.*

Proof Let v be a vector in $\mathbb{R}^\Omega$. Then

$$
\begin{aligned}
v \in V_F \cap V_G &\iff v \text{ is constant on each component of the blue graph and } v \text{ is constant on each component of the red graph} \\
&\iff v \text{ is constant on each component of the red-and-blue graph} \\
&\iff v \in V_{F \vee G}. \quad \blacksquare
\end{aligned}
$$

6.2 Orthogonal partitions

Definition Let F and G be partitions of Ω. Then F is *orthogonal* to G if V_F is geometrically orthogonal to V_G, that is, if $P_F P_G = P_G P_F$.

Note that this implies that F is orthogonal to G if $F \preccurlyeq G$, because $P_F P_G = P_G P_F = P_G$ in that case. In particular, F is orthogonal to itself.

Lemma 6.4 *If F is orthogonal to G, then*

(i) $P_F P_G = P_{F \vee G}$;
(ii) $V_F \cap (V_{F \vee G})^\perp$ is orthogonal to V_G.

Proof These both follow immediately from Lemma 2.2, using the fact that $V_F \cap V_G = V_{F \vee G}$. $\blacksquare$

Corollary 6.5 *Partitions F and G are orthogonal to each other if and only if*

(i) within each class of $F \vee G$, each F-class meets every G-class, and
(ii) for each element ω of Ω,

$$\frac{|F\text{-class containing } \omega|}{|F \vee G\text{-class containing } \omega|} = \frac{|F \wedge G\text{-class containing } \omega|}{|G\text{-class containing } \omega|}.$$

(When we say that one set 'meets' another, we mean that their intersection is not empty.)

Corollary 6.6 *If F is orthogonal to G and if F, G and $F \vee G$ are all uniform then*

(i) $F \wedge G$ is also uniform;
(ii) $k_{F\wedge G}k_{F\vee G} = k_F k_G$;
(iii) $R_F R_G = k_{F\wedge G} R_{F\vee G}$.

Example 6.1 revisited Here rows are orthogonal to columns even though no row meets every column. ■

Example 6.2 revisited As in Example 1.6, write R, C and L for R_{rows}, R_{columns} and R_{letters}. Proposition 6.1 shows that $P_{\text{rows}} = n^{-1}R$, $P_{\text{columns}} = n^{-1}C$, $P_{\text{letters}} = n^{-1}L$ and $P_U = n^{-2}J$. In Example 1.6 we saw that $RC = CR = RL = LR = CL = LR = J$. Thus the partitions into rows, columns and letters are pairwise orthogonal. ■

Theorem 6.7 *Let $\mathcal{F}$ be a set of pairwise orthogonal partitions of Ω which is closed under $\vee$. For F in $\mathcal{F}$, put*

$$W_F = V_F \cap \left(\sum_{G \succ F} V_G \right)^{\perp}.$$

Then

(i) the spaces W_F and W_G are orthogonal to each other whenever F and G are different partitions in $\mathcal{F}$;
(ii) for each F in $\mathcal{F}$,

$$V_F = \bigoplus_{G \succcurlyeq F} W_G.$$

Proof (i) If $F \neq G$ then $F \vee G$ must be different from at least one of F and G. Suppose that $F \vee G \neq F$. Then $F \vee G \succ F$ and $F \vee G \in \mathcal{F}$ so

$$W_F \leqslant V_F \cap V_{F\vee G}^{\perp},$$

while $W_G \leqslant V_G$. Lemma 6.4(ii) shows that $V_F \cap V_{F\vee G}^{\perp}$ is orthogonal to V_G, because F is orthogonal to G. Hence W_F is orthogonal to W_G.

(ii) Since the spaces W_G, for $G \succcurlyeq F$, are pairwise orthogonal, their vector space sum is direct, so it suffices to prove that

$$V_F = \sum_{G \succcurlyeq F} W_G. \tag{6.2}$$

We do this by induction. If there is no G such that $G \succ F$ then $W_F = V_F$ and Equation (6.2) holds. The definition of W_F shows that

$$V_F = W_F + \sum_{H \succ F} V_H.$$

If the inductive hypothesis is true for every H with $H \succ F$ then

$$\sum_{H \succ F} V_H = \sum_{H \succ F} \sum_{G \succcurlyeq H} W_G.$$

If $G \succcurlyeq H \succ F$ then $G \succ F$, and if $G \succ F$ then $G \succcurlyeq G \succ F$ so

$$\sum_{H \succ F} \sum_{G \succcurlyeq H} W_G = \sum_{G \succ F} W_G.$$

Thus

$$V_F = W_F + \sum_{G \succ F} W_G = \sum_{G \succcurlyeq F} W_G. \quad \blacksquare$$

Definition An *orthogonal array of strength* 2 on a set Ω is a set $\mathcal{F}$ of uniform pairwise orthogonal partitions of Ω which contains U and E and for which $F \vee G = U$ whenever F and G are distinct elements of $\mathcal{F} \setminus \{E\}$.

Given an orthogonal array of strength 2, suppose that its non-trivial partitions are $F_1, \ldots, F_m$, with $n_1, \ldots, n_m$ classes of size $k_1, \ldots, k_m$ respectively. Theorem 6.7 shows that $\mathbb{R}^\Omega = W_U \oplus W_1 \oplus \cdots \oplus W_m \oplus W_E$, where the projector onto W_i is

$$\frac{1}{k_i} R_i - \frac{1}{|\Omega|} J$$

and the projector onto W_E is

$$I - \sum_{i=1}^{m} \frac{1}{k_i} R_i + \frac{m-1}{|\Omega|} J.$$

The dimension of W_i is $n_i - 1$; the dimension of W_E is $|\Omega| - 1 - \sum_i (n_i - 1)$. In particular,

$$\sum_{i=1}^{m} (n_i - 1) \leqslant |\Omega| - 1.$$

Example 6.3 Figure 6.3 shows an orthogonal array of strength 2 on a set Ω of size 12. The elements of Ω are the columns of the array. The non-trivial partitions F_1, F_2, F_3 and F_4 are the rows of the array. In row F_i, the different symbols denote the different classes of F_i: thus $n_1 = n_2 = n_3 = 2$ while $n_4 = 3$. ■

	ω_1	ω_2	ω_3	ω_4	ω_5	ω_6	ω_7	ω_8	ω_9	ω_{10}	ω_{11}	ω_{12}
F_1	1	1	2	1	1	1	2	2	2	1	2	2
F_2	2	1	1	2	1	1	1	2	2	2	1	2
F_3	1	2	1	1	2	1	1	1	2	2	2	2
F_4	1	1	2	2	2	3	3	1	2	3	1	3

Fig. 6.3. Orthogonal array in Example 6.3

Although orthogonal arrays have interesting combinatorial properties, they do not, in general, give us association schemes. In Example 6.3, the obvious choice for associate classes are relations such as 'in the same class of F_1 but different classes of F_2, F_3 and F_4'. But ω_1 has one such associate (ω_5) while ω_6 has none. The extra condition that we need is closure under $\wedge$, as we show in the next section. If we attempt to extend $\mathcal{F}$ in Example 6.3 by including infima then we lose orthogonality, for $F_3 \wedge F_4$ is not orthogonal to F_2.

6.3 Orthogonal block structures

Definition An *orthogonal block structure* on a set Ω is a set $\mathcal{F}$ of pairwise orthogonal uniform partitions of Ω which is closed under $\wedge$ and $\vee$ (in particular, $\mathcal{F}$ contains U and E).

Thus, in Example 6.1, the set $\{E$, rows, columns, rectangles, $U\}$ is an orthogonal block structure. The set $\{E$, rows, columns, letters, $U\}$ in Example 6.2 is also an orthogonal block structure.

Note that, if all the other conditions for an orthogonal block structure are met, then orthogonality between partitions F and G is easy to check: whenever $\{\alpha, \beta\}$ is contained in an F-class and $\{\beta, \gamma\}$ is contained in a G-class then there must be some element δ such that $\{\alpha, \delta\}$ is contained in a G-class and $\{\delta, \gamma\}$ is contained in an F-class. That is, wherever can be reached in the two-colour graph for $F \vee G$ by a red edge followed by a blue edge can also be reached by a blue edge followed by a red edge.

Theorem 6.8 *Let $\mathcal{F}$ be an orthogonal block structure on Ω. For F in $\mathcal{F}$, define the subset $\mathcal{C}_F$ of $\Omega \times \Omega$ by*

$$(\alpha, \beta) \in \mathcal{C}_F \quad \textit{if} \quad F = \bigwedge \{G \in \mathcal{F} : \alpha \textit{ and } \beta \textit{ are in the same } G\textit{-class}\}.$$

Then $\{\mathcal{C}_F : F \in \mathcal{F},\ \mathcal{C}_F \neq \varnothing\}$ forms an association scheme on Ω with valencies a_F, where

$$a_F = \sum_{G \in \mathcal{F}} \mu(G, F) k_G.$$

Proof The non-empty $\mathcal{C}_F$ do form a partition of $\Omega \times \Omega$, because $\mathcal{F}$ is closed under $\wedge$. They are symmetric. The equality partition E is in $\mathcal{F}$ and $\mathcal{C}_E = \mathrm{Diag}(\Omega)$.

Let A_F be the adjacency matrix for $\mathcal{C}_F$. Then

α and β are in the same F-class $\iff$ there is some $G \preccurlyeq F$ with $(\alpha, \beta) \in \mathcal{C}_G$,

so

$$R_F = \sum_{G \preccurlyeq F} A_G = \sum_{G \in \mathcal{F}} \zeta(G, F) A_G = \sum_{G \in \mathcal{F}} \zeta'(F, G) A_G. \tag{6.3}$$

This is true for all F in $\mathcal{F}$. The inverse of the matrix ζ' is μ', so we can invert Equation (6.3) (this is called *Möbius inversion*) to give

$$A_F = \sum_{G \in \mathcal{F}} \mu'(F, G) R_G = \sum_{G \in \mathcal{F}} \mu(G, F) R_G. \tag{6.4}$$

Taking row sums of Equation (6.4) gives

$$a_F = \sum_{G \in \mathcal{F}} \mu'(F, G) k_G = \sum_{G \in \mathcal{F}} \mu(G, F) k_G. \tag{6.5}$$

As usual, let $\mathcal{A}$ be the subspace of $\mathbb{R}^{\Omega \times \Omega}$ spanned by $\{A_F : F \in \mathcal{F}\}$. We must show that $\mathcal{A}$ is closed under multiplication. Equations (6.3) and (6.4) show that $\mathcal{A} = \mathrm{span}\{R_F : F \in \mathcal{F}\}$. All the partitions are uniform, so Proposition 6.1 shows that $\mathcal{A} = \mathrm{span}\{P_F : F \in \mathcal{F}\}$. The partitions are pairwise orthogonal, and $\mathcal{F}$ is closed under $\vee$, so we can apply Lemma 6.4(i) and deduce that span $\{P_F : F \in \mathcal{F}\}$ is closed under multiplication. ■

Theorem 6.9 *Let $\mathcal{F}$ be an orthogonal block structure on Ω. For each F in $\mathcal{F}$, put*

$$W_F = V_F \cap \left(\sum_{G \succ F} V_G \right)^{\perp}.$$

Then the non-zero spaces W_F, for F in $\mathcal{F}$, are the strata for the association scheme. Their projectors S_F satisfy

$$S_F = \sum_{G \in \mathcal{F}} \mu(F,G) P_G,$$

and their dimensions d_F satisfy

$$d_F = \sum_{G \in \mathcal{F}} \mu(F,G) n_G.$$

Proof Theorem 6.7 shows that, for all F in $\mathcal{F}$,

$$V_F = \bigoplus_{G \succcurlyeq F} W_G$$

and the summands are orthogonal. So if S_F is the projector onto W_F then

$$P_F = \sum_{G \succcurlyeq F} S_G = \sum_{G \in \mathcal{F}} \zeta(F,G) S_G. \tag{6.6}$$

Möbius inversion gives

$$S_F = \sum_{G \in \mathcal{F}} \mu(F,G) P_G. \tag{6.7}$$

Thus each S_F is in the Bose–Mesner algebra $\mathcal{A}$ of the association scheme. But each S_F is idempotent, so Lemma 2.7 shows that each S_F is a sum of zero or more stratum projectors. If $F \neq G$ then $W_F \perp W_G$, so S_F and S_G cannot contain any stratum projectors in common. Thus no linear combination of $\{S_F : F \in \mathcal{F}\}$ projects onto any non-zero proper subspace of W_G, for any G in $\mathcal{F}$. But Equations (6.6) and (6.7) show that $\mathcal{A} = \operatorname{span}\{S_F : F \in \mathcal{F}\}$, so the non-zero spaces W_F must be precisely the strata.

Taking the trace of both sides of Equation (6.7) gives

$$d_F = \sum_{G \in \mathcal{F}} \mu(F,G) n_G, \tag{6.8}$$

because $\operatorname{tr} P_G = \dim V_G = n_G$. ■

The character table of the association scheme follows immediately from the work done so far. From Equations (6.4) and (6.6) and Proposition 6.1 we have

$$\begin{aligned} A_F &= \sum_G \mu'(F,G)R_G \\ &= \sum_G \mu'(F,G)k_G P_G \\ &= \sum_G \mu'(F,G)k_G \sum_H \zeta(G,H)S_H, \end{aligned}$$

so

$$C = \mu' \operatorname{diag}(k)\zeta. \tag{6.9}$$

Inversion then gives

$$D = C^{-1} = \mu \operatorname{diag}(k)^{-1}\zeta'. \tag{6.10}$$

We can check that this agrees with the results found in Chapter 2. From Equations (6.5) and (6.8) we obtain the (symmetric) diagonal matrices of valencies and dimensions as

$$\operatorname{diag}(a) = \mu' \operatorname{diag}(k) = \operatorname{diag}(k)\mu$$

and

$$\operatorname{diag}(d) = \mu \operatorname{diag}(n).$$

Applying Corollary 2.13 to Equation (6.9) gives

$$\begin{aligned} D &= \frac{1}{|\Omega|} \operatorname{diag}(d)C' \operatorname{diag}(a)^{-1} \\ &= \frac{1}{|\Omega|}\mu \operatorname{diag}(n)\zeta' \operatorname{diag}(k)\mu\mu^{-1} \operatorname{diag}(k)^{-1} \\ &= \frac{1}{|\Omega|}\mu \operatorname{diag}(n)\zeta' \\ &= \mu \operatorname{diag}(k)^{-1}\zeta' \end{aligned}$$

because $n_F k_F = |\Omega|$ for all F.

6.4 Calculations

Although the existence of the Möbius function is useful for proving general results, such as Theorems 6.8 and 6.9, it is rarely used in explicit calculations. Equation (6.3) can be rewritten as

$$A_F = R_F - \sum_{G \prec F} A_G$$

and therefore

$$a_F = k_F - \sum_{G \prec F} a_G.$$

Hence the adjacency matrices and their valencies can be calculated recursively, starting at the bottom of the Hasse diagram. Likewise,

$$S_F = P_F - \sum_{G \succ F} S_G$$

and

$$d_F = n_F - \sum_{G \succ F} d_G$$

and so the stratum projectors and the dimensions of the strata are calculated recursively, starting at the top of the Hasse diagram.

Example 6.1 revisited Write R, C and B respectively for R_{rows}, R_{columns} and $R_{\text{rectangles}}$. We have $k_E = 1$, $k_{\text{rows}} = 4$, $k_{\text{columns}} = 2$, $k_{\text{rectangles}} = 8$ and $k_U = 24$. Working from the bottom of the Hasse diagram in Figure 6.1, we start with $A_E = I$. Then $A_{\text{rows}} = R - I$ and $A_{\text{columns}} = C - I$. Hence

$$\begin{aligned} A_{\text{rectangles}} &= B - A_{\text{rows}} - A_{\text{columns}} - A_E \\ &= B - (R - I) - (C - I) - I \\ &= B - R - C + I. \end{aligned}$$

Continuing in this way, we obtain

$$\begin{aligned} A_E &= I & a_E &= 1 \\ A_{\text{rows}} &= R - I & a_{\text{rows}} &= 3 \\ A_{\text{columns}} &= C - I & a_{\text{columns}} &= 1 \\ A_{\text{rectangles}} &= B - R - C + I & a_{\text{rectangles}} &= 3 \\ A_U &= J - B & a_U &= 16. \end{aligned}$$

To obtain the strata, we start at the top of the Hasse diagram. For the dimensions we use the fact that $n_U = 1$, $n_{\text{rectangles}} = 3$, $n_{\text{columns}} = 12$, $n_{\text{rows}} = 6$ and $n_E = 24$. For example, one stage in the calculation is

$$d_{\text{rows}} = n_{\text{rows}} - d_{\text{rectangles}} - d_U = 6 - 2 - 1 = 3.$$

We obtain

$$
\begin{aligned}
S_U &= \frac{1}{24}J & d_U &= 1\\
S_{\text{rectangles}} &= \frac{1}{8}B - \frac{1}{24}J & d_{\text{rectangles}} &= 2\\
S_{\text{columns}} &= \frac{1}{2}C - \frac{1}{8}B & d_{\text{columns}} &= 9\\
S_{\text{rows}} &= \frac{1}{4}R - \frac{1}{8}B & d_{\text{rows}} &= 3\\
S_E &= I - \frac{1}{4}R - \frac{1}{2}C + \frac{1}{8}B & d_E &= 9. \quad \blacksquare
\end{aligned}
$$

It is often quick and convenient to summarize the information about an orthogonal block structure by tabulating the numbers n_F, k_F, a_F and d_F next to the Hasse diagram, as in Figure 6.4.

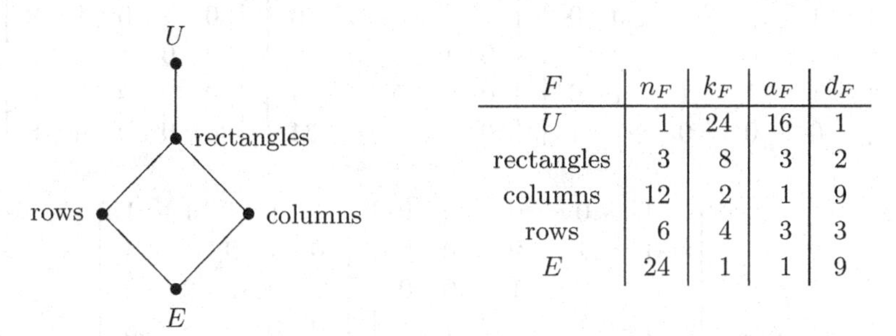

F	n_F	k_F	a_F	d_F
U	1	24	16	1
rectangles	3	8	3	2
columns	12	2	1	9
rows	6	4	3	3
E	24	1	1	9

Fig. 6.4. Summary for Example 6.1

To multiply two adjacency matrices, express them both in terms of the R matrices and multiply them using Corollary 6.6(iii). Then re-express the R matrices in terms of the adjacency matrices.

Example 6.1 revisited One such product is obtained as

$$
\begin{aligned}
A_{\text{rows}}A_{\text{rectangles}} &= (R-I)(B-R-C+I)\\
&= RB - R^2 - RC + R - B + R + C - I\\
&= 4B - 4R - B + 2R - B + C - I\\
&= 2B - 2R + C - I\\
&= 2(A_{\text{rectangles}} + A_{\text{rows}} + A_{\text{columns}} + I)\\
&\qquad - 2(A_{\text{rows}} + I) + (A_{\text{columns}} + I) - I\\
&= 2A_{\text{rectangles}} + 3A_{\text{columns}}. \quad \blacksquare
\end{aligned}
$$

The character table is calculated either by expressing each adjacency matrix in terms of the stratum projectors, or by using Equation (6.9).

Example 6.1 revisited Direct calculation of, say, A_{columns} gives

$$\begin{aligned}
A_{\text{columns}} &= C - I \\
&= 2P_{\text{columns}} - I \\
&= 2(S_{\text{columns}} + S_{\text{rectangles}} + S_U) \\
&\qquad - (S_E + S_{\text{rows}} + S_{\text{columns}} + S_{\text{rectangles}} + S_U) \\
&= -S_E - S_{\text{rows}} + S_{\text{columns}} + S_{\text{rectangles}} + S_U.
\end{aligned}$$

On the other hand, if we keep the elements of $\mathcal{F}$ in the order

$$E \quad \text{rows} \quad \text{columns} \quad \text{rectangles} \quad U,$$

then we obtain C as

$$\begin{bmatrix} 1 & 0 & 0 & 0 & 0 \\ -1 & 1 & 0 & 0 & 0 \\ -1 & 0 & 1 & 0 & 0 \\ 1 & -1 & -1 & 1 & 0 \\ 0 & 0 & 0 & -1 & 1 \end{bmatrix} \begin{bmatrix} 1 & 0 & 0 & 0 & 0 \\ 0 & 4 & 0 & 0 & 0 \\ 0 & 0 & 2 & 0 & 0 \\ 0 & 0 & 0 & 8 & 0 \\ 0 & 0 & 0 & 0 & 24 \end{bmatrix} \begin{bmatrix} 1 & 1 & 1 & 1 & 1 \\ 0 & 1 & 0 & 1 & 1 \\ 0 & 0 & 1 & 1 & 1 \\ 0 & 0 & 0 & 1 & 1 \\ 0 & 0 & 0 & 0 & 1 \end{bmatrix}$$

$$= \begin{bmatrix} 1 & 0 & 0 & 0 & 0 \\ -1 & 1 & 0 & 0 & 0 \\ -1 & 0 & 1 & 0 & 0 \\ 1 & -1 & -1 & 1 & 0 \\ 0 & 0 & 0 & -1 & 1 \end{bmatrix} \begin{bmatrix} 1 & 1 & 1 & 1 & 1 \\ 0 & 4 & 0 & 4 & 4 \\ 0 & 0 & 2 & 2 & 2 \\ 0 & 0 & 0 & 8 & 8 \\ 0 & 0 & 0 & 0 & 24 \end{bmatrix}$$

$$= \begin{bmatrix} 1 & 1 & 1 & 1 & 1 \\ -1 & 3 & -1 & 3 & 3 \\ -1 & -1 & 1 & 1 & 1 \\ 1 & -3 & -1 & 3 & 3 \\ 0 & 0 & 0 & -8 & 16 \end{bmatrix}.$$

Notice that the entries 1 come in the *first* row, while the valencies come in the *last* column.

Likewise we calculate $24D$ as

$$\begin{bmatrix} 1 & -1 & -1 & 1 & 0 \\ 0 & 1 & 0 & -1 & 0 \\ 0 & 0 & 1 & -1 & 0 \\ 0 & 0 & 0 & 1 & -1 \\ 0 & 0 & 0 & 0 & 1 \end{bmatrix} \begin{bmatrix} 24 & 0 & 0 & 0 & 0 \\ 0 & 6 & 0 & 0 & 0 \\ 0 & 0 & 12 & 0 & 0 \\ 0 & 0 & 0 & 3 & 0 \\ 0 & 0 & 0 & 0 & 1 \end{bmatrix} \begin{bmatrix} 1 & 0 & 0 & 0 & 0 \\ 1 & 1 & 0 & 0 & 0 \\ 1 & 0 & 1 & 0 & 0 \\ 1 & 1 & 1 & 1 & 0 \\ 1 & 1 & 1 & 1 & 1 \end{bmatrix}$$

$$= \begin{bmatrix} 1 & -1 & -1 & 1 & 0 \\ 0 & 1 & 0 & -1 & 0 \\ 0 & 0 & 1 & -1 & 0 \\ 0 & 0 & 0 & 1 & -1 \\ 0 & 0 & 0 & 0 & 1 \end{bmatrix} \begin{bmatrix} 24 & 0 & 0 & 0 & 0 \\ 6 & 6 & 0 & 0 & 0 \\ 12 & 0 & 12 & 0 & 0 \\ 3 & 3 & 3 & 3 & 0 \\ 1 & 1 & 1 & 1 & 1 \end{bmatrix}$$

$$= \begin{bmatrix} 9 & -3 & -9 & 3 & 0 \\ 3 & 3 & -3 & -3 & 0 \\ 9 & -3 & 9 & -3 & 0 \\ 2 & 2 & 2 & 2 & -1 \\ 1 & 1 & 1 & 1 & 1 \end{bmatrix}.$$

Here the dimensions come in the first column, while the 1s come in the last row. ■

This example displays a difficulty which was alluded to at the end of Sections 2.2 and 3.4. In the association scheme of an orthogonal block structure, the associate classes and the strata are both naturally labelled by $\mathcal{F}$, so we normally have $\mathcal{K} = \mathcal{E} = \mathcal{F}$. However, the diagonal associate class, usually called $\mathcal{C}_0$, is here $\mathcal{C}_E$, while the all-1s stratum, usually called W_0, is here W_U. In other words, the special associate class and the special stratum correspond to the two different trivial partitions. That is why the special entries in the matrices C and D are not all in the first row and column.

On the other hand, the foregoing example also demonstrates an advantage that orthogonal block structures have over association schemes in general. There are explicit, straightforward formulae for the strata and character table, so there is no need to resort to any of the techniques in Section 2.4.

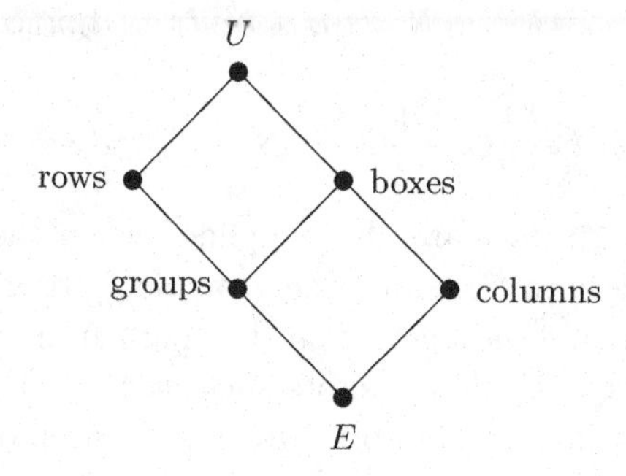

Fig. 6.5. Hasse diagram for Example 6.4

Example 6.4 (Example 5.13 continued) The association scheme $\underline{\underline{2}} \times (\underline{\underline{9}}/\underline{\underline{2}})$ is shown in Figure 5.1. Its Hasse diagram is in Figure 6.5.

We have $k_E = 1$, $k_{\text{groups}} = 2$, $k_{\text{rows}} = 18$, $k_{\text{columns}} = 2$, $k_{\text{boxes}} = 4$ and $k_U = 36$ while $n_E = 36$, $n_{\text{groups}} = 18$, $n_{\text{rows}} = 2$, $n_{\text{columns}} = 18$, $n_{\text{boxes}} = 9$ and $n_U = 1$.

As in Example 5.13, write G, R, C and B for R_{groups}, R_{rows}, R_{columns} and R_{boxes} respectively. Then

$$\begin{array}{lcllcl} A_E & = & I & a_E & = & 1 \\ A_{\text{groups}} & = & G - I & a_{\text{groups}} & = & 1 \\ A_{\text{rows}} & = & R - G & a_{\text{rows}} & = & 16 \\ A_{\text{columns}} & = & C - I & a_{\text{columns}} & = & 1 \\ A_{\text{boxes}} & = & B - G - C + I & a_{\text{boxes}} & = & 1 \\ A_U & = & J - R - B + G & a_U & = & 16. \end{array}$$

The stratum projectors and their dimensions are as follows.

$$\begin{array}{lcllcl} S_U & = & \frac{1}{36}J & d_U & = & 1 \\ S_{\text{boxes}} & = & \frac{1}{4}B - \frac{1}{36}J & d_{\text{boxes}} & = & 8 \\ S_{\text{columns}} & = & \frac{1}{2}C - \frac{1}{4}B & d_{\text{columns}} & = & 9 \\ S_{\text{rows}} & = & \frac{1}{18}R - \frac{1}{36}J & d_{\text{rows}} & = & 1 \\ S_{\text{groups}} & = & \frac{1}{2}G - \frac{1}{18}R - \frac{1}{4}B + \frac{1}{36}J & d_{\text{groups}} & = & 8 \\ S_E & = & I - \frac{1}{2}G - \frac{1}{2}C + \frac{1}{4}B & d_E & = & 9. \end{array}$$ ■

Calculations with data are also simplified when the relevant association scheme is derived from an orthogonal block structure. To estimate treatment effects and variances from the data from an experiment in blocks, we need to calculate vectors such as $SX'QY$, where Y is the data vector, Q is the matrix of orthogonal projection onto $V_B^\perp$, X is the design matrix and S is a projector onto an eigenspace of $X'QX$: see Sections 4.3 and 4.5. Section 5.2 shows that if the block design is partially balanced with respect to an association scheme $\mathcal{Q}$ then S may be taken to be a stratum projector for $\mathcal{Q}$. If $\mathcal{Q}$ is defined by an orthogonal block structure then Equation (6.7) shows that $SX'QY$ can be calculated very simply, because each map of the form $X'QY \mapsto P_F X'QY$ replaces each entry in the vector by its average over that F-class.

Example 6.5 (Example 5.10 continued) The association scheme on the treatments comes from the orthogonal block structure $\{E, R, C, U\}$, where the classes of R and C are rows and columns respectively. In the notation of Example 4.3, $P_3 = S_R + S_C = (P_R - P_U) + (P_C - P_U)$. We noted in Section 4.3 that $QXP_U = 0$, so the entries of $P_3X'QY$ are obtained from $X'QY$ by adding the row- and column-averages. Then $P_4X'QY$ is obtained as $X'QY - P_3X'QY$. ■

6.5 Orthogonal block structures from Latin squares

If any of the $\mathcal{C}_F$ is empty then $\mathcal{K}$ is a proper subset of $\mathcal{F}$. Theorem 2.6 shows that $|\mathcal{E}| = |\mathcal{K}|$, so $\mathcal{E}$ is also a proper subset of $\mathcal{F}$; in fact, the number of F such that $W_F = 0$ is the same as the number of F such that $\mathcal{C}_F = \varnothing$. However, the partitions which give zero subspaces are not usually the same as those which give empty subsets of $\Omega \times \Omega$.

Example 6.2 revisited Although this association scheme is defined by a single Latin square, it is *not* of the Latin square type $\mathrm{L}(3, n)$ described in Example 1.6; the former first associates have been separated into those in the same row, those in the same column and those in the same letter, so that we have the association scheme in Exercise 1.12. If we write R, C and L for R_{rows}, R_{columns} and R_{letters}, we obtain

$$
\begin{array}{lcllcl}
A_E & = & I & a_E & = & 1 \\
A_{\text{rows}} & = & R - I & a_{\text{rows}} & = & n-1 \\
A_{\text{columns}} & = & C - I & a_{\text{columns}} & = & n-1 \\
A_{\text{letters}} & = & L - I & a_{\text{letters}} & = & n-1 \\
A_U & = & J - R - C - L + 2I & a_U & = & (n-1)(n-2)
\end{array}
$$

and

$$
\begin{array}{lcllcl}
S_U & = & \dfrac{1}{n^2}J & d_U & = & 1 \\[2ex]
S_{\text{rows}} & = & \dfrac{1}{n}R - \dfrac{1}{n^2}J & d_{\text{rows}} & = & n-1
\end{array}
$$

$$
\begin{array}{llll}
S_{\text{columns}} = \frac{1}{n}C - \frac{1}{n^2}J & & d_{\text{columns}} = & n-1 \\
S_{\text{letters}} = \frac{1}{n}L - \frac{1}{n^2}J & & d_{\text{letters}} = & n-1 \\
S_E = I - \frac{1}{n}R - \frac{1}{n}C & & & \\
\quad - \frac{1}{n}L + \frac{2}{n^2}J & & d_E = & (n-1)(n-2).
\end{array}
$$

If $n = 2$ then $\mathcal{C}_U = \varnothing$ and $W_E = 0$. ■

More generally, if we have $c - 2$ mutually orthogonal $n \times n$ Latin squares then we can create an orthogonal block structure with c non-trivial partitions of a set of size n^2 into n classes of size n. It is also an orthogonal array. For $i = 1, \ldots, c$ we have $A_i = R_i - I$, where R_i is the relation matrix of the i-th non-trivial partition and A_i is the adjacency matrix of the corresponding associate class; the stratum projector S_i is given by $S_i = n^{-1}R_i - n^{-2}J$. Moreover, for $i \neq j$ we have $R_iR_j = J$ and so $A_iA_j = J - A_i - A_j + I$. If $c = n+1$ then $\mathcal{C}_U = \varnothing$ and $W_E = 0$. For definiteness, let us say that this association scheme is *square* of type $\mathrm{S}(c, n)$.

A popular generalization of square lattice designs uses this association scheme. Given $c - 2$ mutually orthogonal $n \times n$ Latin squares, follow the construction procedure in Section 5.5.2 but use b_i copies of the i-th component for $1 \leqslant i \leqslant c$, where $\sum b_i = r$. Then $\Lambda = rI + \sum b_iA_i = \sum b_iR_i$. Therefore

$$L = rI - \frac{1}{n}\sum_{i=1}^{c} b_iR_i = rS_E + \sum_{i=1}^{c}(r - b_i)S_i.$$

The canonical efficiency factors are 1, with multiplicity $(n+1-c)(n-1)$, and $(r - b_i)/r$, with multiplicity $n - 1$, for $1 \leqslant i \leqslant c$. Moreover,

$$
\begin{aligned}
L^- &= \frac{1}{r}S_E + \sum_{i=1}^{c}\frac{1}{r-b_i}S_i \\
&= \frac{1}{r}\left[I - \sum_{i=1}^{c}\frac{1}{n}R_i + \frac{(c-1)}{n^2}J\right] + \sum_{i=1}^{c}\frac{1}{r-b_i}\left[\frac{1}{n}R_i - \frac{1}{n^2}J\right].
\end{aligned}
$$

Thus the variance of the estimator of $\tau(\theta) - \tau(\eta)$ is equal to

$$2\left(\frac{1}{r} - \frac{c}{rn} + \frac{1}{n}\sum_{i=1}^{c}\frac{1}{r-b_i}\right)\sigma^2$$

if $(\theta, \eta) \in \mathcal{C}_U$, and to

$$2\left(\frac{1}{r} - \frac{c-1}{rn} + \frac{1}{n}\sum_{j\neq i}\frac{1}{r-b_j}\right)\sigma^2$$

if $(\theta, \eta) \in \mathcal{C}_i$.

6.6 Crossing and nesting orthogonal block structures

Any partition F of Ω can be identified with an equivalence relation on Ω, that is, with a subset of $\Omega \times \Omega$. Thus we regard the pair (α, β) as belonging to F if and only if α and β are in the same F-class. The relation matrix R_F is now seen as the adjacency matrix of this subset F. The conditions for an equivalence relation are

(i) (**reflexivity**) $\mathrm{Diag}(\Omega) \subseteq F$;
(ii) (**symmetry**) $F' = F$;
(iii) (**transitivity**) if $(\alpha, \beta) \in F$ and $(\beta, \gamma) \in F$ then $(\alpha, \gamma) \in F$.

From this viewpoint, the two trivial partitions of Ω may be expressed as

$$\begin{aligned} E &= \mathrm{Diag}(\Omega) \\ U &= \Omega \times \Omega. \end{aligned}$$

Moreover, if F_1 and F_2 are two partitions of Ω then $F_1 \wedge F_2 = F_1 \cap F_2$.

Now let F be a partition of Ω_1 and G a partition of Ω_2. As in Section 3.2, we can define the subset $F \times G$ of $(\Omega_1 \times \Omega_2) \times (\Omega_1 \times \Omega_2)$:

$$\begin{aligned} F \times G &= \{((\alpha_1, \alpha_2), (\beta_1, \beta_2)) : (\alpha_1, \beta_1) \in F \text{ and } (\alpha_2, \beta_2) \in G\} \\ &= \left\{ \begin{array}{r} ((\alpha_1, \alpha_2), (\beta_1, \beta_2)) : \alpha_1 \text{ is in the same } F\text{-class as } \beta_1 \text{ and} \\ \alpha_2 \text{ is in the same } G\text{-class as } \beta_2 \end{array} \right\}. \end{aligned}$$

Lemma 6.10 *Let F and G be partitions of Ω_1 and Ω_2 respectively. Then $F \times G$ is a partition of $\Omega_1 \times \Omega_2$. Moreover, $R_{F\times G} = R_F \otimes R_G$ and $P_{F\times G} = P_F \otimes P_G$. The number of classes of $F \times G$ is $n_F n_G$. If F and G are both uniform then $F \times G$ is uniform.*

Proof Our "twisted" definition of products of subsets gives

$$\mathrm{Diag}(\Omega_1 \times \Omega_2) = \mathrm{Diag}(\Omega_1) \times \mathrm{Diag}(\Omega_2) \subseteq F \times G$$

so $F \times G$ is reflexive. Also $(F \times G)' = F' \times G'$, as in the proof of Theorem 3.3, so $F \times G$ is symmetric. If $((\alpha_1, \alpha_2), (\beta_1, \beta_2)) \in F \times G$ and $((\beta_1, \beta_2), (\gamma_1, \gamma_2)) \in F \times G$ then $(\alpha_1, \beta_1) \in F$ and $(\beta_1, \gamma_1) \in F$ so $(\alpha_1, \gamma_1) \in F$; also $(\alpha_2, \beta_2) \in G$ and $(\beta_2, \gamma_2) \in G$ so $(\alpha_2, \gamma_2) \in G$: therefore $((\alpha_1, \alpha_2), (\gamma_1, \gamma_2)) \in F \times G$ and so $F \times G$ is transitive. Thus $F \times G$ is a partition of $\Omega_1 \times \Omega_2$.

Equation (3.1) shows that $R_{F\times G} = R_F \otimes R_G$, because the matrices R_F, R_G and $R_{F\times G}$ are the adjacency matrices of the subsets F, G and $F \times G$ respectively.

Let Δ be the F-class containing ω_1 in Ω_1 and let Γ be the G-class containing ω_2 in Ω_2. Then the $(F \times G)$-class containing (ω_1, ω_2) in $\Omega_1 \times \Omega_2$ is just $\Delta \times \Gamma$. Therefore $n_{F\times G} = n_F n_G$. In particular, if all classes of F have size k_F and all classes of G have size k_G then $F \times G$ is uniform with class size $k_F k_G$.

If $v \in V_F$ and $w \in V_G$ then $v \otimes w$ is constant on each class of $F \times G$. Hence $V_F \otimes V_G \leqslant V_{F\times G}$. A basis of $V_{F\times G}$ consists of the characteristic vectors of the classes of $F \times G$. But $\chi_{\Delta\times\Gamma} = \chi_\Delta \otimes \chi_\Gamma \in V_F \otimes V_G$. Thus $V_{F\times G} = V_F \otimes V_G$. Corollary 3.2 shows that $P_{F\times G} = P_F \otimes P_G$. ■

Lemma 6.11 *Let F_1 and G_1 be partitions of Ω_1, and F_2 and G_2 be partitions of Ω_2. Then*

(i) $(F_1 \times F_2) \wedge (G_1 \times G_2) = (F_1 \wedge G_1) \times (F_2 \wedge G_2)$;
(ii) $(F_1 \times F_2) \vee (G_1 \times G_2) = (F_1 \vee G_1) \times (F_2 \vee G_2)$;
(iii) if $F_1 \perp G_1$ *and* $F_2 \perp G_2$ *then* $(F_1 \times F_2) \perp (G_1 \times G_2)$.

Proof (i) Since $(F_1 \times F_2) \wedge (G_1 \times G_2) = (F_1 \times F_2) \cap (G_1 \times G_2)$,

$$\begin{aligned}
&((\alpha_1, \beta_1), (\alpha_2, \beta_2)) \in (F_1 \times F_2) \wedge (G_1 \times G_2) \\
&\iff (\alpha_1, \beta_1) \in F_1,\ (\alpha_2, \beta_2) \in F_2,\ (\alpha_1, \beta_1) \in G_1 \\
&\qquad\quad \text{and } (\alpha_2, \beta_2) \in G_2 \\
&\iff (\alpha_1, \beta_1) \in F_1 \cap G_1 \text{ and } (\alpha_2, \beta_2) \in F_2 \cap G_2 \\
&\iff ((\alpha_1, \beta_1), (\alpha_2, \beta_2)) \in (F_1 \cap G_1) \times (F_2 \cap G_2) \\
&\iff ((\alpha_1, \beta_1), (\alpha_2, \beta_2)) \in (F_1 \wedge G_1) \times (F_2 \wedge G_2).
\end{aligned}$$

(ii) Let $E_1 = \mathrm{Diag}(\Omega_1)$ and $E_2 = \mathrm{Diag}(\Omega_2)$. Observe that $F_1 \times F_2 = (F_1 \times E_2) \vee (E_1 \times F_2)$ and $G_1 \times G_2 = (G_1 \times E_2) \vee (E_1 \times G_2)$. Also,

$(F_1 \times E_2) \vee (G_1 \times E_2) = (F_1 \vee G_1) \times E_2$ and $(E_1 \times F_2) \vee (E_1 \times G_2) = E_1 \times (F_2 \vee G_2)$. Now, $\vee$ is commutative and associative, so

$$\begin{aligned}
&(F_1 \times F_2) \vee (G_1 \times G_2) \\
&\quad = (F_1 \times E_2) \vee (E_1 \times F_2) \vee (G_1 \times E_2) \vee (E_1 \times G_2) \\
&\quad = (F_1 \times E_2) \vee (G_1 \times E_2) \vee (E_1 \times F_2) \vee (E_1 \times G_2) \\
&\quad = ((F_1 \vee G_1) \times E_2) \vee (E_1 \times (F_2 \vee G_2)) \\
&\quad = (F_1 \vee G_1) \times (F_2 \vee G_2).
\end{aligned}$$

(iii) From Lemma 6.10, $P_{F_1 \times F_2} = P_{F_1} \otimes P_{F_2}$ and $P_{G_1 \times G_2} = P_{G_1} \otimes P_{G_2}$. If $F_1 \perp G_1$ and $F_2 \perp G_2$ then $P_{F_1} P_{G_1} = P_{G_1} P_{F_1}$ and $P_{F_2} P_{G_2} = P_{G_2} P_{F_2}$: hence

$$\begin{aligned}
P_{F_1 \times F_2} P_{G_1 \times G_2} &= (P_{F_1} \otimes P_{F_2})(P_{G_1} \otimes P_{G_2}) \\
&= P_{F_1} P_{G_1} \otimes P_{F_2} P_{G_2} = P_{G_1} P_{F_1} \otimes P_{G_2} P_{F_2} \\
&= (P_{G_1} \otimes P_{G_2})(P_{F_1} \otimes P_{F_2}) = P_{G_1 \times G_2} P_{F_1 \times F_2}
\end{aligned}$$

and therefore $(F_1 \times F_2) \perp (G_1 \times G_2)$. ■

If $\mathcal{F}_1$ is an orthogonal block structure on Ω_1 and $\mathcal{F}_2$ is an orthogonal block structure on Ω_2 then the direct product $\mathcal{F}_1 \times \mathcal{F}_2$ is defined as in Section 3.2. It is a set of subsets of $\Omega_1 \times \Omega_2$.

Theorem 6.12 *For $t = 1, 2$, let $\mathcal{F}_t$ be an orthogonal block structure on a set Ω_t. Then $\mathcal{F}_1 \times \mathcal{F}_2$ is an orthogonal block structure on $\Omega_1 \times \Omega_2$.*

Proof Lemma 6.10 shows that every element of $\mathcal{F}_1 \times \mathcal{F}_2$ is a uniform partition of $\Omega_1 \times \Omega_2$. Lemma 6.11 shows that these partitions are pairwise orthogonal, and that $\mathcal{F}_1 \times \mathcal{F}_2$ is closed under $\wedge$ and $\vee$.

For $t = 1, 2$, let $E_t = \mathrm{Diag}(\Omega_t)$ and $U_t = \Omega_t \times \Omega_t$. Then

$$\mathrm{Diag}(\Omega_1 \times \Omega_2) = E_1 \times E_2 \in \mathcal{F}_1 \times \mathcal{F}_2$$

and

$$(\Omega_1 \times \Omega_2) \times (\Omega_1 \times \Omega_2) = U_1 \times U_2 \in \mathcal{F}_1 \times \mathcal{F}_2.$$

Therefore $\mathcal{F}_1 \times \mathcal{F}_2$ is an orthogonal block structure on $\Omega_1 \times \Omega_2$. ■

The partial order on $\mathcal{F}_1 \times \mathcal{F}_2$ is obtained as the direct product of the partial orders on $\mathcal{F}_1$ and $\mathcal{F}_2$ (considered as subsets of $\mathcal{F}_1 \times \mathcal{F}_1$ and $\mathcal{F}_2 \times \mathcal{F}_2$ respectively), for

$$F_1 \times F_2 \preccurlyeq G_1 \times G_2 \iff F_1 \preccurlyeq G_1 \quad \text{and} \quad F_2 \preccurlyeq G_2.$$

Thus the zeta function for $\mathcal{F}_1 \times \mathcal{F}_2$ is the tensor product of the zeta functions for $\mathcal{F}_1$ and $\mathcal{F}_2$.

Theorem 6.13 *For $t = 1, 2$, let $\mathcal{Q}_t$ be the association scheme of an orthogonal block structure $\mathcal{F}_t$ on a set Ω_t. Let $\mathcal{Q}$ be the association scheme of $\mathcal{F}_1 \times \mathcal{F}_2$. Then $\mathcal{Q} = \mathcal{Q}_1 \times \mathcal{Q}_2$.*

Proof For $t = 1, 2$, let ζ_t and μ_t be the zeta and Möbius functions of $\mathcal{F}_t$. Then the zeta function ζ of $\mathcal{F}_1 \times \mathcal{F}_2$ is $\zeta_1 \otimes \zeta_2$, whose inverse μ is $\mu_1 \otimes \mu_2$. For $t = 1, 2$, the adjacency matrices of $\mathcal{Q}_t$ are

$$A_{F_t} = \sum_{G_t} \mu'_t(F_t, G_t) R_{G_t}$$

for F_t in $\mathcal{F}_t$. Those of the association scheme $\mathcal{Q}$ are

$$\begin{aligned}
A_{F_1 \times F_2} &= \sum_{G_1,\, G_2} \mu'(F_1 \times F_2, G_1 \times G_2) R_{G_1 \times G_2} \\
&= \sum_{G_1,\, G_2} (\mu'_1 \otimes \mu'_2)(F_1 \times F_2, G_1 \times G_2) R_{G_1} \otimes R_{G_2} \\
&= \sum_{G_1,\, G_2} \mu'_1(F_1, G_1) \mu'_2(F_2, G_2) R_{G_1} \otimes R_{G_2} \\
&= \left(\sum_{G_1} \mu'_1(F_1, G_1) R_{G_1} \right) \otimes \left(\sum_{G_2} \mu'_2(F_2, G_2) R_{G_2} \right) \\
&= A_{F_1} \otimes A_{F_2},
\end{aligned}$$

which are precisely the adjacency matrices of $\mathcal{Q}_1 \times \mathcal{Q}_2$. ■

Since every orthogonal block structure contains the trivial partition E, we can also define the wreath product of orthogonal block structures as in Section 3.4.

Theorem 6.14 *For $t = 1, 2$, let $\mathcal{F}_t$ be an orthogonal block structure on a set Ω_t. Then $\mathcal{F}_1 / \mathcal{F}_2$ is an orthogonal block structure on $\Omega_1 \times \Omega_2$.*

Proof As in the proof of Theorem 6.12, the elements of $\mathcal{F}_1/\mathcal{F}_2$ are uniform partitions of $\Omega_1 \times \Omega_2$ and they are pairwise orthogonal. They are $F \times U_2$, for F in $\mathcal{F}_1 \setminus \{E_1\}$, and $E_1 \times G$, for G in $\mathcal{F}_2$, where $E_1 = \mathrm{Diag}(\Omega_1)$ and $U_2 = \Omega_2 \times \Omega_2$. Now

$$(F_1 \times U_2) \wedge (G_1 \times U_2) = (F_1 \wedge G_1) \times U_2;$$

$$(F_1 \times U_2) \vee (G_1 \times U_2) = (F_1 \vee G_1) \times U_2;$$

$$(F \times U_2) \wedge (E_1 \times G) = (F \wedge E_1) \times (U_2 \wedge G) = E_1 \times G;$$

$$(F \times U_2) \vee (E_1 \times G) = (F \vee E_1) \times (U_2 \vee G) = F \times U_2;$$

$$(E_1 \times F_2) \wedge (E_1 \times G_2) = E_1 \times (F_2 \wedge G_2);$$

and

$$(E_1 \times F_2) \vee (E_1 \times G_2) = E_1 \times (F_2 \vee G_2).$$

Since $E_1 \times U_2$ is indeed in $\mathcal{F}_1/\mathcal{F}_2$, this shows that $\mathcal{F}_1/\mathcal{F}_2$ is closed under $\wedge$ and $\vee$. Let $E_2 = \mathrm{Diag}(\Omega_2)$ and $U_1 = \Omega_1 \times \Omega_1$. Then $U_1 \in \mathcal{F}_1$ so $U_1 \times U_2 \in \mathcal{F}_1/\mathcal{F}_2$, and $E_2 \in \mathcal{F}_2$ so $E_1 \times E_2 \in \mathcal{F}_1/\mathcal{F}_2$. Therefore $\mathcal{F}_1/\mathcal{F}_2$ is an orthogonal block structure on $\Omega_1 \times \Omega_2$. ■

The partial order on $\mathcal{F}_1/\mathcal{F}_2$ is obtained by putting the whole partial order for $\mathcal{F}_1$ above the whole partial order for $\mathcal{F}_2$, identifying the equality trivial partition E_1 of $\mathcal{F}_1$ with the universal trivial partition U_2 of $\mathcal{F}_2$. See Figure 6.6.

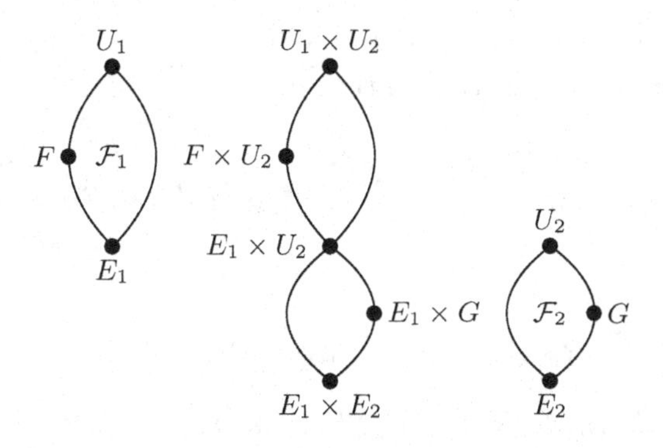

Fig. 6.6. The partial order on $\mathcal{F}_1/\mathcal{F}_2$

Theorem 6.15 *For $t = 1, 2$, let $\mathcal{Q}_t$ be the association scheme of an orthogonal block structure $\mathcal{F}_t$ on a set Ω_t. Let $\mathcal{Q}$ be the association scheme of $\mathcal{F}_1/\mathcal{F}_2$. Then $\mathcal{Q} = \mathcal{Q}_1/\mathcal{Q}_2$.*

Proof The relation and adjacency matrices for $\mathcal{F}_t$ and $\mathcal{Q}_t$ satisfy

$$R_{F_t} = \sum_{G_t} \zeta_t'(F_t, G_t) A_{G_t}$$

for $t = 1, 2$. Let F_2 be any partition in $\mathcal{F}_2$. Then, in $\mathcal{Q}$, we have

$$R_{E_1 \times F_2} = \sum_{G_2} \zeta'(E_1 \times F_2, E_1 \times G_2) A_{E_1 \times G_2} = \sum_{G_2} \zeta_2'(F_2, G_2) A_{E_1 \times G_2}.$$

This is because $\zeta'(E_1 \times F_2, H \times U_2) = 0$ if $H \in \mathcal{F}_1 \setminus \{E_1\}$. But

$$R_{E_1 \times F_2} = R_{E_1} \otimes R_{F_2} = I_{\Omega_1} \otimes R_{F_2}$$

so Möbius inversion in $\mathcal{F}_2$ gives

$$\begin{aligned} A_{E_1 \times F_2} &= \sum_{G_2} \mu_2'(F_2, G_2) R_{E_1 \times G_2} \\ &= I_{\Omega_1} \otimes \sum_{G_2} \mu_2'(F_2, G_2) R_{G_2} \\ &= I_{\Omega_1} \otimes A_{F_2} \end{aligned}$$

for F_2 in $\mathcal{F}_2$.

Now let F_1 be any partition in $\mathcal{F}_1$. In $\mathcal{Q}$ we have

$$R_{F_1 \times U_2} = \sum_{G_1} \zeta'(F_1 \times U_2, G_1 \times U_2) A_{G_1 \times U_2} + \sum_{H \in \mathcal{F}_2} A_{E_1 \times H} - A_{E_1 \times U_2}.$$

Here we have used the fact that $\zeta'(F_1 \times U_2, E_1 \times H) = 1$ for all H in $\mathcal{F}_2$, and we have removed one copy of the term $A_{E_1 \times U_2}$ which occurs in both sums. Now,

$$\sum_H A_{E_1 \times H} = I_{\Omega_1} \otimes \sum_H A_H = I_{\Omega_1} \otimes J_{\Omega_2} = R_{E_1 \times U_2}$$

and

$$R_{F_1 \times U_2} = R_{F_1} \otimes R_{U_2} = R_{F_1} \otimes J_{\Omega_2}.$$

Define $T_F = A_{F \times U_2}$ for F in $\mathcal{F}_1 \setminus \{E_1\}$ and $T_{E_1} = I_{\Omega_1} \otimes J_{\Omega_2}$. Then

$$R_{F_1} \otimes J_{\Omega_2} = \sum_{G_1} \zeta_1'(F_1, G_1) T_{G_1}$$

for all F_1 in $\mathcal{F}_1$. Möbius inversion in $\mathcal{F}_1$ gives

$$T_{F_1} = \sum_{G_1} \mu_1'(F_1, G_1) R_{G_1} \otimes J_{\Omega_2} = A_{F_1} \otimes J_{\Omega_2}$$

for F_1 in $\mathcal{F}_1$. So the adjacency matrices of $\mathcal{Q}$ are $I_{\Omega_1} \otimes A_G$ for G in $\mathcal{F}_2$ and $A_F \otimes J_{\Omega_2}$ for F in $\mathcal{F}_1 \setminus \{E_1\}$, and these are precisely the adjacency matrices of $\mathcal{Q}_1/\mathcal{Q}_2$. ■

Theorems 6.13 and 6.15 justify the use of notation such as $\underline{\underline{3}}/(\underline{\underline{2}} \times \underline{\underline{4}})$ to refer both to an orthogonal block structure and to its corresponding association scheme.

Exercises

6.1 Draw the Hasse diagram for each of the following sets $\mathcal{F}$ of partitions. Find the zeta function and Möbius function in each case.

(a) The set consists of bk elements, grouped into b blocks of size k, and $\mathcal{F} = \{E, \text{blocks}, U\}$.
(b) The set is an $m \times n$ rectangle, and $\mathcal{F} = \{E, \text{rows}, \text{columns}, U\}$.
(c) The set consists of the 90 cows' ears in Example 3.9, and $\mathcal{F} = \{E, \text{cows}, \text{herds}, U\}$.

6.2 Let $(\mathcal{F}, \preccurlyeq)$ be a poset. An element of $\mathcal{F}$ is said to be *maximal* if there is no G in $\mathcal{F}$ such that $F \prec G$.

(a) Prove that if $\mathcal{F}$ is finite then $\mathcal{F}$ has at least one maximal element.
(b) Prove that if $\mathcal{F}$ contains m elements then there is a bijection π from $\mathcal{F}$ to $\{1, \dots, m\}$ such that $F \prec G \Rightarrow \pi(F) < \pi(G)$.

6.3 For each of the following sets $\mathcal{G}$ of partitions, find the smallest set $\mathcal{F}$ which contains $\mathcal{G}$, which is closed under $\wedge$ and $\vee$, and contains E and U. Then draw the Hasse diagram for $\mathcal{F}$ and find its zeta and Möbius functions.

(a) The set consists of the 24 mushroom boxes in Example 3.8 and $\mathcal{G}$ consists of the three partitions into 2-dimensional pieces, that is, rows, columns and shelves.
(b) The set consists of the 48 sheep-months in Example 3.12, and $\mathcal{G} = \{\text{sheep, houses, months}\}$.
(c) The set consists of the 96 vacuum-cleaner tests in Example 3.10, and $\mathcal{G} = \{\text{weeks, housewives}\}$.
(d) The set consists of the 48 vacuum-cleaner tests in Example 3.11, and $\mathcal{G} = \{\text{weeks, housewives}\}$.

6.4 Prove Corollary 6.5.

6.5 For each set $\mathcal{F}$ in Exercises 6.1 and 6.3, find the adjacency matrices and stratum projectors in terms of the relation matrices R_F. Also find the valencies and the dimensions of the strata.

6.6 Let Ω be an $n \times rn$ rectangle. An allocation of n letters to the elements of Ω is a *Latin rectangle* if every letter occurs once in each column and r times in each row. Under what condition is there an orthogonal block structure $\mathcal{F}$ containing $\{\text{rows, columns, letters}\}$?

Draw two Latin rectangles with $n = 5$ and $r = 3$, one of which gives rise to an orthogonal block structure and one of which does not.

Draw the Hasse diagram for the one that is an orthogonal block structure.

For which values of n and r (if any) are any of the matrices A_F equal to zero in the orthogonal block structure?

6.7 A cheese-tasting experiment is planned for the five days Monday–Friday. Four students have volunteered to taste cheese. On each of the five days, all four students will go to the tasting room between 10 a.m. and 11 a.m. Each student will be given three cheeses to taste within that hour (leaving time for any after-taste to disappear). Let Ω be the set of the 60 cheese-tastings.

Let U and E be the two trivial partitions of Ω. Let V be the partition of Ω into volunteers, and let D be the partition of Ω into days.

Let $\mathcal{F} = \{U, E, V, D, V \wedge D\}$. You may assume that $\mathcal{F}$ is an orthogonal block structure.

(a) Explain in words what each of U, E and $V \wedge D$ is, and state how many classes each has.
(b) Draw the Hasse diagram for $\mathcal{F}$. Hence calculate its Möbius function.
(c) Express the adjacency matrices A_V, A_D, $A_{V\wedge D}$, A_U and A_E (of the association scheme defined by $\mathcal{F}$) as linear combinations of R_V, R_D, $R_{V\wedge D}$, R_U and R_E.
(d) Calculate the product $A_V A_D$ in terms of the adjacency matrices.

6.8 Let $\mathcal{L}$ be the orthogonal block structure on a set of size 16 defined by the rows, columns and letters of a 4×4 Latin square. Describe each of the following orthogonal block structures in words. Draw their Hasse diagrams and find their adjacency matrices.

(a) $\underline{\underline{3}} \times \mathcal{L}$
(b) $\underline{\underline{3}}/\mathcal{L}$
(c) $\mathcal{L}/\underline{\underline{3}}$
(d) $\mathcal{L} \times \mathcal{L}$
(e) $\mathcal{L}/\mathcal{L}$

6.9 Construct orthogonal block structures so that

(a) there is a non-trivial partition F with $A_F = O$;
(b) there is a non-trivial partition F with $W_F = 0$.

7
Designs for structured sets

7.1 Designs on orthogonal block structures

In this chapter $\mathcal{F}$ is an orthogonal block structure on a set Ω of experimental units, Θ is a set of treatments, and, as in Chapter 4, $\psi\colon \Omega \to \Theta$ is a design in which treatment θ has replication r_θ. Strictly speaking, the design is the quadruple $(\Omega, \mathcal{F}, \Theta, \psi)$.

As in Chapter 4, define the design matrix X by

$$X(\omega,\theta) = \begin{cases} 1 & \text{if } \psi(\omega) = \theta \\ 0 & \text{otherwise.} \end{cases}$$

For F in $\mathcal{F}$, define the F-concurrence $\Lambda_F(\theta,\eta)$ of treatments θ and η (or the concurrence of θ and η in F-classes) to be

$$\left|\{(\alpha,\beta) \in \Omega\times\Omega : \alpha,\ \beta \text{ in the same } F\text{-class},\ \psi(\alpha)=\theta,\ \psi(\beta)=\eta\}\right|.$$

As in Lemma 4.1,

$$\Lambda_F = X' R_F X. \tag{7.1}$$

In particular,

$$\Lambda_E = X' R_E X = X' I_\Omega X = \operatorname{diag}(r);$$

$$\Lambda_U(\theta,\eta) = (X' R_U X)(\theta,\eta) = (X' J_\Omega X)(\theta,\eta) = r_\theta r_\eta.$$

If $\mathcal{F}$ contains a single non-trivial partition then ψ is just a block design, as in Chapter 4. There are special names for designs on a handful of orthogonal block structures.

7.1.1 Row-column designs

If $(\Omega, \mathcal{F})$ has type $\underline{\underline{n}} \times \underline{\underline{m}}$ the design is called a *row-column design.* The archetypal row-column design is the Latin square, viewed as a design for n treatments in the orthogonal block structure $\underline{\underline{n}} \times \underline{\underline{n}}$.

Example 7.1 A design for the 10 treatments 0, 1, ..., 9 in the orthogonal block structure $\underline{\underline{6}} \times \underline{\underline{5}}$ is shown in Figure 7.1. It is equi-replicate with $r = 3$.

7	0	1	5	2
8	7	6	9	5
0	8	5	4	3
3	4	7	2	9
9	6	4	1	0
1	2	3	8	6

Fig. 7.1. The row-column design in Example 7.1

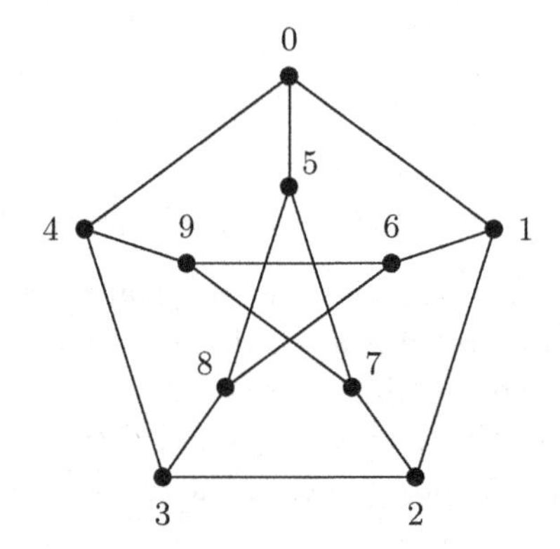

Fig. 7.2. The Petersen graph on the treatments in Example 7.1

Let A be the adjacency matrix for edges in the Petersen graph in Figure 7.2. Then

$$\Lambda_{\text{rows}} = 3I + 2A + (J - A - I)$$

and

$$\Lambda_{\text{columns}} = 3I + A + 2(J - A - I). \quad \blacksquare$$

7.1.2 Nested blocks

If the orthogonal block structure has type $\underline{\underline{n}}/\underline{\underline{m}}/\underline{\underline{l}}$ the design is called either a *nested block design* or a *split-plot design.* The classes of the non-trivial partitions are called large blocks and small blocks in the first case, blocks and whole plots in the second.

Example 7.2 (Example 3.9 continued) The experimental units are cows' ears. The experiment uses both ears from each of 15 cows from each of three herds. The nine treatments a, b, ..., i are allocated as shown in Figure 7.3. $\blacksquare$

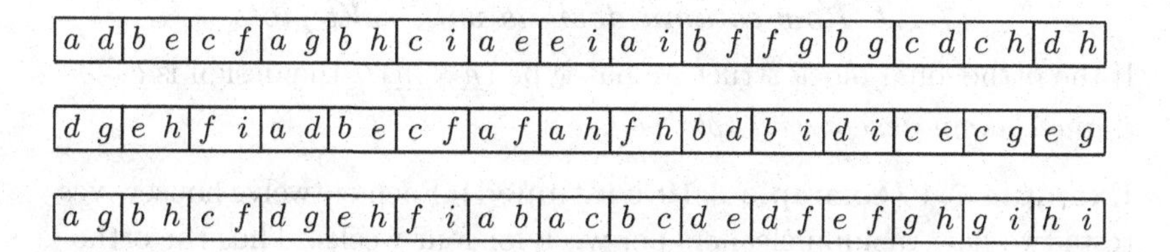

Fig. 7.3. Allocation of nine treatments to cows' ears in Example 7.2

7.1.3 Nested row-column designs

If the orthogonal block structure has type $\underline{\underline{n}}/(\underline{\underline{m}} \times \underline{l})$ the design is called a *nested row-column design.*

Example 7.3 A nested row-column design for six treatments is shown in Figure 7.4. The orthogonal block structure is $\underline{\underline{3}}/(\underline{\underline{2}} \times \underline{\underline{4}})$.

1	2	4	5
5	4	1	2

4	5	3	6
5	3	6	4

1	2	3	6
3	6	2	1

Fig. 7.4. Nested row-column design in Example 7.3

We have

$$\Lambda_{\text{blocks}} = \begin{bmatrix} 8 & 8 & 4 & 4 & 4 & 4 \\ 8 & 8 & 4 & 4 & 4 & 4 \\ 4 & 4 & 8 & 4 & 4 & 8 \\ 4 & 4 & 4 & 8 & 8 & 4 \\ 4 & 4 & 4 & 8 & 8 & 4 \\ 4 & 4 & 8 & 4 & 4 & 8 \end{bmatrix}, \qquad \Lambda_{\text{rows}} = \begin{bmatrix} 4 & 4 & 2 & 2 & 2 & 2 \\ 4 & 4 & 2 & 2 & 2 & 2 \\ 2 & 2 & 4 & 2 & 2 & 4 \\ 2 & 2 & 2 & 4 & 4 & 2 \\ 2 & 2 & 2 & 4 & 4 & 2 \\ 2 & 2 & 4 & 2 & 2 & 4 \end{bmatrix}$$

and

$$\Lambda_{\text{columns}} = \begin{bmatrix} 4 & 0 & 1 & 1 & 1 & 1 \\ 0 & 4 & 1 & 1 & 1 & 1 \\ 1 & 1 & 4 & 0 & 1 & 1 \\ 1 & 1 & 0 & 4 & 1 & 1 \\ 1 & 1 & 1 & 1 & 4 & 0 \\ 1 & 1 & 1 & 1 & 0 & 4 \end{bmatrix}. \quad ■$$

7.1.4 Row-column designs with split plots

If the orthogonal block structure has type $(\underline{\underline{n}} \times \underline{\underline{m}})/\underline{\underline{l}}$ the design is called a *row-column design with split plots*.

Example 7.4 (Example 3.10 continued) Each of twelve housewives tests two new vacuum cleaners per week for four weeks. Thus the orthogonal block structure is $(\underline{\underline{4}} \times \underline{\underline{12}})/\underline{\underline{2}}$. The twelve vacuum cleaners are labelled A, B, C, D, α, β, γ, δ, a, b, c, d. They are allocated as in Figure 7.5. ■

$A\ \alpha$	$B\ \beta$	$C\ \gamma$	$D\ \delta$	$A\ a$	$B\ b$	$C\ c$	$D\ d$	$\alpha\ a$	$\beta\ b$	$\gamma\ c$	$\delta\ d$
$B\ \gamma$	$A\ \delta$	$D\ \alpha$	$C\ \beta$	$B\ d$	$A\ c$	$D\ b$	$C\ a$	$\gamma\ d$	$\delta\ c$	$\alpha\ b$	$\beta\ a$
$C\ \delta$	$D\ \gamma$	$A\ \beta$	$B\ \alpha$	$C\ b$	$D\ a$	$A\ d$	$B\ c$	$\delta\ b$	$\gamma\ a$	$\beta\ d$	$\alpha\ c$
$D\ \beta$	$C\ \alpha$	$B\ \delta$	$A\ \gamma$	$D\ c$	$C\ d$	$B\ a$	$A\ b$	$\beta\ c$	$\alpha\ d$	$\delta\ a$	$\gamma\ b$

Fig. 7.5. Allocation of vacuum cleaners in a row-column design with split plots: see Example 7.4

A row-column design with split plots is called a *semi-Latin rectangle* if every treatment occurs equally often in each row and every treatment occurs equally often in each column. Two semi-Latin rectangles are shown in Figures 7.6 and 7.7. The design in Figure 7.6 was used for an experiment on leaves of tobacco plants: columns represent plants, rows represent heights, there being a pair of leaves at each height.

1 2	2 3	3 4	4 5	5 6	6 7	7 0	0 1
5 7	6 0	7 1	0 2	1 3	2 4	3 5	4 6
3 6	4 7	5 0	6 1	7 2	0 3	1 4	2 5
0 4	1 5	2 6	3 7	4 0	5 1	6 2	7 3

Fig. 7.6. Semi-Latin rectangle for an experiment on leaves: eight treatments in $(\underline{\underline{4}} \times \underline{\underline{8}})/\underline{\underline{2}}$

7.1.5 Other orthogonal block structures

There is essentially no limit to the type of orthogonal block structure that may be used for an experiment, but there are no special names for the designs on most of these structures.

5	2	6	4	5	0	8	7	2	3	1	7	6	9	1	8	4	9	3	0
5	0	8	7	5	9	3	1	9	6	4	7	3	2	4	8	1	2	6	0
2	3	1	7	9	6	4	7	2	9	0	8	0	5	4	1	8	5	6	3
6	9	1	8	3	2	4	8	0	5	4	1	6	3	0	7	7	5	2	9
4	9	3	0	1	2	6	0	8	5	6	3	7	5	2	9	4	1	8	7

Fig. 7.7. Semi-Latin rectangle for ten treatments in $(\underline{\underline{5}} \times \underline{\underline{5}})/\underline{\underline{4}}$

Example 7.5 (Example 3.12 continued) Twelve sheep are used in a feeding trial for four months. The experimental unit is one sheep for one month. The sheep are housed in three houses, four per house. Sixteen feed treatments are allocated as in Figure 7.8. ■

1	2	3	4	16	15	14	13	12	11	10	9
5	6	7	8	1	2	3	4	14	13	16	15
9	10	11	12	6	5	8	7	1	2	3	4
13	14	15	16	11	12	9	10	7	8	5	6

Fig. 7.8. Allocation of sixteen feed treatments in Example 7.5

Other structures for the plots in a design are in Examples 3.8 and 7.7.

7.2 Overall partial balance

The Petersen graph is strongly regular. In Example 7.1, both of the non-trivial concurrence matrices can be expressed in terms of the adjacency matrix of this graph. More generally, we seek designs where all of the concurrence matrices are linear combinations of the adjacency matrices of an association scheme on the treatments.

We thus need notation for an association scheme on Θ as well as the association scheme on Ω defined by $\mathcal{F}$. As a balance between familiar letters and cumbersome notation, we shall write B_F and Q_F for the adjacency matrices and stratum projectors on Ω, leaving A_i and S_e free for those on Θ, and otherwise distinguish where necessary by using the familiar notation but with superscripts (Ω) and (Θ), as shown in Table 7.1. Note that everything in the association scheme on Ω is indexed by $\mathcal{F}$, while the association scheme on Θ, which can be quite general, needs the two indexing sets $\mathcal{K}$ and $\mathcal{E}$ (which have the same size, by Theorem 2.6).

	experimental units		treatments	
set	Ω		Θ	
elements	$\alpha, \beta, \ldots$		$\theta, \eta, \ldots$	
size of set	n_E		t	
associate classes	$\mathcal{C}_F^{(\Omega)}$	for F in $\mathcal{F}$	$\mathcal{C}_i^{(\Theta)}$	for i in $\mathcal{K}$
adjacency matrices	B_F	for F in $\mathcal{F}$	A_i	for i in $\mathcal{K}$
valencies	not used		a_i	for i in $\mathcal{K}$
relation matrices	R_F	for F in $\mathcal{F}$		
partition subspaces	V_F	for F in $\mathcal{F}$		
partition projectors	P_F	for F in $\mathcal{F}$		
stratum projectors	Q_F	for F in $\mathcal{F}$	S_e	for e in $\mathcal{E}$
strata	$W_F^{(\Omega)}$	for F in $\mathcal{F}$	$W_e^{(\Theta)}$	for e in $\mathcal{E}$
dimensions	$d_F^{(\Omega)}$	for F in $\mathcal{F}$	$d_e^{(\Theta)}$	for e in $\mathcal{E}$
Bose–Mesner algebra	not used		$\mathcal{A}$	
character table	not used		C	

Table 7.1. *Notation for association schemes on the experimental units and on the treatments*

Definition Let $\mathcal{Q}$ be an association scheme on a set Θ with associate classes etc. as shown in Table 7.1, let $\mathcal{F}$ be an orthogonal block structure on a set Ω, and let ψ be a function from Ω to Θ with concurrence matrices Λ_F for F in $\mathcal{F}$. The quadruple $(\Omega, \mathcal{F}, \Theta, \psi)$ is *partially balanced overall* with respect to $\mathcal{Q}$ if Λ_F is in the Bose–Mesner algebra $\mathcal{A}$ of $\mathcal{Q}$ for every F in $\mathcal{F}$; that is, there are integers λ_{Fi} for F in $\mathcal{F}$ and i in $\mathcal{K}$ such that

$$\Lambda_F = \sum_{i \in \mathcal{K}} \lambda_{Fi} A_i \qquad \text{for } F \text{ in } \mathcal{F}.$$

As in Chapter 5, partial balance implies equal replication, because the only diagonal matrices in $\mathcal{A}$ are the scalar matrices. Put $r = \lambda_{E0}$; then

$$\Lambda_E = rI_\Theta = rA_0$$

and

$$\Lambda_U = r^2 J_\Theta = r^2 \sum_{i \in \mathcal{K}} A_i.$$

Lemma 4.1(ii) shows that the row sums of Λ_F are all equal to rk_F, so

$$rk_F = \sum_{i \in \mathcal{K}} \lambda_{Fi} a_i \tag{7.2}$$

for all F in $\mathcal{F}$.

For F in $\mathcal{F}$, let $\mathcal{G}_F$ be the group-divisible association scheme on Ω defined by the orthogonal block structure $\{E, F, U\}$. Then overall partial balance for $(\Omega, \mathcal{F}, \Theta, \psi)$ is equivalent to partial balance for all the block designs $(\Omega, \mathcal{G}_F, \Theta, \psi)$ with respect to the *same* association scheme on Θ. It is possible for each of the individual block designs $(\Omega, \mathcal{G}_F, \Theta, \psi)$ to be partially balanced (with respect to different association schemes on Θ) without $(\Omega, \mathcal{F}, \Theta, \psi)$ being partially balanced overall. A test for overall partial balance is commutativity of the concurrence matrices, for if Λ_F and Λ_G are in the same Bose–Mesner algebra then $\Lambda_F \Lambda_G = \Lambda_G \Lambda_F$. However, commutativity does not guarantee partial balance.

Example 7.1 revisited The Petersen graph and its complement are equivalent to the triangular association scheme T(5). Thus the design in Figure 7.1 has overall partial balance. In fact, it is a triangular row-column design: 'triangular' refers to the association scheme on Θ while 'row-column' describes the association scheme on Ω.

Figure 7.9 redisplays this design with the treatments relabelled by the 2-subsets of $\{1, 2, 3, 4, 5\}$. This shows more transparently that the design has overall partial balance with respect to T(5). ■

$\{4,5\}$	$\{3,4\}$	$\{2,5\}$	$\{1,2\}$	$\{1,3\}$
$\{3,5\}$	$\{4,5\}$	$\{1,4\}$	$\{2,3\}$	$\{1,2\}$
$\{3,4\}$	$\{3,5\}$	$\{1,2\}$	$\{1,5\}$	$\{2,4\}$
$\{2,4\}$	$\{1,5\}$	$\{4,5\}$	$\{1,3\}$	$\{2,3\}$
$\{2,3\}$	$\{1,4\}$	$\{1,5\}$	$\{2,5\}$	$\{3,4\}$
$\{2,5\}$	$\{1,3\}$	$\{2,4\}$	$\{3,5\}$	$\{1,4\}$

Fig. 7.9. Row-column design in Example 7.1 with the treatments relabelled

Example 7.6 Consider the two row-column designs in Figure 7.10. Each is for the six treatments 1, ..., 6 in the orthogonal block structure $\underline{\underline{3}} \times \underline{\underline{4}}$.

1	2	3	4
2	6	5	1
5	4	6	3

Design ψ_1

1	2	3	4
6	1	2	5
3	4	5	6

Design ψ_2

Fig. 7.10. Two designs in Example 7.6

In each design the block design in rows is group-divisible with groups $1,2 \parallel 3,4 \parallel 5,6$ and

$$\Lambda_{\text{rows}} = \begin{array}{c} \\ 1 \\ 2 \\ 3 \\ 4 \\ 5 \\ 6 \end{array} \begin{array}{c} \begin{array}{cccccc} 1 & 2 & 3 & 4 & 5 & 6 \end{array} \\ \left[\begin{array}{cccccc} 2 & 2 & 1 & 1 & 1 & 1 \\ 2 & 2 & 1 & 1 & 1 & 1 \\ 1 & 1 & 2 & 2 & 1 & 1 \\ 1 & 1 & 2 & 2 & 1 & 1 \\ 1 & 1 & 1 & 1 & 2 & 2 \\ 1 & 1 & 1 & 1 & 2 & 2 \end{array} \right] \end{array}.$$

The block designs in columns are also group-divisible; the groups are

$$1,6 \parallel 2,3 \parallel 4,5 \quad \text{in} \quad \psi_1$$

and

$$1,5 \parallel 2,6 \parallel 3,4 \quad \text{in} \quad \psi_2.$$

In design ψ_1

$$\Lambda_{\text{columns}} = \left[\begin{array}{cccccc} 2 & 1 & 1 & 1 & 1 & 0 \\ 1 & 2 & 0 & 1 & 1 & 1 \\ 1 & 0 & 2 & 1 & 1 & 1 \\ 1 & 1 & 1 & 2 & 0 & 1 \\ 1 & 1 & 1 & 0 & 2 & 1 \\ 0 & 1 & 1 & 1 & 1 & 2 \end{array} \right]$$

so

$$\Lambda_{\text{rows}}\Lambda_{\text{columns}} = \left[\begin{array}{cccccc} 9 & 9 & 7 & 8 & 8 & 7 \\ 9 & 9 & 7 & 8 & 8 & 7 \\ 8 & 7 & 9 & 9 & 7 & 8 \\ 8 & 7 & 9 & 9 & 7 & 8 \\ 7 & 8 & 8 & 7 & 9 & 9 \\ 7 & 8 & 8 & 7 & 9 & 9 \end{array} \right].$$

As their product is not symmetric, Λ_{rows} and $\Lambda_{\mathrm{columns}}$ do not commute, so design ψ_1 is not partially balanced overall.

In design ψ_2

$$\Lambda_{\mathrm{columns}} = \begin{bmatrix} 2 & 1 & 1 & 1 & 0 & 1 \\ 1 & 2 & 1 & 1 & 1 & 0 \\ 1 & 1 & 2 & 0 & 1 & 1 \\ 1 & 1 & 0 & 2 & 1 & 1 \\ 0 & 1 & 1 & 1 & 2 & 1 \\ 1 & 0 & 1 & 1 & 1 & 2 \end{bmatrix}$$

so

$$\Lambda_{\mathrm{rows}}\Lambda_{\mathrm{columns}} = \begin{bmatrix} 9 & 9 & 8 & 8 & 7 & 7 \\ 9 & 9 & 8 & 8 & 7 & 7 \\ 8 & 8 & 8 & 8 & 8 & 8 \\ 8 & 8 & 8 & 8 & 8 & 8 \\ 7 & 7 & 8 & 8 & 9 & 9 \\ 7 & 7 & 8 & 8 & 9 & 9 \end{bmatrix}.$$

This is symmetric, so the concurrence matrices Λ_{rows} and $\Lambda_{\mathrm{columns}}$ commute. However, their product does not have constant diagonal entries, so they cannot belong to the same Bose–Mesner algebra, so design ψ_2 is not partially balanced either. ■

Example 7.2 revisited The block design in herds is partially balanced with respect to the group-divisible association scheme $\mathcal{Q}_1$ with groups $a, b, c \parallel d, e, f \parallel g, h, i$; and the block design in cows is partially balanced with respect to the group-divisible association scheme $\mathcal{Q}_2$ with groups $a, d, g \parallel b, e, h \parallel c, f, i$. Although these two association schemes are different, the whole design has overall partial balance with respect to the rectangular association scheme R(3, 3) defined by the rectangle

a	b	c
d	e	f
g	h	i

.

Label the associate classes of R(3, 3) as in Example 1.3, so that $\mathcal{C}_1^{(\Theta)}$ consists of unequal pairs in the same row, $\mathcal{C}_2^{(\Theta)}$ consists of unequal pairs in the same column and $\mathcal{C}_3^{(\Theta)}$ contains the other unequal pairs. Then

$$\begin{aligned}\Lambda_U &= 100J_\Theta \\ \Lambda_{\text{herds}} &= 34(I_\Theta + A_1) + 33(A_2 + A_3) \\ \Lambda_{\text{cows}} &= 10I_\Theta + 2A_2 + (A_1 + A_3) \\ \Lambda_E &= 10I_\Theta.\end{aligned}$$

The apparent contradiction is explained by the remark on page 111: the block design in herds is partially balanced with respect to more than one association scheme, and we only need to find one of these with respect to which the block design in cows is partially balanced. ■

Example 7.4 revisited Call one person for one week a *block*. The design is group-divisible with groups

$$A, B, C, D \parallel \alpha, \beta, \gamma, \delta \parallel a, b, c, d.$$

As in Exercise 2.2, let G be the relation matrix for groups. Then

$$\begin{aligned}\Lambda_U &= 64J \\ \Lambda_{\text{weeks}} &= 16J \\ \Lambda_{\text{people}} &= 8G + 4(J - G) = 4G + 4J \\ \Lambda_{\text{blocks}} &= 8I + (J - G) \\ \Lambda_E &= 8I.\end{aligned}$$ ■

Example 7.5 revisited Write the 16 treatments in a square array as in Figure 7.11(a). A pair of mutually orthogonal 4×4 Latin squares is in Figure 7.11(b). Now $\{E$, rows, columns, Latin, Greek, $U\}$ is an

1	2	3	4
5	6	7	8
9	10	11	12
13	14	15	16

(a) rows and columns

A	B	C	D
B	A	D	C
C	D	A	B
D	C	B	A

α	β	γ	δ
γ	δ	α	β
δ	γ	β	α
β	α	δ	γ

(b) Latin and Greek letters

Fig. 7.11. Four partitions on the set of treatments in Example 7.5

orthogonal block structure on the set Θ of treatments: it defines a square association scheme $\mathcal{Q}$ on Θ of type S(4, 4). Then

$$\begin{aligned}
\Lambda_U &= 9J \\
\Lambda_{\text{months}} &= I + A_{\text{row}} + 2J \\
\Lambda_{\text{houses}} &= 3J \\
\Lambda_{\text{house-months}} &= 3(I + A_{\text{row}}) \\
\Lambda_{\text{sheep}} &= 3I + A_{\text{column}} + A_{\text{Latin}} + A_{\text{Greek}} \\
\Lambda_E &= 3I
\end{aligned}$$

and so the design has overall partial balance with respect to $\mathcal{Q}$. ■

Example 7.7 A design for nine treatments in the orthogonal block structure $(\underline{\underline{2}}/\underline{\underline{3}}) \times (\underline{\underline{2}}/\underline{\underline{3}})$ is shown in Figure 7.12. The Hasse diagram for the orthogonal block structure is in Figure 7.13.

1	2	3	5	6	4
5	6	4	7	8	9
9	7	8	3	1	2
2	8	5	6	9	3
6	3	9	1	4	7
7	4	1	8	2	5

Fig. 7.12. Design in Example 7.7

Let A_1 and A_2 be the adjacency matrices for the non-diagonal classes of the Hamming scheme H(2, 3) on the treatments defined by the following square array:

1	2	3
4	5	6
7	8	9

that is, $A_1(\theta, \eta) = 1$ if θ and η are distinct treatments in the same row

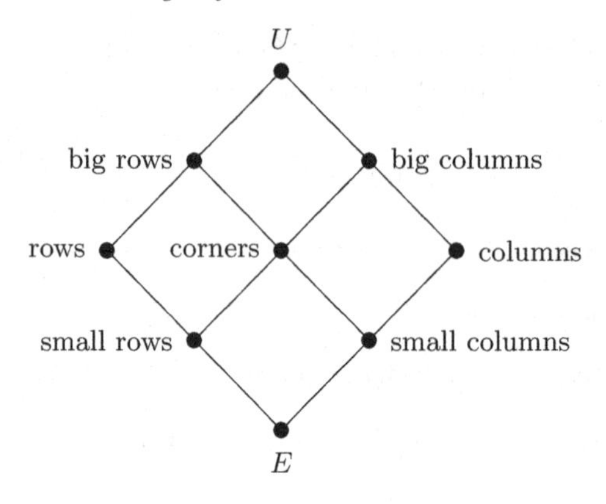

Fig. 7.13. Hasse diagram for Example 7.7

or column. Then

$$
\begin{aligned}
\Lambda_U &= 16J \\
\Lambda_{\text{big rows}} &= 8J \\
\Lambda_{\text{big columns}} &= 8J \\
\Lambda_{\text{corners}} &= 4J \\
\Lambda_{\text{rows}} &= 4I + 3A_1 + 2A_2 \\
\Lambda_{\text{columns}} &= 4I + 2A_1 + 3A_2 \\
\Lambda_{\text{small rows}} &= 4I + 2A_1 \\
\Lambda_{\text{small columns}} &= 4I + 2A_2 \\
\Lambda_E &= 4I.
\end{aligned}
$$

This design has overall partial balance with respect to $\mathrm{H}(2,3)$. ■

7.3 Fixed effects

As in Section 4.3, let the random variable Y be the response on Ω when our design ψ is used for an experiment. There are two widely used models for Y when $(\Omega, \mathcal{F})$ is an orthogonal block structure.

In the *fixed-effects model*, the assumptions are that

$$\mathbb{E}(Y) = X\tau + \sum_{\substack{F \in \mathcal{F} \\ F \neq E}} h_F, \tag{7.3}$$

where τ is an unknown vector in $\mathbb{R}^\Theta$ and, for F in $\mathcal{F} \setminus \{E\}$, h_F is an unknown vector in V_F, and

$$\mathrm{Cov}(Y) = I\sigma^2. \tag{7.4}$$

In other words, every class of every partition in $\mathcal{F}$ except E makes a contribution (fixed effect) to the expectation, but $\mathcal{F}$ has no effect on the variances and covariances. The model given in Section 4.3 is the fixed-effects model for the orthogonal block structure $\{E, \text{blocks}, U\}$; there was no need to write

$$h_{\text{blocks}} + h_U$$

because $h_U \in V_U \leqslant V_{\text{blocks}}$.

Because all of the h_F are unknown, the only way to estimate τ from Y is to project Y onto a subspace which is orthogonal to V_F for all F in $\mathcal{F} \setminus \{E\}$. The largest such subspace is the stratum $W_E^{(\Omega)}$, often called the *bottom stratum*. Thus we estimate τ from $Q_E Y$, where

$$\mathbb{E}(Q_E Y) = Q_E\, \mathbb{E}(Y) = Q_E X\tau.$$

Put $L_E = X'Q_E X$, which is called the *information matrix* in the bottom stratum. The following theorem is proved just like Theorems 4.6 and 4.8.

Theorem 7.1 *Suppose that Equations (7.3) and (7.4) hold. If there is a vector z in $\mathbb{R}^\Theta$ with $L_E z = x$ then $z'X'Q_E Y$ is an unbiased estimator for $x'\tau$ and its variance is $z'L_E z\sigma^2$, which is $x'L_E^- x\sigma^2$. Moreover, this estimator has the smallest variance among all linear unbiased estimators for $x'\tau$.*

The design is said to be *connected* if $\ker L_E$ is the one-dimensional space $W_0^{(\Theta)}$. Using Corollary 4.10, this generalizes the concept of connectedness in Section 4.1, but cannot always be related to connectedness in a graph.

The properties of the design now follow exactly as in Chapters 4 and 5, with L_E in place of L.

Note that

$$\begin{aligned} L_E = X'Q_E X &= X'\left(\sum_{F\in\mathcal{F}} \mu(E,F) P_F\right) X \\ &= \sum_{F\in\mathcal{F}} \mu(E,F) \frac{1}{k_F} X'R_F X \\ &= \sum_{F\in\mathcal{F}} \mu(E,F) \frac{1}{k_F} \Lambda_F \end{aligned}$$

so that if the design has overall partially balance with respect to an association scheme $\mathcal{Q}$ on Θ then L_E is in the Bose–Mesner algebra $\mathcal{A}$ of $\mathcal{Q}$ and all the results of Chapter 5 apply.

Example 7.1 revisited Label the associate classes as in Example 1.4, so that $\mathcal{C}_1^{(\Theta)}$ contains edges and $\mathcal{C}_2^{(\Theta)}$ non-edges. Then

$$\begin{aligned} L_E &= \frac{1}{k_E}\Lambda_E - \frac{1}{k_{\text{rows}}}\Lambda_{\text{rows}} - \frac{1}{k_{\text{columns}}}\Lambda_{\text{columns}} + \frac{1}{k_U}\Lambda_U \\ &= 3I - \frac{1}{5}(3I + 2A_1 + A_2) - \frac{1}{6}(3I + A_1 + 2A_2) + \frac{9}{30}J \\ &= \frac{1}{30}(57I - 17A_1 - 16A_2 + 9J) \\ &= \frac{1}{30}(74I + A_2 - 8J). \end{aligned}$$

From the character table of T(5) in Example 2.5,

$$\begin{aligned} I &= S_0 + S_1 + S_2 \\ A_2 &= 6S_0 + S_1 - 2S_2 \\ J &= 10S_0 \end{aligned}$$

so

$$L_E = \frac{1}{30}(75S_1 + 72S_2) = \frac{1}{10}(25S_1 + 24S_2).$$

Since $r = 3$, the canonical efficiency factors are $5/6$ and $4/5$, with multiplicities 4 and 5 respectively. Also

$$\begin{aligned} L_E^- &= 10\left(\frac{1}{25}S_1 + \frac{1}{24}S_2\right) \\ &= \frac{1}{25}\left(4I - \frac{8}{3}A_1 + \frac{2}{3}A_2\right) + \frac{1}{24}\left(5I + \frac{5}{3}A_1 - \frac{5}{3}A_2\right). \end{aligned}$$

If θ and η are first associates then the variance of the estimator of the treatment difference $\tau(\theta) - \tau(\eta)$ is

$$2\left(\frac{4}{25} + \frac{5}{24} + \frac{8}{75} - \frac{5}{72}\right)\sigma^2 = \frac{73}{90}\sigma^2;$$

otherwise it is

$$2\left(\frac{4}{25} + \frac{5}{24} - \frac{2}{75} + \frac{5}{72}\right)\sigma^2 = \frac{74}{90}\sigma^2. \quad \blacksquare$$

Of course, it is not necessary to have every Λ_F in $\mathcal{A}$ in order to have L_E in $\mathcal{A}$. Thus the design is defined to be *partially balanced in the bottom stratum* if L_E is in the Bose–Mesner algebra of some association scheme on Θ, and to be *balanced in the bottom stratum* if L_E is a linear combination of I_Θ and J_Θ.

If $\mathcal{F} = \{E, F, U\}$ then

$$\begin{aligned} L_U &= \frac{1}{n_E}\Lambda_U = \frac{r^2}{n_E}J_\Theta \\ L_F &= \frac{1}{k_F}\Lambda_F - \frac{r^2}{n_E}J_\Theta \\ L_E &= rI_\Theta - \frac{1}{k_F}\Lambda_F \end{aligned}$$

so that being partially balanced in the bottom stratum is the same as being partially balanced overall. However, for more complicated orthogonal block structures, it is possible to have partial balance in the bottom stratum without overall partial balance.

Example 7.3 revisited The design in Figure 7.4 does not have overall partial balance, because $\Lambda_{\mathrm{rows}}\Lambda_{\mathrm{columns}}$ does not have constant diagonal (as with design ψ_2 in Example 7.6). However,

$$\begin{aligned} L_E &= 4I - \frac{1}{2}\Lambda_{\mathrm{columns}} - \frac{1}{4}\Lambda_{\mathrm{rows}} + \frac{1}{8}\Lambda_{\mathrm{blocks}} \\ &= 4I - \frac{1}{2}\Lambda_{\mathrm{columns}} \end{aligned}$$

because $\Lambda_{\mathrm{blocks}} = 2\Lambda_{\mathrm{rows}}$. Thus the design is partially balanced (in fact, group-divisible) in the bottom stratum. ■

7.4 Random effects

The fixed-effects model is often appropriate for designs in fairly simple orthogonal block structures, such as block designs (Chapters 4–5) and row-column designs (Examples 7.1 and 7.6). However, it may not be appropriate in more complicated cases. In Example 7.2 there is no need for the unknown vector h_{herds} in Equation (7.3); since $V_{\mathrm{herds}} \leqslant V_{\mathrm{cows}}$ the vector h_{herds} can be incorporated into the unknown vector h_{cows}. Nor is V_{herds} needed for the analysis, because

$$\begin{aligned} L_E &= X'P_EX - X'P_{\mathrm{cows}}X \\ &= 10I - \frac{1}{2}\Lambda_{\mathrm{cows}}. \end{aligned}$$

Thus if the fixed-effects model is assumed then the way that treatments are allocated to herds is immaterial, and there is no point in putting any constraints on Λ_{herds}. Similarly, if the fixed-effects model is assumed in Example 7.5 then the allocation of treatments to months is immaterial, and both weeks and people can be ignored in Example 7.4.

It is sometimes more realistic to assume that each partition F contributes not to the expectation of Y but to its pattern of covariance. Thus in the *random-effects* model the assumptions are that

$$\mathbb{E}(Y) = X\tau, \tag{7.5}$$

where τ is an unknown vector in $\mathbb{R}^\Theta$, and that there are unknown real numbers γ_F, for F in $\mathcal{F}$, such that

$$\operatorname{cov}(Y_\alpha, Y_\beta) = \gamma_F \qquad \text{if } (\alpha, \beta) \in \mathcal{C}_F^{(\Omega)};$$

in other words,

$$\operatorname{Cov}(Y) = \sum_{F \in \mathcal{F}} \gamma_F B_F. \tag{7.6}$$

Because $\operatorname{span}\{B_F : F \in \mathcal{F}\} = \operatorname{span}\{Q_F : F \in \mathcal{F}\}$, Equation (7.6) can be re-expressed as

$$\operatorname{Cov}(Y) = \sum_{F \in \mathcal{F}} \xi_F Q_F \tag{7.7}$$

for unknown constants ξ_F, which must be non-negative because $\operatorname{Cov}(Y)$ is non-negative definite. The constant ξ_F is called the *stratum variance* for stratum $W_F^{(\Omega)}$.

Projecting the data onto stratum $W_F^{(\Omega)}$, for any F in $\mathcal{F}$, gives

$$\mathbb{E}(Q_F Y) = Q_F X \tau$$

and

$$\operatorname{Cov}(Q_F Y) = Q_F \left(\sum_{G \in \mathcal{F}} \xi_G Q_G \right) Q_F = \xi_F Q_F,$$

which is scalar on $W_F^{(\Omega)}$. Applying Theorems 4.6 and 4.8 to $Q_F Y$ gives the following.

Theorem 7.2 *Let x be in $\mathbb{R}^\Theta$ and let F be in $\mathcal{F}$. Put $L_F = X' Q_F X$. If $x \notin \operatorname{Im} L_F$ then there is no linear function of $Q_F Y$ which is an unbiased estimator for $x'\tau$. If there is a vector z in $\mathbb{R}^\Theta$ such that $L_F z = x$ then $z' X' Q_F Y$ is an unbiased estimator for $x'\tau$ with variance $z' L_F z \xi_F$; moreover, this variance is minimum among variances of unbiased estimators for $x'\tau$ which are linear functions of $Q_F Y$.*

The matrix L_F is called the information matrix in stratum $W_F^{(\Omega)}$. If $x \notin \operatorname{Im} L_F$ then we say that there is *no information* on $x'\tau$ in stratum $W_F^{(\Omega)}$.

For F in $\mathcal{F}$, we have

$$
\begin{aligned}
L_F = X'Q_FX &= X'\left(\sum_{G\in\mathcal{F}} \mu(F,G)P_G\right)X \\
&= \sum_{G\in\mathcal{F}} \mu(F,G)\frac{1}{k_G}X'R_GX \\
&= \sum_{G\in\mathcal{F}} \mu(F,G)\frac{1}{k_G}\Lambda_G. \qquad (7.8)
\end{aligned}
$$

Thus if the design has overall partial balance with respect to some association scheme $\mathcal{Q}$ on Θ then all of the information matrices are in the Bose–Mesner algebra $\mathcal{A}$ of $\mathcal{Q}$. Hence the treatment strata $W_e^{(\Theta)}$, for e in $\mathcal{E}$, are mutual eigenspaces of all the information matrices. In other words, for all e in $\mathcal{E}$, every non-zero vector in the treatment stratum $W_e^{(\Theta)}$ is a basic contrast in every plot stratum $W_F^{(\Omega)}$ in $\mathbb{R}^\Omega$.

For e in $\mathcal{E}$, let $r\varepsilon_{Fe}$ be the eigenvalue of L_F on $W_e^{(\Theta)}$, so that if $x \in W_e^{(\Theta)}$ then $L_Fx = r\varepsilon_{Fe}x$. If $\varepsilon_{Fe} = 0$ then we cannot estimate $x'\tau$ from Q_FY; otherwise $(r\varepsilon_{Fe})^{-1}x'X'Q_FY$ is a linear unbiased estimator for $x'\tau$ with variance $x'x/(r\varepsilon_{Fe})$. Thus the canonical efficiency factor for x in stratum $W_F^{(\Omega)}$ is

$$
\frac{x'x}{r} \Big/ \frac{x'x}{r\varepsilon_{Fe}},
$$

which is ε_{Fe}.

Theorem 7.3 *If $e \in \mathcal{E}$ then $\sum_{F\in\mathcal{F}} \varepsilon_{Fe} = 1$.*

Proof Let $x \in W_e^{(\Theta)} \setminus \{0\}$. Then

$$
\begin{aligned}
\left(\sum_{F\in\mathcal{F}} \varepsilon_{Fe}\right)x &= \sum_{F\in\mathcal{F}} (\varepsilon_{Fe}x) = \frac{1}{r}\sum_{F\in\mathcal{F}} (L_Fx) \\
&= \frac{1}{r}\left(\sum_{F\in\mathcal{F}} L_F\right)x = \frac{1}{r}\left(\sum_{F\in\mathcal{F}} X'Q_FX\right)x \\
&= \frac{1}{r}X'\left(\sum_{F\in\mathcal{F}} Q_F\right)Xx = \frac{1}{r}X'I_\Omega Xx \\
&= \frac{1}{r}rI_\Theta x = x
\end{aligned}
$$

and so $\sum_{F\in\mathcal{F}} \varepsilon_{Fe} = 1$. ■

The canonical efficiency factor ε_{Fe} is therefore sometimes called the *proportion of information* on x in stratum $W_F^{(\Omega)}$, if $x \in W_e^{(\Theta)}$. This name is inexact, because the information is proportional to ε_{Fe}/ξ_F.

Theorem 7.4 *For F in $\mathcal{F}$ and e in $\mathcal{E}$,*

$$\varepsilon_{Fe} = \frac{1}{r}\sum_{G\in\mathcal{F}} \mu(F,G)\frac{1}{k_G}\sum_{i\in\mathcal{K}} \lambda_{G_i} C(i,e),$$

where C is the character table of $\mathcal{Q}$.

Proof From Equation (7.8),

$$\begin{aligned} L_F &= \sum_{G\in\mathcal{F}} \mu(F,G)\frac{1}{k_G}\Lambda_G \\ &= \sum_{G\in\mathcal{F}} \mu(F,G)\frac{1}{k_G}\sum_{i\in\mathcal{K}} \lambda_{Gi} A_i \\ &= \sum_{G\in\mathcal{F}} \mu(F,G)\frac{1}{k_G}\sum_{i\in\mathcal{K}} \lambda_{Gi}\sum_{e\in\mathcal{E}} C(i,e) S_e. \end{aligned}$$

Thus the eigenvalue of L_F on $W_e^{(\Theta)}$ is

$$\sum_{G\in\mathcal{F}} \mu(F,G)\frac{1}{k_G}\sum_{i\in\mathcal{K}} \lambda_{G_i} C(i,e). \quad ■$$

It is convenient to present the canonical efficiency factors of a partially balanced design in a $\mathcal{F}\times\mathcal{E}$ table. To simplify the calculations, we apply Möbius inversion to Equation (7.8) and obtain

$$\frac{1}{k_F}\Lambda_F = \sum_{G\in\mathcal{F}} \zeta(F,G) L_G.$$

Hence we calculate information matrices recursively, starting at the top of the Hasse diagram, using

$$L_F = \frac{1}{k_F}\Lambda_F - \sum_{G\succ F} L_G.$$

Example 7.1 revisited Using the character table from Example 2.5:

$$
\begin{aligned}
\frac{1}{30}\Lambda_U &= \frac{9}{30}J_\Theta = \frac{3}{10}J_\Theta = 3S_0;\\
\frac{1}{6}\Lambda_{\text{columns}} &= \frac{1}{6}\left(3I + A_1 + 2A_2\right)\\
&= \frac{1}{6}(3(S_0+S_1+S_2) + (3S_0 - 2S_1 + S_2)\\
&\qquad + 2(6S_0 + S_1 - 2S_2))\\
&= 3S_0 + \frac{1}{2}S_1;\\
\frac{1}{5}\Lambda_{\text{rows}} &= \frac{1}{5}\left(3I + 2A_1 + A_2\right)\\
&= \frac{1}{5}(3(S_0+S_1+S_2) + 2(3S_0 - 2S_1 + S_2)\\
&\qquad + (6S_0 + S_1 - 2S_2))\\
&= 3S_0 + \frac{3}{5}S_2;\\
\Lambda_E &= 3I = 3S_0 + 3S_1 + 3S_2.
\end{aligned}
$$

Therefore

$$
\begin{aligned}
L_U &= \frac{1}{30}\Lambda_U = 3S_0;\\
L_{\text{columns}} &= \frac{1}{6}\Lambda_{\text{columns}} - L_U = \frac{1}{2}S_1;\\
L_{\text{rows}} &= \frac{1}{5}\Lambda_{\text{rows}} - L_U = \frac{3}{5}S_2;\\
L_E &= \Lambda_E - L_{\text{rows}} - L_{\text{columns}} - L_U = \frac{5}{2}S_1 + \frac{12}{5}S_2,
\end{aligned}
$$

as already shown. The replication is 3, so dividing by 3 gives the canonical efficiency factors, which are shown in Table 7.2. The stratum dimensions are shown in parentheses. ∎

strata for Ω		strata for Θ 0 (1)	1 (4)	2 (5)
U	(1)	1	0	0
columns	(4)	0	$\frac{1}{6}$	0
rows	(5)	0	0	$\frac{1}{5}$
E	(20)	0	$\frac{5}{6}$	$\frac{4}{5}$

Table 7.2. *Canonical efficiency factors in Example 7.1*

Example 7.4 revisited The non-trivial stratum projectors are

$$S_1 = \frac{1}{4}G - \frac{1}{12}J \quad \text{and} \quad S_2 = I - \frac{1}{4}G.$$

Expressing the concurrence matrices in terms of these gives

$$\begin{aligned}
\frac{1}{96}\Lambda_U &= \frac{64}{96}J = 8S_0 \\
\frac{1}{24}\Lambda_{\text{weeks}} &= \frac{16}{24}J = 8S_0 \\
\frac{1}{8}\Lambda_{\text{people}} &= \frac{1}{8}(4G + 4J) = \frac{1}{8}(16S_1 + 64S_0) = 2S_1 + 8S_0 \\
\frac{1}{2}\Lambda_{\text{blocks}} &= \frac{1}{2}(8I + J - G) = \frac{1}{2}(8S_2 + 4S_1 + 16S_0) \\
&= 4S_2 + 2S_1 + 8S_0 \\
\Lambda_E &= 8I = 8(S_0 + S_1 + S_2).
\end{aligned}$$

Hence

$$\begin{aligned}
L_U &= \frac{1}{96}\Lambda_U = 8S_0 \\
L_{\text{weeks}} &= \frac{1}{24}\Lambda_{\text{weeks}} - L_U = 0
\end{aligned}$$

$$L_{\text{people}} = \frac{1}{8}\Lambda_{\text{people}} - L_U = 2S_1$$

$$L_{\text{blocks}} = \frac{1}{2}\Lambda_{\text{blocks}} - L_{\text{people}} - L_{\text{weeks}} - L_U = 4S_2$$

$$L_E = \Lambda_E - L_{\text{blocks}} - L_{\text{people}} - L_{\text{weeks}} - L_U = 6S_1 + 4S_2.$$

The canonical efficiency factors are in Table 7.3. ∎

strata for Ω		strata for Θ 0 (1)	1 (2)	2 (9)
U	(1)	1	0	0
weeks	(3)	0	0	0
people	(11)	0	$\frac{1}{4}$	0
blocks	(33)	0	0	$\frac{1}{2}$
E	(48)	0	$\frac{3}{4}$	$\frac{1}{2}$

Table 7.3. *Canonical efficiency factors in Example 7.4*

Example 7.5 revisited We use the stratum projectors found in Section 6.5.

$$\begin{aligned}
\frac{1}{48}\Lambda_U &= \frac{9}{48}J = 3S_0 \\
\frac{1}{12}\Lambda_{\text{months}} &= \frac{1}{12}(4(S_{\text{rows}} + S_0) + 32S_0) = \frac{1}{3}S_{\text{rows}} + 3S_0 \\
\frac{1}{16}\Lambda_{\text{houses}} &= \frac{3}{16}J = 3S_0 \\
\frac{1}{4}\Lambda_{\text{house-months}} &= \frac{3}{4}(4(S_{\text{rows}} + S_0)) = 3S_{\text{rows}} + 3S_0
\end{aligned}$$

$$\frac{1}{4}\Lambda_{\text{sheep}} = \frac{1}{4}(4(S_{\text{columns}} + S_{\text{Latin}} + S_{\text{Greek}} + 3S_0))$$

$$= S_{\text{columns}} + S_{\text{Latin}} + S_{\text{Greek}} + 3S_0$$

$$\Lambda_E = 3I = 3(S_0 + S_{\text{rows}} + S_{\text{columns}} + S_{\text{Latin}} + S_{\text{Greek}} + S_E).$$

Thus

$$L_U = \frac{1}{48}\Lambda_U = 3S_0$$

$$L_{\text{months}} = \frac{1}{12}\Lambda_{\text{months}} - L_U = \frac{1}{3}S_{\text{rows}}$$

$$L_{\text{houses}} = \frac{1}{16}\Lambda_{\text{houses}} - L_U = 0$$

$$L_{\text{house-months}} = \frac{1}{4}\Lambda_{\text{house-months}} - L_{\text{houses}} - L_{\text{months}} - L_U = \frac{8}{3}S_{\text{rows}}$$

$$L_{\text{sheep}} = \frac{1}{4}\Lambda_{\text{sheep}} - L_{\text{houses}} - L_U = S_{\text{columns}} + S_{\text{Latin}} + S_{\text{Greek}}$$

$$L_E = \Lambda_E - L_{\text{sheep}} - L_{\text{house-months}} - L_{\text{houses}} - L_{\text{months}} - L_U$$

$$= 2(S_{\text{columns}} + S_{\text{Latin}} + S_{\text{Greek}}) + 3S_E.$$

The canonical efficiency factors are in Table 7.4. ■

For e in $\mathcal{E}$, put $\mathcal{F}_e = \{F \in \mathcal{F} : \varepsilon_{Fe} \neq 0\}$. If $x \in W_e^{(\Theta)}$ we now have estimates for $x'\tau$ from all the strata $W_F^{(\Omega)}$ with F in $\mathcal{F}_e$. Consider the linear combination

$$Z = \sum_{F \in \mathcal{F}_e} c_F \frac{1}{r\varepsilon_{Fe}}(x'X'Q_FY).$$

Then $\mathbb{E}(Z) = \sum_{F \in \mathcal{F}_e} c_F x'\tau$ so we need $\sum c_F = 1$ for an unbiased estimator for $x'\tau$.

Lemma 4.3 shows that

$$\operatorname{Var} Z = \operatorname{Var}\left(\left(\sum_{F \in \mathcal{F}_e} \frac{c_F}{r\varepsilon_{Fe}} x'X'Q_F\right) Y\right)$$

strata for Ω		strata for Θ 0 (1)	rows (3)	columns (3)	Latin (3)	Greek (3)	E (3)
U	(1)	1	0	0	0	0	0
months	(3)	0	$\frac{1}{9}$	0	0	0	0
houses	(2)	0	0	0	0	0	0
house-months	(6)	0	$\frac{8}{9}$	0	0	0	0
sheep	(9)	0	0	$\frac{1}{3}$	$\frac{1}{3}$	$\frac{1}{3}$	0
E	(27)	0	0	$\frac{2}{3}$	$\frac{2}{3}$	$\frac{2}{3}$	1

Table 7.4. *Canonical efficiency factors in Example 7.5*

$$
\begin{aligned}
&= \left(\sum_{F\in\mathcal{F}_e} \frac{c_F}{r\varepsilon_{Fe}} x'X'Q_F\right)\left(\sum_{H\in\mathcal{F}} \xi_H Q_H\right)\left(\sum_{G\in\mathcal{F}_e} \frac{c_G}{r\varepsilon_{Ge}} Q_G X x\right)\\
&= \sum_{F\in\mathcal{F}_e} \frac{c_F^2}{(r\varepsilon_{Fe})^2} x'L_F x \xi_F\\
&= \sum_{F\in\mathcal{F}_e} c_F^2 \frac{\xi_F}{r\varepsilon_{Fe}} x'x.
\end{aligned}
$$

For fixed $\sum c_F$, this variance is minimized when c_F is proportional to $r\varepsilon_{Fe}/\xi_F$, so we combine the separate estimators for $x'\tau$ into the estimator

$$
\sum_{F\in\mathcal{F}_e} \left(\frac{r\varepsilon_{Fe}}{\hat{\xi}_F}\right)\left(\frac{1}{r\varepsilon_{Fe}} x'X'Q_F Y\right) \Big/ \sum_{F\in\mathcal{F}_e} \frac{r\varepsilon_{Fe}}{\hat{\xi}_F}.
$$

This combined estimator requires estimates of $\hat{\xi}_F$ for F in $\mathcal{F}_e$.

For F in $\mathcal{F}$, put $\mathcal{E}_F = \{e \in \mathcal{E} : \varepsilon_{Fe} \neq 0\}$. Replacing $V_B^\perp$ in Section 4.5 by $W_F^{(\Omega)}$, we obtain the following analogue of Corollary 4.16.

Proposition 7.5 *Put* $d_F^* = d_F^{(\Omega)} - \sum_{e \in \mathcal{E}_F} d_e^{(\Theta)}$. *If* $d_F^* \neq 0$ *then*

$$\left(\langle Q_F Y, Q_F Y \rangle - \sum_{e \in \mathcal{E}_F} \frac{1}{r \varepsilon_{Fe}} \langle S_e X' Q_F Y, S_e X' Q_F Y \rangle \right) \Big/ d_F^*$$

is an unbiased estimator for ξ_F.

We use Proposition 7.5 to obtain estimates for the stratum variances and hence to obtain combined estimates for contrasts. What to do if d_F^* is zero is beyond the scope of this book.

Example 7.8 (Example 4.3 continued) As rewritten with respect to the rectangular association scheme in Example 5.10, the table of canonical efficiency factors is as shown in Table 7.5. Here B indicates the partition into blocks. In Example 4.3 we estimated the treatment effect $\tau(1) - \tau(2) - \tau(5) + \tau(6)$ as -325 with variance ξ_E, where $\hat{\xi}_E = 319$. This estimation used information from $W_E^{(\Omega)}$ only. Since $\chi_1 - \chi_2 - \chi_5 + \chi_6 \in W_E^{(\Theta)}$ and $\varepsilon_{BE} = 0$ there is no further information on this contrast in the blocks stratum $W_B^{(\Omega)}$.

strata for Ω		strata for Θ U (1)	R (1)	C (3)	E (3)
U	(1)	1	0	0	0
B	(7)	0	$\frac{1}{4}$	$\frac{1}{4}$	0
E	(24)	0	$\frac{3}{4}$	$\frac{3}{4}$	1

Table 7.5. *Canonical efficiency factors in Example 7.8*

To calculate $X'Q_BY$ we sum the entries of P_BY over each treatment and subtract four times the overall mean of Y. This is shown in Table 7.6. From this, $P_RX'Q_BY$ and $P_CX'Q_BY$ are calculated from row means and column means. Since $r\varepsilon_{BR} = r\varepsilon_{BC} = 1$, we calculate estimates of treatment effects directly from $P_RX'Q_BY$ and $P_CX'Q_BY$. For example, $\chi_1 - \chi_2 + \chi_5 - \chi_6 \in W_C^{(\Theta)}$ so we estimate $\tau(1) - \tau(2) + \tau(5) - \tau(6)$ as $-13.875 - 33.625 - 13.875 - 33.625 = -95$, with variance $4\xi_B/1$.

Θ	$X'Q_BY$	$P_RX'Q_BY$	$P_CX'Q_BY$
1	−10.125	3.750	−13.875
2	37.375	3.750	33.625
3	−18.125	3.750	−21.875
4	5.875	3.750	2.125
5	−17.625	−3.750	−13.875
6	29.875	−3.750	33.625
7	−25.625	−3.750	−21.875
8	−1.625	−3.750	2.125

Table 7.6. *Vectors in* $\mathbb{R}^\Theta$ *in Example 7.8*

Summing squares gives

vector	sum of squares	multiply by	
Q_BY	4498.969	1	4498.969
$P_RX'Q_BY/1$	112.500	−1	−112.500
$P_CX'Q_BY/1$	3612.375	−1	−3612.375
			774.094

so $\hat{\xi}_B = 774.094/(7-1-3) = 258$.

In Example 4.3 we estimated $\tau(1) - \tau(2) + \tau(5) - \tau(6)$ from Q_EY as -53 with estimated variance $4\hat{\xi}_E/3$, that is 425. The combined estimate of $\tau(1) - \tau(2) + \tau(5) - \tau(6)$ is

$$\frac{1032 \times -53 + 425 \times -95}{1032 + 425} = -65$$

with variance 301.

This example is atypical in one way. When blocks are well chosen then ξ_B should be bigger than ξ_E.

The Genstat output in Figure 4.2 shows the calculation of 258 as the estimate of ξ_B. The information summary in Figure 4.2 can be interpreted as part of the table of canonical efficiency factors shown in Table 7.5. Figure 7.14 shows the further output given by Genstat when combined estimates are requested. Here the combined estimate of $\tau(1) - \tau(2) + \tau(5) - \tau(6)$ is given as $106.2 - 300.2 + 451.9 - 320.3$, which is -62.4. This differs from our estimate of -65 because Genstat uses a slightly different method from the one described here. ■

```
***** Tables of combined means *****

Variate: yield

 fertiliz         1        2        3        4        5        6
     frow         1        1        1        1        2        2
     fcol         1        2        3        4        1        2
              106.2    300.2    417.9    349.6    451.9    320.3

 fertiliz         7        8
     frow         2        2
     fcol         3        4
              106.3    280.2


*** Standard errors of differences of combined means ***

Table              fertiliz
rep.                      4
s.e.d.                12.89
Except when comparing means with the same level(s) of
 frow                 12.83
 fcol                 12.78
```

Fig. 7.14. Part of Genstat output for Example 7.8

7.5 Special classes of design

7.5.1 Orthogonal designs

If ψ is a design on an orthogonal block structure $(\Omega, \mathcal{F})$ then ψ induces a partition T on Ω whose classes are the sets of plots which receive the same treatment. If ψ is equi-replicate then T is uniform. Let V_T be the partition subspace of $\mathbb{R}^\Theta$ defined by T. Lemma 4.7 shows that the projector P_T onto V_T is equal to $r^{-1}XX'$ when the design is equi-replicate.

Definition A design on an orthogonal block structure $(\Omega, \mathcal{F})$ with treatment partition T is *orthogonal* if T is orthogonal to F for all F in $\mathcal{F}$.

Thus

$$\begin{aligned} \text{the design is orthogonal} \quad &\iff \quad P_T P_F = P_F P_T \quad \text{for all } F \text{ in } \mathcal{F} \\ &\iff \quad P_T Q_F = Q_F P_T \quad \text{for all } F \text{ in } \mathcal{F}. \end{aligned}$$

Lemma 7.6 *Let X be the design matrix of a design which has equal replication r, and let $L = X'QX$, where Q is any orthogonal projector* *in* $\mathbb{R}^{\Omega\times\Omega}$. *Then* XX' *commutes with* Q *if and only if all the eigenvalues of* $r^{-1}L$ *are in* $\{0,1\}$.

Proof By assumption, $X'X = rI$. If XX' commutes with Q then $(r^{-1}L)^2 = r^{-2}X'QXX'QX = r^{-2}X'QQXX'X = r^{-2}X'QX(X'X) = r^{-2}L(rI) = r^{-1}L$, so $r^{-1}L$ is idempotent and all its eigenvalues are in $\{0,1\}$.

Conversely, suppose that all of the eigenvalues of $r^{-1}L$ are in $\{0,1\}$. Let $x \in \mathbb{R}^\Theta$. If $Lx = 0$ then $QXx = 0$, by Lemma 4.7. If $r^{-1}Lx = x$ then $\langle QXx, QXx\rangle = x'Lx = rx'x = x'(rI)x = x'X'Xx = \langle Xx, Xx\rangle$; since Q is a projector this shows that $QXx = Xx$. Therefore $QX = r^{-1}XL$. Hence $QXX' = r^{-1}XLX'$, which is symmetric, so $QXX' = (QXX')' = XX'Q$. ∎

Thus a partially balanced design for the orthogonal block structure $(\Omega, \mathcal{F})$ is orthogonal if every canonical efficiency factor is equal to either 0 or 1; that is, if $|\mathcal{F}_e| = 1$ for all e in $\mathcal{E}$.

In an orthogonal design each basic contrast is estimated in precisely one stratum, and there is no need to combine estimates. Moreover, if $\varepsilon_{Fe} = 1$ then the proof of Lemma 7.6 shows that $S_eX'Q_F = S_eX'$: thus the quantity $\langle S_eX'Q_FY, S_eX'Q_FY\rangle/(r\varepsilon_{Fe})$ in Proposition 7.5 can be simplified to $\langle S_eX'Y, S_eX'Y\rangle/r$.

Orthogonal designs are extremely common in practice, especially when there are constraints on how treatments may be allocated.

Example 7.9 In an experiment on beans, the treatments had structure $\underline{\underline{3}} \times \underline{\underline{2}} \times \underline{\underline{2}}$, being all combinations of three cultivars, two seeding rates, and presence or absence of molybdenum. The experimental area was divided into four blocks. Each block was divided into three strips in one direction: cultivars were applied to strips. In the perpendicular direction, each block was divided into two halves, which were further divided into two quarters: seeding rates were applied to halves, and molybdenum to one quarter per half.

The Hasse diagram for the orthogonal block structure on Ω is shown in Figure 7.15. Table 7.7 lists the strata $W_F^{(\Omega)}$. For each F the dimension $d_F^{(\Omega)}$ is shown, followed by those treatment strata $W_e^{(\Theta)}$ for which $\varepsilon_{Fe} = 1$ and their dimensions; d_F^* is calculated by subtraction. ∎

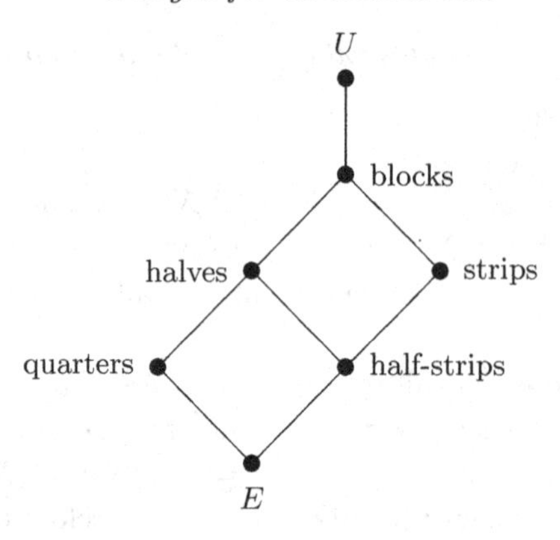

Fig. 7.15. The orthogonal block structure in Example 7.9

strata in $\mathbb{R}^\Omega$		strata in $\mathbb{R}^\Theta$		d_F^*
F	$\dim W_F^{(\Omega)}$	e	$\dim W_e^{(\Theta)}$	
U_Ω	1	U_Θ	1	0
blocks	3	–	–	3
strips	8	cultivars	2	6
halves	4	rates	1	3
half-strips	8	cultivars $\wedge$ rates	2	6
quarters	8	molybdenum	1	
		rates $\wedge$ molybdenum	1	6
E_Ω	16	cultivars $\wedge$ molybdenum	2	
		E_Θ	2	12

Table 7.7. *Allocation to strata in Example 7.9*

7.5.2 Balanced designs

Definition Suppose that $(\Omega, \mathcal{F}, \Theta, \psi)$ is a design on an orthogonal block structure. The design is *balanced* if it has overall partial balance with respect to the trivial association scheme $\underline{\underline{t}}$ on Θ.

Latin squares are both orthogonal and balanced.

Example 7.10 A balanced design for nine treatments in the orthogonal block structure $(\underline{\underline{4}} \times \underline{\underline{9}})/\underline{\underline{2}}$ is shown in Figure 7.16. It has

$$\Lambda_U = 64J$$

$$
\begin{aligned}
\Lambda_{\text{rows}} &= 4J \\
\Lambda_{\text{columns}} &= 8I + 7(J - I) \\
\Lambda_{\text{blocks}} &= 8I + (J - I) \\
\Lambda_E &= 8I. \quad \blacksquare
\end{aligned}
$$

1 8	2 0	3 1	4 2	5 3	6 4	7 5	8 6	0 7
2 7	3 8	4 0	5 1	6 2	7 3	8 4	0 5	1 6
3 6	4 7	5 8	6 0	7 1	8 2	0 3	1 4	2 5
4 5	5 6	6 7	7 8	8 0	0 1	1 2	2 3	3 4

Fig. 7.16. Balanced design for a row-column structure with split plots: see Example 7.10

Example 7.11 Figure 7.17 shows a balanced semi-Latin rectangle for seven treatments. ■

0 0	3 4	6 1	2 5	5 2	1 6	4 3
5 4	1 1	4 5	0 2	3 6	6 3	2 0
3 1	6 5	2 2	5 6	1 3	4 0	0 4
1 5	4 2	0 6	3 3	6 0	2 4	5 1
6 2	2 6	5 3	1 0	4 4	0 1	3 5
4 6	0 3	3 0	6 4	2 1	5 5	1 2
2 3	5 0	1 4	4 1	0 5	3 2	6 6

Fig. 7.17. Balanced semi-Latin rectangle: see Example 7.11

In a balanced design there are integers λ_{F0} and λ_{F1}, for F in $\mathcal{F}$, such that

$$\Lambda_F = \lambda_{F0} I_\Theta + \lambda_{F1}(J_\Theta - I_\Theta).$$

Equation (7.2) gives

$$rk_F = \lambda_{F0} + (t - 1)\lambda_{F1},$$

so

$$\begin{aligned}\Lambda_F &= (\lambda_{F0} - \lambda_{F1})\left(I_\Theta - \frac{1}{t}J_\Theta\right) + \frac{rk_F}{t}J_\Theta \\ &= (\lambda_{F0} - \lambda_{F1})\,S_1 + rk_F S_0.\end{aligned}$$

Therefore

$$\begin{aligned}L_U &= rS_0 \\ L_F &= \sum_{G\in\mathcal{F}} \mu(F,G)\frac{1}{k_G}(\lambda_{G0} - \lambda_{G1})\,S_1 \qquad \text{if } F \neq U.\end{aligned}$$

In other words,

$$\begin{aligned}\varepsilon_{F1} &= \frac{1}{r}\sum_{G\in\mathcal{F}} \mu(F,G)\frac{1}{k_G}(\lambda_{G0} - \lambda_{G1}) \\ &= \sum_{G\in\mathcal{F}} \mu(F,G)\left(1 - \frac{t\lambda_{G1}}{rk_G}\right).\end{aligned}$$

Theorem 7.7 (Generalization of Fisher's Inequality) *Let F be a partition in an orthogonal block structure upon which there is a balanced design for t treatments. If $F \neq U$ and $L_F \neq 0$ then $d_F^{(\Omega)} \geqslant t-1$.*

Proof If $F \neq U$ then $L_F = r\varepsilon_{F1}S_1$. If $L_F \neq 0$ then $\varepsilon_{F1} \neq 0$ so the proof of Lemma 4.14(i) gives a bijection from $W_1^{(\Theta)}$ to a subspace of $W_F^{(\Omega)}$. ∎

Corollary 7.8 *In a balanced design for t treatments in the row-column structure $\underline{\underline{n}} \times \underline{\underline{m}}$, either $t \leqslant n$ or t divides m.*

Proof The dimension of $W_{\text{rows}}^{(\Omega)}$ is $n-1$, so either $t-1 \leqslant n-1$ or $L_{\text{rows}} = 0$, which occurs if and only if all treatments occur equally often in each row. ∎

7.5.3 Cyclic designs

Definition A design $(\Omega, \mathcal{F}, \mathbb{Z}_t, \psi)$ is *weakly cyclic* if $(\Omega, \mathcal{G}_F, \mathbb{Z}_t, \psi)$ is cyclic for each F in $\mathcal{F}$. It is *strongly cyclic* if there is a permutation π of Ω which

(i) preserves the partition F, for all F in $\mathcal{F}$, in the sense that α and β are in the same F-class if and only if $\pi(\alpha)$ and $\pi(\beta)$ are in the same F-class, and
(ii) satisfies $\psi(\pi(\omega)) = \psi(\omega) + 1$ for all ω in Ω.

Thus the design is weakly cyclic if, for each F in $\mathcal{F}$, the F-classes can be constructed from one or more initial F-classes by successively adding 1 modulo t. Strongly cyclic designs are weakly cyclic, but there exist weakly cyclic designs that are not strongly cyclic. However, all weakly cyclic designs have overall partial balance with respect to the association scheme ⓣ.

Example 7.12 To construct a strongly cyclic nested row-column design for the structure $\underline{t}/(\underline{n}\times\underline{m})$, allocate elements of $\mathbb{Z}_t$ to one $n\times m$ rectangle and then translate this rectangle as in Section 5.4. An example with $n = m = 2$ and $t = 6$ is shown in Figure 7.18. ■

0	2		1	3		2	4		3	5		4	0		5	1
4	1		5	2		0	3		1	4		2	5		3	0

Fig. 7.18. Strongly cyclic nested row-column design in Example 7.12

To construct a strongly cyclic design for t treatments in either of the structures $\underline{n}\times\underline{t}$ or $(\underline{n}\times\underline{t})/\underline{m}$, allocate treatments to the first column and form the other columns by translation. Example 7.10 was constructed in this way.

Example 7.13 Figure 7.19 shows a design for treatment set $\mathbb{Z}_8$ in the orthogonal block structure $(\underline{4}\times\underline{4})/\underline{2}$. Every row and every column contains each treatment once, and the design in blocks is generated cyclically from the initial blocks $\{0,1\}$ and $\{0,3\}$. Thus the design is weakly cyclic.

Suppose that the design ψ is strongly cyclic under a permutation π with $\psi(\pi(\omega)) = \psi(\omega)+1$. Then π must take the block with $\{1,4\}$ to the block with $\{2,5\}$, so it fixes the last row. However, π must also take the

0 1	2 7	4 3	6 5
2 3	0 5	6 1	4 7
4 5	6 3	0 7	2 1
6 7	4 1	2 5	0 3

Fig. 7.19. Cyclic row-column design with split plots in Example 7.13

block with $\{6,7\}$ to the block with $\{7,0\}$, so it moves the last row to the third row. This contradiction shows that there can be no such π.

In fact, there is not even a way of relabelling the treatments to make the design strongly cyclic. The pairs of treatments in the blocks are all different, so any relabelling of the treatments must produce at least one block δ which contains 0 and an odd number. This block has eight distinct translates, which must be equally distributed among some rows permuted in a cycle by π. Either these translates form all the blocks in two rows interchanged by π, or π permutes all the rows in a 4-cycle. In either case π^4 fixes every row. Similarly, π^4 fixes every column. Therefore π^4 must fix every block. However, $\psi(\pi^4(\omega)) = \psi(\omega) + 4$, so π^4 does not fix δ. ■

Let Υ_1 and Υ_2 be subsets of $\mathbb{Z}_t$. Suppose that, for $\{i,j\} = \{1,2\}$, the translates

$$\Upsilon_i + \upsilon \qquad \text{for } \upsilon \text{ in } \Upsilon_j$$

consist of all distinct translates of Υ_i equally often. Then the following construction gives a weakly cyclic row-column design: label the rows by Υ_1 and the columns by Υ_2, and put treatment $\upsilon_1 + \upsilon_2$ into the cell in row υ_1 and column υ_2.

Example 7.14 In $\mathbb{Z}_{10}$ we may take $\Upsilon_1 = \{0,5,1,6,2,7\}$ and $\Upsilon_2 = \{0,2,4,6,8\}$. This gives the design in Figure 7.20, which is strongly cyclic. ■

0	2	4	6	8
5	7	9	1	3
1	3	5	7	9
6	8	0	2	4
2	4	6	8	0
7	9	1	3	5

Fig. 7.20. Cyclic row-column design in Example 7.14

Theorem 7.9 *In a cyclic design on an orthogonal block structure the canonical efficiency factors in stratum* $W_F^{(\Omega)}$ *are*

$$\sum_{G\in\mathcal{F}} \mu(F,G)\frac{1}{rk_G}\sum_{\theta\in\mathbb{Z}_t} \lambda_{G\theta}\epsilon^{\theta}$$

for complex t-th roots of unity ϵ.

Proof Use Theorems 7.4 and 2.18. ■

Example 7.12 revisited The first row of each concurrence matrix is shown below. On the right are calculated the eigenvalues of each concurrence matrix on the treatment strata defined by ϵ_6^0, $\epsilon_6^{\pm 1}$, $\epsilon_6^{\pm 2}$ and ϵ_6^3, where ϵ_6 is a primitive sixth root of unity (compare with Example 2.6).

F	0	1	2	3	4	5	ϵ_6^0	$\epsilon_6^{\pm 1}$	$\epsilon_6^{\pm 2}$	ϵ_6^3	rk_F
U	16	16	16	16	16	16	96	0	0	0	96
squares	4	2	3	2	3	2	16	1	1	4	16
rows	4	0	1	2	1	0	8	1	5	4	8
columns	4	1	1	0	1	1	8	4	2	4	8
E	4	0	0	0	0	0	4	4	4	4	4

Subtraction gives the canonical efficiency factors in Table 7.8. ■

strata for Ω		strata for Θ			
		ϵ_6^0	$\epsilon_6^{\pm 1}$	$\epsilon_6^{\pm 2}$	ϵ_6^3
		(1)	(2)	(2)	(1)
U	(1)	1	0	0	0
squares	(5)	0	$\frac{1}{16}$	$\frac{1}{16}$	$\frac{1}{4}$
rows	(6)	0	$\frac{1}{16}$	$\frac{9}{16}$	$\frac{1}{4}$
columns	(6)	0	$\frac{7}{16}$	$\frac{3}{16}$	$\frac{1}{4}$
E	(6)	0	$\frac{7}{16}$	$\frac{3}{16}$	$\frac{1}{4}$

Table 7.8. *Canonical efficiency factors in Example 7.12*

Example 7.15 Contrast the two cyclic designs for 13 treatments in the nested row-column structure $\underline{\underline{13}}/(\underline{\underline{2}} \times \underline{\underline{3}})$ whose initial blocks are shown in Figure 7.21.

Design ψ_1 is partially balanced with respect to the cyclic association scheme $\mathcal{Q}$ defined by the blueprint in Example 1.8. Putting

$$A_1(\theta, \eta) = 1 \quad \text{if } \theta - \eta \in \{\pm 1, \pm 3, \pm 4\},$$

0	1	4
4	0	1

initial block of ψ_1

1	2	4
5	8	3

initial block of ψ_2

Fig. 7.21. Initial blocks of two cyclic designs in Example 7.15

we find that

$$\begin{aligned}\Lambda_{\text{blocks}} &= 12I + 4A_1\\ \Lambda_{\text{rows}} &= 6I + 2A_1\\ \Lambda_{\text{columns}} &= 6I + A_1.\end{aligned}$$

Exercise 2.16 shows that $\mathcal{Q}$ has two 6-dimensional strata and that

$$A_1 = 6S_0 + \left(\frac{-1+\sqrt{13}}{2}\right) S_1 + \left(\frac{-1-\sqrt{13}}{2}\right) S_2.$$

Hence we can calculate the canonical efficiency factors, which are shown in Table 7.9.

strata for Ω		strata for Θ 0 (1)	1 (6)	2 (6)
U	(1)	1	0	0
blocks	(12)	0	$\frac{5+\sqrt{13}}{18} \approx 0.478$	$\frac{5-\sqrt{13}}{18} \approx 0.077$
rows	(13)	0	0	0
columns	(26)	0	$\frac{13-\sqrt{13}}{72} \approx 0.130$	$\frac{13+\sqrt{13}}{72} \approx 0.231$
E	(26)	0	$\frac{13-\sqrt{13}}{24} \approx 0.391$	$\frac{13+\sqrt{13}}{24} \approx 0.692$

Table 7.9. *Canonical efficiency factors for design ψ_1 in Example 7.15*

In design ψ_2, the first rows of the concurrence matrices are as follows.

	0	1	2	3	4	5	6	7	8	9	10	11	12
U	36	36	36	36	36	36	36	36	36	36	36	36	36
blocks	6	4	3	3	2	1	2	2	1	2	3	3	4
rows	6	1	2	2	0	1	0	0	1	0	2	2	1
columns	6	1	0	0	1	0	1	1	0	1	0	0	1
E	6	0	0	0	0	0	0	0	0	0	0	0	0

Hence

$$\begin{aligned} L_E &= 6I - \frac{1}{2}\Lambda_{\text{columns}} - \frac{1}{3}\Lambda_{\text{rows}} + \frac{1}{6}\Lambda_{\text{blocks}} \\ &= \frac{13}{6}\left(I - \frac{1}{13}J\right) \end{aligned}$$

so the design is balanced in the bottom stratum with canonical efficiency factor $13/36 \approx 0.361$.

However, in design ψ_1 the canonical efficiency factors in the bottom stratum are approximately equal to 0.391 and 0.692, with harmonic mean 1/2. Thus if the fixed-effects model is assumed then ψ_1 is better than ψ_2 for every contrast, and considerably better overall, with relative efficiency 18/13. It is somewhat counter-intuitive that the design which is balanced in the bottom stratum is worse, but this can be explained by noting that in ψ_1 there is no information in the rows stratum and rather little in the columns stratum.

strata for Ω		strata for Θ						
		ϵ_{13}^{0}	$\epsilon_{13}^{\pm 1}$	$\epsilon_{13}^{\pm 2}$	$\epsilon_{13}^{\pm 3}$	$\epsilon_{13}^{\pm 4}$	$\epsilon_{13}^{\pm 5}$	$\epsilon_{13}^{\pm 6}$
		(1)	(2)	(2)	(2)	(2)	(2)	(2)
U	(1)	1	0	0	0	0	0	0
blocks	(12)	0	0.289	0.094	0.019	0.133	0.017	0.031
rows	(13)	0	0.212	0.021	0.096	0.083	0.485	0.186
columns	(26)	0	0.137	0.524	0.524	0.422	0.137	0.422
E	(26)	0	0.361	0.361	0.361	0.361	0.361	0.361

Table 7.10. *Canonical efficiency factors for design ψ_2 in Example 7.15*

The picture changes if the random-effects model is assumed. Table 7.10 shows all the canonical efficiency factors for design ψ_2, which are calculated using a primitive 13-th root ϵ_{13} of unity and given to three decimal places. In typical use of a nested row-column design ξ_{blocks} would be so much bigger than ξ_E that there would be little point in including the estimates from the blocks stratum, but ξ_{rows} and ξ_{columns} might be only two or three times as big as ξ_E. Thus design ψ_2, which has little information in the blocks stratum, is preferable to ψ_1. ■

Example 7.15 illustrates a general difficulty about optimality for complicated block structures. In a block design the variance of the combined estimator is

$$\frac{\xi_E \xi_{\text{blocks}}}{r(\varepsilon_{Ee}\xi_{\text{blocks}} + (1-\varepsilon_{Ee})\xi_E)}$$

if we know the ratio $\xi_{\text{blocks}} :: \xi_E$ in advance. Since we expect ξ_{blocks} to be bigger than ξ_E, a design which is good for the fixed-effects model (having high values of ε_{Ee}) is likely to be good for the random-effects model also. However, with more complicated structures such as the nested row-column design this is no longer true: a design that is optimal for one model may be seriously worse than optimal for the other model.

7.5.4 Lattice squares

Lattice squares are designs for n^2 treatments in the nested row-column structure $\underline{\underline{r}}/(\underline{\underline{n}} \times \underline{\underline{n}})$. The association scheme on the treatments is square of type $\mathrm{S}(c, n)$, derived from $c-2$ mutually orthogonal $n \times n$ Latin squares, where $c \geqslant 3$. There are c non-trivial partitions into n classes of size n; if $F_i \neq F_j$ then $F_i \wedge F_j = E$ and $F_i \vee F_j = U$. Together with E and U these form an orthogonal block structure $\mathcal{L}$, which defines the association scheme.

For each block, choose any two distinct non-trivial partitions in $\mathcal{L}$ and use these to make the rows and columns of the block. For $i = 1, \ldots, c$ let b_{1i} and b_{2i} be the numbers of blocks in which F_i is used for the rows and columns respectively. Thus $\sum b_{1i} = \sum b_{2i} = r$ and $b_{1i} + b_{2i} \leqslant r$ for each i. Now

$$
\begin{aligned}
\frac{1}{rn^2}\Lambda_U &= \frac{r}{n^2}J = rS_0 \\
\frac{1}{n^2}\Lambda_{\text{blocks}} &= \frac{r}{n^2}J = rS_0 \\
\frac{1}{n}\Lambda_{\text{rows}} &= \frac{1}{n}\left(rI + \sum b_{1i}A_i\right) = \sum b_{1i}(S_0 + S_i) = rS_0 + \sum b_{1i}S_i
\end{aligned}
$$

$$\frac{1}{n}\Lambda_{\text{columns}} = \frac{1}{n}\left(rI + \sum b_{2i}A_i\right) = \sum b_{2i}(S_0 + S_i) = rS_0 + \sum b_{2i}S_i$$

$$\Lambda_E \qquad = rI$$

and so $\varepsilon_{\text{rows},i} = b_{1i}/r$, $\varepsilon_{\text{columns},i} = b_{2i}/r$ and $\varepsilon_{Ei} = (r - b_{1i} - b_{2i})/r$.

7.6 Valid randomization

Some statisticians think that the fixed-effects model and the random-effects model are both too specific. Instead, they assume that

$$Y = X_\psi \tau + Z \tag{7.9}$$

where Z is an unknown random vector with finite expectation and variance. Here we write X_ψ to indicate that the design matrix depends on the design ψ. At first sight it seems hopeless to estimate anything from the vague model in Equation (7.9). The solution is to *randomize*; that is, choose ψ at random from a set Ψ of designs. Then the matrix X_ψ changes with ψ but the stratum projectors Q_F do not. We assume that the random choice of ψ is independent of the distribution of Z.

Definition Randomization by random choice from Ψ is *valid* for a given orthogonal block structure if

(i) the estimators for $x'\tau$ given in Theorem 7.2 are unbiased;
(ii) the variance of each of those estimators is estimated in an unbiased fashion by the estimators implied in Proposition 7.5;
(iii) the covariance between estimators from different strata in $\mathbb{R}^\Omega$ is zero.

Here all expectations are taken over both the distribution of Z and the random choice from Ψ.

What is estimable in stratum $W_F^{(\Omega)}$, and with what variance, depends on the information matrix L_F. Thus it is desirable that Ψ satisfy the following condition:

for each F in $\mathcal{F}$, there is a matrix L_F in $\mathbb{R}^{\Theta\times\Theta}$ such that $X'_\psi Q_F X_\psi = L_F$ for all ψ in Ψ. (7.10)

Equivalently, the concurrence matrices Λ_F do not depend on ψ, or the set of matrices $X'_\psi B_F X_\psi$ do not depend on ψ. In particular, since

$\Lambda_E = \text{diag}(r)$, the replication r_θ of treatment θ is the same for all ψ in Ψ.

If $\psi(\omega) = \theta$ for all ψ in Ψ then $\widehat{\tau(\theta)}$ will always involve the unknown Z_ω and there is no hope of an unbiased estimator. We need the collection of designs in Ψ to be sufficiently general to ensure that the treatments are, in some sense, randomly allocated to plots. The estimators of variance in Proposition 7.5 involve products $Y(\alpha)Y(\beta)$, so validity actually requires a condition on *pairs* of plots. It turns out that a satisfactory condition is the following:

> for each F in $\mathcal{F}$ and each pair (θ, η) in $\Theta \times \Theta$ there is an integer $q(F; \theta, \eta)$ such that, for all pairs (α, β) in $\mathcal{C}_F^{(\Omega)}$, there are exactly $q(F; \theta, \eta)$ designs ψ in Ψ for which $\psi(\alpha) = \theta$ and $\psi(\beta) = \eta$. (7.11)

Suppose that condition (7.11) holds. Then $q(E; \theta, \eta) = 0$ if $\theta \neq \eta$ and $q(E; \theta, \theta)$ is equal to the number of designs ψ in Ψ for which $\psi(\alpha) = \theta$, irrespective of α. If condition (7.10) also holds then a counting argument shows that

$$|\Psi|\, r_\theta = |\Omega|\, q(E; \theta, \theta).$$

If condition (7.10) holds and any one design in Ψ has overall partial balance with respect to $\mathcal{Q}$ then every design in Ψ has overall partial balance with respect to $\mathcal{Q}$. If condition (7.11) also holds then $q(F; \theta_1, \eta_1) = q(F; \theta_2, \eta_2)$ whenever (θ_1, η_1) and (θ_2, η_2) are in the same associate class of $\mathcal{Q}$.

Lemma 7.10 *Let M be a matrix in $\mathbb{R}^{\Theta \times \Theta}$ and let F be in $\mathcal{F}$. If condition (7.11) holds then there is a scalar $\kappa(M)$ such that*

$$\sum_{\psi \in \Psi} X_\psi M X_\psi' Q_F = |\Psi|\, \kappa(M) Q_F.$$

Proof Put $\tilde{M} = \sum_{\psi \in \Psi} X_\psi M X_\psi'$. Condition (7.11) implies that

$$\tilde{M}(\alpha, \beta) = \sum_{\theta \in \Theta} \sum_{\eta \in \Theta} q(F; \theta, \eta) M(\theta, \eta)$$

if $(\alpha, \beta) \in \mathcal{C}_F^{(\Omega)}$. Hence

$$\tilde{M} = \sum_{F \in \mathcal{F}} \sum_{\theta \in \Theta} \sum_{\eta \in \Theta} q(F; \theta, \eta) M(\theta, \eta) B_F,$$

and so $\tilde{M}$ is also a linear combination of $\{Q_G : G \in \mathcal{F}\}$. Therefore $\tilde{M} Q_F$ is a scalar multiple of Q_F. ■

Theorem 7.11 *Let Ψ be a set of designs for treatment set Θ in the orthogonal block structure $(\Omega, \mathcal{F})$. If Ψ satisfies both conditions (7.10) and (7.11) then randomization by random choice from Ψ is valid.*

Proof Write $\mathbb{E}_1$ for expectation over the distribution of Z and $\mathbb{E}_\Psi$ for expectation over the random choice from Ψ. Let $F \in \mathcal{F} \setminus \{U\}$.

(i) First we consider estimation of contrasts in stratum $W_F^{(\Omega)}$. Let x be in $\mathbb{R}^\Theta$. Suppose that there is a vector z in $\mathbb{R}^\Theta$ such that $L_F z = x$. Then

$$\begin{aligned} z'X'_\psi Q_F Y &= z'X'_\psi Q_F X_\psi \tau + z'X'_\psi Q_F Z \\ &= z'L_F\tau + z'X'_\psi Q_F Z, \qquad \text{by condition (7.10)}, \\ &= x'\tau + z'X'_\psi Q_F Z. \end{aligned}$$

Therefore

$$\mathbb{E}(z'X'_\psi Q_F Y) = x'\tau + z'\,\mathbb{E}_\Psi(X'_\psi) Q_F\, \mathbb{E}_1(Z)$$

by the independence of Z and the choice of ψ. Condition (7.11) shows that the θ-row of $\mathbb{E}_\Psi(X'_\psi)$ is equal to $(q(E;\theta,\theta)/|\Psi|)u'$, where u is the all-1 vector in $\mathbb{R}^\Omega$. Since $F \neq U$, $u'Q_F = 0$ and so $z'\,\mathbb{E}_\Psi(X'_\psi)Q_F\,\mathbb{E}_1(Z) = 0$. Thus the estimator $z'X'_\psi Q_F Y$ is unbiased for $x'\tau$.

(ii) Secondly we consider the variance of the estimator in part (i). Since $\mathbb{E}(z'X'_\psi Q_F Y) = x'\tau$,

$$\begin{aligned} \mathrm{Var}(z'X'_\psi Q_F Y) &= \mathbb{E}((z'X'_\psi Q_F Y - x'\tau)^2) \\ &= \mathbb{E}(Z'Q_F X_\psi z z' X'_\psi Q_F Z). \end{aligned}$$

Lemma 7.10 shows that

$$\mathbb{E}_\Psi(Q_F X_\psi z z' X'_\psi Q_F) = \kappa(zz')Q_F.$$

Thus

$$\begin{aligned} \kappa(zz')d_F^{(\Omega)} &= \mathrm{tr}(\kappa(zz')Q_F) \\ &= \mathbb{E}_\Psi(\mathrm{tr}(Q_F X_\psi z z' X'_\psi Q_F)) \\ &= \mathbb{E}_\Psi(z'X'_\psi Q_F Q_F X_\psi z) \\ &= \mathbb{E}_\Psi(z'L_F z) \\ &= z'L_F z. \end{aligned}$$

Therefore

$$\begin{aligned}\mathrm{Var}(z'X'_\psi Q_F Y) &= \mathbb{E}_1(Z'\kappa(zz')Q_F Z)\\ &= \frac{z'L_F z}{d_F^{(\Omega)}}\mathbb{E}_1(Z'Q_F Z).\end{aligned}$$

However, we estimate this variance by $z'L_F z\hat{\xi}_F$, so we need to show that $\mathbb{E}(\hat{\xi}_F) = \mathbb{E}_1(Z'Q_F Z)/d_F^{(\Omega)}$.

According to Proposition 7.5,

$$\hat{\xi}_F = \frac{1}{d_F^*}\left(Y'Q_F Y - \sum_{e\in\mathcal{E}_F}\frac{1}{r\varepsilon_{Fe}}Y'Q_F X_\psi S_e X'_\psi Q_F Y\right),$$

where $d_F^* = d_F^{(\Omega)} - \sum_{e\in\mathcal{E}_F} d_e^{(\Theta)}$. Now,

$$Y'Q_F Y = \tau'X'_\psi Q_F X_\psi\tau + Z'Q_F X_\psi\tau + \tau'X'_\psi Q_F Z + Z'Q_F Z$$

so

$$\mathbb{E}_\Psi(Y'Q_F Y) = \tau'L_F\tau + Z'Q_F Z$$

and so

$$\mathbb{E}(Y'Q_F Y) = \tau'L_F\tau + \mathbb{E}_1(Z'Q_F Z).$$

Suppose that $e\in\mathcal{E}_F$. Then

$$\begin{aligned}&Y'Q_F X_\psi S_e X'_\psi Q_F Y\\ &\quad= \tau'X'_\psi Q_F X_\psi S_e X'_\psi Q_F X_\psi\tau + \tau'X'_\psi Q_F X_\psi S_e X'_\psi Q_F Z\\ &\qquad + Z'Q_F X_\psi S_e X'_\psi Q_F X_\psi\tau + Z'Q_F X_\psi S_e X'_\psi Q_F Z\\ &\quad= \tau'L_F S_e L_F\tau + \tau'L_F S_e X'_\psi Q_F Z\\ &\qquad + Z'Q_F X_\psi S_e L_F\tau + Z'Q_F X_\psi S_e X'_\psi Q_F Z\end{aligned}$$

so

$$\begin{aligned}&\mathbb{E}_\Psi(Y'Q_F X_\psi S_e X'_\psi Q_F Y)\\ &\quad= r\varepsilon_{Fe}\tau'L_F S_e\tau + \mathbb{E}_\Psi(Z'Q_F X_\psi S_e X'_\psi Q_F Z).\end{aligned}$$

Lemma 7.10 shows that

$$\mathbb{E}_\Psi(Q_F X_\psi S_e X'_\psi Q_F) = \kappa(S_e)Q_F.$$

By Lemma 4.14, $(r\varepsilon_{Fe})^{-1}Q_F X_\psi S_e X'_\psi Q_F$ is the orthogonal projector onto $\mathrm{Im}\, Q_F X_\psi S_e$, which has dimension $d_e^{(\Theta)}$. Therefore

$$\kappa(S_e)d_F^{(\Omega)} = \mathrm{tr}(\kappa(S_e)Q_F)$$

$$= \mathbb{E}_\Psi(\mathrm{tr}(Q_F X_\psi S_e X'_\psi Q_F))$$
$$= r\varepsilon_{Fe} d_e^{(\Theta)}.$$

Consequently,

$$\mathbb{E}_\Psi(Z'Q_F X_\psi S_e X'_\psi Q_F Z) = \frac{r\varepsilon_{Fe} d_e^{(\Theta)}}{d_F^{(\Omega)}} Z'Q_F Z$$

and so

$$\mathbb{E}\left(\frac{1}{r\varepsilon_{Fe}} Y'Q_F X_\psi S_e X'_\psi Q_F Y\right) = \tau' L_F S_e \tau + \frac{d_e^{(\Theta)}}{d_F^{(\Omega)}} \mathbb{E}_1(Z'Q_F Z).$$

Now, $\sum_{e\in\mathcal{E}_F} L_F S_e = L_F$ and $\sum_{e\in\mathcal{E}_F} d_e^{(\Theta)} = d_F^{(\Omega)} - d_F^*$, so $\mathbb{E}(\hat{\xi}_F) = \mathbb{E}_1(Z'Q_F Z)/d_F^{(\Omega)}$.

(iii) Estimators from different strata have the form $z'X'_\psi Q_F Y$ and $w'X'_\psi Q_G Y$ for some vectors z and w in $\mathbb{R}^\Theta$ and distinct partitions F and G in $\mathcal{F}$. From part (i),

$$\begin{aligned} \mathrm{cov}(z'X'_\psi Q_F Y, w'X'_\psi Q_G Y) &= \mathbb{E}((z'X'_\psi Q_F Z)(w'X'_\psi Q_G Z)) \\ &= \mathbb{E}(Z'Q_G X_\psi w z' X'_\psi Q_F Z). \end{aligned}$$

Lemma 7.10 shows that $\mathbb{E}_\Psi(X_\psi wz'X'_\psi)Q_F = \kappa(wz')Q_F$. Since $Q_G Q_F = 0$ we have $\mathbb{E}_\Psi(Z'Q_G X_\psi wz'X'_\psi Q_F Z) = 0$. ■

Note that Theorem 7.11 does not require that the designs in Ψ be partially balanced, or even equi-replicate.

Example 7.16 Suppose that $(\Omega, \mathcal{F}) = \underline{\underline{n}} \times \underline{\underline{n}}$ and we want ψ to be a Latin square. Then each of the following is a valid method of randomization.

- Choose at random from among all $n \times n$ Latin squares.
- Fix one Latin square and randomly permute its rows and columns independently.
- Choose at random one Latin square from a set of $n-1$ mutually orthogonal $n \times n$ Latin squares, then randomly permute its letters. ■

Example 7.17 The following randomization is valid for all nested row-column designs. Randomly permute the blocks. Within each block, randomly permute rows and randomly permute columns. ■

7.7 Designs on association schemes

7.7.1 General theory

In the majority of experiments, the structure on the experimental units is an orthogonal block structure. However, there are some experiments with other types of association scheme on the plots. The structure of a field trial is typically $\underline{\underline{n}} \times \underline{\underline{m}}$; however, if $n = m$ and the plots are square and neither direction corresponds to a potential influence such as prevailing wind or direction of ploughing, then the association scheme $\mathrm{L}(2, n)$ may be appropriate. The triangular scheme is appropriate for experiments where the experimental units are unordered pairs: pairs of people sharing a task, or hybrid plants when the gender of the parent is immaterial. For ordered pairs, $\mathrm{Pair}(n)$ is appropriate: for example, if the two people play different roles, such as sending and receiving a message, or if the gender of the parents of the hybrid plants does matter. The association scheme $\underline{\underline{m}}/\textcircled{n}$ occurs in serology and marine biology, when treatments are arranged around the circumference of m petri dishes or cylindrical tanks.

In this section we extend the previous results in this chapter to general association schemes, as far as possible. Let $\mathcal{P}$ be the association scheme on Ω. For convenience, let its adjacency matrices be B_F for F in some index set $\mathcal{F}$, and its stratum projectors Q_G for G in some index set $\mathcal{G}$. Since there may be no relation matrices for $\mathcal{P}$, we cannot define a concurrence matrix. However, when $\mathcal{P}$ is the association scheme defined by an orthogonal block structure then $\mathrm{span}\{X'R_FX : F \in \mathcal{F}\} = \mathrm{span}\{X'B_FX : F \in \mathcal{F}\}$ and so we may define partial balance directly in terms of the matrices $X'B_FX$.

Definition Let $\mathcal{P}$ be an association scheme on a set Ω with adjacency matrices B_F for F in $\mathcal{F}$, and let $\mathcal{Q}$ be an association scheme on a set Θ with associate classes $\mathcal{C}_i^{(\Theta)}$ for i in $\mathcal{K}$. Let ψ be a function from Ω to Θ with design matrix X. The design ψ is *partially balanced* for $\mathcal{P}$ with respect to $\mathcal{Q}$ if $X'B_FX$ is in the Bose–Mesner algebra of $\mathcal{Q}$ for all F in $\mathcal{F}$; that is, if there are integers ν_{Fi} for (F, i) in $\mathcal{F} \times \mathcal{K}$ such that

$$\left|\left\{(\alpha, \beta) \in \mathcal{C}_F^{(\Omega)} : \psi(\alpha) = \theta,\ \psi(\beta) = \eta\right\}\right| = \nu_{Fi}$$

whenever $(\theta, \eta) \in \mathcal{C}_i^{(\Theta)}$. The design is *balanced* for $\mathcal{P}$ if it is partially balanced for $\mathcal{P}$ with respect to the trivial scheme on Θ.

If $\mathcal{P}$ is trivial then every equi-replicate design is balanced for $\mathcal{P}$.

Example 7.18 Let $(\Omega, \mathcal{P})$ be ⑥ and $(\Theta, \mathcal{Q})$ be $\underline{\underline{3}}$. The design in Figure 7.22 is balanced with the values of ν_{Fi} shown in the table. ■

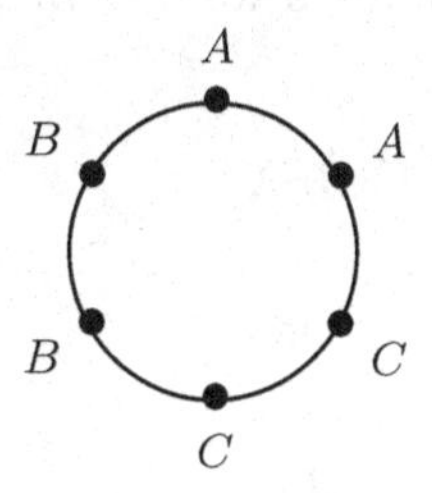

	equal	different
0	2	0
1	2	1
2	0	2
3	0	1

Fig. 7.22. Balanced design on the 6-circuit

Example 7.19 Let $(\Omega, \mathcal{P})$ be $\mathrm{L}(2,4)$ and $(\Theta, \mathcal{Q})$ be $\mathrm{GD}(4,2)$ with groups $A, B \parallel C, D \parallel E, F \parallel G, H$. The design in Figure 7.23 is partially balanced with the values of ν_{Fi} shown in the table. ■

A	C	E	G
C	B	G	F
F	H	A	D
H	E	D	B

	equal	same group	other
0	2	0	0
1	0	0	2
2	2	4	2

Fig. 7.23. Design partially balanced for $\mathrm{L}(2,4)$ with respect to $\mathrm{GD}(4,2)$

Example 7.20 If the structure in Figure 7.12 is regarded as $\mathrm{L}(2,6)$ then the design shown there is balanced. ■

The lattice squares in Section 7.5.4 are partially balanced for $\underline{\underline{r}}/\,\mathrm{L}(2,n)$ with respect to $\mathrm{S}(c,n)$.

If there are no partition subspaces of $\mathbb{R}^\Omega$ then there is no obvious way to define the fixed-effects model in general. However, the random-effects model can be defined by Equation (7.5) and

$$\mathrm{Cov}(Y) = \sum_{G \in \mathcal{G}} \xi_G Q_G.$$

The only difference between this and Equation (7.7) is that we can no longer use the same set to index both the adjacency matrices and the strata of $\mathcal{P}$. The information matrix L_G in stratum $W_G^{(\Omega)}$ is defined, as before, by $L_G = X'Q_GX$. Estimation, variances, canonical efficiency factors and combined estimators all work just as they did in Section 7.4, except that formulas for the canonical efficiency factors involve the explicit inverse $\tilde{D}$ of the character table for $\mathcal{P}$. In place of Equation (7.8) and Theorem 7.4 we have

$$\begin{aligned} L_G &= X'Q_GX = X' \sum_{F\in\mathcal{F}} \tilde{D}(G,F)B_FX \\ &= \sum_{F\in\mathcal{F}} \tilde{D}(G,F) \sum_{i\in\mathcal{K}} \nu_{Fi}A_i \\ &= \sum_{F\in\mathcal{F}} \tilde{D}(G,F) \sum_{i\in\mathcal{K}} \nu_{Fi} \sum_{e\in\mathcal{E}} C(i,e)S_e, \end{aligned}$$

so the canonical efficiency factor ε_{Ge} for treatment stratum $W_e^{(\Theta)}$ in plot stratum $W_G^{(\Omega)}$ is

$$\frac{1}{r} \sum_{F\in\mathcal{F}} \tilde{D}(G,F) \sum_{i\in\mathcal{K}} \nu_{Fi}C(i,e).$$

Moreover,

$$\sum_{e\in\mathcal{E}} d_e\varepsilon_{Ge} = \frac{1}{r} \operatorname{tr} L_G = \frac{t}{r} \sum_{F\in\mathcal{F}} \tilde{D}(G,F)\nu_{F0}.$$

Note that $X'Q_0X = rS_0$ always, because

$$X'Q_0X = \frac{1}{|\Omega|} X'J_\Omega X = \frac{r^2}{|\Omega|} J_\Theta = \frac{r^2t}{rt} S_0 = rS_0.$$

If $\mathcal{P}$ is a general association scheme on Ω then we cannot define an orthogonal design in terms of partitions. However, Lemma 7.6 shows that if we define a design to be orthogonal if all the canonical efficiency factors are in $\{0,1\}$ then this definition does specialize to the one given in Section 7.5.1 if $\mathcal{P}$ is defined by an orthogonal block structure. Thus a design is orthogonal if there is a function $e \mapsto e'$ from $\mathcal{E}$ to $\mathcal{G}$ such that $\varepsilon_{Ge} = 1$ if $G = e'$ and all other canonical efficiency factors are zero. Equivalent conditions are that the treatment subspace V_T is geometrically orthogonal to every plot stratum $W_G^{(\Omega)}$; and that the treatment projector P_T commutes with every adjacency matrix B_F.

For a balanced design, Fisher's inequality still holds: that is, for each plot stratum $W_G^{(\Omega)}$ either $L_G = 0$ or $d_G^{(\Omega)} \geqslant t-1$.

Corollary 7.12 *In a balanced design on the cyclic association scheme* ⓝ, *the number of treatments is at most three.*

Proof The information matrices cannot all be zero, and no stratum for Ⓝ has dimension bigger than two. ■

Similarly, the results about valid randomization also extend to more general association schemes. The only change needed is that the information matrices in Condition (7.10) are labelled by the strata in $\mathcal{P}$ while Condition (7.11) refers to the associate classes of $\mathcal{P}$.

Example 7.21 Suppose that we want a balanced design for 13 treatments in $\underline{\underline{26}}/⑥$. The following randomization is valid. Form the 13 translates of each circle below. Randomly allocate the resulting 26 circles to the circles in the experiment. For each circle independently, rotate it through a random integer multiple of 60° and, with probability one half, reflect it in the vertical axis. ■

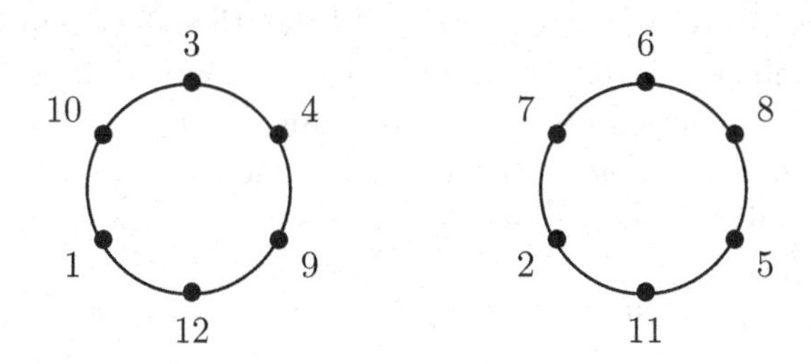

7.7.2 Composite designs

If $\psi: \Omega \to \Theta$ and $\phi: \Theta \to \Delta$ are functions then we can form the composite function $\phi\psi: \Omega \to \Delta$. If both ψ and ϕ are designs then so is $\phi\psi$, so we call it the *composite design*. If ψ and ϕ are equi-replicate with replications r_1 and r_2 then $\phi\psi$ is equi-replicate with replication $r_1 r_2$.

Theorem 7.13 *Let $\mathcal{P}$, $\mathcal{Q}$ and $\mathcal{R}$ be association schemes on Ω, Θ and Δ respectively. Let $\psi: \Omega \to \Theta$ and $\phi: \Theta \to \Delta$ be designs. If ψ is partially balanced for $\mathcal{P}$ with respect to $\mathcal{Q}$ and ϕ is partially balanced for $\mathcal{Q}$ with respect to $\mathcal{R}$ then $\phi\psi$ is partially balanced for $\mathcal{P}$ with respect to $\mathcal{R}$.*

In addition, if ψ satisfies

$$\left\{X'_\psi B_F X_\psi : F \in \mathcal{F}\right\} \text{ spans the Bose–Mesner algebra of } \mathcal{Q}, \qquad (7.12)$$

where the adjacency matrices of $\mathcal{P}$ are B_F for F in $\mathcal{F}$, then the converse is true in the sense that if $\phi\psi$ is partially balanced then so is ϕ.

Proof Let the adjacency matrices of $\mathcal{Q}$ be A_i for i in $\mathcal{K}$. Fix F in $\mathcal{F}$. Because ψ is partially balanced, there are scalars ν_{Fi}, for i in $\mathcal{K}$, such that $X'_\psi B_F X_\psi = \sum_{i\in\mathcal{K}} \nu_{Fi} A_i$. For each i in $\mathcal{K}$, $X'_\phi A_i X_\phi$ is in the Bose–Mesner algebra of $\mathcal{R}$, because ϕ is partially balanced. Technique 1.1 shows that $X_{\phi\psi} = X_\psi X_\phi$, so

$$X'_{\phi\psi} B_F X_{\phi\psi} = X'_\phi X'_\psi B_F X_\psi X_\phi = \sum_{i\in\mathcal{K}} \nu_{Fi} X'_\phi A_i X_\phi,$$

which is in the Bose–Mesner algebra of $\mathcal{R}$.

Conversely, suppose that ψ satisfies Condition (7.12), which implies that ψ is partially balanced. If A_i is an adjacency matrix for $\mathcal{Q}$ then $A_i = X'_\psi B X_\psi$ for some B in the Bose–Mesner algebra of $\mathcal{P}$. Now $X'_\phi A_i X_\phi = X'_{\phi\psi} B X_{\phi\psi}$, which is the Bose–Mesner algebra of $\mathcal{R}$ if $\phi\psi$ is partially balanced. ■

If Θ is a relatively simple association scheme then Theorem 7.13 can be used to combine a straightforward design ϕ on Θ with a design ψ on Ω to create a new design $\phi\psi$ on Ω. For example, if ψ is balanced and ϕ is equi-replicate then $\phi\psi$ is balanced: in other words, in a balanced design any merging of treatments into classes of equal size gives another balanced design.

On the other hand, if Ω is relatively simple then a known partially balanced design ψ can be used to test a design $\phi\colon \Theta \to \Delta$ for partial balance. If $\mathcal{P}$ and $\mathcal{Q}$ both have two associate classes then ψ satisfies Condition (7.12) so long as it is partially balanced but not balanced. This is particularly useful when $\mathcal{P}$ is group-divisible, for then ψ and $\phi\psi$ are both block designs.

Theorem 7.14 *Let $\mathcal{P}$, $\mathcal{Q}$, $\mathcal{R}$, ψ and ϕ be as in Theorem 7.13. Suppose that ψ is partially balanced for $\mathcal{P}$ with respect to $\mathcal{Q}$ and that ϕ is partially balanced for $\mathcal{Q}$ with respect to $\mathcal{R}$; that the strata for $\mathcal{P}$, $\mathcal{Q}$, and $\mathcal{R}$ are labelled by $\mathcal{G}$, $\mathcal{E}$ and $\mathcal{X}$ respectively; that the canonical efficiency factors for ψ are $\varepsilon^{(1)}_{Ge}$ for (G,e) in $\mathcal{G}\times\mathcal{E}$ and those for ϕ are $\varepsilon^{(2)}_{ex}$ for (e,x) in $\mathcal{E}\times\mathcal{X}$. Then the canonical efficiency factors for the composite design $\phi\psi$ are ε_{Gx}, where*

$$\varepsilon_{Gx} = \sum_{e\in\mathcal{E}} \varepsilon^{(1)}_{Ge}\varepsilon^{(2)}_{ex}$$

for (G,x) in $\mathcal{G}\times\mathcal{X}$.

Proof Let Q_G, S_e and T_x be the stratum projectors for G in $\mathcal{G}$, e in $\mathcal{E}$

and x in $\mathcal{X}$. Let r_1 and r_2 be the replications of ψ and ϕ respectively. By definition of the canonical efficiency factors,

$$X'_\psi Q_G X_\psi = r_1 \sum_{e\in\mathcal{E}} \varepsilon^{(1)}_{Ge} S_e$$

for G in $\mathcal{G}$, and

$$X'_\phi S_e X_\phi = r_2 \sum_{x\in\mathcal{X}} \varepsilon^{(2)}_{ex} T_x$$

for e in $\mathcal{E}$. Therefore

$$X'_{\phi\psi} Q_G X_{\phi\psi} = r_1 r_2 \sum_{x\in\mathcal{X}} \sum_{e\in\mathcal{E}} \varepsilon^{(1)}_{Ge} \varepsilon^{(2)}_{ex} T_x.$$

Since $\phi\psi$ has replication $r_1 r_2$, this proves the result. ■

7.7.3 Designs on triangular schemes

Theorem 7.13 gives a simple way of recognizing partially balanced designs on triangular schemes. Let Γ be an n-set. Let Ω consist of the 2-subsets of Γ, with the triangular association scheme $\mathrm{T}(n)$. Let Φ be a set of size $n(n-1)$, with association scheme $\mathrm{GD}(n, n-1)$. Let $\rho\colon \Phi \to \Omega$ be the design whose blocks have the form

$$\{\{\gamma, \delta\} : \delta \in \Gamma \setminus \{\gamma\}\}$$

for γ in Γ. Let Θ be a set of treatments with association scheme $\mathcal{Q}$ and let $\psi\colon \Omega \to \Theta$ be a design. Then ψ is partially balanced for $\mathrm{T}(n)$ with respect to $\mathcal{Q}$ if and only if $\psi\rho$ is partially balanced for $\mathrm{GD}(n, n-1)$ with respect to $\mathcal{Q}$. To visualize the block design $\psi\rho$, write the treatments in a square array with the diagonal missing, putting $\psi(\{i,j\})$ into cells (i,j) and (j,i). The rows of this array are the blocks of $\psi\rho$.

Label the blocks and bottom strata for $\mathrm{GD}(n, n-1)$ by B and E respectively. Exercise 5.7 shows that the canonical efficiency factors for ρ are

$$\varepsilon_{B,\text{parents}} = \frac{n-2}{2(n-1)} \qquad \varepsilon_{B,\text{offspring}} = 0$$

$$\varepsilon_{E,\text{parents}} = \frac{n}{2(n-1)} \qquad \varepsilon_{E,\text{offspring}} = 1.$$

Theorem 7.14 relates the canonical efficiency factors for the block design $\psi\rho$ to those for ψ as follows:

$$\varepsilon_{Be} = \frac{(n-2)}{2(n-1)} \varepsilon_{\text{parents},e} \tag{7.13}$$

and

$$\varepsilon_{Ee} = \frac{n}{2(n-1)}\varepsilon_{\text{parents},e} + \varepsilon_{\text{offsping},e}$$

for each treatment stratum $W_e^{(\Theta)}$.

Although general association schemes do not have fixed-effects models, for designs on the triangular scheme it is plausible to assume either a fixed-effects model in which

$$\mathbb{E}(Y) = X\tau + h$$

for some h in $W_{\text{parents}}^{(\Omega)}$, or the random-effects model

$$\text{Cov}(Y) = \xi_0 Q_0 + \xi_{\text{parents}} Q_{\text{parents}} + \xi_{\text{offspring}} Q_{\text{offspring}}$$

with $\xi_{\text{offspring}}$ smaller than ξ_{parents}. In either case we want the canonical efficiency factors in the parents stratum to be small. Equation (7.13) shows that ψ is efficient if and only if $\psi\rho$ is an efficient block design.

Definition A Latin square is *unipotent* if a single letter occurs throughout the main diagonal; it is *idempotent* if each letter occurs once on the main diagonal.

Given a symmetric $n \times n$ Latin square Π, there is an obvious way of turning it into a design ψ on $\text{T}(n)$: let $\psi(\{i,j\})$ be the letter in row i and column j. If n is even then there is a symmetric unipotent square Π: then $t = n-1$ and $r = n/2$. Now $\psi\rho$ is a complete-block design with $\varepsilon_B = 0$ for all treatment contrasts. Hence $\varepsilon_{\text{parents}} = 0$ for all treatment contrasts and so the design is orthogonal.

On the other hand, if n is odd then there is a symmetric idempotent Latin square Π, which gives a design ψ in which $t = n$ and $r = (n-1)/2$. Now $\psi\rho$ is a balanced incomplete-block design with block size $n-1$, so $\varepsilon_{E1} = n(n-2)/(n-1)^2$ and $\varepsilon_{B1} = 1/(n-1)^2$ for the trivial scheme on the letters. Hence ψ is also balanced and

$$\varepsilon_{\text{parents},1} = \frac{2}{(n-1)(n-2)}, \quad \varepsilon_{\text{offspring},1} = \frac{n(n-3)}{(n-1)(n-2)}.$$

Example 7.22 Figure 7.24 shows two balanced designs for 7 treatments in $\text{T}(7)$. Design ψ_1 is constructed from a symmetric idempotent Latin square so it has $\varepsilon_{\text{offspring},1} = 14/15$. However, the composite design $\psi_2\rho$ is not binary: it has concurrence matrix $12I_\Theta + 4(J_\Theta - I_\Theta)$ and so $\varepsilon_{\text{offspring},1} = 7/15$. ■

1	2	3	4	5	6	7	
	B	C	D	E	F	G	1
		D	E	F	G	A	2
			F	G	A	B	3
				A	B	C	4
					C	D	5
						E	6
							7

Design ψ_1

1	2	3	4	5	6	7	
	A	G	A	E	E	G	1
		B	A	B	F	F	2
			C	B	C	G	3
				D	C	D	4
					E	D	5
						F	6
							7

Design ψ_2

Fig. 7.24. Two balanced designs for 7 treatments in T(7)

7.7.4 Designs on Latin-square schemes

There is a similar straightforward analysis of designs ψ for an association scheme $(\Omega, \mathcal{P})$ of Latin-square type $\mathrm{L}(c,n)$. Now let Φ be a set of size cn^2, with association scheme $\mathrm{GD}(cn,n)$, and let $\rho\colon \Phi \to \Omega$ be a square lattice design. Assume that $c \leqslant n$, so that ρ is not balanced. Then Theorem 7.13 shows that ψ is partially balanced if and only if $\psi\rho$ is. Label the strata of $\mathcal{P}$ so that $W_1^{(\Omega)}$ has dimension $c(n-1)$ and $W_2^{(\Omega)}$ has dimension $(n+1-c)(n-1)$. Section 5.5.2 shows that the canonical efficiency factors for ρ are

$$\varepsilon_{B1} = \frac{1}{c} \qquad \varepsilon_{B2} = 0$$

$$\varepsilon_{E1} = \frac{c-1}{c} \qquad \varepsilon_{E2} = 1.$$

Theorem 7.14 gives

$$\varepsilon_{Be} = \frac{1}{c}\varepsilon_{1e} \quad \text{and} \quad \varepsilon_{Ee} = \frac{(c-1)}{c}\varepsilon_{1e} + \varepsilon_{2e} \tag{7.14}$$

for every treatment stratum $W_e^{(\Theta)}$.

Example 7.19 revisited In $\psi\rho$, each block contains one treatment from each group, so groups are orthogonal to blocks and $\varepsilon_{E,\mathrm{between}} = 1$. Technique 5.4 gives

$$3\varepsilon_{E,\mathrm{between}} + 4\varepsilon_{E,\mathrm{within}} = (8 \times 3)/4,$$

so $\varepsilon_{E,\mathrm{within}} = 3/4$. Therefore $\varepsilon_{B,\mathrm{between}} = 0$ and $\varepsilon_{B,\mathrm{within}} = 1/4$. Since $c = 2$, we obtain

$$\begin{array}{ll} \varepsilon_{1,\mathrm{between}} = 0 & \varepsilon_{1,\mathrm{within}} = \frac{1}{2} \\ \varepsilon_{2,\mathrm{between}} = 1 & \varepsilon_{2,\mathrm{within}} = \frac{1}{2}. \end{array} \quad \blacksquare$$

For designs on an association scheme of Latin-square type it is plausible to assume either a fixed-effects model with

$$\mathbb{E}(Y) = X\tau + h$$

for some h in $W_1^{(\Omega)}$ or the random-effects model with ξ_1 bigger than ξ_2. In either case, Equation (7.14) shows that ψ is a good design if and only if $\psi\rho$ is.

7.7.5 Designs on pair schemes

Suppose that Ω has the Pair(n) association scheme, with the adjacency matrices I, A, B, M and G given on page 134. Exercise 5.24 shows that the non-trivial strata for Pair(n) are as follows.

stratum	description	dimension	projector
$W_1^{(\Omega)}$	symmetric row-column	$n-1$	Q_1
$W_2^{(\Omega)}$	antisymmetric row-column	$n-1$	Q_2
$W_3^{(\Omega)}$	rest of symmetric	$\frac{n(n-3)}{2}$	Q_3
$W_4^{(\Omega)}$	rest of antisymmetric	$\frac{(n-1)(n-2)}{2}$	Q_4

Moreover,

$$\begin{aligned} Q_1 &= \frac{1}{2(n-2)}(2(I+M) + A + B + 4n^{-1}J) \\ Q_2 &= \frac{1}{2n}(2(I-M) + A - B) \\ Q_3 &= \frac{1}{2(n-2)}((n-4)(I+M) - A - B + 2(n-1)^{-1}J) \\ Q_4 &= \frac{1}{2n}((n-2)(I-M) - A + B). \end{aligned}$$

For brevity, we consider only designs ψ defined by unipotent $n \times n$

Latin squares: thus $t = n - 1$ and $r = n$. For such a design,

$$\begin{aligned} X'IX &= nI \\ X'AX &= 2n(J - I) \\ X'(2M + B)X &= 2nJ \\ X'JX &= n^2J. \end{aligned}$$

Therefore $X'Q_1X = X'Q_2X = O$.

Now, $X'(I + M)X$ is the concurrence matrix Λ for the block design in mirror-image pairs, so this block design is partially balanced with respect to an association scheme $\mathcal{Q}$ on the treatments if and only if $(\Omega, \mathrm{Pair}(n), \Theta, \psi)$ is. Moreover,

$$X'Q_3X = \frac{1}{2}\Lambda - rS_0 \quad \text{and} \quad X'Q_4X = rI - \frac{1}{2}\Lambda$$

so the canonical efficiency factors in strata $W_3^{(\Omega)}$ and $W_4^{(\Omega)}$ are the same as those of the block design in the blocks and bottom strata.

If n is an odd integer there is an idempotent Latin square Π in which the $n - 1$ treatments are split into groups of two in such a way that mirror-image pairs in Ω always have the two treatments in a group. For example, take the set of treatments (letters) to be $\mathbb{Z}_n$ and put letter $i - j$ in cell (i, j). An example is shown in Figure 7.25.

	A	C	D	B
B		A	C	D
D	B		A	C
C	D	B		A
A	C	D	B	

Fig. 7.25. Group-divisible design on Pair(5)

	A	B	C	D	E
C		E	A	B	D
A	B		D	E	C
D	E	A		C	B
B	C	D	E		A
E	D	C	B	A	

Fig. 7.26. Balanced design on Pair(6)

Such a Latin square gives a block design in which the between-groups contrasts are between-blocks contrasts while the within-groups contrasts are orthogonal to blocks. Therefore the design on Pair(n) is partially balanced for the association scheme $\mathrm{GD}((n-1)/2, 2)$: in fact it is orthogonal with

$$\varepsilon_{3,\text{between}} = 1 \qquad \varepsilon_{3,\text{within}} = 0$$

$$\varepsilon_{4,\text{between}} = 0 \qquad \varepsilon_{4,\text{within}} = 1.$$

On the other hand, we may be able to choose Π so that the mirror images of the n replications of each treatment consist of every other treatment just once and itself twice. An example is shown in Figure 7.26. Then $\Lambda = (n+1)I + J$, so the block design is balanced with $\varepsilon_{B1} = (n+1)/2n$ and $\varepsilon_{E1} = (n-1)/2n$. Therefore the design on Pair(n) is also balanced, with canonical efficiency factors $\varepsilon_{31} = (n+1)/2n$ and $\varepsilon_{41} = (n-1)/2n$.

Finally, if Π is a symmetric Latin square then $\Lambda = 2nI$ so $X'Q_3X = rI - rS_0$ and $X'Q_4X = O$. Now the design is orthogonal with all estimation in stratum $W_3^{(\Omega)}$.

Exercises

7.1 Here are two designs for nested blocks. Show that one is partially balanced and the other is not. Find the table of canonical efficiency factors for the design which is partially balanced.

(a) Eight treatments in eight large blocks, each containing two small blocks of size two.

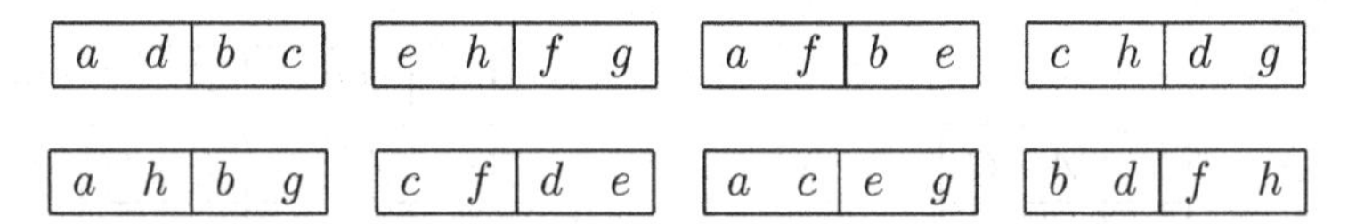

(b) Nine treatments in three large blocks, each containing three small blocks of size four.

a c g i	b d h a	f h c e
c e i b	g i d f	h a e g
d f a c	e g b d	i b f h

7.2 Consider the semi-Latin rectangles in Figures 7.6 and 7.7. For each one, decide whether it is partially balanced. If it is partially balanced, find its table of canonical efficiency factors.

7.3 Here are two semi-Latin rectangles for twelve treatments in the orthogonal block structure $(\underline{\underline{6}} \times \underline{\underline{6}})/\underline{\underline{2}}$. Show that one is partially balanced and the other is not. Find the table of canonical efficiency factors for the design which is partially balanced.

(a) First semi-Latin square.

<table>
<tr><td>a l</td><td>f k</td><td>c h</td><td>b g</td><td>d i</td><td>e j</td></tr>
<tr><td>c i</td><td>b j</td><td>e f</td><td>h l</td><td>g k</td><td>a d</td></tr>
<tr><td>e k</td><td>h i</td><td>d g</td><td>a f</td><td>j l</td><td>b c</td></tr>
<tr><td>d j</td><td>a e</td><td>i l</td><td>c k</td><td>b f</td><td>g h</td></tr>
<tr><td>f g</td><td>c d</td><td>a b</td><td>i j</td><td>e h</td><td>k l</td></tr>
<tr><td>b h</td><td>g l</td><td>j k</td><td>d e</td><td>a c</td><td>f i</td></tr>
</table>

(b) Second semi-Latin square.

<table>
<tr><td>a c</td><td>e g</td><td>l i</td><td>k j</td><td>d h</td><td>b f</td></tr>
<tr><td>k i</td><td>a d</td><td>b g</td><td>c e</td><td>j f</td><td>l h</td></tr>
<tr><td>j e</td><td>i f</td><td>a k</td><td>b h</td><td>c l</td><td>d g</td></tr>
<tr><td>f h</td><td>c b</td><td>d j</td><td>a l</td><td>k g</td><td>i e</td></tr>
<tr><td>d b</td><td>l j</td><td>e h</td><td>f g</td><td>a i</td><td>c k</td></tr>
<tr><td>l g</td><td>k h</td><td>c f</td><td>d i</td><td>b e</td><td>a j</td></tr>
</table>

7.4 Here is a design for a variant of the orthogonal block structure in Example 7.5, with only three months and only twelve treatments. Show that it is partially balanced and find its canonical efficiency factors.

<table>
<tr><td>a</td><td>d</td><td>i</td><td>l</td><td>j</td><td>h</td><td>e</td><td>c</td><td>k</td><td>f</td><td>b</td><td>g</td></tr>
<tr><td>b</td><td>f</td><td>g</td><td>k</td><td>d</td><td>a</td><td>l</td><td>i</td><td>e</td><td>j</td><td>h</td><td>c</td></tr>
<tr><td>c</td><td>e</td><td>h</td><td>j</td><td>g</td><td>k</td><td>b</td><td>f</td><td>i</td><td>a</td><td>d</td><td>l</td></tr>
</table>

7.5 Find the canonical efficiency factors for the designs in Figures 7.3, 7.12, 7.16, 7.17, 7.19 and 7.20.

7.6 Here are two designs for 15 treatments in a 6×10 row-column array. Show that they are both partially balanced, and find their canonical efficiency factors. Which do you think is better if the fixed-effects model is assumed?

(a) The treatments are all 2-subsets of $\{1, \ldots, 6\}$.

$\{2,3\}$	$\{2,4\}$	$\{2,5\}$	$\{2,6\}$	$\{3,4\}$	$\{3,5\}$	$\{3,6\}$	$\{4,5\}$	$\{4,6\}$	$\{5,6\}$
$\{1,3\}$	$\{1,4\}$	$\{1,5\}$	$\{1,6\}$	$\{5,6\}$	$\{4,6\}$	$\{4,5\}$	$\{3,6\}$	$\{3,5\}$	$\{3,4\}$
$\{1,2\}$	$\{5,6\}$	$\{4,6\}$	$\{4,5\}$	$\{1,4\}$	$\{1,5\}$	$\{1,6\}$	$\{2,6\}$	$\{2,5\}$	$\{2,4\}$
$\{5,6\}$	$\{1,2\}$	$\{3,6\}$	$\{3,5\}$	$\{1,3\}$	$\{2,6\}$	$\{2,5\}$	$\{1,5\}$	$\{1,6\}$	$\{2,3\}$
$\{4,6\}$	$\{3,6\}$	$\{1,2\}$	$\{3,4\}$	$\{2,6\}$	$\{1,3\}$	$\{2,4\}$	$\{1,4\}$	$\{2,3\}$	$\{1,6\}$
$\{4,5\}$	$\{3,5\}$	$\{3,4\}$	$\{1,2\}$	$\{2,5\}$	$\{2,4\}$	$\{1,3\}$	$\{2,3\}$	$\{1,4\}$	$\{1,5\}$

(b) The treatments are the integers modulo 15.

0	1	3	4	6	7	9	10	12	13
2	3	5	6	8	9	11	12	14	0
5	6	8	9	11	12	14	0	2	3
7	8	10	11	13	14	1	2	4	5
10	11	13	14	1	2	4	5	7	8
12	13	0	1	3	4	6	7	9	10

7.7 For $i = 1, 2$, let Δ_i be the collection of b_i blocks of size k_i in a balanced incomplete-block design for treatment-set Θ_i of size t_i. Form a nested row-column design of size $\underline{\underline{b_1 b_2}}/(\underline{\underline{k_1}} \times \underline{\underline{k_2}})$ as follows: the set of treatments is $\Theta_1 \times \Theta_2$; for δ_1 in Δ_1 and δ_2 in Δ_2 there is a product block $\delta_1 \times \delta_2$ (as in Section 4.6.4) in which the rows have the form $\delta_1 \times \{\theta_2\}$ for each θ_2 in δ_2 and the columns have the form $\{\theta_1\} \times \delta_2$ for each θ_1 in δ_1.

Show that this design is partially balanced with respect to the rectangular association scheme $\mathrm{R}(t_1, t_2)$ on $\Theta_1 \times \Theta_2$.

7.8 In an experiment on watering chicory, the plants were grown in 27 boxes arranged in nine stacks of three. Water came from three tanks, each having three outflow taps. The taps were adjusted to give three different flow-rates. There was one stack of boxes per tap. The sides of the boxes were also adjustable, giving three different depths to which water would be retained in a box before overflowing. The treatments consisted of all combinations of flow-rate and depth, as follows.

depth	flow-rate 1	2	3
1	a	b	c
2	d	e	f
3	g	h	i

Treatments were allocated to boxes as shown below.

	tank 1			tank 2			tank 3		
	tap 1	tap 2	tap 3	tap 4	tap 5	tap 6	tap 7	tap 8	tap 9
height 3	a	i	e	g	f	b	c	d	h
height 2	d	c	h	a	i	e	f	g	b
height 1	g	f	b	d	c	h	i	a	e

Show that this design is orthogonal and partially balanced. Allocate treatment strata to plot strata as in Table 7.7.

7.9 Show that, if n is an odd integer greater than 1, there is a balanced semi-Latin rectangle for n treatments in the orthogonal block structure $(\underline{\underline{n}} \times \underline{\underline{n}})/\underline{\underline{2}}$. Is the same true for even n?

7.10 Construct balanced semi-Latin rectangles for seven treatments in the orthogonal block structures $(\underline{\underline{7}} \times \underline{\underline{7}})/\underline{\underline{k}}$ when $k = 3$, 4, 5 and 6.

7.11 Consider the following design for ten treatments in nested blocks.

2 6 8 9	0 3 4 5	2 3 5 9	1 6 7 8
1 2 7 9	3 4 6 8	4 8 0 1	2 5 6 7
4 5 7 1	3 8 9 0	3 4 9 1	5 6 8 0
6 0 2 3	4 7 8 9	6 7 9 3	5 0 1 2
5 6 1 3	7 8 0 2	8 2 4 5	6 9 0 1
8 9 1 5	7 2 3 4	7 8 3 5	9 0 2 4
0 4 6 7	8 1 2 3	0 1 3 7	9 4 5 6
9 0 5 7	1 2 4 6		

Investigate whether it is balanced, weakly cyclic or strongly cyclic. Show that it is partially balanced, and find its table of canonical efficiency factors.

7.12 Is the design in Figure 7.17 (a) strongly cyclic (b) weakly cyclic?

7.13 Let ψ be a semi-Latin rectangle for t treatments in the orthogonal block structure $(\underline{\underline{n}} \times \underline{\underline{m}})/\underline{\underline{k}}$. Prove that if $t > (n-1)(m-1)+1$ then there are at least $t-(n-1)(m-1)-1$ canonical efficiency factors in the bottom stratum equal to 1.

7.14 Show that, if n is an integer greater than 1, there is a balanced cyclic design for $2n+1$ treatments in the orthogonal block structure $(\underline{\underline{n}} \times \underline{\underline{2n+1}})/\underline{\underline{2}}$.

7.15 Two cyclic designs for seven treatments in the orthogonal block structure $\underline{\underline{7}}/(\underline{\underline{2}} \times \underline{\underline{3}})$ have initial blocks

1	2	4
6	5	3

and

1	2	4
2	4	1

.

Show that both designs are balanced. Which is better for the fixed-effects model? Which is better for the random-effects model?

7.16 There is a cyclic nested row-column design for 13 treatments with initial block

1	9	3
4	10	12

.

Compare this with the two designs in Example 7.15.

7.17 Explain why the methods of randomization in Examples 7.16, 7.17 and 7.21 satisfy both conditions (7.10) and (7.11). Is the randomization in Example 7.21 valid for $\underline{\underline{26}}/\underline{\underline{6}}$?

7.18 Show that if n is odd then there is a symmetric idempotent $n \times n$ Latin square and that if n is even then there is a symmetric unipotent $n \times n$ Latin square.

7.19 Construct an orthogonal design ψ_1 for five treatments on T(6) by using a symmetric unipotent 6×6 Latin square.

Let ψ_2 be the block design with 15 blocks of size 3 in Example 5.12. Construct the composite design $\psi_1\psi_2$.

7.20 Construct a balanced design for five treatments on the Petersen graph.

7.21 Investigate the designs on Pair(n) defined by idempotent Latin squares.

8
Groups

8.1 Blueprints

In this section we generalize the idea of a cyclic association scheme (Section 1.4.5) to arbitrary finite groups.

Let Ω be a finite group, written multiplicatively, with identity element 1_Ω. Convolution in $\mathbb{R}^\Omega$ now has to be defined by

$$(f * g)(\omega) = \sum_{\alpha \in \Omega} f(\alpha) g(\alpha^{-1}\omega)$$

for f, g in $\mathbb{R}^\Omega$; in particular,

$$\chi_\alpha * \chi_\beta = \chi_{\alpha\beta}$$

for α, β in Ω. In this chapter we use $*$ to denote convolution to distinguish it from the pointwise product of functions from Ω to a field.

Let Δ be a partition of Ω into classes Δ_0, Δ_1, ..., Δ_s. In the notation of Section 6.1, V_Δ is the subspace of $\mathbb{R}^\Omega$ spanned by the characteristic functions χ_{Δ_i} of the classes Δ_i.

Definition A partition Δ of the group Ω is a *blueprint* for Ω if

(i)‴ $\{1_\Omega\}$ is a class of Δ;
(ii)‴ for each element ω of Ω and each class Δ_i of Δ, if $\omega \in \Delta_i$ then $\omega^{-1} \in \Delta_i$;
(iii)‴ V_Δ is closed under convolution.

Example 8.1 For any group Ω, let Inverse be the partition of Ω whose classes are $\{\omega, \omega^{-1}\}$ for ω in Ω. If Ω is Abelian then Inverse is a blueprint. Property (iii)‴ holds because

$$\begin{aligned}(\chi_\alpha + \chi_{\alpha^{-1}}) * (\chi_\beta + \chi_{\beta^{-1}}) \\ = \quad \chi_{\alpha\beta} + \chi_{\alpha^{-1}\beta^{-1}} + \chi_{\alpha\beta^{-1}} + \chi_{\alpha^{-1}\beta}\end{aligned}$$

$$= \left(\chi_{\alpha\beta} + \chi_{\beta^{-1}\alpha^{-1}}\right) + \left(\chi_{\alpha\beta^{-1}} + \chi_{\beta\alpha^{-1}}\right)$$
$$= \left(\chi_{\alpha\beta} + \chi_{(\alpha\beta)^{-1}}\right) + \left(\chi_{\alpha\beta^{-1}} + \chi_{(\alpha\beta^{-1})^{-1}}\right). \quad \blacksquare$$

If Δ is a blueprint then we always label its classes so that $\Delta_0 = \{1_\Omega\}$. Define $\varphi\colon \mathbb{R}^\Omega \to \mathbb{R}^{\Omega\times\Omega}$ by

$$\varphi(\chi_\omega)(\alpha,\beta) = \begin{cases} 1 & \text{if } \alpha^{-1}\beta = \omega \\ 0 & \text{otherwise,} \end{cases}$$

extended linearly. Technique 1.1 shows that

$$\begin{aligned} \left(\varphi(\chi_\theta)\varphi(\chi_\phi)\right)(\alpha,\beta) &= \left|\left\{\gamma \in \Omega : \alpha^{-1}\gamma = \theta \text{ and } \gamma^{-1}\beta = \phi\right\}\right| \\ &= \begin{cases} 1 & \text{if } \alpha^{-1}\beta = \theta\phi \\ 0 & \text{otherwise,} \end{cases} \end{aligned}$$

so $\varphi(\chi_\theta)\varphi(\chi_\phi) = \varphi(\chi_\theta * \chi_\phi)$. In fact, φ also preserves addition and scalar multiplication, so it is an algebra isomorphism.

Theorem 8.1 *Let Δ be a partition of Ω into classes Δ_0, Δ_1, ..., Δ_s. For $i = 0$, ..., s, define the subset $\mathcal{C}_i$ of $\Omega \times \Omega$ by*

$$\mathcal{C}_i = \left\{(\alpha,\beta) : \alpha^{-1}\beta \in \Delta_i\right\}.$$

Let $\mathcal{Q}(\Delta)$ be the partition of $\Omega\times\Omega$ into $\mathcal{C}_0$, ..., $\mathcal{C}_s$. Then Δ is a blueprint if and only if $\mathcal{Q}(\Delta)$ is an association scheme.

Proof The three defining properties of an association scheme match one by one the three defining properties of a blueprint.

(i) If Δ is a blueprint then $\Delta_0 = \{1_\Omega\}$ so $\mathcal{C}_0 = \mathrm{Diag}(\Omega)$. If $\mathcal{Q}(\Delta)$ is an association scheme then $\mathcal{C}_0 = \mathrm{Diag}(\Omega)$ so $\Delta_0 = \{1_\Omega\}$.

(ii) Suppose that Δ is a blueprint and $(\alpha,\beta) \in \mathcal{C}_i$. Then $\alpha^{-1}\beta \in \Delta_i$ so $(\alpha^{-1}\beta)^{-1} \in \Delta_i$, by (ii)$''$. But $(\alpha^{-1}\beta)^{-1} = \beta^{-1}\alpha$ so $(\beta,\alpha) \in \mathcal{C}_i$ and so $\mathcal{C}_i$ is symmetric. Conversely, suppose that $\mathcal{Q}(\Delta)$ is an association scheme and that $\omega \in \Delta_i$. Then $(1_\Omega,\omega) \in \mathcal{C}_i$ and $\mathcal{C}_i$ is symmetric so $(\omega, 1_\Omega) \in \mathcal{C}_i$ and so $\omega^{-1} \in \Delta_i$.

(iii) If A_i is the adjacency matrix of $\mathcal{C}_i$ then $A_i = \varphi(\chi_{\Delta_i})$. Since φ is an isomorphism, A_iA_j is a linear combination of A_0, ..., A_s if and only if $\chi_{\Delta_i} * \chi_{\Delta_j}$ is a linear combination of χ_{Δ_0}, ..., χ_{Δ_s}. In other words, condition (iii) for the association scheme is equivalent to condition (iii)$'''$ for the blueprint. $\blacksquare$

The association scheme $\mathcal{Q}(\Delta)$ derived from a blueprint Δ is called a *group scheme*, or an *Abelian-group scheme* if Ω is Abelian.

Corollary 8.2 *If Δ is a blueprint for Ω then the characteristic functions of its classes commute under convolution.*

Proof Let $\mathcal{A}$ be the Bose–Mesner algebra of $\mathcal{Q}(\Delta)$. The restriction of φ to V_Δ gives an isomorphism between V_Δ and $\mathcal{A}$. Multiplication in $\mathcal{A}$ is commutative, so convolution in V_Δ is commutative. ∎

To check that the third condition for a blueprint is satisfied, construct an $(s+1) \times (s+1)$ table: in row i and column j write all products $\alpha\beta$ with α in Δ_i and β in Δ_j. Then Δ is a blueprint if and only if there are integers p_{ij}^k such that every element of Δ_k occurs p_{ij}^k times in cell (i,j).

Example 8.2 Take Ω to be the group S_3 of all permutations on three points, which is not Abelian. Put $\Delta_0 = \{1\}$, $\Delta_1 = \{(1\,2\,3),(1\,3\,2)\}$, $\Delta_2 = \{(1\,2)\}$ and $\Delta_3 = \{(1\,3),(2\,3)\}$. The table of products is:

		Δ_0	Δ_1		Δ_2	Δ_3	
		1	(1 2 3),	(1 3 2)	(1 2)	(1 3),	(2 3)
Δ_0	1	1	(1 2 3),	(1 3 2)	(1 2)	(1 3),	(2 3)
Δ_1	(1 2 3)	(1 2 3)	(1 3 2),	1	(2 3)	(1 2),	(1 3)
	(1 3 2)	(1 3 2)	1,	(1 2 3)	(1 3)	(2 3),	(1 2)
Δ_2	(1 2)	(1 2)	(1 3),	(2 3)	1	(1 2 3),	(1 3 2)
Δ_3	(1 3)	(1 3)	(2 3),	(1 2)	(1 3 2)	1,	(1 2 3)
	(2 3)	(2 3)	(1 2),	(1 3)	(1 2 3)	(1 3 2),	1

This shows that Δ is a blueprint and that its convolution table is as follows.

	χ_{Δ_0}	χ_{Δ_1}	χ_{Δ_2}	χ_{Δ_3}
χ_{Δ_0}	χ_{Δ_0}	χ_{Δ_1}	χ_{Δ_2}	χ_{Δ_3}
χ_{Δ_1}	χ_{Δ_1}	$2\chi_{\Delta_0} + \chi_{\Delta_1}$	χ_{Δ_3}	$2\chi_{\Delta_2} + \chi_{\Delta_3}$
χ_{Δ_2}	χ_{Δ_2}	χ_{Δ_3}	χ_{Δ_0}	χ_{Δ_1}
χ_{Δ_3}	χ_{Δ_3}	$2\chi_{\Delta_2} + \chi_{\Delta_3}$	χ_{Δ_1}	$2\chi_{\Delta_0} + \chi_{\Delta_1}$

In fact, $\mathcal{Q}(\Delta) \cong \underline{\underline{3}} \times \underline{\underline{2}}$, so this association scheme is not new even though it demonstrates a non-trivial blueprint in a non-Abelian group. ∎

8.2 Characters

The technique for finding the character table of a cyclic association scheme (Technique 2.7) extends with no difficulty to all Abelian-group schemes.

If Ω is an Abelian group then ϕ is defined to be an *irreducible character* of Ω if ϕ is a non-zero homomorphism from Ω to the complex numbers under multiplication. A good account of these characters can be found in [159, Chapter 2]. However, all we need here are:

(i) if Ω is any finite Abelian group then Ω is the direct product $\Omega_1 \times \cdots \times \Omega_m$ of cyclic groups Ω_i;

(ii) if $\Omega = \Omega_1 \times \cdots \times \Omega_m$ and ϕ is an irreducible character of Ω then there are irreducible characters ϕ_i of Ω_i for $i = 1, \ldots, m$ such that $\phi(\omega_1, \ldots, \omega_m) = \phi_1(\omega_1)\phi_2(\omega_2)\cdots\phi_m(\omega_m)$;

(iii) if Ω_i is cyclic with generator α_i of order n_i then the irreducible characters of Ω_i are $\phi_{i1}, \ldots, \phi_{in_i}$ where

$$\phi_{ij}(\alpha_i^k) = \epsilon^{jk}$$

and ϵ is a fixed primitive complex n_i-th root of unity;

(iv) there is an irreducible character ϕ_0, called the *trivial* character, for which $\phi_0(\omega) = 1_{\mathbb{C}}$ for all ω in Ω;

(v) if ϕ is an irreducible character of Ω then so is its complex conjugate $\bar{\phi}$, and $\bar{\phi}(\omega) = (\phi(\omega))^{-1}$.

Theorem 8.3 *Let Δ be a blueprint for an Abelian group Ω. Let A_i be the adjacency matrix of $\mathcal{Q}(\Delta)$ corresponding to the class Δ_i of Δ. Let ϕ be an irreducible character of Ω. Then the real functions $\phi + \bar{\phi}$ and $\mathrm{i}(\phi - \bar{\phi})$ are eigenvectors of A_i with eigenvalue $\sum_{\beta\in\Delta_i}\phi(\beta)$.*

Proof For ω in Ω, we have

$$\begin{aligned}(A_i\phi)(\omega) &= \sum_{\alpha\in\Omega} A_i(\omega,\alpha)\phi(\alpha)\\ &= \sum\{\phi(\alpha) : \omega^{-1}\alpha \in \Delta_i\}\\ &= \sum\{\phi(\omega\beta) : \beta\in\Delta_i\}\\ &= \sum_{\beta\in\Delta_i}\phi(\omega)\phi(\beta),\end{aligned}$$

because ϕ is a homomorphism, so $A_i\phi = \left(\sum_{\beta\in\Delta_i}\phi(\beta)\right)\phi$.

Similarly, $\bar{\phi}$ is an eigenvector of A_i with eigenvalue $\sum_{\beta\in\Delta_i}\bar{\phi}(\beta)$. But

$$\bar{\phi}(\beta) = (\phi(\beta))^{-1} = \phi(\beta^{-1}),$$

so property (ii)$'''$ for a blueprint shows that

$$\sum_{\beta\in\Delta_i}\bar{\phi}(\beta) = \sum_{\beta\in\Delta_i}\phi(\beta^{-1}) = \sum_{\beta\in\Delta_i}\phi(\beta).$$

Thus ϕ and $\bar{\phi}$ have the same eigenvalue $\sum_{\beta\in\Delta_i}\phi(\beta)$, which is necessarily real because A_i is symmetric. Hence the real vectors $\phi+\bar{\phi}$ and $\mathrm{i}(\phi-\bar{\phi})$ are eigenvectors of A_i with eigenvalue $\sum_{\beta\in\Delta_i}\phi(\beta)$. ■

Since the number of irreducible characters of Ω is equal to the size of Ω, an eigenvector basis of the association scheme arises in this way.

Technique 8.1 Given an Abelian-group association scheme, calculate $\sum_{\beta\in\Delta_i}\phi(\beta)$ for each class Δ_i and one of each pair $\{\phi,\bar{\phi}\}$ of complex conjugate irreducible characters of the group. Form the character table of the association scheme by amalgamating those irreducible characters which have the same eigenvalue on every adjacency matrix.

If Δ is the Inverse partition on Ω then the columns of the character table of $\mathcal{Q}(\Delta)$ are precisely the pairs of complex conjugate irreducible characters of Ω (or singletons when $\phi = \bar{\phi}$), so the double use of the word 'character' is not entirely arbitrary.

Example 8.3 Let $\Omega = \mathbb{Z}_3 \times \mathbb{Z}_3 = \langle a, b : a^3 = b^3 = 1,\ ab = ba\rangle$. The irreducible characters of Ω are given in the following table, where ϵ_3 is a primitive cube root of unity in $\mathbb{C}$.

	1	a	a^2	b	b^2	ab	a^2b^2	ab^2	a^2b
ϕ_0	1	1	1	1	1	1	1	1	1
ϕ_1	1	ϵ_3	ϵ_3^2	1	1	ϵ_3	ϵ_3^2	ϵ_3	ϵ_3^2
$\bar{\phi}_1$	1	ϵ_3^2	ϵ_3	1	1	ϵ_3^2	ϵ_3	ϵ_3^2	ϵ_3
ϕ_2	1	1	1	ϵ_3	ϵ_3^2	ϵ_3	ϵ_3^2	ϵ_3^2	ϵ_3
$\bar{\phi}_2$	1	1	1	ϵ_3^2	ϵ_3	ϵ_3^2	ϵ_3	ϵ_3	ϵ_3^2
ϕ_3	1	ϵ_3	ϵ_3^2	ϵ_3	ϵ_3^2	ϵ_3^2	ϵ_3	1	1
$\bar{\phi}_3$	1	ϵ_3^2	ϵ_3	ϵ_3^2	ϵ_3	ϵ_3	ϵ_3^2	1	1
ϕ_4	1	ϵ_3	ϵ_3^2	ϵ_3^2	ϵ_3	1	1	ϵ_3^2	ϵ_3
$\bar{\phi}_4$	1	ϵ_3^2	ϵ_3	ϵ_3	ϵ_3^2	1	1	ϵ_3	ϵ_3^2

If Δ = Inverse then the character table of $\mathcal{Q}(\Delta)$ is as follows.

$$\begin{array}{cc} & \\ & \\ \{1\} & (1) \\ \{a,a^2\} & (2) \\ \{b,b^2\} & (2) \\ \{ab,a^2b^2\} & (2) \\ \{ab^2,a^2b\} & (2) \end{array} \begin{array}{c} \phi_0 \quad \phi_1 \quad \phi_2 \quad \phi_3 \quad \phi_4 \\ (1) \quad (2) \quad (2) \quad (2) \quad (2) \\ \left[\begin{array}{rrrrr} 1 & 1 & 1 & 1 & 1 \\ 2 & -1 & 2 & -1 & -1 \\ 2 & 2 & -1 & -1 & -1 \\ 2 & -1 & -1 & -1 & 2 \\ 2 & -1 & -1 & 2 & -1 \end{array}\right] \end{array}$$ ■

Example 8.4 Let $\Omega = \mathbb{Z}_4 \times \mathbb{Z}_2 = \langle a, b : a^4 = b^2 = 1,\ ab = ba \rangle$. The irreducible characters are given in the following table.

	1	a	a^2	a^3	b	ab	a^2b	a^3b
ϕ_0	1	1	1	1	1	1	1	1
ϕ_1	1	1	1	1	−1	−1	−1	−1
ϕ_2	1	i	−1	−i	1	i	−1	−i
ϕ_3	1	i	−1	−i	−1	−i	1	i
ϕ_4	1	−1	1	−1	1	−1	1	−1
ϕ_5	1	−1	1	−1	−1	1	−1	1
$\bar{\phi}_2$	1	−i	−1	i	1	−i	−1	i
$\bar{\phi}_3$	1	−i	−1	i	−1	i	1	−i

Hence we obtain the character table for the association scheme defined by the blueprint in Exercise 8.1.

$$\begin{array}{cc} & \\ & \\ \{1\} & (1) \\ \{a,a^3\} & (2) \\ \{a^2\} & (1) \\ \{ab,a^3b\} & (2) \\ \{b,a^2b\} & (2) \end{array} \begin{array}{c} \phi_0 \quad \phi_1 \quad \phi_2,\ \phi_3 \quad \phi_4 \quad \phi_5 \\ (1) \quad (1) \quad (4) \quad (1) \quad (1) \\ \left[\begin{array}{rrrrr} 1 & 1 & 1 & 1 & 1 \\ 2 & 2 & 0 & -2 & -2 \\ 1 & 1 & -1 & 1 & 1 \\ 2 & -2 & 0 & -2 & 2 \\ 2 & -2 & 0 & 2 & -2 \end{array}\right] \end{array}$$ ■

(The reader who knows about the character theory of arbitrary finite groups will ask how this technique generalizes to non-Abelian groups. The answer is not straightforward. For instance, the association scheme in Example 8.2 has three associate classes and hence four characters, but it is derived from a blueprint for S_3, which has only three irreducible characters.)

8.3 Crossing and nesting blueprints

Here we extend the ideas of Chapter 3 to blueprints. Let Ω_1 and Ω_2 be finite groups. Multiplication in their direct product $\Omega_1 \times \Omega_2$ is coordinatewise. For $t = 1, 2$ let f_t and g_t be in $\mathbb{R}^{\Omega_t}$. Then we can form the convolution of $f_1 \otimes f_2$ with $g_1 \otimes g_2$, both of which are functions from $\Omega_1 \times \Omega_2$ to $\mathbb{R}$, as in Section 3.1. Now,

$$\begin{aligned}
&((f_1 \otimes f_2) * (g_1 \otimes g_2))\,(\omega_1, \omega_2) \\
&= \sum_{(\alpha_1,\alpha_2)\in\Omega_1\times\Omega_2} (f_1 \otimes f_2)(\alpha_1, \alpha_2) \times (g_1 \otimes g_2)(\alpha_1^{-1}\omega_1, \alpha_2^{-1}\omega_2) \\
&= \sum_{(\alpha_1,\alpha_2)\in\Omega_1\times\Omega_2} f_1(\alpha_1) f_2(\alpha_2) g_1(\alpha_1^{-1}\omega_1) g_2(\alpha_2^{-1}\omega_2) \\
&= \sum_{\alpha_1\in\Omega_1} f_1(\alpha_1) g_1(\alpha_1^{-1}\omega_1) \sum_{\alpha_2\in\Omega_2} f_2(\alpha_2) g_2(\alpha_2^{-1}\omega_2) \\
&= (f_1 * g_1)\,(\omega_1)\,(f_2 * g_2)\,(\omega_2) \\
&= ((f_1 * g_1) \otimes (f_2 * g_2))\,(\omega_1, \omega_2)
\end{aligned}$$

and so

$$(f_1 \otimes f_2) * (g_1 \otimes g_2) = (f_1 * g_1) \otimes (f_2 * g_2)\,. \tag{8.1}$$

Now we apply this result to characteristic functions of subsets. Let Δ_1 and Δ_2 be subsets of Ω_1, and Γ_1 and Γ_2 be subsets of Ω_2. Then

$$\chi_{\Delta_t \times \Gamma_t} = \chi_{\Delta_t} \otimes \chi_{\Gamma_t} \tag{8.2}$$

so

$$\begin{aligned}
\chi_{\Delta_1\times\Gamma_1} * \chi_{\Delta_2\times\Gamma_2} &= (\chi_{\Delta_1} \otimes \chi_{\Gamma_1}) * (\chi_{\Delta_2} \otimes \chi_{\Gamma_2}) \\
&= (\chi_{\Delta_1} * \chi_{\Delta_2}) \otimes (\chi_{\Gamma_1} * \chi_{\Gamma_2}),
\end{aligned} \tag{8.3}$$

by Equation (8.1).

Let Δ be a partition of Ω_1 into classes Δ_i for i in $\mathcal{K}_1$ and let Γ be a partition of Ω_2 into classes Γ_x for x in $\mathcal{K}_2$. As in Section 3.2, we define the *direct product* $\Delta \times \Gamma$ of Δ and Γ to be the partition of $\Omega_1 \times \Omega_2$ into the sets $\Delta_i \times \Gamma_x$ for (i, x) in $\mathcal{K}_1 \times \mathcal{K}_2$.

Theorem 8.4 *If Δ and Γ are blueprints for Ω_1 and Ω_2 respectively then $\Delta \times \Gamma$ is a blueprint for $\Omega_1 \times \Omega_2$.*

Proof (i)‴ $\Delta_0 = \{1_{\Omega_1}\}$ and $\Gamma_0 = \{1_{\Omega_2}\}$ so $\Delta_0 \times \Gamma_0 = \{(1_{\Omega_1}, 1_{\Omega_2})\}$ and $(1_{\Omega_1}, 1_{\Omega_2})$ is the identity element of $\Omega_1 \times \Omega_2$.

(ii)‴ Let $(\omega_1, \omega_2) \in \Delta_i \times \Gamma_x$. Then $\omega_1^{-1} \in \Delta_i$ and $\omega_2^{-1} \in \Gamma_x$, and so $(\omega_1, \omega_2)^{-1} = (\omega_1^{-1}, \omega_2^{-1}) \in \Delta_i \times \Gamma_x$.

(iii)‴ Both V_Δ and V_Γ are closed under convolution, so closure of $V_{\Delta\times\Gamma}$ under convolution follows from Equations (8.3) and (8.2). ■

We can also define wreath products in a manner similar to that in Section 3.4 but with Δ_0 playing the role previously played by $\mathrm{Diag}(\Omega_1)$. Specifically, if $\Delta_0 = \{1_{\Omega_1}\}$ then we define the *wreath product* Δ/Γ to be the partition of $\Omega_1 \times \Omega_2$ whose classes are

$$\begin{array}{ll} \Delta_i \times \Omega_2 & \text{for } i \text{ in } \mathcal{K}_1 \setminus \{0\} \\ \Delta_0 \times \Gamma_x & \text{for } x \text{ in } \mathcal{K}_2. \end{array}$$

Theorem 8.5 *If Δ and Γ are blueprints for Ω_1 and Ω_2 respectively then Δ/Γ is a blueprint for $\Omega_1 \times \Omega_2$.*

Proof The first two conditions are straightforward. For the third, there are three cases to consider. Suppose that

$$\chi_{\Delta_i} * \chi_{\Delta_j} = \sum_k p_{ij}^k \chi_{\Delta_k}$$

and

$$\chi_{\Gamma_x} * \chi_{\Gamma_y} = \sum_z q_{xy}^z \chi_{\Gamma_z}.$$

(a) If $i, j \in \mathcal{K}_1 \setminus \{0\}$ then $\chi_{\Delta_i\times\Omega_2} * \chi_{\Delta_j\times\Omega_2}$

$$\begin{aligned} &= (\chi_{\Delta_i} * \chi_{\Delta_j}) \otimes (\chi_{\Omega_2} * \chi_{\Omega_2}) \\ &= \sum_{k\in\mathcal{K}_1} p_{ij}^k \chi_{\Delta_k} \otimes |\Omega_2|\,\chi_{\Omega_2} \\ &= |\Omega_2| \sum_{k\in\mathcal{K}_1\setminus\{0\}} p_{ij}^k \chi_{\Delta_k\times\Omega_2} + |\Omega_2|\, p_{ij}^0 \sum_{x\in\mathcal{K}_2} \chi_{\Delta_0\times\Gamma_x}. \end{aligned}$$

(b) If $i \in \mathcal{K}_1 \setminus 0$ and $x \in \mathcal{K}_2$ then

$$\begin{aligned} \chi_{\Delta_i\times\Omega_2} * \chi_{\Delta_0\times\Gamma_x} &= (\chi_{\Delta_i} * \chi_{\Delta_0}) \otimes (\chi_{\Omega_2} * \chi_{\Gamma_x}) \\ &= \chi_{\Delta_i} \otimes |\Gamma_x|\,\chi_{\Omega_2} \\ &= |\Gamma_x|\,\chi_{\Delta_i\times\Omega_2}. \end{aligned}$$

(c) If $x, y \in \mathcal{K}_2$ then

$$\begin{aligned} \chi_{\Delta_0\times\Gamma_x} * \chi_{\Delta_0\times\Gamma_y} &= (\chi_{\Delta_0} * \chi_{\Delta_0}) \otimes (\chi_{\Gamma_x} * \chi_{\Gamma_y}) \\ &= \chi_{\Delta_0} \otimes \sum_{z\in\mathcal{K}_2} q_{xy}^z \chi_{\Gamma_z} \\ &= \sum_{z\in\mathcal{K}_2} q_{xy}^z \chi_{\Delta_0\times\Gamma_z}. \quad ■ \end{aligned}$$

Proposition 8.6 *Let Δ and Γ be blueprints for Ω_1 and Ω_2 respectively. Then $\mathcal{Q}(\Delta \times \Gamma) = \mathcal{Q}(\Delta) \times \mathcal{Q}(\Gamma)$ and $\mathcal{Q}(\Delta/\Gamma) = \mathcal{Q}(\Delta)/\mathcal{Q}(\Gamma)$.*

8.4 Abelian-group block-designs

In Section 5.4 we considered incomplete-block designs whose treatment sets can be identified with cyclic groups. Once again, we generalize to arbitrary finite groups as far as possible, so some definitions need reworking.

Let Θ be a finite group of order t, written multiplicatively. If $\Upsilon \subseteq \Theta$, a *left translate* of Υ is a set of the form

$$\theta\Upsilon = \{\theta\upsilon : \upsilon \in \Upsilon\}.$$

The *stabilizer* of Υ is the set $\{\theta \in \Theta : \theta\Upsilon = \Upsilon\}$. This is a subgroup of Θ, and its index l in Θ is equal to the number of distinct translates of Υ. This is the justification for calling l the 'index of Υ'.

Definition An incomplete-block design with treatment set the group Θ is a *thin group block-design* if there is some subset Υ of Θ such that the blocks are all the distinct translates of Υ. It is a *group block-design* if its set of blocks can be partitioned into subsets of blocks such that each subset is a thin block-design. If Θ is Abelian then a group block-design for Θ is called an *Abelian-group block-design.*

Theorem 8.7 *Let Θ be a group of order t, and let Υ be a subset of Θ of index l. For θ in Θ, put*

$$m_\theta(\Upsilon) = \left|\{(\upsilon_1, \upsilon_2) \in \Upsilon \times \Upsilon : \upsilon_1^{-1}\upsilon_2 = \theta\}\right|.$$

In the thin group block-design generated by Υ,

$$\Lambda(1_\Theta, \theta) = m_\theta(\Upsilon) \times \frac{l}{t}$$

and

$$\Lambda(\eta, \zeta) = \Lambda(1_\Theta, \eta^{-1}\zeta). \tag{8.4}$$

Proof Treatments 1_Θ and θ concur in the translate $\rho\Upsilon$ if and only if there are elements υ_1 and υ_2 in Υ such that $\rho\upsilon_1 = 1_\Theta$ and $\rho\upsilon_2 = \theta$, that is, $\rho = \upsilon_1^{-1}$ and $\theta = \upsilon_1^{-1}\upsilon_2$. The $m_\theta(\Upsilon)$ such pairs give $m_\theta(\Upsilon)$ translates, but each translate which occurs in this way occurs t/l times. Hence the concurrence in the thin design is $(l/t)m_\theta(\Upsilon)$.

Treatments 1_Θ and θ concur in $\rho\Upsilon$ if and only if η and $\eta\theta$ concur in $\eta\rho\Upsilon$, so

$$\Lambda(\eta, \eta\theta) = \Lambda(1_\Theta, \theta);$$

in other words,

$$\Lambda(\eta, \zeta) = \Lambda(1_\Theta, \eta^{-1}\zeta). \quad \blacksquare$$

Corollary 8.8 *Every Abelian-group block-design is partially balanced with respect to the association scheme $\mathcal{Q}$ defined by the blueprint* Inverse.

Proof Every class of $\mathcal{Q}$ has the form

$$\mathcal{C}_\theta = \left\{(\eta, \zeta) \in \Theta \times \Theta : \eta^{-1}\zeta \in \{\theta, \theta^{-1}\}\right\}$$

for some θ in Θ. Since Equation (8.4) holds in each component of the design, there are constants λ_θ such that $\Lambda(\eta, \zeta) = \lambda_{\eta^{-1}\zeta}$. The concurrence matrix is symmetric, so

$$\lambda_\theta = \Lambda(1_\Theta, \theta) = \Lambda(\theta, 1_\Theta) = \lambda_{\theta^{-1}}$$

and hence $\Lambda(\eta, \zeta) = \lambda_\theta$ for all (η, ζ) in $\mathcal{C}_\theta$. $\blacksquare$

The above argument generalizes to any group for which Inverse is a blueprint, such as the quaternion group Q_8. However, in general, group block-designs constructed from non-Abelian groups are not partially balanced.

Example 8.2 revisited In S_3, the block design generated by $\{1, (1\,2)\}$ and two copies of $\{1, (1\,3)\}$ has $r = \lambda_1 = 3$, $\lambda_{(1\,2)} = 1$, $\lambda_{(1\,3)} = 2$ and $\lambda_\theta = 0$ otherwise. If the design were partially balanced then $\varphi(\chi_{(1\,2)})$ and $\varphi(\chi_{(1\,3)})$ would have to be adjacency matrices of the association scheme. These matrices do not commute, so the design cannot be partially balanced. $\blacksquare$

To calculate the λ_θ for a thin design, form the *table of quotients* for Υ. This is the $\Upsilon \times \Upsilon$ table with element $v_1^{-1}v_2$ in row v_1 and column v_2. Then $m_\theta(\Upsilon)$ is the number of occurrences of θ in this table.

Example 8.3 revisited In $\mathbb{Z}_3 \times \mathbb{Z}_3$, the block $\{1, a, b\}$ gives the following table of quotients.

	1	a	b
1	1	a	b
a	a^2	1	a^2b
b	b^2	ab^2	1

In the thin design generated by this block, $r = \lambda_1 = 3$, $\lambda_{ab} = \lambda_{a^2b^2} = 0$ and $\lambda_\theta = 1$ otherwise. Therefore the design is group-divisible with groups $1, ab, a^2b^2 \parallel a, a^2b, b^2 \parallel a^2, b, ab^2$. ■

Definition A subset Υ of a group Θ is a *perfect difference set* for Θ if there is an integer λ such that every non-identity element of Θ occurs λ times in the table of quotients for Υ.

(Some authors call perfect difference sets just *difference sets*. Unfortunately, the word 'difference' is used even for groups written multiplicatively.)

Theorem 8.9 *The thin group block-design generated by Υ is balanced if and only if Υ is a perfect difference set.*

Example 8.5 Let $\Theta = \mathbb{Z}_2 \times \mathbb{Z}_2 \times \mathbb{Z}_2 \times \mathbb{Z}_2$, the Abelian group generated by elements a, b, c, d of order 2. Let $\Upsilon = \{1, a, b, c, d, abcd\}$. The table of quotients follows.

	1	a	b	c	d	$abcd$
1	1	a	b	c	d	$abcd$
a	a	1	ab	ac	ad	bcd
b	b	ab	1	bc	bd	acd
c	c	ac	bc	1	cd	abd
d	d	ad	bd	cd	1	abc
$abcd$	$abcd$	bcd	acd	abd	abc	1

Thus Υ is a perfect difference set. It generates a balanced incomplete-block design with $\lambda = 2$. ■

Theorem 8.10 *In an Abelian-group block-design with treatment set Θ and $\Lambda(1_\Theta, \theta) = \lambda_\theta$, the canonical efficiency factors are*

$$1 - \frac{1}{rk} \sum_{\theta \in \Theta} \lambda_\theta \phi(\theta)$$

for the irreducible characters ϕ of Θ, except for the trivial irreducible character.

Proof Consider the association scheme $\mathcal{Q}(\mathsf{Inverse})$, as in Corollary 8.8. From Theorem 8.3, the eigenvalue of $\lambda_\theta A_\theta$ on ϕ is $\lambda_\theta \phi(\theta)$ if $\theta = \theta^{-1}$; otherwise it is $\lambda_\theta(\phi(\theta) + \phi(\theta^{-1}))$, which is equal to $\lambda_\theta \phi(\theta) + \lambda_{\theta^{-1}} \phi(\theta^{-1})$. Hence the eigenvalue of Λ on ϕ is $\sum_{\theta \in \Theta} \lambda_\theta \phi(\theta)$. ■

Example 8.4 revisited Let $\Upsilon = \{1, a, b\}$ in $\mathbb{Z}_4 \times \mathbb{Z}_2$. Then, in the thin design generated by Υ, the λ_θ are as in the following table.

θ	1	a	a^2	a^3	b	ab	a^2b	a^3b
λ_θ	3	1	0	1	2	1	0	1

Since $r = k = 3$, we use Theorem 8.10 and the characters on page 232 to obtain the canonical efficiency factors as follows.

ϕ	$\sum \lambda_\theta \phi(\theta)$	canonical efficiency factor
ϕ_1	1	$\frac{8}{9}$
$\phi_2, \bar{\phi}_2$	5	$\frac{4}{9}$
$\phi_3, \bar{\phi}_3$	1	$\frac{8}{9}$
ϕ_4	1	$\frac{8}{9}$
ϕ_5	1	$\frac{8}{9}$

■

Abelian-group block-designs with the same group and the same block size can be juxtaposed. As the component designs have the same basic contrasts, the canonical efficiency factors of the new design can be calculated from those of its components as in Section 4.6.1.

8.5 Abelian-group designs on structured sets

The cyclic designs on orthogonal block structures in Section 7.5.3 can also be generalized to an arbitrary Abelian group. Let $(\Omega, \mathcal{F})$ be an orthogonal block structure and let Θ be an Abelian group. A design $(\Omega, \mathcal{F}, \Theta, \psi)$ is defined to be a *weak* Abelian-group design if the block design $(\Omega, \mathcal{G}_F, \Theta, \psi)$ is an Abelian-group block-design for each F in $\mathcal{F}$. It is a *strong* Abelian-group design if there are permutations π_θ of Ω for θ in Θ such that

(i) $\pi_\theta(\pi_\eta(\omega)) = \pi_{\theta\eta}(\omega)$ for all ω in Ω and all θ, η in Θ;

(ii) for all F in $\mathcal{F}$ and all θ in Θ, α and β are in the same F-class if and only if $\pi_\theta(\alpha)$ and $\pi_\theta(\beta)$ are in the same F-class;

(iii) $\psi(\pi_\theta(\omega)) = (\psi(\omega))\theta$ for all ω in Ω and all θ in Θ.

Let $|\Theta| = t$. Strong Abelian-group designs for the structure $\underline{\underline{t}}/(\underline{\underline{n}} \times \underline{\underline{m}})$ can be constructed by translating an initial block, while strong Abelian-group designs for the structures $\underline{\underline{n}} \times \underline{\underline{t}}$ and $(\underline{\underline{n}} \times \underline{\underline{t}})/\underline{\underline{m}}$ can be constructed by translating an initial column.

Example 8.6 Let $\Theta = \mathbb{Z}_2 \times \mathbb{Z}_2 \times \mathbb{Z}_2 =$

$$\langle a, b, c : a^2 = b^2 = c^2 = 1,\ ab = ba,\ ac = ca,\ bc = cb \rangle.$$

In the nested block design in Figure 8.1, each subgroup Υ of Θ of order 4 gives a pair of large blocks, which are the cosets of that subgroup. These are divided into small blocks, which are the cosets of a subgroup of Υ of order 2, in such a way that all subgroups of Θ of order 2 are used. Hence this is a strong Abelian-group design. In fact, it is a balanced design. ■

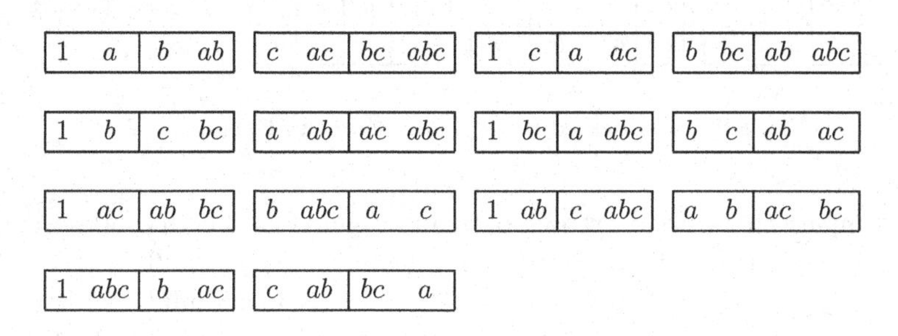

Fig. 8.1. Abelian-group design in nested blocks in Example 8.6

The results of Section 7.5.3 all extend to weak Abelian-group designs. In particular, every weak Abelian-group design is partially balanced with respect to the association scheme defined by the blueprint Inverse, and the canonical efficiency factors in stratum $W_F^{(\Omega)}$ are

$$\sum_{G \in \mathcal{F}} \mu(F, G) \frac{1}{rk_G} \sum_{\theta \in \Theta} \lambda_{G\theta} \phi(\theta)$$

for irreducible characters ϕ of Θ (compare with Theorem 7.9).

Example 8.7 The irreducible characters of the group Θ in Example 8.6 are given in Table 8.1. Figure 8.2 shows a nested row-column design which is strong for Θ. The first row of each concurrence matrix is shown on the left of Table 8.2, with its eigenvalues on the irreducible characters on the right. Subtraction gives the canonical efficiency factors in Table 8.3.

	1	a	b	ab	c	ac	bc	abc
ϕ_0	1	1	1	1	1	1	1	1
ϕ_1	1	-1	1	-1	1	-1	1	-1
ϕ_2	1	1	-1	-1	1	1	-1	-1
ϕ_3	1	-1	-1	1	1	-1	-1	1
ϕ_4	1	1	1	1	-1	-1	-1	-1
ϕ_5	1	-1	1	-1	-1	1	-1	1
ϕ_6	1	1	-1	-1	-1	-1	1	1
ϕ_7	1	-1	-1	1	-1	1	1	-1

Table 8.1. *Character table of* $\mathbb{Z}_2 \times \mathbb{Z}_2 \times \mathbb{Z}_2$

c	b	1	bc
b	1	bc	c

ac	ab	a	abc
ab	a	abc	ac

a	c	ac	1
1	a	c	ac

ab	bc	abc	b
b	ab	bc	abc

c	ab	1	abc
abc	c	ab	1

ac	b	a	bc
bc	ac	b	a

ab	a	b	1
abc	ac	bc	c

Fig. 8.2. First nested row-column design in Example 8.7

Compare this with the strong design in Figure 8.3. The second design is balanced, with canonical efficiency factors $\varepsilon_{\text{blocks},1} = 0$, $\varepsilon_{\text{rows},1} = 1/7$, $\varepsilon_{\text{columns},1} = 3/7$ and $\varepsilon_{E1} = 3/7$. Just as in Example 7.15, the balanced design is appreciably worse than the other for the fixed-effects model. ■

8.6 Group block structures

Groups give an interesting subfamily of orthogonal block structures.

Let Ω be a finite group, written multiplicatively, and let Υ be a subgroup of Ω. Then the left translates of Υ are called *left cosets* of Υ and

b	c	bc	1
abc	a	ab	ac

ab	bc	ac	1
b	abc	c	a

c	ac	a	1
ab	b	bc	abc

bc	a	abc	1
c	ab	ac	b

ac	abc	b	1
bc	c	a	ab

a	b	ab	1
ac	bc	abc	c

abc	ab	c	1
a	ac	b	bc

Fig. 8.3. Balanced nested row-column design in Example 8.7

1	a	b	ab	c	ac	bc	abc	ϕ_0	ϕ_1	ϕ_2	ϕ_3	ϕ_4	ϕ_5	ϕ_6	ϕ_7	rk_F
49	49	49	49	49	49	49	49	392	0	0	0	0	0	0	0	392
13	5	5	5	13	5	5	5	56	16	16	16	0	0	0	0	56
7	3	3	3	6	2	2	2	28	8	8	8	4	0	0	0	28
7	1	1	1	1	1	1	1	14	6	6	6	6	6	6	6	14
7	0	0	0	0	0	0	0	7	7	7	7	7	7	7	7	7
first row of concurrence matrix								eigenvalues on characters								

Table 8.2. *Calculations in Example 8.7: the partitions F are in the order U, blocks, rows, columns, E*

strata for Ω		strata for Θ: ϕ_0 (1)	ϕ_1, ϕ_2, ϕ_3 (3)	ϕ_4 (1)	ϕ_5, ϕ_6, ϕ_7 (3)
U	(1)	1	0	0	0
blocks	(6)	0	$\frac{2}{7}$	0	0
rows	(7)	0	0	$\frac{1}{7}$	0
columns	(21)	0	$\frac{1}{7}$	$\frac{3}{7}$	$\frac{3}{7}$
E	(21)	0	$\frac{4}{7}$	$\frac{3}{7}$	$\frac{4}{7}$

Table 8.3. *Canonical efficiency factors for first design in Example 8.7*

they form a uniform partition $F(\Upsilon)$ of Ω. If Ξ is another subgroup then $\Upsilon \cap \Xi$ is also a subgroup of Ω and $F(\Upsilon) \wedge F(\Xi) = F(\Upsilon \cap \Xi)$.

The subset $\Upsilon\Xi$ of Ω is defined by

$$\Upsilon\Xi = \{\upsilon\xi : \upsilon \in \Upsilon,\ \xi \in \Xi\}.$$

Subgroups Υ and Ξ are said to *commute* with each other if $\Upsilon\Xi = \Xi\Upsilon$. A standard piece of group theory shows that $\Upsilon\Xi$ is a subgroup of Ω if and only if $\Upsilon\Xi = \Xi\Upsilon$.

Theorem 8.11 *If $\Upsilon\Xi = \Xi\Upsilon$ then $F(\Upsilon)$ is orthogonal to $F(\Xi)$ and $F(\Upsilon\Xi) = F(\Upsilon) \vee F(\Xi)$; otherwise $F(\Upsilon)$ and $F(\Xi)$ are not orthogonal to each other.*

Proof For ω in Ω, let

$$m(\omega) = |\{(\upsilon,\xi) \in \Upsilon \times \Xi : \upsilon\xi = \omega\}|,$$

and define $m'(\omega)$ similarly with $\Upsilon \times \Xi$ replaced by $\Xi \times \Upsilon$. Then the (α,β)-element of $R_{F(\Upsilon)}R_{F(\Xi)}$ is equal to $m(\alpha^{-1}\beta)$ while the (α,β)-element of $R_{F(\Xi)}R_{F(\Upsilon)}$ is equal to $m'(\alpha^{-1}\beta)$. If $\Upsilon\Xi \neq \Xi\Upsilon$ then there is some pair (α,β) for which $m(\alpha^{-1}\beta) \neq m'(\alpha^{-1}\beta)$. Thus $R_{F(\Upsilon)}R_{F(\Xi)} \neq R_{F(\Xi)}R_{F(\Upsilon)}$ so $P_{F(\Upsilon)}P_{F(\Xi)} \neq P_{F(\Xi)}P_{F(\Upsilon)}$ and, by definition, $F(\Upsilon)$ is not orthogonal to $F(\Xi)$.

If $\upsilon_1\xi_1 = \upsilon_2\xi_2 = \omega$ with υ_1, υ_2 in Υ and ξ_1, ξ_2 in Ξ then $\upsilon_2^{-1}\upsilon_1 = \xi_2^{-1}\xi_1$. Thus $m(\omega)$ is equal to either zero or $|\Upsilon \cap \Xi|$. In particular, $|\Upsilon\Xi| = |\Upsilon|\,|\Xi| \,/\, |\Upsilon \cap \Xi|$. Therefore

$$\left(P_{F(\Upsilon)}P_{F(\Xi)}\right)(\alpha,\beta) = \frac{m(\alpha^{-1}\beta)}{|\Upsilon|\,|\Xi|} = \begin{cases} 0 & \text{if } \alpha^{-1}\beta \notin \Upsilon\Xi \\ \dfrac{1}{|\Upsilon\Xi|} & \text{otherwise.} \end{cases}$$

If $\Upsilon\Xi = \Xi\Upsilon$ then $\Upsilon\Xi$ is a subgroup and

$$P_{F(\Upsilon)}P_{F(\Xi)} = P_{F(\Upsilon\Xi)} = P_{F(\Xi)}P_{F(\Upsilon)}.$$

Therefore $F(\Upsilon)$ is orthogonal to $F(\Xi)$, and Lemma 6.4 shows that $F(\Upsilon\Xi) = F(\Upsilon) \vee F(\Xi)$. ■

Corollary 8.12 *Any set of pairwise commuting subgroups of a group Ω which is closed under intersection and product defines an orthogonal block structure on Ω.*

Such an orthogonal block structure is called a *group block structure*.

If at least one of Υ, Ξ is a normal subgroup of Ω then $\Upsilon\Xi = \Xi\Upsilon$. So the set of all normal subgroups of Ω always gives a group block structure. If Ω is Abelian then all subgroups are normal and we obtain the following special blueprint, called Subgroup.

Theorem 8.13 *Let Ω be an Abelian group. Let Υ_0, ..., Υ_s be the cyclic subgroups of Ω, where $\Upsilon_0 = \langle 1_\Omega \rangle$. For $i = 0$, ..., s, let $\Delta_i = \{\omega \in \Omega : \langle\omega\rangle = \Upsilon_i\}$. Then Δ_0, Δ_1, ..., Δ_s form a blueprint for Ω.*

Proof Consider the association scheme $\mathcal{Q}$ of the group block structure defined by the set of all subgroups of Ω. For any subgroup Ξ, elements α and β are in the same class of $F(\Xi)$ if and only if $\alpha^{-1}\beta \in \Xi$. The definition of $\mathcal{Q}$ in Theorem 6.8 shows that $\mathcal{C}_{F(\Xi)}$ is empty if Ξ is not cyclic: in fact,

$$\mathcal{C}_{F(\Xi)} = \left\{(\alpha,\beta) : \langle\alpha^{-1}\beta\rangle = \Xi\right\}.$$

Thus the non-empty classes are $\mathcal{C}_0, \ldots, \mathcal{C}_s$, where

$$\mathcal{C}_i = \left\{(\alpha, \beta) : \langle \alpha^{-1}\beta \rangle = \Upsilon_i\right\}.$$

Now $\mathcal{Q} = \mathcal{Q}(\Delta)$ where $\Delta_i = \{\omega \in \Omega : \langle \omega \rangle = \Upsilon_i\}$. Theorem 8.1 shows that Δ is a blueprint for Ω. ∎

Theorem 8.14 *Let Ξ be a subgroup of an Abelian group Θ. The thin block design whose blocks are the cosets of Ξ is partially balanced with respect to $\mathcal{Q}(\mathsf{Subgroup})$. If ϕ is an irreducible character of Θ then the canonical efficiency factor for ϕ is*

$$\begin{cases} 0 & \text{if } \phi(\xi) = 1_{\mathbb{C}} \text{ for all } \xi \text{ in } \Xi \\ 1 & \text{otherwise.} \end{cases}$$

Proof In Theorems 8.7 and 8.10 we have $t = |\Theta|$, $k = |\Xi|$, $r = 1$, $l = t/k$ and

$$m_\theta(\Xi) = \begin{cases} k & \text{if } \theta \in \Xi \\ 0 & \text{otherwise.} \end{cases}$$

Thus

$$\lambda_\theta = \begin{cases} 1 & \text{if } \theta \in \Xi \\ 0 & \text{otherwise.} \end{cases}$$

If θ_1 and θ_2 are in the same class of $\mathsf{Subgroup}$ then either both or neither of θ_1 and θ_2 are in Ξ, so the design is partially balanced with respect to $\mathcal{Q}(\mathsf{Subgroup})$.

Theorem 8.10 shows that the canonical efficiency factor ε for ϕ is

$$1 - \frac{1}{k}\sum_{\xi \in \Xi} \phi(\xi).$$

If $\phi(\xi) = 1_{\mathbb{C}}$ for all ξ in Ξ then $\varepsilon = 0$; otherwise, Theorem 2.1 of [159] shows that $\sum_{\xi \in \Xi} \phi(\xi) = 0$ and so $\varepsilon = 1$. ∎

Statisticians call such a block design a *confounded* design, in which ϕ is confounded if $\phi(\xi) = 1_{\mathbb{C}}$ for all ξ in Ξ. Juxtaposing designs formed from r subgroups $\Xi_1, \ldots, \Xi_r$ of the same size gives a block design in which the canonical efficiency factor for ϕ is

$$1 - \frac{|\{i : \phi(\xi) = 1_{\mathbb{C}} \text{ for all } \xi \text{ in } \Xi_i\}|}{r}.$$

Non-Abelian groups also admit some non-trivial group block structures.

Example 8.8 Let $\Omega = \langle a, b : a^9 = b^3 = 1,\ b^{-1}ab = a^4 \rangle$. This is not Abelian, but its subgroups commute pairwise. The lattice of subgroups is shown in Figure 8.4: here

$$
\begin{array}{lll}
\Psi = \langle b, a^3 \rangle & & \\
\Xi_1 = \langle a \rangle & \Xi_2 = \langle ab \rangle & \Xi_3 = \langle a^2 b \rangle \\
\Phi = \langle a^3 \rangle & & \\
\Upsilon_1 = \langle b \rangle & \Upsilon_2 = \langle a^3 b \rangle & \Upsilon_3 = \langle a^6 b \rangle.
\end{array}
$$

This is identical to the Hasse diagram for the orthogonal block structure.

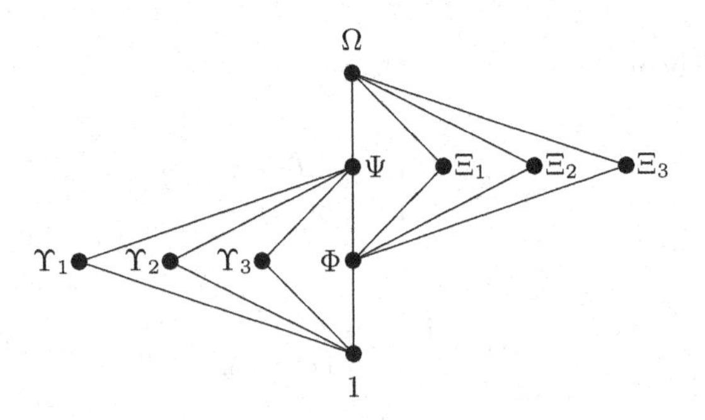

Fig. 8.4. Lattice of all subgroups in Example 8.8

All subgroups are cyclic except Ω and Ψ: the "associate classes" which correspond to these two are empty. Calculating stratum dimensions from Equation (6.8) shows that the zero "strata" are those corresponding to Φ and 1. In fact, the zeta and Möbius functions are

$$
\zeta = \begin{array}{c}
 \\ 1 \\ \Upsilon_1 \\ \Upsilon_2 \\ \Upsilon_3 \\ \Phi \\ \Psi \\ \Xi_1 \\ \Xi_2 \\ \Xi_3 \\ \Omega
\end{array}
\begin{array}{c}
\begin{array}{cccccccccc} 1 & \Upsilon_1 & \Upsilon_2 & \Upsilon_3 & \Phi & \Psi & \Xi_1 & \Xi_2 & \Xi_3 & \Omega \end{array} \\
\left[\begin{array}{cccccccccc}
1 & 1 & 1 & 1 & 1 & 1 & 1 & 1 & 1 & 1 \\
0 & 1 & 0 & 0 & 0 & 1 & 0 & 0 & 0 & 1 \\
0 & 0 & 1 & 0 & 0 & 1 & 0 & 0 & 0 & 1 \\
0 & 0 & 0 & 1 & 0 & 1 & 0 & 0 & 0 & 1 \\
0 & 0 & 0 & 0 & 1 & 1 & 1 & 1 & 1 & 1 \\
0 & 0 & 0 & 0 & 0 & 1 & 0 & 0 & 0 & 1 \\
0 & 0 & 0 & 0 & 0 & 0 & 1 & 0 & 0 & 1 \\
0 & 0 & 0 & 0 & 0 & 0 & 0 & 1 & 0 & 1 \\
0 & 0 & 0 & 0 & 0 & 0 & 0 & 0 & 1 & 1 \\
0 & 0 & 0 & 0 & 0 & 0 & 0 & 0 & 0 & 1
\end{array}\right]
\end{array}
$$

and

$$\mu = \begin{array}{c} \\ 1 \\ \Upsilon_1 \\ \Upsilon_2 \\ \Upsilon_3 \\ \Phi \\ \Psi \\ \Xi_1 \\ \Xi_2 \\ \Xi_3 \\ \Omega \end{array} \begin{array}{c} \begin{array}{cccccccccc} 1 & \Upsilon_1 & \Upsilon_2 & \Upsilon_3 & \Phi & \Psi & \Xi_1 & \Xi_2 & \Xi_3 & \Omega \end{array} \\ \left[\begin{array}{cccccccccc} 1 & -1 & -1 & -1 & -1 & 3 & 0 & 0 & 0 & 0 \\ 0 & 1 & 0 & 0 & 0 & -1 & 0 & 0 & 0 & 0 \\ 0 & 0 & 1 & 0 & 0 & -1 & 0 & 0 & 0 & 0 \\ 0 & 0 & 0 & 1 & 0 & -1 & 0 & 0 & 0 & 0 \\ 0 & 0 & 0 & 0 & 1 & -1 & -1 & -1 & -1 & 3 \\ 0 & 0 & 0 & 0 & 0 & 1 & 0 & 0 & 0 & -1 \\ 0 & 0 & 0 & 0 & 0 & 0 & 1 & 0 & 0 & -1 \\ 0 & 0 & 0 & 0 & 0 & 0 & 0 & 1 & 0 & -1 \\ 0 & 0 & 0 & 0 & 0 & 0 & 0 & 0 & 1 & -1 \\ 0 & 0 & 0 & 0 & 0 & 0 & 0 & 0 & 0 & 1 \end{array}\right] \end{array}.$$

Equation (6.9) gives the character table as follows.

$$\begin{array}{cc} \\ \\ 1 & (1) \\ \Upsilon_1 & (2) \\ \Upsilon_2 & (2) \\ \Upsilon_3 & (2) \\ \Phi & (2) \\ \Psi & (0) \\ \Xi_1 & (6) \\ \Xi_2 & (6) \\ \Xi_3 & (6) \\ \Omega & (0) \end{array} \begin{array}{c} \begin{array}{cccccccccc} 1 & \Upsilon_1 & \Upsilon_2 & \Upsilon_3 & \Phi & \Psi & \Xi_1 & \Xi_2 & \Xi_3 & \Omega \\ (0) & (6) & (6) & (6) & (0) & (2) & (2) & (2) & (2) & (1) \end{array} \\ \left[\begin{array}{cccccccccc} 1 & 1 & 1 & 1 & 1 & 1 & 1 & 1 & 1 & 1 \\ -1 & 2 & -1 & -1 & -1 & 2 & -1 & -1 & -1 & 2 \\ -1 & -1 & 2 & -1 & -1 & 2 & -1 & -1 & -1 & 2 \\ -1 & -1 & -1 & 2 & -1 & 2 & -1 & -1 & -1 & 2 \\ -1 & -1 & -1 & -1 & 2 & 2 & 2 & 2 & 2 & 2 \\ 3 & 0 & 0 & 0 & 0 & 0 & 0 & 0 & 0 & 0 \\ 0 & 0 & 0 & 0 & -3 & -3 & 6 & -3 & -3 & 6 \\ 0 & 0 & 0 & 0 & -3 & -3 & -3 & 6 & -3 & 6 \\ 0 & 0 & 0 & 0 & -3 & -3 & -3 & -3 & 6 & 6 \\ 0 & 0 & 0 & 0 & 9 & 0 & 0 & 0 & 0 & 0 \end{array}\right] \end{array}$$

Of course, two rows and two columns should be deleted from this matrix! ■

8.7 Automorphism groups

In Section 3.3 we introduced weak and strong automorphisms of association schemes. A thorough investigation of these requires more knowledge of permutation group theory than is assumed in this book, so this section develops only those ideas that are most relevant to other parts of this book, specifically Chapters 3 and 11.

Suppose that $\mathcal{P}$ is a partition of $\Omega \times \Omega$, where Ω is any finite set. Let π be a permutation of Ω. For each class $\mathcal{C}$ of $\mathcal{P}$, put $\pi(\mathcal{C}) = \{(\pi(\alpha), \pi(\beta)) : (\alpha, \beta) \in \mathcal{C}\}$. Then the classes $\pi(\mathcal{C})$ form a partition $\pi(\mathcal{P})$

of $\Omega \times \Omega$, which is an association scheme if $\mathcal{P}$ is. If $\pi(\mathcal{P}) = \mathcal{P}$ then π induces a permutation of the classes of $\mathcal{P}$: by a slight abuse of the terminology in Section 3.3, we call π a *weak automorphism* of $\mathcal{P}$. If $\pi(\mathcal{C}) = \mathcal{C}$ for every class $\mathcal{C}$ of $\mathcal{P}$ then π is a strong automorphism of $\mathcal{P}$.

Lemma 8.15 *Let ϕ be a group automorphism of a finite group Ω. Let Δ be a blueprint for Ω with classes Δ_0, Δ_1, ..., Δ_s. If there is a permutation σ of $\{0,\ldots,s\}$ such that $\{\phi(\omega) : \omega \in \Delta_i\} = \Delta_{\sigma(i)}$ for $i = 0$, ..., s then ϕ is a weak automorphism of $\mathcal{Q}(\Delta)$.*

Proof By definition,

$$\phi(\mathcal{C}_i) = \{(\phi(\alpha), \phi(\beta)) : (\alpha, \beta) \in \mathcal{C}_i\} = \left\{(\phi(\alpha), \phi(\beta)) : \alpha^{-1}\beta \in \Delta_i\right\}$$

so the existence of σ ensures that

$$\phi(\mathcal{C}_i) = \left\{(\phi(\alpha), \phi(\beta)) : \phi(\alpha^{-1}\beta) \in \Delta_{\sigma(i)}\right\}.$$

Since ϕ is a group automorphism of Ω, $\phi(\alpha^{-1}\beta) = (\phi(\alpha))^{-1}\phi(\beta)$ and so

$$\phi(\mathcal{C}_i) = \left\{(\phi(\alpha), \phi(\beta)) : (\phi(\alpha))^{-1}\phi(\beta) \in \Delta_{\sigma(i)}\right\} = \mathcal{C}_{\sigma(i)}. \quad \blacksquare$$

Proposition 3.5 is a special case of Lemma 8.15.

Proposition 8.16 *Let $\mathcal{Q}$ be an association scheme on a set Ω. The set of all weak automorphisms of $\mathcal{Q}$ forms a group under composition. So does the set of all strong automorphisms. The latter is a subgroup of the former.*

For $t = 1$, 2, let $\mathcal{Q}_t$ be an association scheme on Ω_t and let Φ_t be a group of automorphisms of Ω_t (either weak or strong, so long as both groups are of the same type). For ϕ_t in Φ_t define the permutation (ϕ_1, ϕ_2) of $\Omega_1 \times \Omega_2$ by

$$(\phi_1, \phi_2)\colon (\alpha, \beta) \mapsto (\phi_1(\alpha), \phi_2(\beta)).$$

Then (ϕ_1, ϕ_2) is an automorphism of $\mathcal{Q}_1 \times \mathcal{Q}_2$ of the same type. The set of all such permutations is called the *direct product* of Φ_1 and Φ_2.

Now let ϕ be in Φ_1 and, for all α in Ω_1, let ψ_α be in Φ_2. Define the permutation π of $\Omega_1 \times \Omega_2$ by

$$\pi\colon (\alpha, \beta) \mapsto (\phi(\alpha), \psi_\alpha(\beta)).$$

Then π is an automorphism of $\mathcal{Q}_1/\mathcal{Q}_2$ of the same type. The set of all such permutations is called the *wreath product* of Φ_1 and Φ_2.

Theorem 8.17 *Let $\mathcal{Q}$ be an association scheme on a set Ω. Let Φ be a group of weak automorphisms of $\mathcal{Q}$. Form the partition $\mathcal{P}$ of $\Omega \times \Omega$ by merging classes $\mathcal{C}_i$ and $\mathcal{C}_j$ of $\mathcal{Q}$ whenever there is some ϕ in Φ with $\phi(\mathcal{C}_i) = \mathcal{C}_j$. Then $\mathcal{P}$ is an association scheme on Ω and Φ is a group of strong automorphisms of $\mathcal{P}$.*

Proof We check the partition conditions for an association scheme.

(i) For every ϕ in Φ, $\phi(\mathrm{Diag}(\Omega)) = \mathrm{Diag}(\Omega)$, so $\mathrm{Diag}(\Omega)$ is a class of $\mathcal{P}$.

(ii) Since every class of $\mathcal{Q}$ is symmetric, so is every class of $\mathcal{P}$.

(iii) Write $[i] = \{j : \exists\phi \in \Phi \text{ with } \pi(\mathcal{C}_i) = \mathcal{C}_j\}$, and let $\mathcal{D}_{[i]}$ be the class of $\mathcal{P}$ obtained by merging the $\mathcal{C}_j$ for j in $[i]$. For (α, β) in $\mathcal{D}_{[k]}$, we need to find the size $p([i], [j], \alpha, \beta)$ of $\mathcal{D}_{[i]}(\alpha) \cap \mathcal{D}_{[j]}(\beta)$. This set is the disjoint union

$$\bigcup_{m\in[i]} \bigcup_{n\in[j]} (\mathcal{C}_m(\alpha) \cap \mathcal{C}_n(\beta))$$

so its size is $\sum_{m\in[i]} \sum_{n\in[j]} p^k_{mn}$ if $(\alpha, \beta) \in \mathcal{C}_k$.

If $l \in [k]$ then there is some ϕ in Φ such that $\phi(\mathcal{C}_k) = \mathcal{C}_l$. Since $(\phi(\mathcal{D}))(\phi(\alpha)) = \phi(\mathcal{D}(\alpha))$ for every subset $\mathcal{D}$ of $\Omega \times \Omega$, and ϕ fixes both $\mathcal{D}_{[i]}$ and $\mathcal{D}_{[j]}$,

$$\mathcal{D}_{[i]}(\phi(\alpha)) \cap \mathcal{D}_{[j]}(\phi(\beta)) = \phi\left(\mathcal{D}_{[i]}(\alpha) \cap \mathcal{D}_{[j]}(\beta)\right)$$

and so $p([i], [j], \alpha, \beta) = p([i], [j], \phi(\alpha), \phi(\beta))$. Therefore

$$\sum_{m\in[i]} \sum_{n\in[j]} p^k_{mn} = \sum_{m\in[i]} \sum_{n\in[j]} p^l_{mn}$$

and so $p([i], [j], \alpha, \beta)$ is constant for (α, β) in $\mathcal{D}_{[k]}$, as required.

By construction of $\mathcal{P}$, each ϕ in Φ fixes each class of $\mathcal{P}$ and so is a strong automorphism of $\mathcal{P}$. ■

If Ω is a group, we can define a *permutation action* of Ω on itself. For ω in Ω, define the permutation π_ω of Ω by

$$\pi_\omega(\alpha) = \omega\alpha \qquad \text{for } \alpha \text{ in } \Omega.$$

Then, for ω_1, ω_2 in Ω, we have

$$\pi_{\omega_1}(\pi_{\omega_2}(\alpha)) = \omega_1\omega_2\alpha = \pi_{\omega_1\omega_2}(\alpha)$$

so the map $\omega \mapsto \pi_\omega$ is indeed a homomorphism, which is what is meant by π being an 'action'. This is called the *left regular action*.

(More information about permutation actions may be found in books such as [71, 186]. However, the formulae there are superficially different from those here, because permutation-group theorists usually write functions on the right of their arguments.)

Definition A group Φ of permutations of Ω is *transitive* if, for all elements α, β in Ω, there is at least one element ϕ in Φ with $\phi(\alpha) = \beta$; it is *regular* if, for all elements α, β in Ω, there is exactly one element ϕ in Φ with $\phi(\alpha) = \beta$.

Lemma 8.18 *Let Ω be a finite group and let $\mathcal{P}$ be a partition of $\Omega \times \Omega$. Then there is a partition Δ of Ω such that $\mathcal{P} = \mathcal{Q}(\Delta)$ if and only if π_ω is a strong automorphism of $\mathcal{P}$ for every element ω of Ω.*

Proof Suppose that $\mathcal{P} = \mathcal{Q}(\Delta)$ and that $\mathcal{C}$ corresponds to the class Δ_i of Δ. If $(\alpha, \beta) \in \mathcal{C}$ then $\alpha^{-1}\beta \in \Delta_i$. Thus

$$[\pi_\omega(\alpha)]^{-1}\,\pi_\omega(\beta) = (\omega\alpha)^{-1}(\omega\beta) = \alpha^{-1}\omega^{-1}\omega\beta = \alpha^{-1}\beta \in \Delta_i$$

and so $(\pi_\omega(\alpha), \pi_\omega(\beta)) \in \mathcal{C}$. Therefore $\pi_\omega(\mathcal{C}) = \mathcal{C}$.

Conversely, suppose that $\mathcal{C} = \pi_\omega(\mathcal{C})$ for all ω in Ω, and that $(\alpha, \beta) \in \mathcal{C}$. If $\alpha_1^{-1}\beta_1 = \alpha^{-1}\beta$ then $(\alpha_1, \beta_1) = (\omega\alpha, \omega\beta)$ for $\omega = \alpha_1\alpha^{-1} = \beta_1\beta^{-1}$ and so $(\alpha_1, \beta_1) \in \mathcal{C}$. If this is true for all classes $\mathcal{C}$ of $\mathcal{P}$ then $\mathcal{P} = \mathcal{Q}(\Delta)$ for some partition Δ of Ω. ■

Theorem 8.19 *An association scheme has a regular group of strong automorphisms if and only if it is a group scheme.*

Proof First, suppose that Ω is a group with a blueprint Δ. For each ω in Ω, Lemma 8.18 shows that π_ω is a strong automorphism of $\mathcal{Q}(\Delta)$. Thus $\{\pi_\omega : \omega \in \Omega\}$ is a group of strong automorphisms of $\mathcal{Q}(\Delta)$. It is regular because, given any pair α, β of elements of Ω, there is a unique one of these automorphisms, namely $\pi_{\beta\alpha^{-1}}$, taking α to β.

Conversely, suppose that Φ is a regular group of strong automorphisms of an association scheme $\mathcal{Q}$ on a set Ω with classes $\mathcal{C}_i$. Choose a fixed element ω_0 of Ω. For $\alpha \in \Omega$, write $\bar{\alpha} = \phi$ where ϕ is the unique element of Φ such that $\phi(\omega_0) = \alpha$. Then $\alpha \mapsto \bar{\alpha}$ is a bijection which identifies Ω with Φ.

Suppose that $\bar{\alpha_1}^{-1}\bar{\beta_1} = \bar{\alpha_2}^{-1}\bar{\beta_2}$. Put $\bar{\omega} = \bar{\alpha_2}\bar{\alpha_1}^{-1} = \bar{\beta_2}\bar{\beta_1}^{-1}$. Then $\bar{\omega}\bar{\alpha_1} = \bar{\alpha_2}$ so $\bar{\omega}(\alpha_1) = \bar{\omega}\,(\bar{\alpha_1}(\omega_0)) = \bar{\alpha_2}(\omega_0) = \alpha_2$; similarly $\bar{\omega}(\beta_1) = \beta_2$, and so $(\alpha_2, \beta_2) = (\bar{\omega}(\alpha_1), \bar{\omega}(\beta_1))$. Since $\bar{\omega}$ is a strong automorphism, (α_1, β_1) and (α_2, β_2) are in the same class of $\mathcal{Q}$. Thus $\mathcal{Q} = \mathcal{Q}(\Delta)$ for some partition Δ of Ω. Theorem 8.1 shows that Δ is a blueprint. ■

A	B	C	D	E
B	A	D	E	C
C	E	A	B	D
D	C	E	A	B
E	D	B	C	A

Fig. 8.5. A Latin square of order 5

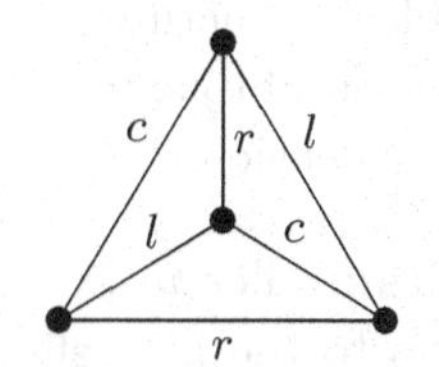

Fig. 8.6. Associate classes among four points

Regular groups of automorphisms are often called *translation groups*, so association schemes derived from blueprints are often called *translation schemes*.

There are association schemes which have groups of strong automorphisms which are transitive but not regular. For example, the triangular scheme T(5) has the symmetric group S_5 acting on the unordered pairs of a 5-set in the natural way. This is a group of strong automorphisms. None of its subgroups is regular, so T(5) is not a translation scheme.

Automorphisms also provide a way of showing that not every orthogonal block structure is a group block structure. Consider the square association scheme $\mathcal{Q}$ defined as in Section 6.5 by the Latin square in Figure 8.5. The point in the top left-hand corner is contained in four 2×2 Latin subsquares, hence in four quadrangles of the type shown in Figure 8.6, where the labels r, c, l denote associates by A_{rows}, A_{columns} and A_{letters}. The point in the top right-hand corner is in only one such quadrangle, so there can be no strong automorphism of $\mathcal{Q}$ taking it to the point in the top left-hand corner. Hence $\mathcal{Q}$ is not a translation scheme and so $\mathcal{F}$ cannot be a group block structure.

8.8 Latin cubes

Let G be a finite group. The *Cayley table* of G is defined to be the Latin square with rows, columns and letters labelled by the elements of G such that the letter in row g and column h is gh. As in Example 6.2, this Latin square defines an orthogonal block structure on $G \times G$. If G is Abelian then the orthogonal block structure is a group block structure: the non-trivial partitions are into the cosets of the subgroups $1 \times G$, $G \times 1$ and $\{(g, g^{-1}) : g \in G\}$.

Most Latin squares are not Cayley tables but they still define orthogonal block structures.

It is natural to extend the ideas of Latin squares and Cayley tables to higher dimensions.

Definition For $m \geqslant 3$, a *Latin m-cube* of order n is an allocation of n letters to the n^m cells of a set Γ^m, where $|\Gamma| = n$, such that if any $m-1$ of the coordinates are fixed at any of the possible n^{m-1} values then each letter occurs once among the cells obtained by varying the remaining coordinate. A Latin 3-cube is called simply a Latin cube.

Given any Latin square Π of order n, we can construct a Latin cube of order n by repeating Π in every layer but permuting the letters by an n-cycle as we move upwards.

Each cell of a Latin m-cube may be identified with the $(m+1)$-tuple

$$(\text{1st coordinate, 2nd coordinate, } \ldots, \text{ } m\text{-th coordinate, letter}).$$

A set of such $(m+1)$-tuples is a Latin m-cube if and only if, for each of the $(m+1)$ positions, each m-tuple occurs exactly once in the remaining positions. This characterization shows that the role of the letters is no different from the role of any of the coordinate positions.

Thus a Latin m-cube of order n is a set of size n^m together with $m+1$ uniform partitions into n classes, such that any m of those partitions, together with the infimum of every subset of them, form an orthogonal block structure isomorphic to $(\underline{\underline{n}})^m$. Unlike a Latin square, a Latin m-cube for $m \geqslant 3$ is not itself an orthogonal block structure, because its set of partitions is not closed under taking infima. However, we can ask when its set of partitions is *contained* in an orthogonal block structure. To answer this question, we need to generalize the idea of Cayley tables.

Definition For $m \geqslant 3$, an m-dimensional *extended Cayley table* for the group G is the Latin m-cube whose coordinates and letters are all labelled by the elements of G in such a way that the letter in cell $(g_1, g_2, \ldots, g_m)$ is $g_1 g_2 \cdots g_m$.

Theorem 8.20 *Let G be an Abelian group. Put $\Omega = G^m$. For $i = 1$, $\ldots$, m let F_i be the partition of Ω according to the value of the i-th coordinate, and let F_{m+1} be the partition of Ω according to the letters of the extended Cayley table of G. Put $\mathcal{F} = \{F_1, \ldots, F_{m+1}\}$. Then there is an orthogonal block structure $\mathcal{H}$ on Ω with $\mathcal{F} \subset \mathcal{H}$.*

Proof For $i = 1, \ldots, m$, let H_i be the subgroup $H_{i1} \times H_{i2} \times \cdots \times H_{im}$ of Ω, where $H_{ii} = 1$ and $H_{ij} = G$ if $j \neq i$. Let

$$H_{m+1} = \{(g_1, g_2, \ldots, g_m) \in G^m : g_1 g_2 \cdots g_m = 1\}.$$

Then H_{m+1} is a subgroup of G and, for $i = 1, \ldots, m+1$, the classes of F_i are the cosets of H_i.

Let $\mathcal{H}$ be the set of coset partitions defined by all subgroups of Ω. Then $\mathcal{F} \subset \mathcal{H}$. Because G is Abelian, $\mathcal{H}$ is an orthogonal block structure. ■

For simplicity, I shall prove the converse of Theorem 8.20 just for the case $m = 3$. This is enough to indicate that dimension 2 is very different from higher dimensions: while every Latin square is an orthogonal block structure, most Latin cubes are not even contained in orthogonal block structures.

Theorem 8.21 *Suppose that the partitions F_1, F_2, F_3, F_4 form a Latin cube on a set Ω. Put $\mathcal{F} = \{F_1, F_2, F_3, F_4\}$. If there is an orthogonal block structure $\mathcal{H}$ on Ω with $\mathcal{F} \subseteq \mathcal{H}$ then $(\Omega, \mathcal{F})$ is isomorphic to the extended Cayley table of an Abelian group.*

Proof Let G be the set of letters occurring in the cube; that is, G is the set of F_4-classes. Choose an element in G and call it 1. Let ω_0 be an element of Ω which is in F_4-class 1. For $i = 1, \ldots, 3$, label the F_i-class containing ω_0 by 1. Since each element of Ω is uniquely specified by any three of the classes containing it, we may uniquely label the remaining F_i-classes, for $i = 1, \ldots, 3$, in such a way that Ω contains the quadruples

$$(1, 1, g, g), \quad (1, g, 1, g) \quad \text{and} \quad (g, 1, 1, g)$$

for all g in G.

Suppose that (a, b, c, d), (a, b, e, f) and (g, h, e, f) are three quadruples in Ω. Then (a, b, c, d) and (a, b, e, f) are in the same class of $F_1 \wedge F_2$ while (a, b, e, f) and (g, h, e, f) are in the same class of $F_3 \wedge F_4$. If $\mathcal{F}$ is contained in an orthogonal block structure $\mathcal{H}$ then $F_1 \wedge F_2$ and $F_3 \wedge F_4$ are in $\mathcal{H}$ and their relation matrices commute. Hence Ω contains an element ω which is in the same $(F_3 \wedge F_4)$-class as (a, b, c, d) and the same $(F_1 \wedge F_2)$-class as (g, h, e, f): then ω must be (g, h, c, d). In other words, if Ω contains (a, b, c, d), (a, b, e, f) and (g, h, e, f) then it contains (g, h, c, d); and similarly for the other two partitions of $\mathcal{F}$ into two pairs. Call this the Quadrangle Rule.

Let $(a, b, 1, c)$ be an element of Ω. Since Ω contains $(1, b, 1, b)$ and $(1, 1, b, b)$, the Quadrangle Rule shows that Ω also contains $(a, 1, b, c)$.

Similarly, if 1 occurs in any of the first three positions of a quadruple in Ω, we obtain another quadruple in Ω if we interchange that 1 with either of the other letters in the first three positions. Call this the Rule of Three.

For each (a, b) in G^2 there is a unique quadruple in Ω of the form $(a, b, 1, ?)$, so we may define a closed binary operation $\circ$ on G by putting $a \circ b = c$ if Ω contains $(a, b, 1, c)$. We shall show that G is an Abelian group under the operation $\circ$.

Since Ω contains $(g, 1, 1, g)$ and $(1, g, 1, g)$ for all g in G, the element 1 is an identity for G. For all g in G, Ω contains quadruples of the form $(g, ?, 1, 1)$ and $(?, g, 1, 1)$, so g has an inverse under $\circ$.

If $a \circ b = c$ then Ω contains $(a, b, 1, c)$. The Rule of Three shows that Ω contains $(a, 1, b, c)$, $(1, a, b, c)$ and $(b, a, 1, c)$. Therefore $b \circ a = c$, and so the operation $\circ$ is commutative.

Let a, b, c be any elements of G. Put $x = a \circ b$, $y = b \circ c$ and $z = x \circ c$. Then Ω contains $(a, b, 1, x)$, $(b, c, 1, y)$ and $(x, c, 1, z)$. By the Rule of Three, Ω contains $(x, 1, c, z)$. Applying the Quadrangle Rule to the elements $(a, b, 1, x)$, $(x, 1, 1, x)$ and $(x, 1, c, z)$ shows that Ω contains (a, b, c, z). Similarly, $(b, c, 1, y)$ gives $(1, b, c, y)$ via the Rule of Three and then (a, b, c, z), $(1, b, c, y)$ and $(1, y, 1, y)$ give $(a, y, 1, z)$ via the Quadrangle Rule. Therefore $a \circ y = z$. Thus $a \circ (b \circ c) = a \circ y = z = x \circ c = (a \circ b) \circ c$ and so $\circ$ is associative. This completes the proof that G is an Abelian group under the operation $\circ$.

Finally, since $z = a \circ b \circ c$, we have just shown that Ω contains the quadruple $(a, b, c, a \circ b \circ c)$ for all (a, b, c) in G^3, and so $(\Omega, \mathcal{F})$ is an extended Cayley table for G. ■

Exercises

8.1 Let $\Omega = \mathbb{Z}_4 \times \mathbb{Z}_2 = \langle a, b : a^4 = b^2 = 1,\ ab = ba \rangle$. Show that $\{1\}$, $\{a, a^3\}$, $\{a^2\}$, $\{ab, a^3b\}$, $\{b, a^2b\}$ is a blueprint for $\mathbb{Z}_4 \times \mathbb{Z}_2$ and that the corresponding association scheme is isomorphic to $(\underline{\underline{2}} \times \underline{\underline{2}})/\underline{\underline{2}}$.

8.2 Let $\Omega = \mathbb{Z}_4 \times \mathbb{Z}_2$. Show that $\mathcal{Q}(\mathsf{Inverse})$ is isomorphic to ④ $\times\ \underline{\underline{2}}$. Generalize.

8.3 Find all the blueprints for the non-Abelian groups S_3, D_8 and Q_8. Here D_8 is the dihedral group $\langle a, b : a^4 = b^2 = 1,\ bab = a^3 \rangle$ and Q_8 is the quaternion group $\langle a, b : a^4 = 1,\ b^2 = a^2,\ b^{-1}ab = a^3 \rangle$.

8.4 Let Δ be a blueprint for a group Ω. Find a condition on Δ which is necessary and sufficient for $\mathcal{Q}(\Delta)$ to be the association scheme defined by a distance-regular graph.

8.5 Show that $\{1\}$, $\{a^2\}$, $\{b\}$, $\{a^2b\}$, $\{a, a^3, ab, a^3b\}$ is a blueprint for $\mathbb{Z}_4 \times \mathbb{Z}_2$. Find the character table of its association scheme.

8.6 Let $\Omega = \mathbb{Z}_3 \times \mathbb{Z}_3 \times \mathbb{Z}_3 =$

$$\langle a, b, c : a^3 = b^3 = c^3 = 1,\ ab = ba,\ ac = ca,\ bc = cb \rangle.$$

Put

$$\begin{aligned}
\Delta_0 &= \{1\}, \\
\Delta_1 &= \{a, a^2, b, b^2, c, c^2, abc, a^2b^2c^2\}, \\
\Delta_2 &= \{ab, a^2b^2, ac, a^2c^2, bc, b^2c^2\}, \\
\Delta_3 &= \{ab^2, a^2b, ac^2, a^2c, bc^2, b^2c, abc^2, a^2b^2c, ab^2c, a^2bc^2, a^2bc, ab^2c^2\}.
\end{aligned}$$

Prove that Δ is a blueprint for $\mathbb{Z}_3 \times \mathbb{Z}_3 \times \mathbb{Z}_3$. Find the character table of its association scheme.

8.7 Let Ω be an Abelian group. Prove that the set of irreducible characters of Ω forms a group Φ under pointwise multiplication.

8.8 Let Ω be an Abelian group and let Φ be the group of irreducible characters defined in Question 8.7. Let Δ be a blueprint for Ω.

(a) Let Γ be the partition of Φ formed by putting ϕ_1 and ϕ_2 in the same class if and only if they lie in the same stratum for $\mathcal{Q}(\Delta)$. Prove that Γ is a blueprint for Φ.

(b) Find Γ for the blueprint Δ in Example 8.4. Also find the character table for $\mathcal{Q}(\Gamma)$.

(c) Elements of Ω can be identified with irreducible characters of Φ by putting

$$\omega(\phi) = \phi(\omega).$$

With this identification, prove that the labelled character table for $\mathcal{Q}(\Gamma)$ is the inverse of the labelled character table for $\mathcal{Q}(\Delta)$, multiplied by $|\Omega|$.

8.9 Prove Proposition 8.6.

8.10 Let Ξ be a subgroup of index l in a group Ω. Let Δ be a blueprint for Ξ. Extend Δ to a partition Γ of Ω by adjoining $\Omega \setminus \Xi$ to Δ. Prove that Γ is a blueprint for Ω and that $\mathcal{Q}(\Gamma) \cong \underline{l}/\mathcal{Q}(\Delta)$.

8.11 Find a perfect difference set of size 6 for the group $\mathbb{Z}_4 \times \mathbb{Z}_4$.

8.12 Use the group $\mathbb{Z}_3 \times \mathbb{Z}_3$ to construct a partially balanced incomplete-block design for nine treatments in nine blocks of size 4. Calculate its canonical efficiency factors.

8.13 Construct the Subgroup blueprint for $\mathbb{Z}_5 \times \mathbb{Z}_5$ and for $\mathbb{Z}_6 \times \mathbb{Z}_6$.

8.14 Under what conditions on the groups Ω_1 and Ω_2 is

$$\mathsf{Subgroup}(\Omega_1) \times \mathsf{Subgroup}(\Omega_2) = \mathsf{Subgroup}(\Omega_1 \times \Omega_2)?$$

8.15 Let $\mathcal{Q}$ be the association scheme of the orthogonal block structure $\mathcal{F}$ defined by the Cayley table of a finite group G. Show that $\mathcal{Q}$ is a translation scheme. When is $\mathcal{F}$ a group block structure?

8.16 Construct a Latin cube which is not an extended Cayley table.

8.17 Let $\mathcal{Q}$ be an association scheme in which $a_i = 1$ for $0 \leqslant i \leqslant s$. Prove that there is an integer m such that $|\Omega| = 2^m$ and that $\mathcal{Q} = \mathcal{Q}(\mathsf{Inverse})$ for the elementary Abelian group $\mathbb{Z}_2^m$.

8.18 Find a blueprint Δ for $\mathbb{Z}_2 \times \mathbb{Z}_2 \times \mathbb{Z}_2 \times \mathbb{Z}_2$ which has classes of size 1, 5, 5 and 5. Show that there is a weak automorphism of $\mathcal{Q}(\Delta)$ which permutes the non-diagonal classes of $\mathcal{Q}(\Delta)$ transitively.

9

Posets

9.1 Product sets

Throughout this chapter, $\Omega = \Omega_1 \times \cdots \times \Omega_m$ where $|\Omega_i| \geqslant 2$ for $i = 1$, ..., m. Put $M = \{1, \ldots, m\}$. We shall use structure on M to derive structure on Ω.

Let L be a subset of M. This defines an equivalence relation $\sim_L$ on Ω by:

$$\alpha \sim_L \beta \qquad \text{if and only if } \alpha_i = \beta_i \text{ for all } i \text{ in } L.$$

This in turn defines a partition $F(L)$ whose classes are the equivalence classes of $\sim_L$. The relation matrix $R_{F(L)}$ defined in Section 6.1 is given by

$$R_{F(L)} = \bigotimes_{i \in L} I_i \otimes \bigotimes_{i \in M \setminus L} J_i, \tag{9.1}$$

where we are abbreviating I_{Ω_i} to I_i, and similarly for J.

Let $n_i = |\Omega_i|$ for i in M and $n = \prod_{i=1}^m n_i$. Then $F(L)$ has $n_{F(L)}$ classes of size $k_{F(L)}$, where

$$n_{F(L)} = \prod_{i \in L} n_i$$

and

$$k_{F(L)} = \prod_{i \in M \setminus L} n_i = n/n_{F(L)}.$$

Thus $F(L)$ is uniform and its projection matrix $P_{F(L)}$ satisfies

$$P_{F(L)} = \frac{1}{k_{F(L)}} R_{F(L)} \tag{9.2}$$

by Proposition 6.1.

Lemma 9.1 *Let K and L be subsets of M. Then*

(i) *if $K \subseteq L$ then $F(L) \preccurlyeq F(K)$;*
(ii) *$F(K) \wedge F(L) = F(K \cup L)$;*
(iii) *$R_{F(K)}R_{F(L)} = \left(\prod_{i \notin K \cup L} n_i\right) R_{F(K \cap L)}$;*
(iv) *$F(K)$ is orthogonal to $F(L)$;*
(v) *$F(K) \vee F(L) = F(K \cap L)$.*

Proof Let α, β be in Ω.

(i) If $K \subseteq L$ and $\alpha \sim_L \beta$ then $\alpha \sim_K \beta$.
(ii) Elements α and β are in the same class of $F(K) \wedge F(L)$ if and only if $\alpha \sim_K \beta$ and $\alpha \sim_L \beta$, that is, if and only if $\alpha \sim_{K \cup L} \beta$.
(iii) Technique 1.1 shows that $R_{F(K)}R_{F(L)}(\alpha, \beta) = |\Psi(\alpha, \beta)|$, where $\Psi(\alpha, \beta) = \{\gamma \in \Omega : \alpha \sim_K \gamma \text{ and } \gamma \sim_L \beta\}$. If there is any i in $K \cap L$ for which $\alpha_i \neq \beta_i$ then $\Psi(\alpha, \beta) = \varnothing$. If $\alpha \sim_{K \cap L} \beta$ then

$$\Psi(\alpha, \beta) = \prod_{i \in K} \{\alpha_i\} \times \prod_{i \in L \setminus K} \{\beta_i\} \times \prod_{i \notin K \cup L} \Omega_i$$

and so $R_{F(K)}R_{F(L)}(\alpha, \beta) = \prod_{i \notin K \cup L} n_i$.
(iv) Part (iii) and Equation (9.2) show that the matrices $P_{F(K)}$ and $P_{F(L)}$ commute with each other, so $F(K)$ is orthogonal to $F(L)$.
(v) Part (iii) shows that

$$k_{F(K)}k_{F(L)}P_{F(K)}P_{F(L)} = \left(\prod_{i \notin K \cup L} n_i\right) k_{F(K \cap L)}P_{F(K \cap L)}.$$

By inclusion-exclusion,

$$k_{F(K)}k_{F(L)} = \left(\prod_{i \notin K \cup L} n_i\right) k_{F(K \cap L)}$$

and so $P_{F(K)}P_{F(L)} = P_{F(K \cap L)}$. From part (i) of Lemma 6.4, $F(K) \vee F(L) = F(K \cap L)$. ■

9.2 A partial order on subscripts

Let $\sqsubseteq$ be a partial order on M. To avoid confusion with $\preccurlyeq$, I shall draw the Hasse diagram for $\sqsubseteq$ using unfilled dots, and say that j *dominates* i if $i \sqsubseteq j$. (Statisticians often say that 'i is nested in j', which invites confusion with $\preccurlyeq$.) The pair $(M, \sqsubseteq)$ is now a partially ordered set, or *poset* for short.

Definition A subset L of M is *ancestral* (with respect to $\sqsubseteq$) if $j \in L$ whenever $i \in L$ and $i \sqsubseteq j$.

Denote by $\mathcal{S}(\sqsubseteq)$ the set of subsets of M which are ancestral with respect to $\sqsubseteq$. Where I need to emphasize the set as well as the partial order, I shall write $\mathcal{S}(M, \sqsubseteq)$.

(Some authors, such as Aigner [9], call ancestral subsets *filters*. Others, such as [87], call them *up-sets* or *up-ideals*.)

Proposition 9.2 *If K and L are ancestral subsets then so are $K \cap L$ and $K \cup L$.*

The idea behind the dominating partial order is the following: if $i \sqsubset j$ and $\alpha_j \neq \beta_j$ then we do not care whether or not $\alpha_i = \beta_i$. For example, the set of plots in a block design can be regarded as $\Omega_1 \times \Omega_2$ where $|\Omega_1| = b$ and $|\Omega_2| = k$. Then equality of the first subscript indicates plots in the same block, but equality of the second subscript is irrelevant if the first subscripts differ: in other words $2 \sqsubset 1$. Thus we are interested in the partitions $F(L)$ only for ancestral subsets L of M.

Example 9.1 Over a period of n_1 weeks, samples of milk are tested for bacteria. There are n_2 laboratories involved in the testing, each employing n_3 technicians for this task. Each week n_4 samples of milk are sent to each laboratory, where they are split into n_3 subsamples, one of which is tested by each technician. The relevant poset is shown in Figure 9.1.

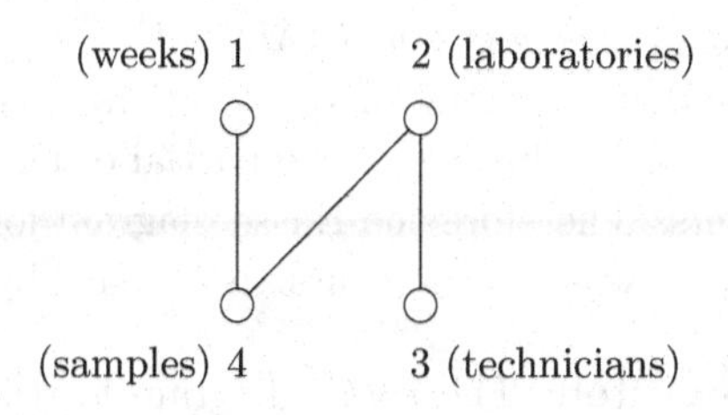

Fig. 9.1. The dominating poset in Example 9.1

The subsamples can be labelled by the elements of $\Omega_1 \times \Omega_2 \times \Omega_3 \times \Omega_4$, where $\Omega_i = \{1, \ldots, n_i\}$ for $i = 1, \ldots, 4$. The element of Ω_1 indicates the week, the element of Ω_2 the laboratory, the element of Ω_3 the technician within the given laboratory and the element of Ω_4 the sample within the given week and laboratory. The ancestral subsets, and their corresponding equivalence relations on the set of $n_1 n_2 n_3 n_4$ subsamples, are as follows.

ancestral L	(class of) $\sim_L$
$\varnothing$	universe (everything equivalent to everything else)
$\{1\}$	same week
$\{2\}$	same laboratory
$\{1,2\}$	same laboratory in same week
$\{2,3\}$	same technician
$\{1,2,4\}$	same sample
$\{1,2,3\}$	same technician in same week
$\{1,2,3,4\}$	same subsample

■

Definition Given a product set $\Omega = \prod_{i \in M} \Omega_i$ for which $|\Omega_i| \geqslant 2$ for i in M, a *poset block structure* on Ω is the set of partitions $F(L)$ given by the ancestral subsets L of some partial order $\sqsubseteq$ on M.

Theorem 9.3 *Poset block structures are orthogonal block structures.*

Proof Every partition $F(L)$ is uniform. The partitions are orthogonal, by Lemma 9.1(iv). Proposition 9.2 and parts (ii) and (v) of Lemma 9.1 show that the set of partitions given by the ancestral subsets is closed under $\wedge$ and $\vee$. ■

The two trivial partitions come from the two trivial subsets, which are both ancestral: $F(\varnothing) = U$ and $F(M) = E$.

By Lemma 9.1(i), the partial order $\preccurlyeq$ on the poset block structure $\mathcal{F}(\sqsubseteq)$ defined by $\sqsubseteq$ is the inverse of the partial order $\subseteq$ on the set $\mathcal{S}(\sqsubseteq)$ of ancestral subsets. Therefore the Hasse diagram for $\mathcal{F}(\sqsubseteq)$ appears to be the wrong way up when its elements are labelled by ancestral subsets.

Example 9.1 revisited The Hasse diagram for the partitions in the poset block structure on milk subsamples is shown in Figure 9.2. ■

Theorem 9.4 *Every poset block structure is a group block structure.*

Proof Identify Ω_i with any group of order n_i. For L in $\mathcal{S}(\sqsubseteq)$, put $\Omega_L = \prod_{i \in L} \{1_{\Omega_i}\} \times \prod_{i \notin L} \Omega_i$. Then Ω_L is a normal subgroup of Ω and $F(L)$ is the partition of Ω into cosets of Ω_L. Moreover, $\Omega_K \cap \Omega_L = \Omega_{K \cup L}$ and $\Omega_K \Omega_L = \Omega_{K \cap L}$. ■

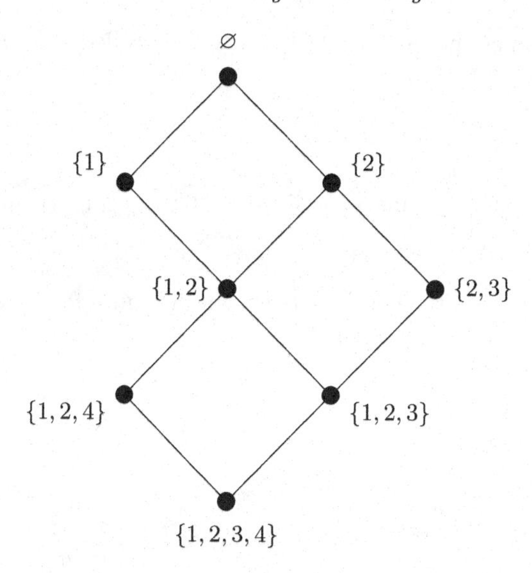

Fig. 9.2. Hasse diagram for the poset block structure defined by the poset in Figure 9.1

9.3 Crossing and nesting

In Section 6.6 we showed that if $\mathcal{F}_1$ and $\mathcal{F}_2$ are orthogonal block structures then so are $\mathcal{F}_1 \times \mathcal{F}_2$ and $\mathcal{F}_1/\mathcal{F}_2$. Here we specialize this result to poset block structures.

For $t = 1, 2$, let $\sqsubseteq_t$ be a partial order on a non-empty set M_t. The partial order $\sqsubseteq_t$ is often identified with the subset $\{(i,j) : i \sqsubseteq_t j\}$ of $M_t \times M_t$. Thus if $M_1 \cap M_2 = \varnothing$ we can define a new partial order $\sqsubseteq$ on $M_1 \cup M_2$ by putting

$$\sqsubseteq = \sqsubseteq_1 \cup \sqsubseteq_2.$$

Then $i \sqsubseteq j$ if *either* $\{i,j\} \subseteq M_1$ and $i \sqsubseteq_1 j$ *or* $\{i,j\} \subseteq M_2$ and $i \sqsubseteq_2 j$. The Hasse diagram for $(M_1 \cup M_2, \sqsubseteq)$ consists of those for $(M_1, \sqsubseteq_1)$ and $(M_2, \sqsubseteq_2)$ side by side. The new poset $(M_1 \cup M_2, \sqsubseteq)$ is called the *cardinal sum* of $(M_1, \sqsubseteq_1)$ and $(M_2, \sqsubseteq_2)$.

Alternatively, we can define a new partial order $\sqsubseteq_1/\sqsubseteq_2$ on $M_1 \cup M_2$ by putting

$$\sqsubseteq_1/\sqsubseteq_2 = \sqsubseteq_1 \cup \sqsubseteq_2 \cup (M_2 \times M_1).$$

In this order, $i \sqsubseteq_1/\sqsubseteq_2 j$ if $\{i,j\} \subseteq M_1$ and $i \sqsubseteq_1 j$ *or* $\{i,j\} \subseteq M_2$ and $i \sqsubseteq_2 j$ *or* $i \in M_2$ and $j \in M_1$. The Hasse diagram for $(M_1 \cup M_2, \sqsubseteq_1/\sqsubseteq_2)$ consists of the Hasse diagram for $(M_1, \sqsubseteq_1)$ above the Hasse diagram for

$(M_2, \sqsubseteq_2)$. The new poset $(M_1 \cup M_2, \sqsubseteq_1/\sqsubseteq_2)$ is called the *ordinal sum* of $(M_1, \sqsubseteq_1)$ and $(M_2, \sqsubseteq_2)$.

Example 9.2

If $(M_1, \sqsubseteq_1)$ is 1 over 2 (1 above 2, joined by a line) and $(M_2, \sqsubseteq_2)$ is 3 ○ ○ 4 then the above operations give the three new posets shown in Figure 9.3. ■

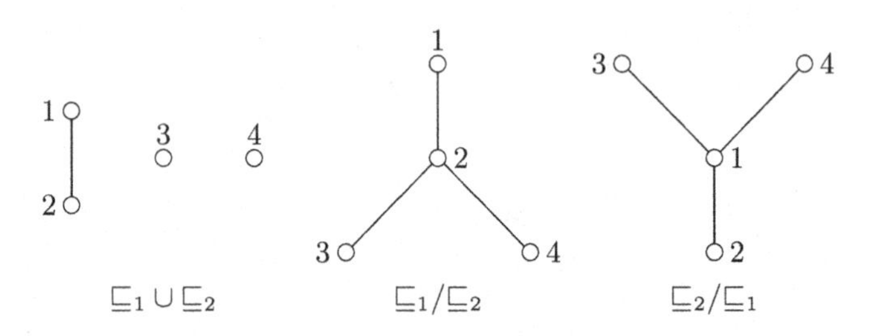

Fig. 9.3. New posets in Example 9.2

Theorem 9.5 *For $t = 1, 2$ let $(M_t, \sqsubseteq_t)$ be a poset, where $M_1 \cap M_2 = \varnothing$. For $t = 1, 2$ and i in M_t let Ω_{ti} be a set of size n_{ti} with $n_{ti} \geqslant 2$; put $\Omega_t = \prod_{i \in M_t} \Omega_{ti}$ and let $\mathcal{F}(\sqsubseteq_t)$ be the poset block structure on Ω_t defined by $\sqsubseteq_t$. Let $\mathcal{F}(\sqsubseteq_1 \cup \sqsubseteq_2)$ and $\mathcal{F}(\sqsubseteq_1/\sqsubseteq_2)$ be the poset block structures defined on $\Omega_1 \times \Omega_2$ by $\sqsubseteq_1 \cup \sqsubseteq_2$ and $\sqsubseteq_1/\sqsubseteq_2$ respectively. Then*

(i) $\mathcal{F}(\sqsubseteq_1) \times \mathcal{F}(\sqsubseteq_2) = \mathcal{F}(\sqsubseteq_1 \cup \sqsubseteq_2)$;
(ii) $\mathcal{F}(\sqsubseteq_1)/\mathcal{F}(\sqsubseteq_2) = \mathcal{F}(\sqsubseteq_1/\sqsubseteq_2)$.

Proof If $L \subseteq M_1 \cup M_2$ then $L = L_1 \cup L_2$ where $L_t \subseteq M_t$. Equation (9.1) shows that

$$R_{F(L)} = R_{F(L_1)} \otimes R_{F(L_2)}$$

and so $F(L) = F(L_1) \times F(L_2)$, by Lemma 6.10.

(i) The subset L of $M_1 \cup M_2$ is ancestral with respect to $\sqsubseteq_1 \cup \sqsubseteq_2$ if and only if L_t is ancestral in M_t for $t = 1, 2$. Thus the partitions in $\mathcal{F}(\sqsubseteq_1 \cup \sqsubseteq_2)$ are precisely those in $\mathcal{F}(\sqsubseteq_1) \times \mathcal{F}(\sqsubseteq_2)$.

(ii) The subset L of $M_1 \cup M_2$ is ancestral with respect to $\sqsubseteq_1/\sqsubseteq_2$ if and only if either

$$\begin{array}{lllll} & L_1 \in \mathcal{S}(M_1, \sqsubseteq_1) & \text{and} & L_2 = \varnothing \\ \text{or} & L_1 = M_1 & \text{and} & L_2 \in \mathcal{S}(M_2, \sqsubseteq_2). \end{array}$$

Now,

$$F(L_1 \cup \varnothing) = F(L_1) \times F(\varnothing) = F(L_1) \times U_2,$$

where $U_2 = \Omega_2 \times \Omega_2$, and

$$F(M_1 \cup L_2) = F(M_1) \times F(L_2) = E_1 \times F(L_2),$$

where $E_1 = \mathrm{Diag}(\Omega_1)$. As noted in the proof of Theorem 6.14, the partitions in $\mathcal{F}(\sqsubseteq_1)/\mathcal{F}(\sqsubseteq_2)$ are precisely those of the form $F(L_1) \times U_2$ for L_1 in $\mathcal{S}(M_1, \sqsubseteq_1)$ and those of the form $E_1 \times F(L_2)$ for L_2 in $\mathcal{S}(M_2, \sqsubseteq_2)$. ■

Corollary 9.6 *Every association scheme constructed from trivial association schemes by (iterated) crossing and nesting is the association scheme of a poset block structure.*

Up to isomorphism, there are five posets with three elements: see Exercise 9.1. The corresponding association schemes appear as examples of iterated crossing and nesting in Examples 3.8–3.12. However, the poset with four elements in Example 9.1 is not expressible as either $\sqsubseteq_1 \cup \sqsubseteq_2$ or $\sqsubseteq_1 / \sqsubseteq_2$ with both $\sqsubseteq_1$ and $\sqsubseteq_2$ being partial orders on smaller sets. Therefore Example 9.1 is not obtainable from trivial structures by crossing and nesting. All other posets with four elements can be obtained from the trivial partial order on singleton sets by taking cardinal and ordinal sums. Thus their poset block structures can be obtained from trivial structures by crossing and nesting: see Exercise 3.13.

Example 9.2 revisited Identifying each singleton set with the unique partial order on that set, the three partial orders in Figure 9.3 are

$$(\{1\} / \{2\}) \cup \{3\} \cup \{4\}, \; \{1\} / \{2\} / (\{3\} \cup \{4\}) \text{ and } (\{3\} \cup \{4\}) / \{1\} / \{2\}.$$

The corresponding association schemes are

$$(\underline{\underline{n_1}}/\underline{\underline{n_2}}) \times \underline{\underline{n_3}} \times \underline{\underline{n_4}}, \; \underline{\underline{n_1}}/\underline{\underline{n_2}}/(\underline{\underline{n_3}} \times \underline{\underline{n_4}}) \text{ and } (\underline{\underline{n_3}} \times \underline{\underline{n_4}})/\underline{\underline{n_1}}/\underline{\underline{n_2}}$$

respectively. ■

There are two special cases of iterated crossing and nesting. Iterated crossing of m trivial posets gives the trivial partial order on M, in which $i \sqsubseteq j$ if and only if $i = j$. At the other extreme, iterated nesting gives the linear order, in which $i \sqsubseteq j$ if and only if $i \geqslant j$.

9.4 Parameters of poset block structures

Like all orthogonal block structures, every poset block structure defines an association scheme. Its parameters can be given rather more explicitly than those in Section 6.3. In this section we find explicit forms for the adjacency matrices, strata and character tables of a poset block structure. This requires some technical manipulation of some subsets of a poset.

Definition Let L be any subset of the poset M. An element i of L is a *maximal* element of L if there is no j in L with $i \sqsubset j$; it is a *minimal* element of L if there is no j in L with $j \sqsubset i$.

Write $\max(L)$ for the set of maximal elements of L and $\min(L)$ for the set of minimal elements of L. The top and bottom of L are defined by $\mathrm{top}(L) = L \setminus \min(L)$ and $\mathrm{bot}(L) = L \setminus \max(L)$.

Theorem 9.7 *Let L be an ancestral subset of M. In the association scheme defined by the poset block structure $\mathcal{F}(\sqsubseteq)$, the adjacency matrix and valency for the associate class corresponding to $F(L)$ are given by:*

$$A_L = \bigotimes_{i\in L} I_i \otimes \bigotimes_{i\in\max(M\setminus L)} (J_i - I_i) \otimes \bigotimes_{i\in\mathrm{bot}(M\setminus L)} J_i \tag{9.3}$$

$$a_L = \prod_{i\in\max(M\setminus L)} (n_i - 1) \times \prod_{i\in\mathrm{bot}(M\setminus L)} n_i.$$

Proof The pair (α, β) is in the associate class defined by $F(L)$ if $F(L)$ is finest subject to $\alpha \sim_L \beta$, that is, if L is largest subject to $\alpha \sim_L \beta$. For i in L we need $\alpha_i = \beta_i$ if $\alpha \sim_L \beta$. If $i \in \max(M \setminus L)$ then $L \cup \{i\}$ is ancestral: in order that α and β are not in the same class of $L \cup \{i\}$ we must have $\alpha_i \neq \beta_i$. If $i \in \mathrm{bot}(M \setminus L)$ then there is some j in $\max(M \setminus L)$ with $i \sqsubset j$; so long as $\alpha_j \neq \beta_j$ then there is no ancestral subset K of M for which $i \in K$ and $\alpha \sim_K \beta$, and so there is no constraint on α_i and β_i. This gives A_L, from which a_L follows by calculating row sums. ■

Corollary 9.8 *(i) In the poset block structure $\mathcal{F}(\sqsubseteq)$, the number of associate classes is equal to $|\mathcal{S}(\sqsubseteq)|$.*

(ii) The set of matrices $\{R_{F(L)} : L \subseteq M\}$ is linearly independent.

Proof (i) Since $|\Omega_i| \geqslant 2$ for each i in M, none of the matrices A_L is zero.

(ii) Apply part (i) to the trivial partial order on M. ■

Equations (9.3) and (9.1) give A_L as a linear combination of some of the relation matrices with coefficients ± 1. Specifically

$$A_L = \sum_{H \subseteq \max(M \setminus L)} (-1)^{|H|} R_{F(L \cup H)}.$$

However, Equation (6.4) gives

$$A_L = \sum_{\text{ancestral } K} \mu(K, L) R_{F(K)},$$

where μ is the Möbius function of the partial order $\supseteq$ on $\mathcal{S}(\sqsubseteq)$ (which corresponds to the partial order $\preccurlyeq$ on the partitions $F(L)$ for L in $\mathcal{S}(\sqsubseteq)$). Since the relation matrices are linearly independent, we can compare the two expressions for A_L and deduce that

$$\mu(K, L) = \begin{cases} (-1)^{|H|} & \text{if } K = L \cup H \text{ and } H \subseteq \max(M \setminus L) \\ 0 & \text{otherwise.} \end{cases} \tag{9.4}$$

Theorem 9.9 *In the association scheme which is defined by the poset block structure $\mathcal{F}(\sqsubseteq)$, the projection matrix and dimension of the stratum corresponding to the ancestral subset L are given by:*

$$\begin{aligned} S_L &= \bigotimes_{i \in \text{top}(L)} I_i \otimes \bigotimes_{i \in \min(L)} \left(I_i - \frac{1}{n_i} J_i \right) \otimes \bigotimes_{i \in M \setminus L} \frac{1}{n_i} J_i \\ d_L &= \prod_{i \in \text{top}(L)} n_i \times \prod_{i \in \min(L)} (n_i - 1). \end{aligned}$$

Proof Theorem 6.9 gives

$$S_L = \sum_{\text{ancestral } K} \mu(L, K) P_{F(K)}.$$

Equation (9.4) shows that if $\mu(L, K) \neq 0$ then $L = K \cup H$ with H contained in $\max(M \setminus K)$; that is, $K = L \setminus H$ and $H \subseteq \min(L)$. Thus

$$\begin{aligned} S_L &= \sum_{H \subseteq \min(L)} (-1)^{|H|} P_{F(L \setminus H)} \\ &= \bigotimes_{i \in \text{top}(L)} I_i \otimes \bigotimes_{i \in M \setminus L} \frac{1}{n_i} J_i \otimes T \end{aligned}$$

where

$$T = \sum_{H \subseteq \min(L)} (-1)^{|H|} \left(\bigotimes_{i \in H} \frac{1}{n_i} J_i \otimes \bigotimes_{i \in \min(L) \setminus H} I_i \right)$$

$$= \bigotimes_{i \in \min(L)} \left(I_i - \frac{1}{n_i} J_i \right).$$

Therefore

$$S_L = \bigotimes_{i \in \mathrm{top}(L)} I_i \otimes \bigotimes_{i \in M \setminus L} \frac{1}{n_i} J_i \otimes \bigotimes_{i \in \min(L)} \left(I_i - \frac{1}{n_i} J_i \right).$$

Taking the trace of each side gives d_L. ■

Theorem 9.10 *Let K and L be ancestral subsets of M. In the character table C for the association scheme defined by the poset block structure $\mathcal{F}(\sqsubseteq)$,*

(i) $C(K, L) = C(K, K \cup L)$;

(ii) if $L \cap \mathrm{bot}(M \setminus K) \neq \varnothing$ *then* $C(K, L) = 0$;

(iii) if $L = K \cup H$ *and* $H \subseteq \max(M \setminus K)$ *then*

$$C(K, L) = (-1)^{|H|} \prod_{i \in \max(M \setminus K) \setminus H} (n_i - 1) \times \prod_{i \in \mathrm{bot}(M \setminus K)} n_i.$$

Proof From Equation (6.9)

$$C = \mu' \operatorname{diag}(k) \zeta$$

so

$$C(K, L) = \sum_{\text{ancestral } G} \mu'(K, G) k_{F(G)} \zeta(G, L).$$

Write $N = \max(M \setminus K)$. Equation (9.4) gives

$$C(K, L) = \sum_{H \subseteq N} (-1)^{|H|} k_{F(K \cup H)} \zeta(K \cup H, L).$$

(i) No matter what subset H is, $K \cup H \supseteq L$ if and only if $K \cup H \supseteq K \cup L$, and so $\zeta(K \cup H, L) = \zeta(K \cup H, K \cup L)$.

(ii) If L contains an element of $\mathrm{bot}(M \setminus K)$ and $H \subseteq N$ then $K \cup H \not\supseteq L$ and so $\zeta(K \cup H, L) = 0$.

(iii) Let $H \subseteq N$. Then

$$C(K, K \cup H) = \sum_{G \subseteq N} (-1)^{|G|} k_{F(K \cup G)} \zeta(K \cup G, K \cup H).$$

For subsets G of N,

$$k_{F(K \cup G)} = \prod_{i \in N \setminus G} n_i \times \prod_{i \in \mathrm{bot}(M \setminus K)} n_i.$$

Consider the poset block structure $\mathcal{P}$ on $\prod_{i\in N}\Omega_i$ defined by the trivial partial order on N in which all subsets are ancestral; denote its character table by $\tilde{C}$ and the size of $F(G)$-classes $\tilde{k}_G$ for G contained in N. Then

$$\begin{aligned}\tilde{C}(\varnothing,H) &= \sum_{G\subseteq N}(-1)^{|G|}\tilde{k}_G\zeta(G,H)\\ &= \sum_{G\subseteq N}(-1)^{|G|}\left(\prod_{i\in N\setminus G} n_i\right)\zeta(K\cup G,K\cup H)\end{aligned}$$

and so

$$C(K,K\cup H)=\tilde{C}(\varnothing,H)\times\prod_{i\in\mathrm{bot}(M\setminus K)} n_i.$$

Now, the association scheme of $\mathcal{P}$ is just the cross of the trivial association schemes $\underline{\underline{n_i}}$ for i in N, whose character tables are given on page 64. Part (ii) of Theorem 3.4 shows that

$$\tilde{C}(\varnothing,H)=(-1)^{|H|}\prod_{i\in N\setminus H}(n_i-1). \quad \blacksquare$$

9.5 Lattice laws

If $\mathcal{F}$ is an orthogonal block structure then it is a *lattice* because it is closed under $\wedge$ and $\vee$. More details about lattices can be found in [9, Chapter II] or [106]. Here we investigate the *modular* and *distributive* laws, which hold in some lattices.

Definition A lattice $\mathcal{F}$ is *modular* if

$$G\wedge(H\vee F)=(G\wedge H)\vee F \tag{9.5}$$

when F, G, H are in $\mathcal{F}$ and $F\preccurlyeq G$.

Notice that we always have

$$G\wedge(H\vee F)\succcurlyeq(G\wedge H)\vee F$$

when $F\preccurlyeq G$. Since $F\preccurlyeq G$ and $F\preccurlyeq H\vee F$ it follows that $F\preccurlyeq G\wedge(H\vee F)$; similarly, since $G\wedge H\preccurlyeq G$ and $G\wedge H\preccurlyeq H\preccurlyeq H\vee F$ we have $G\wedge H\preccurlyeq G\wedge(H\vee F)$; and therefore $(G\wedge H)\vee F\preccurlyeq G\wedge(H\vee F)$.

Theorem 9.11 *Every orthogonal block structure is modular.*

Proof Let $\mathcal{F}$ be an orthogonal block structure on a set Ω, and let F, G, H be partitions in $\mathcal{F}$ such that $F \preccurlyeq G$. We need to show that

$$G \wedge (H \vee F) \preccurlyeq (G \wedge H) \vee F.$$

Suppose that α, β are elements in the same class of $G \wedge (H \vee F)$. Then α and β belong to the same G-class Δ; moreover, there is an element γ such that α and γ belong to the same H-class Γ while γ and β belong to the same F-class Θ. Now, $F \preccurlyeq G$ so $\Theta \subseteq \Delta$ so $\gamma \in \Delta$. Thus α and γ belong to $\Delta \cap \Gamma$, which is a class of $G \wedge H$, while γ and β belong to the same F-class: in other words, α and β belong to the same class of $(G \wedge H) \vee F$. ■

It is not known which modular lattices are isomorphic (as lattices) to orthogonal block structures.

Definition A lattice $\mathcal{F}$ is *distributive* if, for all F, G, H in $\mathcal{F}$,

$$F \wedge (G \vee H) = (F \wedge G) \vee (F \wedge H) \tag{9.6}$$

$$F \vee (G \wedge H) = (F \vee G) \wedge (F \vee H). \tag{9.7}$$

In fact, each of these laws implies the other. If Equation (9.6) holds then

$$\begin{aligned}
(F \vee G) \wedge (F \vee H) &= ((F \vee G) \wedge F) \vee ((F \vee G) \wedge H) \\
&= F \vee ((F \vee G) \wedge H), \qquad \text{because } F \preccurlyeq F \vee G, \\
&= F \vee ((F \wedge H) \vee (G \wedge H)) \\
&= (F \vee (F \wedge H)) \vee (G \wedge H), \qquad \text{because } \vee \text{ is associative,} \\
&= F \vee (G \wedge H),
\end{aligned}$$

because $F \wedge H \preccurlyeq F$, and so Equation (9.7) holds. The dual argument, interchanging $\wedge$ and $\vee$, shows that Equation (9.6) follows from Equation (9.7).

Theorem 9.12 *Poset block structures are distributive.*

Proof This follows from Lemma 9.1(ii) and (v), using the facts that

$$L_1 \cap (L_2 \cup L_3) = (L_1 \cap L_2) \cup (L_1 \cap L_3)$$

and

$$L_1 \cup (L_2 \cap L_3) = (L_1 \cup L_2) \cap (L_1 \cup L_3)$$

for all subsets L_1, L_2, L_3 of M. ■

Not all orthogonal block structures are distributive. In the orthogonal block structure defined by a Latin square (Example 6.2), we have

$$\text{rows} \wedge (\text{columns} \vee \text{letters}) = \text{rows} \wedge U = \text{rows}$$

while

$$(\text{rows} \wedge \text{columns}) \vee (\text{rows} \wedge \text{letters}) = E \vee E = E.$$

Theorem 9.13 *Let $\mathcal{F}$ be an orthogonal block structure. If $\mathcal{F}$ is distributive then it is isomorphic to a poset block structure.*

Proof For F in $\mathcal{F}$, put

$$\bar{F} = \bigwedge \{G \in \mathcal{F} : F \prec G\}$$

and define F to be *irreducible* if $F \neq \bar{F}$. Let M be the set of all irreducible elements of $\mathcal{F}$, and let $\sqsubseteq$ be the restriction of $\preccurlyeq$ to M.

Define $\varphi: \mathcal{S}(\sqsubseteq) \to \mathcal{F}$ by

$$\varphi(L) = \bigwedge \{G \in \mathcal{F} : G \in L\}$$

for ancestral subsets L of M. If $K \in \mathcal{F}$ then $K = \varphi(L)$ where $L = \{G \in M : K \preccurlyeq G\}$, which is ancestral. Hence φ is onto.

Suppose that there are ancestral subsets L_1, L_2 with $\varphi(L_1) = \varphi(L_2)$ and that there is an irreducible F in $L_2 \setminus L_1$. Then $F \succcurlyeq \varphi(L_2) = \varphi(L_1)$ so

$$\begin{aligned} F = F \vee \varphi(L_1) &= F \vee \left(\bigwedge \{G : G \in L_1\}\right) \\ &= \bigwedge \{F \vee G : G \in L_1\} \end{aligned}$$

by the distributive law. Since F is irreducible, there is some G in L_1 for which $F \vee G = F$. Then $G \preccurlyeq F$, so $F \in L_1$, because L_1 is ancestral. This contradiction shows that φ is one-to-one.

It is clear that $\varphi(L_1 \cup L_2) = \varphi(L_1) \wedge \varphi(L_2)$ for all L_1, L_2 in $\mathcal{S}(\sqsubseteq)$. If $F \in M$ and $F \succcurlyeq \varphi(L)$ for some L in $\mathcal{S}(\sqsubseteq)$ then $\varphi(L \cup \{F\}) = \varphi(L)$. Since φ is a bijection, it follows that $F \in L$. Thus $\varphi(L_1) \preccurlyeq \varphi(L_2)$ if and only if $L_1 \supseteq L_2$. Hence $\varphi(L_1) \vee \varphi(L_2) = \varphi(L_1 \cap L_2)$ for all L_1, L_2 in $\mathcal{S}(\sqsubseteq)$.

For F in M, put $m_F = n_F/n_{\bar{F}}$, which is the number of F-classes contained in each $\bar{F}$-class and so $m_F \geqslant 2$. Put $\Omega_F = \{1, 2, \ldots, m_F\}$. Label the F-classes by the elements of Ω_F in such a way that the labels within each $\bar{F}$-class are distinct. For ω in Ω, let ω_F be the label of the F-class containing ω.

For ancestral subsets L of M, define $\vartheta_L: \Omega \to \prod_{F \in L} \Omega_F$ by $\vartheta_L(\omega) =$

$(\omega_F : F \in L)$. For $L = \varnothing$ we take the convention that the empty product is a singleton, so that $\vartheta_\varnothing(\omega)$ takes the same value for all ω in Ω. Then I claim that, for every α in $\prod_{F\in L}\Omega_F$,

$$\{\omega \in \Omega : \vartheta_L(\omega) = \alpha\} \text{ is a class of } \varphi(L). \tag{9.8}$$

This is proved by induction on L. The above convention ensures that it is true for $L = \varnothing$, because $\varphi(\varnothing) = U$.

Suppose that L is a non-empty ancestral subset, and that condition (9.8) holds for all ancestral subsets strictly contained in L. Let $G = \varphi(L)$. If G is irreducible then put $\bar{L} = \varphi^{-1}(\bar{G})$, so that $\bar{L} \subset L$. Then

$$\vartheta_L(\omega) = (\vartheta_{\bar{L}}(\omega), \omega_G).$$

Given α in $\prod_{F\in L}\Omega_F$, put $\bar{\alpha} = (\alpha_F : F \in \bar{L})$. By condition (9.8) for $\bar{L}$, $\{\omega \in \Omega : \vartheta_{\bar{L}}(\omega) = \bar{\alpha}\}$ is a class of $\bar{G}$, so the labelling of G-classes ensures that $\{\omega \in \Omega : \vartheta_L(\omega) = \alpha\}$ is a class of G.

If G is not irreducible then $G = G_1 \wedge G_2$ where $G_i = \varphi(L_i)$ and $L_i \subset L$. Then $G = \varphi(L_1 \cup L_2)$, so $L_1 \cup L_2 = L$. Given α in $\prod_{F\in L}\Omega_F$, put

$$\begin{aligned}
\alpha_1 &= (\alpha_F : F \in L_1) & \Delta_1 &= \{\omega \in \Omega : \vartheta_{L_1}(\omega) = \alpha_1\}\\
\alpha_2 &= (\alpha_F : F \in L_2) & \Delta_2 &= \{\omega \in \Omega : \vartheta_{L_2}(\omega) = \alpha_2\}\\
\alpha_3 &= (\alpha_F : F \in L_1 \cap L_2) & \Delta_3 &= \{\omega \in \Omega : \vartheta_{L_1\cap L_2}(\omega) = \alpha_3\}.
\end{aligned}$$

By condition (9.8) for L_1, L_2 and $L_1 \cap L_2$ respectively, Δ_1, Δ_2 and Δ_3 are classes of G_1, G_2 and $\varphi(L_1 \cap L_2)$ respectively. Moreover, Δ_1 and Δ_2 are contained in Δ_3. Since $\varphi(L_1 \cap L_2) = G_1 \vee G_2$, every G_1-class in Δ_3 meets every G_2-class in Δ_3, so $\Delta_1 \cap \Delta_2$ is not empty. Thus $\Delta_1 \cap \Delta_2$ is a class of $G_1 \wedge G_2$. Since $\Delta_1 \cap \Delta_2 = \{\omega \in \Omega : \vartheta_L(\omega) = \alpha\}$, this proves condition (9.8) for L.

Now, $\varphi(M) = E$, so condition (9.8) for M shows that ϑ_M is a bijection from Ω to $\prod_{F\in M}\Omega_F$. For each ancestral subset L of M, condition (9.8) shows that ϑ_M maps classes of $\varphi(L)$ to classes of $F(L)$ in the poset block structure defined by $\sqsubseteq$ on $\prod_{F\in M}\Omega_F$. Thus $\mathcal{F}$ is isomorphic to this poset block structure. ■

9.6 Poset operators

Theorem 9.5 and Exercise 9.1 show that crossing two trivial association schemes $\underline{\underline{n_1}}$ and $\underline{\underline{n_2}}$ gives the rectangular scheme $\underline{\underline{n_1}} \times \underline{\underline{n_2}}$, which is also the poset block structure defined by the poset in Figure 9.4, and that nesting these two trivial association schemes gives the group-divisible scheme $\underline{\underline{n_1}}/\underline{\underline{n_2}}$, which is also the poset block structure defined by the poset in Figure 9.5. Thus crossing and nesting in some way correspond to the two posets of size two. However, crossing and nesting can be used to combine *any* pair of association schemes. In this section we show that any poset of size m gives an m-ary operator on association schemes, that is, a way of combining m association schemes into a single new association scheme. We call these operators *poset operators* of order m. Thus this section generalizes Chapter 3.

Fig. 9.4. The poset which gives the crossing operator

Fig. 9.5. The poset which gives the nesting operator

For the remainder of this section we use the following notation. For $i = 1, \ldots, m$, $\mathcal{Q}_i$ is an association scheme on a set Ω_i of size n_i greater than 1; its adjacency matrices are $A_k^{(i)}$ for k in $\mathcal{K}_i$, with $|\mathcal{K}_i| = s_i + 1$ and $A_0^{(i)} = I_i$; its set of adjacency matrices is $\mathrm{Adj}(i)$ and $\mathcal{A}_i$ is its Bose–Mesner algebra.

Let $M = \{1, \ldots, m\}$ and let $\sqsubseteq$ be a partial order on M. For subsets K of M, put $\Omega_K = \prod_{i \in K} \Omega_i$. For each ancestral subset L of M put

$$G_L = \prod_{j \in \max(M \setminus L)} (\mathcal{K}_j \setminus \{0\})$$

and, for each element g of G_L, put

$$A_{L,g} = \bigotimes_{i \in L} I_i \otimes \bigotimes_{i \in \max(M \setminus L)} A_{g_i}^{(i)} \otimes \bigotimes_{i \in \mathrm{bot}(M \setminus L)} J_i. \tag{9.9}$$

Let $\mathrm{Adj}(M, \sqsubseteq) = \{A_{L,g} : L \in \mathcal{S}(M, \sqsubseteq),\ g \in G_L\}$ and let $\mathcal{A}$ be the set of real linear combinations of matrices in $\mathrm{Adj}(M, \sqsubseteq)$.

Lemma 9.14 $\sum_{\text{ancestral } L} \sum_{g \in G_L} A_{L,g} = J_{\Omega_M}$.

Proof Let A_L be the matrix defined in Equation (9.3). Then

$$\sum_{g \in G_L} A_{L,g} = A_L.$$

From Theorem 9.7, $\{A_L : L \in \mathcal{S}(\sqsubseteq)\}$ is the set of adjacency matrices of the association scheme defined by the poset block structure $\mathcal{F}(\sqsubseteq)$, and so

$$\sum_{L\in\mathcal{S}(\sqsubseteq)} A_L = J_{\Omega_M}. \quad \blacksquare$$

Define B in $\bigotimes_{i\in M} \mathcal{A}_i$ to be *nice* if, whenever $j \sqsubset i$, either B_j is a scalar multiple of J_j or B_i is a scalar multiple of I_i, and to be *fundamentally nice* if B is nice and $B_i \in \mathrm{Adj}(i) \cup \{J_i\}$ for all i in M. Let $\mathcal{B}$ be the set of nice matrices.

Lemma 9.15 *Every matrix in* $\mathrm{Adj}(M, \sqsubseteq)$ *is nice.*

Proof Consider the matrix $A_{L,g}$ in $\mathrm{Adj}(M, \sqsubseteq)$, where L is an ancestral subset of M and $g \in G_L$. Certainly $A_{L,g} \in \bigotimes_{i\in M} \mathcal{A}_i$. Suppose that $j \sqsubset i$. If $(A_{L,g})_i \neq I_i$ then $i \in M \setminus L$ so $j \in \mathrm{bot}(M \setminus L)$ and therefore $(A_{L,g})_j = J_j$. ■

Lemma 9.16 *The set* $\mathcal{B}$ *is closed under matrix multiplication.*

Proof Suppose that B and C are nice and that $j \sqsubset i$. Now, $(BC)_k = B_k C_k$ for all k in M. If either B_j or C_j is a scalar multiple of J_j then so is $(BC)_j$, because AJ_j is a scalar multiple of J_j for every matrix A in $\mathcal{A}_j$. If neither of B_j and C_j is a scalar multiple of J_j then both B_i and C_i are scalar multiples of I_i and therefore so is $(BC)_i$. ■

Lemma 9.17 *If* B *is nice then* B *is a linear combination of fundamentally nice matrices.*

Proof Let $N(B)$ be the set

$$\{n \in M : B_n \text{ is not a scalar multiple of } J_n \text{ or an adjacency matrix}\}.$$

The proof is by induction on $|N(B)|$.

If $N(B) = \varnothing$ then B is a scalar multiple of a fundamentally nice matrix.

Suppose that $N(B) \neq \varnothing$, and choose n in N. For k in $\mathcal{K}_n$, put

$$C(k) = \bigotimes_{i\in M\setminus\{n\}} B_i \otimes A_k^{(n)}.$$

Then B is a linear combination of the $C(k)$ for k in $\mathcal{K}_n$. If $j \sqsubset n$ then B_j is a scalar multiple of J_j, and if $n \sqsubset j$ then B_j is a scalar multiple of I_j, because B is nice. Hence $C(k)$ is nice for each k in $\mathcal{K}_n$. However, $N(C(k)) = N(B) \setminus \{n\}$, so $|N(C(k))| = |N(B)| - 1$. By induction, B is a linear combination of fundamentally nice matrices. ■

Lemma 9.18 *If B is fundamentally nice then $B \in \mathcal{A}$.*

Proof Suppose that B is fundamentally nice. Put

$$\begin{aligned} L_0 &= \{i \in M : B_i = I_i\} \\ N &= \{i \in M : B_i \neq I_i,\ B_i \neq J_i\} \\ H &= \{i \in M : \exists j \in N \text{ with } i \sqsubset j\} \\ K &= M \setminus L_0 \setminus N \setminus H. \end{aligned}$$

Then L_0 is an ancestral subset of M. Put $\bar{B} = \bigotimes_{i \notin K} B_i$.

For each ancestral subset L_1 of K, put

$$G_{L_1}^{(K)} = \prod_{j \in \max(K \setminus L_1)} (\mathcal{K}_j \setminus \{0\});$$

for each h in $G_{L_1}^{(K)}$, define $A_{L_1,h}^{(K)}$ analogously to $A_{L,g}$.

Suppose that $j \in K$ and $j \sqsubset i$. If $i \in N \cup H$ then there is some k in N with $i \sqsubseteq k$; hence $j \sqsubset k$ and so $j \in H$, which is a contradiction. Therefore $i \in L_0 \cup K$. Thus if L_1 is an ancestral subset of K then $L_0 \cup L_1$ is an ancestral subset of M. Put $L = L_0 \cup L_1$. Then $\max(M \setminus L) = N \cup \max(K \setminus L_1)$. Given any element h in $G_{L_1}^{(K)}$, we can define g in G_L by

$$\begin{aligned} g_j &= h_j && \text{if } j \in \max K \setminus L_1 \\ B_j &= A_{g_j}^{(j)} && \text{if } j \in N. \end{aligned}$$

Then $\bar{B} \otimes A_{L_1,h}^{(K)} = A_{L,g}$, which is in $\mathrm{Adj}(M, \sqsubseteq)$.

Now, Lemma 9.14 for K shows that J_{Ω_K} is the sum of the matrices $A_{L_1,h}^{(K)}$, as L_1 and h take all possible values. Moreover, $B = \bar{B} \otimes J_{\Omega_K}$. Therefore B is a sum of matrices in $\mathrm{Adj}(M, \sqsubseteq)$ and so is in $\mathcal{A}$. ∎

Theorem 9.19 *The set of matrices $\mathrm{Adj}(M, \sqsubseteq)$ forms an association scheme on Ω_M.*

Proof Because $n_i \geqslant 2$ for all i in M, G_L is not empty if $M \setminus L$ is not empty. Thus every matrix $A_{L,g}$ is well defined, and none of its components is zero. Lemma 9.14 shows that the sum of the matrices in $\mathrm{Adj}(M, \sqsubseteq)$ is J_{Ω_M}.

The ancestral subset M itself has $G_M = \varnothing$ and $A_M = I_{\Omega_M}$. Since every matrix in $\mathcal{A}_i$ is symmetric for all i in M, so is every matrix in $\bigotimes_{i \in M} \mathcal{A}_i$, which contains $\mathrm{Adj}(M, \sqsubseteq)$. Lemmas 9.15 and 9.16 show that every product of adjacency matrices is nice. Lemmas 9.17 and 9.18 show that every nice matrix is in $\mathcal{A}$. ∎

The association scheme given in Theorem 9.19 is called the $\sqsubseteq$-*product* of the $\mathcal{Q}_i$.

In an obvious notation, the valency of the associate class which has adjacency matrix $A_{L,g}$ is

$$\prod_{i \in \max(M \setminus L)} a_{g_i}^{(i)} \times \prod_{i \in \mathrm{bot}(M \setminus L)} n_i.$$

It is now natural to refine the definition of 'atomic' given in Section 3.5 and say that an association scheme is *atomic* if it cannot be obtained from association schemes on smaller sets by any poset operator. Thus the association scheme in Example 6.2 is atomic but the one in Example 9.1 is not.

Definition A subset N of M is an *antichain* if whenever i and j are in N and $i \sqsubseteq j$ then $i = j$.

Denote by $\mathcal{N}(M)$ the set of antichains in M.

If K is any subset of M then $\max(K)$ and $\min(K)$ are both antichains. If N is an antichain, put

$$\begin{aligned} \not\downarrow(N) &= \{i \in M : \nexists j \text{ in } N \text{ with } i \sqsubseteq j\} \\ \uparrow(N) &= \{i \in M : \exists j \text{ in } N \text{ with } j \sqsubseteq i\}. \end{aligned}$$

Then both $\not\downarrow(N)$ and $\uparrow(N)$ are ancestral.

Proposition 9.20 *(i) The functions* $\min\colon \mathcal{S}(M, \sqsubseteq) \to \mathcal{N}(M)$ *and* $\uparrow\colon \mathcal{N}(M) \to \mathcal{S}(M, \sqsubseteq)$ *are mutually inverse.*

(ii) The functions

$$K \mapsto \max(M \setminus K) \text{ for } K \text{ in } \mathcal{S}(M, \sqsubseteq)$$

and $\not\downarrow\colon \mathcal{N}(M) \to \mathcal{S}(M, \sqsubseteq)$ *are mutually inverse.*

It was natural to define adjacency matrices of the $\sqsubseteq$-product in terms of ancestral subsets, to show the link with poset block structures. However, Proposition 9.20(ii) shows that there is an alternative labelling of the adjacency matrices by antichains. For each antichain N the corresponding adjacency matrices are those B in $\mathcal{B}$ for which

$$B_i = \begin{cases} I_i & \text{if } i \in \not\downarrow(N) \\ A_k^{(i)} \text{ for some } k \text{ in } \mathcal{K}_i \setminus \{0\} & \text{if } i \in N \\ J_i & \text{otherwise.} \end{cases}$$

Proposition 9.21 *If the association scheme $\mathcal{Q}_i$ has s_i associate classes for i in M then the number of associate classes in the $\sqsubseteq$-product of the $\mathcal{Q}_i$ is equal to*

$$\sum_{N \in \mathcal{N}(M)} \prod_{i \in N} s_i - 1.$$

Example 9.1 revisited Here the antichains are $\varnothing$, $\{1\}$, $\{2\}$, $\{1,2\}$, $\{3\}$, $\{4\}$, $\{1,3\}$ and $\{3,4\}$, so the number of associate classes in a $\sqsubseteq$-product such as the one shown in Figure 9.6 is equal to

$$\begin{aligned} & 1 + s_1 + s_2 + s_1 s_2 + s_3 + s_4 + s_1 s_3 + s_3 s_4 - 1 \\ = \quad & (s_1+1)(s_2+1) + (s_3+1)(s_4+1) + s_1 s_3 - 2 \\ = \quad & s_2(s_1+1) + s_4(s_3+1) + (s_1+1)(s_3+1) - 1. \quad \blacksquare \end{aligned}$$

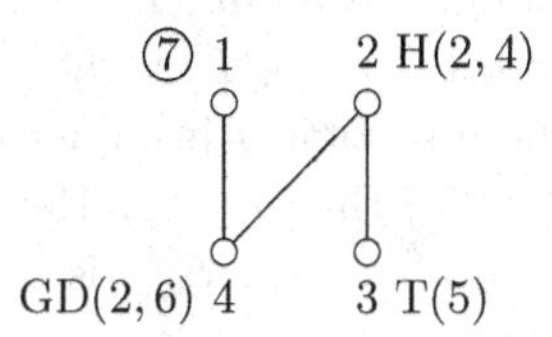

Fig. 9.6. A poset operator combining four association schemes

Theorem 9.22 *For i in M, let the stratum projectors for $\mathcal{Q}_i$ be $S_e^{(i)}$ for e in $\mathcal{E}_i$, with $S_0^{(i)} = n_i^{-1} J_i$. Let $C^{(i)}$ be the character table of $\mathcal{Q}_i$. For each ancestral subset L of M put*

$$\tilde{G}_L = \prod_{j \in \min(L)} (\mathcal{E}_j \setminus \{0\}),$$

and, for each element g of $\tilde{G}_L$, put

$$S_{L,g} = \bigotimes_{i \in \operatorname{top}(L)} I_i \otimes \bigotimes_{i \in \min(L)} S_{g_i}^{(i)} \otimes \bigotimes_{i \in M \setminus L} n_i^{-1} J_i.$$

Then the matrices $S_{L,g}$ are the stratum projectors of the $\sqsubseteq$-product of the $\mathcal{Q}_i$. Moreover, if K is an ancestral subset of M and $f \in G_K$ then the eigenvalue of $A_{K,f}$ on the stratum whose projector is $S_{L,g}$ is equal to

$$\prod_{i \in \min(L) \cap \max(M \setminus K)} C^{(i)}(f_i, g_i) \times \prod_{i \in \max(M \setminus K) \setminus L} C^{(i)}(f_i, 0) \times \prod_{i \in \operatorname{bot}(M \setminus K)} n_i$$

if $\min(L) \subseteq K \cup \max(M \setminus K)$; otherwise it is zero.

Proof Each component of $S_{L,g}$ is idempotent, so $S_{L,g}$ is itself idempotent. Let $W_{L,g}$ be the image of $S_{L,g}$.

Because $n_i \geqslant 2$ for all i in M, $\tilde{G}_L$ is not empty if L is not empty. Thus $S_{L,g}$ is well defined and non-zero, so $W_{L,g}$ is not zero. Note that $\tilde{G}_\varnothing = \varnothing$ and $S_\varnothing = n^{-1}J_{\Omega_M}$.

For a given ancestral subset L, let g and h be distinct elements of $\tilde{G}_L$. There is some i in $\min(L)$ for which $g_i \neq h_i$: then $S^{(i)}_{g_i}S^{(i)}_{h_i} = O$ and so $S_{L,g}S_{L,h} = O$. Hence $W_{L,g}$ is orthogonal to $W_{L,h}$. The sum of the $S_{L,g}$ over g in $\tilde{G}_L$ is equal to the matrix S_L in Theorem 9.9, which is the projector onto the stratum W_L in the association scheme defined by the poset block structure $\mathcal{F}(\sqsubseteq)$. Hence the spaces $W_{L,g}$, as L and g take all possible values, are mutually orthogonal and their direct sum is $\mathbb{R}^{\Omega_M}$.

Now consider the product $A_{K,f}S_{L,g}$. The i-th component of this product is equal to zero if $i \in \mathrm{bot}(M\setminus K)\cap\min(L)$, because then $J_iS^{(i)}_{g_i} = O_i$.

If the product is non-zero then $\min(L) \subseteq K \cup \max(M \setminus K)$. In this case, if $i \in \mathrm{top}(L)$ then there is some j in $\min(L)$ with $j \sqsubset i$: if $j \in K$ then $i \in K$; if $j \in \max(M \setminus K)$ then $i \in K$. Hence $\mathrm{top}(L) \subseteq K$. Thus the i-th component of the product $A_{K,f}S_{L,g}$ is equal to

$$
\begin{array}{ll}
I_i & \text{if } i \in \mathrm{top}(L)\\
S^{(i)}_{g_i} & \text{if } i \in \min(L)\cap K\\
A^{(i)}_{f_i}S^{(i)}_{g_i} & \text{if } i \in \min(L)\cap\max(M\setminus K)\\
n_i^{-1}J_i & \text{if } i \in K\setminus L\\
a^{(i)}_{f_i}n_i^{-1}J_i & \text{if } i \in \max(M\setminus K)\setminus L\\
J_i & \text{if } i \in \mathrm{bot}(M\setminus K).
\end{array}
$$

Now, $A^{(i)}_{f_i}S^{(i)}_{g_i} = C^{(i)}(f_i,g_i)S^{(i)}_{g_i}$ and $a^{(i)}_{f_i} = C^{(i)}(f_i,0)$. This shows that $W_{L,g}$ is an eigenspace of $A_{K,f}$ with the eigenvalue claimed. In particular, $W_{L,g}$ is a subspace of each stratum.

To show that the subspaces $W_{L,g}$ are precisely the strata, we need to show that the number of these subspaces is equal to one more than the number of associate classes, by Theorem 2.6. Proposition 9.20(i) shows that the number of spaces $W_{L,g}$ is equal to

$$\sum_{N\in\mathcal{N}(M)}\prod_{i\in N} s_i,$$

which, by Proposition 9.21, is equal to one more than the number of associate classes. ■

In an obvious notation, the dimension of the stratum $W_{L,g}$ is

$$\prod_{i \in \mathrm{top}(L)} n_i \times \prod_{i \in \min(L)} d_{g_i}^{(i)}.$$

Proposition 9.23 *Poset operators preserve the following families of association schemes, in the sense that if $\mathcal{Q}_i$ is in the family for all i in M then so is their $\sqsubseteq$-product for any partial order $\sqsubseteq$ on M:*

(i) poset block structures;
(ii) orthogonal block structures;
(iii) group block structures;
(iv) group schemes.

Exercises

9.1 For $m \leqslant 3$, find all partial orders on $\{1,\ldots,m\}$ which cannot be obtained from each other by relabelling the elements of $\{1,\ldots,m\}$. For each partial order, write down its ancestral subsets and the Hasse diagram for the partitions which they define. Write down the adjacency matrices and stratum projectors as tensor products of m matrices. Write down the valencies and dimensions in terms of n_1, ..., n_m. Identify each association scheme under another name.

9.2 Draw the Hasse diagram for each of the posets which are implicit in Exercise 3.13.

9.3 Find the character table for Example 9.1.

9.4 Prove the analogue of Theorem 9.10 for the D matrix.

9.5 Prove that the association scheme in Example 6.2 is atomic.

9.6 How many associate classes are there in the association scheme shown in Figure 9.6? Calculate its valencies and stratum dimensions.

9.7 Consider the following association scheme defined by a 5-ary poset operator.

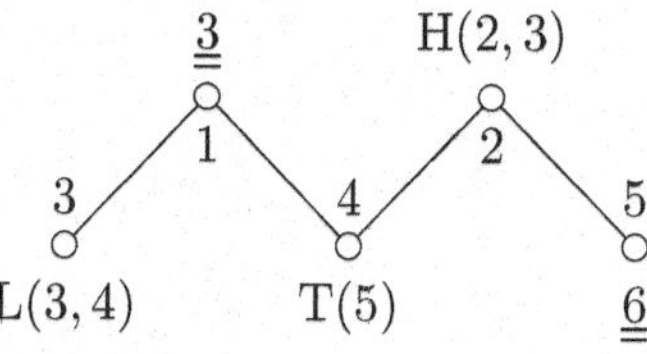

Write down its adjacency matrices and stratum projectors as tensor products of five components. Also calculate its valencies, stratum dimensions and character table.

9.8 Prove the analogue of Theorem 9.22 for the D matrix.

9.9 Prove Proposition 9.23.

10

Subschemes, quotients, duals and products

10.1 Inherent partitions

Definition Let $\mathcal{Q}$ be an association scheme on Ω. A partition F of Ω is *inherent* in $\mathcal{Q}$ if its relation matrix R_F is in the Bose–Mesner algebra of $\mathcal{Q}$.

Clearly, E and U are inherent in every association scheme, because every Bose–Mesner algebra contains I and J. An association scheme with no non-trivial inherent partition is call *primitive*; otherwise it is *imprimitive*. If F is any partition in an orthogonal block structure $\mathcal{F}$ on Ω then F is inherent in the association scheme $\mathcal{Q}(\mathcal{F})$ on Ω defined by $\mathcal{F}$. Hence all orthogonal block structures except $\{E, U\}$ are imprimitive.

Suppose that F is an inherent partition in an association scheme $\mathcal{Q}$ whose associate classes are indexed by $\mathcal{K}$. Then there is a subset $\mathcal{L}$ of $\mathcal{K}$ such that $R_F = \sum_{i \in \mathcal{L}} A_i$. Suppose further that F is non-trivial: then $\mathcal{L} \neq \mathcal{K}$ and $|\mathcal{L}| \geqslant 2$. For each i in $\mathcal{L} \setminus \{0\}$ the graph on Ω whose edges are the pairs in $\mathcal{C}_i$ is disconnected.

Conversely, given any non-empty subset $\mathcal{M}$ of $\mathcal{K} \setminus \{0\}$ we can form the graph $\mathcal{G}$ on Ω whose adjacency matrix is $\sum_{i \in \mathcal{M}} A_i$. Suppose that Γ is a connected component of $\mathcal{G}$ and that $\omega \in \Gamma$. If $\{i, j\} \subseteq \mathcal{M}$ and $p_{ij}^k \neq 0$ then $\mathcal{C}_k(\omega) \subseteq \Gamma$. Let $\mathcal{L}$ be the minimal subset of $\mathcal{K}$ which satsifies

(i) $\mathcal{M} \subseteq \mathcal{L}$, and
(ii) $p_{ij}^k = 0$ if $\{i, j\} \subseteq \mathcal{L}$ and $k \notin \mathcal{L}$.

Then $\sum_{i \in \mathcal{L}} A_i$ is the relation matrix for the partition of Ω into the connected components of $\mathcal{G}$, so this partition is inherent.

Let F be an inherent partition with $R_F = \sum_{i \in \mathcal{L}} A_i$. Then $p_{ij}^k = 0$ if

$\{i, j\} \subseteq \mathcal{L}$ and $k \notin \mathcal{L}$, so span$\{A_i : i \in \mathcal{L}\}$ is a subalgebra $\mathcal{A}[\mathcal{L}]$ of the Bose–Mesner algebra $\mathcal{A}$ of $\mathcal{Q}$. Moreover, $A_i R_F = a_i R_F$ for $i \in \mathcal{L}$.

Since $k_F^{-1} R_F$ is idempotent, Lemma 2.7 shows that there is a subset $\mathcal{H}$ of $\mathcal{E}$ such that $k_F^{-1} R_F = \sum_{e \in \mathcal{H}} S_e$. If $e \in \mathcal{H}$ then $R_F S_e = k_F S_e$; otherwise $R_F S_e = O$.

Conversely, suppose that $R = \sum_{i \in \mathcal{L}} A_i = k \sum_{e \in \mathcal{H}} S_e$ for some $\mathcal{L}$, $\mathcal{E}$ and k. Then R is symmetric with entries in $\{0, 1\}$, so it represents a symmetric relation on Ω. The trace of R is $k \sum_{e \in \mathcal{H}} d_e$, which is non-zero, so $0 \in \mathcal{L}$ and the relation is reflexive. We have $R^2 = kR$, so each entry of R^2 is positive if and only if the corresponding entry of R is, so the relation is transitive. Thus R is the relation matrix of a partition F. Moreover, every class of F has size k, because $R^2 = kR$.

Example 10.1 Consider the association scheme Pair(n). The elements fall into mirror-image pairs (i, j) and (j, i) for $i \neq j$. The partition F into these pairs in inherent in Pair(n). In the notation used on page 134, $R_F = I + M$. ■

Theorem 10.1 *Let $\mathcal{Q}$ be an association scheme on Ω. Then the set $\mathcal{F}$ of inherent partitions in $\mathcal{Q}$ forms an orthogonal block structure on Ω.*

Proof Clearly $\mathcal{F}$ contains E and U. If $F \in \mathcal{F}$ then R_F is in the Bose–Mesner algebra $\mathcal{A}$ of $\mathcal{Q}$, so R_F has constant row-sums and so F is uniform. Moreover, $R_F = \sum_{i \in \mathcal{L}} A_i$ for some subset $\mathcal{L}$ of $\mathcal{K}$. If G is also in $\mathcal{F}$ then there is a subset $\mathcal{M}$ of $\mathcal{K}$ such that $R_G = \sum_{i \in \mathcal{M}} A_i$. Now $R_{F \wedge G} = \sum_{i \in \mathcal{L} \cap \mathcal{M}} A_i \in \mathcal{A}$ and so $F \wedge G \in \mathcal{F}$.

Matrices in $\mathcal{A}$ commute, so $R_F R_G = R_G R_F$. Since F and G are both uniform, this implies that $P_F P_G = P_G P_F$; in other words, F is orthogonal to G. Since $F \wedge G$ is also uniform, Corollary 6.5 shows that $F \vee G$ is uniform. Then Lemma 6.4(i) shows that $R_F R_G$ is a scalar multiple of $R_{F \vee G}$ and hence $F \vee G \in \mathcal{F}$. ■

Proposition 10.2 *Let $\mathcal{Q}_1$ and $\mathcal{Q}_2$ be association schemes on sets Ω_1 and Ω_2. If F and G are inherent partitions in $\mathcal{Q}_1$ and $\mathcal{Q}_2$ respectively then $F \times G$ is inherent in $\mathcal{Q}_1 \times \mathcal{Q}_2$ while $F \times U_2$ and $E_1 \times G$ are inherent in $\mathcal{Q}_1/\mathcal{Q}_2$, where U_2 is the universal partition of Ω_2 and $E_1 = \mathrm{Diag}(\Omega)$.*

Proposition 10.3 *Let Δ be a blueprint for a group Ω. If $\bigcup_{i \in \mathcal{L}} \Delta_i$ is a subgroup of Ω then the partition of Ω into the left cosets of this subgroup is inherent in the group scheme on Ω defined by Δ. Moreover, all inherent partitions in group schemes arise in this way.*

Example 10.2 (Example 8.2 continued) In the symmetric group S_3, $\Delta_0 \cup \Delta_2 = \{1, (1\,2)\}$, which is a subgroup H of S_3. For α, β in S_3,

$$(\alpha, \beta) \in \mathcal{C}_0 \cup \mathcal{C}_2 \iff \alpha^{-1}\beta \in H,$$

so H defines the inherent partition of S_3 whose classes are $\{1, (1\,2)\}$, $\{(1\,2\,3), (2\,3)\}$ and $\{(1\,3\,2), (1\,3)\}$. ■

Note that the subgroup defining an inherent partition in a group scheme is not necessarily a normal subgroup.

10.2 Subschemes

Let $\mathcal{Q}$ be an association scheme on a set Ω with associate classes $\mathcal{C}_i$ for i in $\mathcal{K}$. Let Γ be a subset of Ω. For i in $\mathcal{K}$ put

$$\mathcal{C}_i^\Gamma = \mathcal{C}_i \cap (\Gamma \times \Gamma),$$

which is the restriction of $\mathcal{C}_i$ to Γ. If the non-empty classes $\mathcal{C}_i^\Gamma$ form an association scheme on Γ then this is said to be a *subscheme* of $\mathcal{Q}$.

Example 10.3 In the Hamming scheme $\mathrm{H}(m, 2)$ let Γ consist of those elements of Ω which have exactly n coordinates equal to 1, where $n \leqslant m$. Then Γ can be identified with the set of n-subsets of an m-set. The restrictions of the associate classes of the Hamming scheme to Γ give the Johnson scheme $\mathrm{J}(m, n)$, so $\mathrm{J}(m, n)$ is a subscheme of $\mathrm{H}(m, 2)$. ■

Let A_i^Γ be the adjacency matrix of $\mathcal{C}_i^\Gamma$. For (α, β) in $\Gamma \times \Gamma$,

$$\left(A_i^\Gamma A_j^\Gamma\right)(\alpha, \beta) = \left|\mathcal{C}_i^\Gamma(\alpha) \cap \mathcal{C}_j^\Gamma(\beta)\right|.$$

If Γ is a class of an inherent partition F with $R_F = \sum_{i \in \mathcal{L}} A_i$ then $\mathcal{C}_i^\Gamma = \varnothing$ if $i \neq \mathcal{L}$ while $\mathcal{C}_i(\omega) \subseteq \Gamma$ for all ω in Γ and all i in $\mathcal{L}$. Therefore, for (α, β) in $\Gamma \times \Gamma$ and (i, j) in $\mathcal{L} \times \mathcal{L}$,

$$\left(A_i^\Gamma A_j^\Gamma\right)(\alpha, \beta) = |\mathcal{C}_i(\alpha) \cap \mathcal{C}_j(\beta)| = (A_i A_j)(\alpha, \beta)$$

so

$$A_i^\Gamma A_j^\Gamma = \sum_{k \in \mathcal{K}} p_{ij}^k A_k^\Gamma = \sum_{k \in \mathcal{L}} p_{ij}^k A_k^\Gamma.$$

Therefore $\left\{\mathcal{C}_i^\Gamma : i \in \mathcal{L}\right\}$ is a subscheme of $\mathcal{Q}$ with Bose–Mesner algebra isomorphic to $\mathcal{A}[\mathcal{L}]$. Such a subscheme is called an *algebraic* subscheme.

The algebraic subschemes on different classes of an inherent partition may not be isomorphic to each other. We shall see an example in Section 10.6.1.

Example 10.4 In the Hamming scheme $\mathrm{H}(m,2)$ let Γ consist of those elements which have an even number of coordinates equal to 1, and let F be the partition of Ω into Γ and $\Omega\setminus\Gamma$. Then $R_F = \sum_{\text{even } i} A_i$ and so F is inherent in $\mathrm{H}(m,2)$. The algebraic subschemes are isomorphic to each other and are called the *halves* of the m-cube. ■

Example 10.5 In $\mathrm{H}(m,2)$ put $\bar{\omega}_j = 1-\omega_j$ for $j = 1, \ldots, m$. Let F be the partition of Ω into the antipodal pairs $\{\omega, \bar{\omega}\}$. Then $R_F = A_0 + A_m$ and F is inherent. The algebraic subschemes are all isomorphic to the trivial scheme $\underline{\underline{2}}$. ■

Write $\mathcal{Q}|_\Gamma$ for the algebraic subscheme $\left\{C_i^\Gamma : i \in \mathcal{L}\right\}$, where Γ is a class of the inherent partition F with $R_F = \sum_{i\in\mathcal{L}} A_i$. Since the Bose–Mesner algebra of $\mathcal{Q}|_\Gamma$ is isomorphic to $\mathcal{A}[\mathcal{L}]$, the eigenvalues of A_i^Γ are the same as the eigenvalues of A_i, for i in $\mathcal{L}$. For e, f in $\mathcal{E}$, put $e \approx f$ if $C(i,e) = C(i,f)$ for all i in $\mathcal{L}$. Then $\approx$ is an equivalence relation and its classes correspond to the the strata of $\mathcal{Q}|_\Gamma$, so there are $|\mathcal{L}|$ classes.

Let $\mathcal{H}$ be the subset of $\mathcal{E}$ such that $R_F = k_F \sum_{e\in\mathcal{H}} S_e$. If $e \in \mathcal{H}$ and $i \in \mathcal{L}$ then

$$A_i S_e = A_i \frac{1}{k_F} R_F S_e = a_i \frac{1}{k_F} R_F S_e = a_i S_e.$$

Given f in $\mathcal{E}$, if $A_i S_f = a_i S_f$ for all i in $\mathcal{L}$ then

$$R_F S_f = \sum_{i\in\mathcal{L}} A_i S_f = \sum_{i\in\mathcal{L}} a_i S_f = k_F S_f$$

and so $f \in \mathcal{H}$. Hence $\mathcal{H}$ is the equivalence class of $\approx$ containing 0.

Suppose that $\mathcal{F}$ is an orthogonal block structure on Ω, that F and G are partitions in $\mathcal{F}$ and that Γ is an F-class. Then G^Γ is a uniform partition of Γ and $G^\Gamma = (F \wedge G)^\Gamma$. Hence $\left\{G^\Gamma : G \in \mathcal{F},\ G \preccurlyeq F\right\}$ is an orthogonal block structure on Γ, and the association scheme which it defines is an algebraic subscheme of the association scheme $\mathcal{Q}(\mathcal{F})$.

Since the partitions E and U are inherent in every association scheme, the trivial association scheme $\underline{\underline{1}}$ is an algebraic subscheme of every association scheme and every association scheme is an algebraic subscheme of itself.

If $\mathcal{Q}_3$ and $\mathcal{Q}_4$ are algebraic subschemes of $\mathcal{Q}_1$ and $\mathcal{Q}_2$ respectively then $\mathcal{Q}_3 \times \mathcal{Q}_4$ is an algebraic subscheme of $\mathcal{Q}_1 \times \mathcal{Q}_2$. In particular, $\mathcal{Q}_1$, $\mathcal{Q}_2$, $\mathcal{Q}_3$ and $\mathcal{Q}_4$ are algebraic subschemes of $\mathcal{Q}_1 \times \mathcal{Q}_2$. Moreover, $\mathcal{Q}_3/\mathcal{Q}_2$, $\mathcal{Q}_2$ and $\mathcal{Q}_4$ are algebraic subschemes of $\mathcal{Q}_1/\mathcal{Q}_2$.

10.3 Ideal partitions

As in Section 6.1, every partition $\mathcal{P}$ of $\Omega \times \Omega$ defines a subspace $V_{\mathcal{P}}$ of the space $\mathbb{R}^{\Omega\times\Omega}$. Of course, if $\mathrm{Diag}(\Omega)$ is not a union of classes of $\mathcal{P}$ then I_Ω is not in $V_{\mathcal{P}}$.

Definition Given an association scheme $\mathcal{Q}$ on a set Ω, a partition $\mathcal{P}$ of $\Omega \times \Omega$ is an *ideal partition* for $\mathcal{Q}$ if

(i) $\mathcal{Q} \preccurlyeq \mathcal{P}$;
(ii) $AB \in V_{\mathcal{P}}$ for all A in $V_{\mathcal{Q}}$ and all B in $V_{\mathcal{P}}$.

Note that $V_{\mathcal{Q}}$ is just the Bose–Mesner algebra $\mathcal{A}$ of $\mathcal{Q}$. Thus condition (i) implies that $V_{\mathcal{P}} \leqslant \mathcal{A}$ (using Lemma 6.2) while condition (ii) states that $V_{\mathcal{P}}$ is an *ideal* of $\mathcal{A}$ in the ring-theoretic sense.

Theorem 10.4 *Let $\mathcal{P}$ be an ideal partition for an association scheme $\mathcal{Q}$ and let R be the adjacency matrix for the class of $\mathcal{P}$ which contains* $\mathrm{Diag}(\Omega)$. *Then R is the relation matrix for an inherent partition in $\mathcal{Q}$. Moreover, if B is the adjacency matrix for the class of $\mathcal{P}$ containing the class in $\mathcal{Q}$ whose adjacency matrix is A, then AR is an integer multiple of B.*

Proof Let B_0, B_1, …, B_s be the adjacency matrices of the classes of $\mathcal{P}$, where $B_0 = R$. Let their valencies be b_0, …, b_s. Condition (ii) implies that there are positive integers $m_0, m_1, \ldots, m_s$ such that $AR = m_0B_0 + m_1B_1 + \cdots + m_sB_s$. If $i \neq j$ then the diagonal entries of B_iB_j are zero, while the diagonal entires of B_i^2 are equal to b_i. Therefore the diagonal entries of ARB_j are equal to m_jb_j. Choose i so that $B = B_i$. If $j \neq i$ then the diagonal elements of AB_j are zero, but $AB_j \in V_{\mathcal{P}}$ so $AB_j = r_1B_1 + \cdots + r_sB_s$ for some integers r_1, …, r_s. Then $ARB_j = AB_jR = AB_jB_0$, whose diagonal entries are zero, and so $m_j = 0$. Thus $AR = m_iB$.

There is a subset $\mathcal{L}$ of $\mathcal{K}$ such that $R = \sum_{i\in\mathcal{L}} A_i$ and $0 \in \mathcal{L}$. Then A_iR is an integer multiple of R for all i in $\mathcal{L}$ and so R^2 is an integer multiple of R. Therefore R is the adjacency matrix of a uniform partition of Ω: the partition is inherent because $R \in \mathcal{A}$. ■

Theorem 10.5 *Let $\mathcal{Q}$ be an association scheme with adjacency matrices A_i for i in $\mathcal{K}$. If $\mathcal{Q}$ has an inherent partition F with relation matrix R then there is an ideal partition of $\mathcal{Q}$ whose adjacency matrices are scalar multiples of A_iR for i in $\mathcal{K}$.*

Proof Denote by $F(\alpha)$ the F-class containing α. Let $\mathcal{L}$ be the subset of $\mathcal{K}$ such that $R = \sum_{i\in\mathcal{L}} A_i$. There are positive integers m_{ij} such that

$$RA_i = A_iR = \sum_{j\in\mathcal{K}} m_{ij}A_j.$$

This means that, for all (α,β) in $\mathcal{C}_j$,

$$m_{ij} = (A_iR)(\alpha,\beta) = |\mathcal{C}_i(\alpha)\cap F(\beta)|\,.$$

Put $i \sim j$ if $m_{ij} \neq 0$. Now, $0 \in \mathcal{L}$ so $m_{ii} \geqslant 1$ and so $\sim$ is reflexive. If $m_{ij} \neq 0$ and $(\alpha,\beta) \in \mathcal{C}_j$ then there is some γ in $\mathcal{C}_i(\alpha)$ with $F(\beta) = F(\gamma)$. Thus $\beta \in \mathcal{C}_j(\alpha) \cap F(\gamma)$. This implies that $(A_jR)(\alpha,\gamma) \neq 0$ and so $m_{ji} \neq 0$. Therefore $\sim$ is symmetric.

Suppose that $m_{ij} \neq 0$, $m_{jk} \neq 0$ and $\delta \in \mathcal{C}_k(\alpha)$. Since $m_{jk} \neq 0$ there is some β in $\mathcal{C}_j(\alpha)$ with $F(\beta) = F(\delta)$; then since $m_{ij} \neq 0$ there is some γ in $\mathcal{C}_i(\alpha)$ with $F(\gamma) = F(\beta)$. Then $F(\gamma) = F(\delta)$, because F is a partition, so $m_{ik} \neq 0$. Moreover, $m_{ij} = |\mathcal{C}_i(\alpha)\cap F(\beta)| = |\mathcal{C}_i(\alpha)\cap F(\delta)| = m_{ik}$. Thus $\sim$ is also transitive, and so $\sim$ is an equivalence relation. In addition, for fixed i, all the non-zero values of m_{ij} are equal to a constant, say m_i.

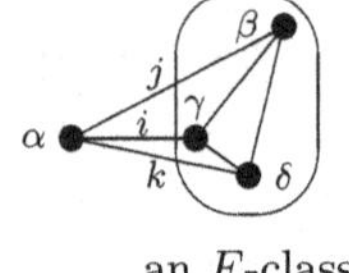

an F-class

Put $[i] = \{j \in \mathcal{K} : j \sim i\}$ and $B_{[i]} = \sum_{j\sim i} A_j$. Then the distinct $B_{[i]}$ are the adjacency matrices of a partition $\mathcal{P}$ of $\Omega\times\Omega$ with $\mathcal{Q} \preccurlyeq \mathcal{P}$. Moreover, $A_iR = m_iB_{[i]}$. Finally,

$$A_jB_{[i]} = \frac{1}{m_i}A_jA_iR = \frac{1}{m_i}\sum_k p_{ji}^kA_kR = \frac{1}{m_i}\sum_k p_{ji}^k m_k B_{[k]},$$

which is in $V_{\mathcal{P}}$. ■

For an ideal partition $\mathcal{P}$, let $\vartheta(\mathcal{P})$ be the inherent partition whose relation matrix is the adjacency matrix for the class of $\mathcal{P}$ containing $\mathrm{Diag}(\Omega)$. For an inherent partition F, let $\varrho(F)$ be the ideal partition whose adjacency matrices are scalar multiples of the matrices A_iR_F. Then R_F is an adjacency matrix of $\varrho(F)$, because $A_0R_F = R_F$: hence $\vartheta(\varrho(F)) = F$. On the other hand, Theorem 10.4 shows that $\mathcal{P} = \varrho(\vartheta(\mathcal{P}))$, so ϱ and ϑ are mutually inverse bijections. Note that $\varrho(E) = \mathcal{Q}$ and $\varrho(U) = U_{\Omega\times\Omega}$.

Corollary 10.6 *An association scheme $\mathcal{Q}$ on a set Ω is imprimitive if and only if it has an ideal partition different from $\mathcal{Q}$ and from $U_{\Omega\times\Omega}$.*

10.4 Quotient schemes and homomorphisms

Let F be an inherent partition for an association scheme $\mathcal{Q}$. Let $\bar{\Omega}$ consist of the classes of F. Continue to use the notation introduced in the proof of Theorem 10.5.

Let Γ and Δ be F-classes and let i and j be in $\mathcal{K}$. If $(\Gamma \times \Delta) \cap \mathcal{C}_i \neq \varnothing$ and $i \sim j$ then $(\Gamma \times \Delta) \cap \mathcal{C}_j \neq \varnothing$; while at most one of $(\Gamma \times \Delta) \cap \mathcal{C}_i$ and $(\Gamma \times \Delta) \cap \mathcal{C}_j$ is non-empty if $i \not\sim j$. Therefore we can define the $\bar{\Omega} \times \bar{\Omega}$ matrix $\bar{A}_{[i]}$ by

$$\bar{A}_{[i]}(\Gamma, \Delta) = \begin{cases} 1 & \text{if } (\Gamma \times \Delta) \cap \mathcal{C}_i \neq \varnothing \\ 0 & \text{otherwise.} \end{cases} \tag{10.1}$$

Then the distinct $\bar{A}_{[i]}$ sum to $J_{\bar{\Omega}}$. Moreover, the $\bar{A}_{[i]}$ are symmetric and $\bar{A}_{[0]} = I_{\bar{\Omega}}$.

Let X be the $\Omega \times \bar{\Omega}$ matrix with

$$X(\omega, \Gamma) = \begin{cases} 1 & \text{if } \omega \in \Gamma \\ 0 & \text{otherwise,} \end{cases}$$

so that $XX' = R$ and $X'X = k_F I_{\bar{\Omega}}$, where R is the relation matrix for F and k_F is the size of the F-classes. Now,

$$B_{[i]}(\alpha, \beta) = \bar{A}_{[i]}(\Gamma, \Delta) \qquad \text{for all } \alpha \text{ in } \Gamma \text{ and all } \beta \text{ in } \Delta.$$

Therefore, $X' B_{[i]} X = k_F^2 \bar{A}_{[i]}$. Hence

$$\begin{aligned} \bar{A}_{[i]} \bar{A}_{[j]} &= \frac{1}{k_F^4} X' B_{[i]} X X' B_{[j]} X \\ &= \frac{1}{k_F^4} X' \frac{1}{m_i} (A_i R) R \frac{1}{m_j} (A_j R) X \\ &= \frac{1}{k_F^2 m_i m_j} X' A_i A_j R X \\ &= \frac{1}{k_F^2 m_i m_j} \sum_{l \in \mathcal{K}} p_{ij}^l X' A_l R X \\ &= \frac{1}{k_F^2 m_i m_j} \sum_{l \in \mathcal{K}} p_{ij}^l X' m_l B_{[l]} X \\ &= \frac{1}{m_i m_j} \sum_{l \in \mathcal{K}} p_{ij}^l m_l \bar{A}_{[l]} \end{aligned} \tag{10.2}$$

and so the distinct $\bar{A}_{[i]}$ are the adjacency matrices of an association scheme $\bar{\mathcal{Q}}$ on $\bar{\Omega}$, which is called the *quotient* association scheme of $\mathcal{Q}$ by F. The valencies a_i, $b_{[i]}$ and $\bar{a}_{[i]}$ satisfy $a_i k_F = m_i b_{[i]}$, $b_{[i]} = \sum_{j \in [i]} a_j$ and $\bar{a}_{[i]} = b_{[i]}/k_F = a_i/m_i$.

(It would be natural to an algebraist to write the quotient scheme as $\mathcal{Q}/F$, but this conflicts with the notation for nesting.)

Example 10.6 In H(4, 2), let F be the partition into antipodal pairs. Then

$$\begin{aligned} R &= A_0 + A_4 && \text{so } [0] = \{0,4\}, \; m_0 = 1 \text{ and } \bar{a}_{[0]} = 1 \\ A_1R &= A_1 + A_3 && \text{so } [1] = \{1,3\}, \; m_1 = 1 \text{ and } \bar{a}_{[1]} = 4 \\ A_2R &= \quad 2A_2 && \text{so } [2] = \{2\}, \quad m_2 = 2 \text{ and } \bar{a}_{[2]} = 3. \end{aligned}$$

Now,

$$A_2^2 = 6(A_0 + A_4) + 4A_2$$

so Equation (10.2) gives

$$(\bar{A}_{[2]})^2 = \frac{1}{4}(6\bar{A}_{[0]} + 6\bar{A}_{[4]} + 4 \times 2\bar{A}_{[2]}) = 3\bar{A}_{[0]} + 2\bar{A}_{[2]}$$

and therefore the quotient scheme is GD(2, 4). ■

In general, the quotient of the Hamming scheme by the antipodal partition is called the *folded cube*.

Let F and G be partitions in an orthogonal block structure $\mathcal{F}$ on Ω, with $F \preccurlyeq G$, and let $\bar{\Omega}$ be the set of F-classes. Then G naturally gives a partition $\bar{G}$ of $\bar{\Omega}$: two F-classes belong to the same class of $\bar{G}$ if they are contained in the same class of G. Moreover, if $\bar{R}_G$ is defined analogously to $\bar{A}_{[i]}$ in Equation (10.1) then $\bar{R}_G$ is the relation matrix $R_{\bar{G}}$ for $\bar{G}$. Hence the quotient of the orthogonal block structure $\mathcal{F}$ by its inherent partition F is the orthogonal block structure $\{\bar{G} : G \in \mathcal{F},\ F \preccurlyeq G\}$.

Subschemes and quotient schemes of orthogonal block structures can be read directly off the Hasse diagram. For example, consider the inherent partition into halves in Example 7.9. All the algebraic subschemes are isomorphic to $\underline{\underline{2}} \times \underline{\underline{3}}$, while the quotient scheme is $\underline{\underline{4}}/\underline{\underline{2}}$.

If $\mathcal{Q} = \mathcal{Q}_1 \times \mathcal{Q}_2$ then both $\mathcal{Q}_1$ and $\mathcal{Q}_2$ are quotients of $\mathcal{Q}$. If $\mathcal{Q} = \mathcal{Q}_1/\mathcal{Q}_2$ then $\mathcal{Q}_1$ is a quotient of $\mathcal{Q}$.

If m is odd then each antipodal pair in H(m, 2) contains exactly one element with an even number of zeros, so

$$\mathrm{H}(m,2) \cong (\mathrm{half}\,\mathrm{H}(m,2)) \times \underline{\underline{2}}.$$

Thus the half cube is both an algebraic subscheme and a quotient scheme. However, this is not true if m is even. For example, the folded cube from H(4, 2) is isomorphic to GD(2, 4) while the half cube is GD(4, 2).

We can use the trick with the incidence matrix X to obtain the strata and character table $\bar{C}$ of $\bar{\mathcal{Q}}$.

Theorem 10.7 *Let $\bar{\mathcal{Q}}$ be the quotient of the association scheme $\mathcal{Q}$ by its inherent partition F whose relation matrix is R. Let $\mathcal{H}$ be the subset of $\mathcal{E}$ such that $R = k_F \sum_{e\in\mathcal{H}} S_e$. Then the stratum projectors for $\bar{\mathcal{Q}}$ are $k_F^{-1}X'S_eX$ for e in $\mathcal{H}$, and the character tables C and $\bar{C}$ for $\mathcal{Q}$ and $\bar{\mathcal{Q}}$ respectively are related by $\bar{C}([i],e) = C(i,e)/m_i$. In particular, if $i \sim j$ and $e \in \mathcal{H}$ then $C(i,e)/m_i = C(j,e)/m_j$.*

Proof If $e \in \mathcal{E}\setminus\mathcal{H}$ then $RS_e = O$ and so $(X'S_eX)^2 = X'S_eXX'S_eX = X'S_eRS_eX = O$ and therefore $X'S_eX = O$. If $e \in \mathcal{H}$ then $RS_e = k_FS_e$. Hence if e and f are in $\mathcal{H}$ then $X'S_eXX'S_fX = X'S_eRS_fX = k_FX'S_eS_fX$ so $\{k_F^{-1}X'S_eX : e \in \mathcal{H}\}$ is a set of orthogonal idempotents. Their sum is $I_{\bar{\Omega}}$, because

$$
\begin{aligned}
\sum_{e\in\mathcal{H}} \frac{1}{k_F} X'S_eX &= \sum_{e\in\mathcal{E}} \frac{1}{k_F} X'S_eX \\
&= \frac{1}{k_F} X' \left(\sum_{e\in\mathcal{E}} S_e \right) X \\
&= \frac{1}{k_F} X' I_\Omega X \\
&= \frac{1}{k_F} k_F I_{\bar{\Omega}}.
\end{aligned}
$$

Put $\bar{S}_e = k_F^{-1}X'S_eX$ for e in $\mathcal{H}$. Then

$$
\begin{aligned}
\bar{A}_{[i]}\bar{S}_e &= \frac{1}{k_F^3} X'B_{[i]}XX'S_eX \\
&= \frac{1}{k_F^3 m_i} X'A_iR^2S_eX \\
&= \frac{1}{k_F m_i} X'C(i,e)S_eX \\
&= \frac{C(i,e)}{m_i} \bar{S}_e
\end{aligned}
$$

and so $\bar{S}_e$ is an eigenprojector for every $\bar{A}_{[i]}$. Moreover, the eigenvalue of $\bar{A}_{[i]}$ on $\operatorname{Im} \bar{S}_e$ is $C(i,e)/m_i$.

Also, if $e \in \mathcal{H}$ then

$$
S_e = \frac{1}{k_F} S_eR = \frac{1}{k_F} \sum_{i\in\mathcal{K}} D(e,i)A_iR = \frac{1}{k_F} \sum_{i\in\mathcal{K}} D(e,i)m_iB_{[i]},
$$

so $\bar{S}_e$ is a linear combination of the $\bar{A}_{[i]}$. This shows that the $\bar{S}_e$, for e in $\mathcal{H}$, are the stratum projectors for the quotient scheme $\bar{\mathcal{Q}}$. ■

Example 10.6 revisited From Exercise 2.7, the character table of H(4, 2) is

$$\begin{array}{cc} & \\ 0 & (1) \\ 1 & (4) \\ 2 & (6) \\ 3 & (4) \\ 4 & (1) \end{array} \begin{array}{c} \begin{array}{ccccc} S_0 & S_1 & S_2 & S_3 & S_4 \end{array} \\ \left[\begin{array}{rrrrr} 1 & 1 & 1 & 1 & 1 \\ 4 & -4 & 2 & -2 & 0 \\ 6 & 6 & 0 & 0 & -2 \\ 4 & -4 & -2 & 2 & 0 \\ 1 & 1 & -1 & -1 & 1 \end{array}\right] \end{array}.$$

Thus $A_0 + A_4 = 2(S_0 + S_1 + S_4)$ so $\mathcal{H} = \{0, 1, 4\}$. Theorem 10.7 shows that the character table for the quotient scheme of H(4, 2) by the inherent partition into its antipodal pairs is

$$\begin{array}{cc} & \\ [0] & (1) \\ [1] & (4) \\ [2] & (3) \end{array} \begin{array}{c} \begin{array}{ccc} \bar{S}_0 & \bar{S}_1 & \bar{S}_4 \end{array} \\ \left[\begin{array}{rrr} 1 & 1 & 1 \\ 4 & -4 & 0 \\ 3 & 3 & -1 \end{array}\right] \end{array}.$$

Since the quotient scheme is GD(2, 4), this agrees with the character table on page 40. ■

Theorem 10.8 *Let D and $\bar{D}$ be the inverses of the character tables of $\mathcal{Q}$ and $\bar{\mathcal{Q}}$ respectively. If $e \in \mathcal{H}$ then $D(e, i) = \bar{D}(e, [i])/k_F$ for all i in $\mathcal{K}$.*

Proof Put $T(e, i) = \bar{D}(e, [i])/k_F$ for e in $\mathcal{H}$ and i in $\mathcal{K}$. It suffices to show that T consists of the $\mathcal{H}$-rows of C^{-1}. First note that

$$\sum_{j \sim i} m_j = \sum_{j \sim i} \frac{a_j k_F}{b_{[i]}} = k_F.$$

If $f \in \mathcal{E} \setminus \mathcal{H}$ and $i \in \mathcal{K}$ then $B_{[i]} S_f = O$ because $R S_f = O$. Hence

$$O = \sum_{j \sim i} A_j S_f = \sum_{j \sim i} \sum_{e \in \mathcal{E}} C(j, e) S_e S_f = \sum_{j \sim i} C(j, f) S_f,$$

and so $\sum_{j \sim i} C(j, f) = 0$. Since $T(e, i) = T(e, j)$ whenever $i \sim j$, this shows that if $e \in \mathcal{H}$ then $\sum_{i \in \mathcal{K}} T(e, i) C(i, f) = 0$.

Now let e and f be in $\mathcal{H}$. Theorem 10.7 shows that

$$\sum_{j \sim i} T(e, j) C(j, f) \quad = \quad \frac{\bar{D}(e, [i])}{k_F} \sum_{j \sim i} m_j \bar{C}([i], f)$$

$$= \bar{D}(e,[i])\bar{C}([i],f).$$

Hence

$$\sum_{i\in\mathcal{K}} T(e,i)C(i,f) = \sum_{\text{classes }[i]} \bar{D}(e,[i])\bar{C}([i],f) = \begin{cases} 1 & \text{if } e=f \\ 0 & \text{otherwise,} \end{cases}$$

because $\bar{D}$ is the inverse of $\bar{C}$. ■

As every algebraist will expect, quotient schemes are intimately bound up with homomorphisms, which generalize isomorphisms.

Definition Let $\mathcal{Q}_1$ be an association scheme on Ω_1 with classes $\mathcal{C}_i$ for i in $\mathcal{K}_1$ and Bose–Mesner algebra $\mathcal{A}_1$, and let $\mathcal{Q}_2$ be an association scheme on Ω_2 with classes $\mathcal{D}_j$ for j in $\mathcal{K}_2$. Let ϕ and π be functions from Ω_1 to Ω_2 and from $\mathcal{K}_1$ to $\mathcal{K}_2$ respectively. The pair (ϕ,π) is a *homomorphism* from $\mathcal{Q}_1$ to $\mathcal{Q}_2$ if

$$(\alpha,\beta)\in\mathcal{C}_i \Longrightarrow (\phi(\alpha),\phi(\beta))\in\mathcal{D}_{\pi(i)}.$$

Given such a homomorphism, let $\mathcal{L} = \{i\in\mathcal{K}_1 : \pi(i)=0\} = \pi^{-1}(0)$ and let $F = \{(\alpha,\beta)\in\Omega_1\times\Omega_1 : \phi(\alpha)=\phi(\beta)\}$. Then F is a partition of Ω_1 and $R_F = \sum_{i\in\mathcal{L}} A_i$ so F is inherent in $\mathcal{Q}_1$.

Further, let $\mathcal{P} = \pi^{-1}(\mathcal{Q}_2)$, so that $\mathcal{P}$ is a partition of $\Omega_1\times\Omega_1$ with classes of the form $\bigcup\{\mathcal{C}_i : \pi(i)=l\}$ for fixed l in $\mathcal{K}_2$. Then $\mathcal{Q}_1 \preccurlyeq \mathcal{P}$. For i, j in $\mathcal{K}_1$, put $i\equiv j$ if $\pi(i)=\pi(j)$. Evidently $\equiv$ is an equivalence relation, and $i\equiv j$ if and only if there exists a quadrangle in Ω of the following form.

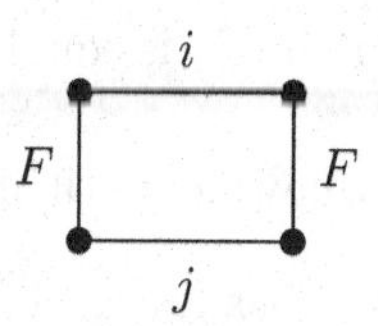

Matrices in $\mathcal{A}_1$ commute, so $R_FA_jR_F = A_jR_F^2 = A_jk_FR_F$, so the existence of such a quadrangle is equivalent to the existence of a triangle of the following form.

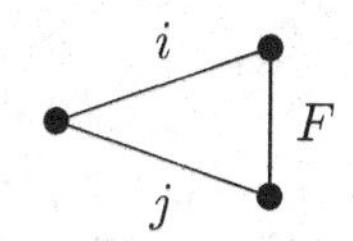

Now the proof of Theorem 10.5 shows that $\mathcal{P}$ is an ideal partition, in fact $\mathcal{P} = \varrho(F)$.

It is now clear that the quotient of $\mathcal{Q}_1$ by F is isomorphic to $\mathcal{Q}_2$.

10.5 Dual schemes

Definition For $t = 1, 2$, let $\mathcal{Q}_t$ be an association scheme on a set of size n, with associate classes labelled by $\mathcal{K}_t$ and strata by $\mathcal{E}_t$, with character table C_t and inverse D_t. Then $\mathcal{Q}_1$ is *formally dual* to $\mathcal{Q}_2$ if there is a bijection * from $\mathcal{K}_1$ to $\mathcal{E}_2$ and $\mathcal{E}_1$ to $\mathcal{K}_2$ such that

$$C_2(e^*, i^*) = nD_1(e, i) \tag{10.3}$$

for all e in $\mathcal{E}_1$ and all i in $\mathcal{K}_1$.

Inverting Equation (10.3) gives

$$nD_2(i^*, e^*) = C_1(i, e). \tag{10.4}$$

Thus if $\mathcal{Q}_1$ is formally dual to $\mathcal{Q}_2$ then $\mathcal{Q}_2$ is also formally dual to $\mathcal{Q}_1$.

Any duality * must interchange the 0-th associate class (diagonal) with the 0-th stratum (constant vectors) because these are the only rows of C and D with all entries equal to 1 and $1/n$ respectively. Thus

$$C_2(e^*, 0) = nD_1(e, 0);$$

that is, the valency of associate class e^* in $\mathcal{Q}_2$ is equal to the dimension of stratum e in $\mathcal{Q}_1$. Equation (10.4) gives

$$nD_2(i^*, 0) = C_1(i, 0),$$

so the dimension of stratum i^* in $\mathcal{Q}_2$ is equal to the valency of associate class i in $\mathcal{Q}_1$.

An association scheme may have more than one formal dual, or it may have none.

Example 10.7 Let $\mathcal{Q}$ be any association scheme of Latin-square type $\mathrm{L}(3, m)$, for some fixed m. Exercise 2.12 shows that the character table C is equal to

$$\begin{array}{cc} & \\ & \\ 0 & (1) \\ 1 & (3(m-1)) \\ 2 & ((m-1)(m-2)) \end{array} \begin{array}{ccc} 0 & 1 & 2 \\ (1) & (3(m-1)) & ((m-1)(m-2)) \\ \left[\begin{array}{c} 1 \\ 3(m-1) \\ (m-1)(m-2) \end{array}\right. & \begin{array}{c} 1 \\ m-3 \\ 2-m \end{array} & \left.\begin{array}{c} 1 \\ -3 \\ 2 \end{array}\right] \end{array}$$

and that m^2D is the same matrix if associate class i is identified with stratum i for $i = 0$, 1, 2. Hence all association schemes of Latin-square type L$(3, m)$ are formally dual to each other. ■

Example 10.8 As shown in Example 2.5, the stratum dimensions for the Petersen graph are 1, 4 and 5. Equation (2.13) shows that the only strongly regular graph with these valencies is GD$(2, 5)$, whose stratum dimensions 1, 1 and 8 are not equal to the valencies of the Petersen graph. Therefore the Petersen graph has no formal dual. ■

Theorem 10.9 *If $\mathcal{Q}_1^*$ is formally dual to $\mathcal{Q}_1$ and $\mathcal{Q}_2^*$ is formally dual to $\mathcal{Q}_2$ then $\mathcal{Q}_1^* \times \mathcal{Q}_2^*$ is formally dual to $\mathcal{Q}_1 \times \mathcal{Q}_2$ and $\mathcal{Q}_2^*/\mathcal{Q}_1^*$ is formally dual to $\mathcal{Q}_1/\mathcal{Q}_2$.*

Proof For the direct product, this follows directly from Theorem 3.4. The wreath product needs a little more care.

We use the notation and result of Theorem 3.9, except that we write $C_{2/1}^*$ for the character table of $\mathcal{Q}_2^*/\mathcal{Q}_1^*$ and $D_{1/2}$ for the inverse of the character table of $\mathcal{Q}_1/\mathcal{Q}_2$, and use $*$ to indicate the index sets of $\mathcal{Q}_1^*$ and $\mathcal{Q}_2^*$. Let $i \in \mathcal{K}_1 \setminus \{0\}$, $x \in \mathcal{K}_2$, $e \in \mathcal{E}_1$ and $f \in \mathcal{E}_2 \setminus \{0\}$. Each bijection * takes the special associate class (diagonal) to the special stratum (constant vectors) so $i^* \in \mathcal{E}_1^* \setminus \{0\}$, $x^* \in \mathcal{E}_2^*$, $e^* \in \mathcal{K}_1^*$ and $f^* \in \mathcal{K}_2^* \setminus \{0\}$. Therefore

$$
\begin{aligned}
C_{2/1}^*(f^*, x^*) &= n_1 C_2^*(f^*, x^*) \\
&= n_1 n_2 D_2(f, x) \\
&= n_1 n_2 D_{1/2}(f, x); \\
C_{2/1}^*(f^*, i^*) &= 0 \\
&= n_1 n_2 D_{1/2}(f, i); \\
C_{2/1}^*(e^*, x^*) &= b_{e^*} \\
&= d_e \\
&= n_1 n_2 D_{1/2}(e, x); \\
C_{2/1}^*(e^*, i^*) &= C_1^*(e^*, i^*) \\
&= n_1 D_1(e, i) \\
&= n_1 n_2 D_{1/2}(e, i). \quad ■
\end{aligned}
$$

Theorem 10.10 *Let $\mathcal{F}$ be an orthogonal block structure on a finite set Ω of size m. Then there is an orthogonal block structure $\mathcal{F}^*$ on a set Ω^* of size m such that the association schemes defined by $(\Omega, \mathcal{F})$ and $(\Omega^*, \mathcal{F}^*)$ are formally dual.*

Proof Putting together orthogonal bases for the strata of the association scheme $\mathcal{Q}$ defined by $(\Omega, \mathcal{F})$ gives an orthogonal basis for $\mathbb{R}^\Omega$ with the property that, for each F in $\mathcal{F}$, n_F elements of the basis lie in V_F and the remainder lie in $V_F^\perp$. Let Ω^* be this basis: then $|\Omega^*| = m$.

For F in $\mathcal{F}$, define a partition F^* on Ω^* by putting basis elements f and g in the same class of F^* if and only if f and g are in the same coset of V_F. Then F^* is uniform with $k_{F^*} = \dim V_F = n_F$ and hence $n_{F^*} = m/n_F = k_F$. Since $\dim V_U = 1$, U^* is the equality partition on Ω^*; since $V_E = \mathbb{R}^\Omega$, E^* is the universal partition on Ω^*. Moreover, if $F \preccurlyeq G$ then $V_G \leqslant V_F$ so $G^* \preccurlyeq F^*$.

From Lemma 6.3, $V_F \cap V_G = V_{F \vee G}$. Thus elements f, g of Ω^* are in the same coset of $V_{F\vee G}$ if and only if they are in the same coset of V_F and in the same coset of V_G; in other words $(F \vee G)^* = F^* \wedge G^*$.

Let Γ be a class of $F^* \vee G^*$. Then Γ consists of at least $k_{G^*}/k_{F^*\wedge G^*}$ classes of F^*, with equality if and only if every F^*-class in Γ meets every G^*-class in Γ. However, $F^* \preccurlyeq (F \wedge G)^*$ and $G^* \preccurlyeq (F \wedge G)^*$ so $F^* \vee G^* \preccurlyeq (F \wedge G)^*$, and

$$k_{(F\wedge G)^*} = n_{F\wedge G} = \frac{m}{k_{F\wedge G}} = \frac{m k_{F\vee G}}{k_F k_G} = \frac{k_{F^*} k_{G^*}}{k_{F^*\wedge G^*}}.$$

Hence $F^* \vee G^* = (F \wedge G)^*$ and F^* is orthogonal to G^*.

Put $\mathcal{F}^* = \{F^* : F \in \mathcal{F}\}$. Then $\mathcal{F}^*$ is an orthogonal block structure on Ω^*. Since $G^* \preccurlyeq F^*$ if $F \preccurlyeq G$, the zeta function ζ^* of $\mathcal{F}^*$ is the transpose of the zeta function ζ of $\mathcal{F}$.

Let C^* be the character table of the association scheme defined by $(\Omega^*, \mathcal{F}^*)$. Equation (6.9) gives

$$\begin{aligned} C^*(F^*, G^*) &= \sum_{H^* \in \mathcal{F}^*} \mu^*(H^*, F^*) k_{H^*} \zeta^*(H^*, G^*) \\ &= \sum_{H \in \mathcal{F}} \mu(F, H) n_H \zeta(G, H) \\ &= m \sum_{H \in \mathcal{F}} \mu(F, H) k_H^{-1} \zeta(G, H) \\ &= m D(F, G) \end{aligned}$$

from Equation (6.10). ■

Theorem 10.11 *Suppose that $\mathcal{Q}$ is an association scheme which has a formal dual $\mathcal{Q}^*$. If F is an inherent partition in $\mathcal{Q}$ then $\mathcal{Q}^*$ has an inherent partition F^* such that, for each class Γ of F^*, the algebraic* *subscheme $\mathcal{Q}^*|_\Gamma$ of $\mathcal{Q}^*$ is formally dual to the quotient scheme $\bar{\mathcal{Q}}$ of $\mathcal{Q}$ by F.*

Proof Let n be the size of the underlying set. Let C and C^* be the character tables of $\mathcal{Q}$ and $\mathcal{Q}^*$ respectively, with inverses D and D^*.

In $\mathcal{Q}$, let $\mathcal{L}$ and $\mathcal{H}$ be the subsets of $\mathcal{K}$ and $\mathcal{E}$ respectively such that $R_F = \sum_{i\in\mathcal{L}} A_i = k_F \sum_{e\in\mathcal{H}} S_e$. Then

$$\sum_{i\in\mathcal{L}} A_i = k_F \sum_{e\in\mathcal{H}} \sum_{j\in\mathcal{K}} D(e,j)A_j$$

so

$$\sum_{e\in\mathcal{H}} D(e,j) = \begin{cases} \frac{1}{k_F} & \text{if } j \in \mathcal{L} \\ 0 & \text{otherwise.} \end{cases}$$

Hence, in $\mathcal{Q}^*$,

$$\begin{aligned} \sum_{e\in\mathcal{H}} A_{e^*} &= \sum_{e\in\mathcal{H}} \sum_{i\in\mathcal{K}} C^*(e^*,i^*)S_{i^*} \\ &= \sum_{i\in\mathcal{K}} \sum_{e\in\mathcal{H}} nD(e,i)S_{i^*} \\ &= \frac{n}{k_F} \sum_{i\in\mathcal{L}} S_{i^*}. \end{aligned}$$

Therefore $\sum_{e\in\mathcal{H}} A_{e^*}$ is the relation matrix of a partition F^* whose classes have size n/k_F. If Γ is any class of F^* then $\mathcal{Q}^*|_\Gamma$ is an association scheme on a set of size n/k_F, as is $\bar{\mathcal{Q}}$.

If $i \sim j$ in $\mathcal{Q}$ then Theorem 10.8 shows that $D(e,i) = D(e,j)$ for all e in $\mathcal{H}$. Therefore $C^*(e^*,i^*) = C^*(e^*,j^*)$ for all e in $\mathcal{H}$, and so $i^* \approx j^*$ in $\mathcal{Q}^*$. Let $[[i^*]]$ be the equivalence class of $\approx$ containing i^* and $[i]$ the equivalence class of $\sim$ containing i. We have shown that the duality * maps $[i]$ into $[[i^*]]$. The number of classes of $\approx$ is equal to the number of strata in $\mathcal{Q}^*|_\Gamma$, which is equal to the number of adjacency matrices in $\mathcal{Q}^*|_\Gamma$, which is $|\mathcal{H}|$. However, $|\mathcal{H}|$ is the number of strata in $\bar{\mathcal{Q}}$, by Theorem 10.7, which is equal to the number of adjacency matrices of $\bar{\mathcal{Q}}$. Hence * induces a bijection from the adjacency matrices of $\bar{\mathcal{Q}}$ to the strata of $\mathcal{Q}^*|_\Gamma$. The strata of $\bar{\mathcal{Q}}$ and the adjacency matrices of $\mathcal{Q}^*|_\Gamma$ are both indexed by $\mathcal{H}$.

Let C^*_Γ be the character table of $\mathcal{Q}^*|_\Gamma$. For e in $\mathcal{H}$ and $i \in \mathcal{K}$ we have

$$C^*_\Gamma(e^*,[[i^*]]) = C^*(e^*,i^*) = nD(e,i) = \frac{n}{k_F}\bar{D}(e,[i]),$$

by Theorem 10.8. This shows that $\mathcal{Q}^*|_\Gamma$ is formally dual to $\bar{\mathcal{Q}}$. ■

Because $\mathcal{Q}^*$ is also formally dual to $\mathcal{Q}$, the algebraic subschemes of $\mathcal{Q}$ on the classes of F are all formally dual to the quotient of $\mathcal{Q}^*$ by F^*.

10.6 New from old

This chapter concludes with some more ways, different from those in Chapters 3 and 9, of combining association schemes.

10.6.1 Relaxed wreath products

In the wreath product $\mathcal{Q}_1/\mathcal{Q}_2$ every copy of Ω_2 is equipped with the *same* association scheme $\mathcal{Q}_2$. However, the i-th associates of an element of any copy of Ω_2 *either* are all in the same copy of of Ω_2 *or* consist of all the elements in certain other copies of Ω_2. So there is no need for all the copies of Ω_2 to have the same association scheme: all that is needed is that they have association schemes with the same parameters.

Thus, let $\mathcal{Q}_1$ be an association scheme on a set Ω_1 of size n. For ω in Ω_1, let $\mathcal{P}_\omega$ be an association scheme on the set Ω_2. Suppose that all of the association schemes $\mathcal{P}_\omega$ have their associate classes indexed by the set $\mathcal{K}_2$, with parameters q_{xy}^z and b_x as shown in Table 3.1. Let the associate classes of $\mathcal{P}_\omega$ be $\mathcal{D}_x^{(\omega)}$ for x in $\mathcal{K}_2$. The *relaxed wreath product* $\mathcal{Q}_1/[\mathcal{P}_1;\mathcal{P}_2;\ldots;\mathcal{P}_n]$ has as its associate classes the following subsets of $\Omega_1\times\Omega_2$:

$$\{((\alpha_1,\alpha_2),(\beta_1,\beta_2)) : (\alpha_1,\beta_1)\in\mathcal{C}\}$$

for each non-diagonal class $\mathcal{C}$ of $\mathcal{Q}_1$, and

$$\left\{((\omega,\alpha),(\omega,\beta)) : \omega\in\Omega_1 \text{ and } (\alpha,\beta)\in\mathcal{D}_x^{(\omega)}\right\}$$

for each x in $\mathcal{K}_2$. The proof that this is an association scheme is a slight modification of the proof of Theorem 3.8.

Example 10.9 Let $\mathcal{Q}_1$ and $\mathcal{Q}_2$ be the association schemes of L(3, 4) type described in Example 3.5. These are not isomorphic to each other. The association scheme $\underline{\underline{2}}/[\mathcal{Q}_1;\mathcal{Q}_2]$ is defined on the set of 32 cells in Figure 10.1. The associate classes are

same cell;
different cell, same row or column or letter, same square;
different square. ■

A	B	C	D
D	A	B	C
C	D	A	B
B	C	D	A

A	B	C	D
B	A	D	C
C	D	A	B
D	C	B	A

Fig. 10.1. Set of 32 cells in Example 10.9

The partition of $\Omega_1 \times \Omega_2$ defined by the first coordinate is inherent in $\mathcal{Q}_1/[\mathcal{P}_1; \mathcal{P}_2; \ldots; \mathcal{P}_n]$. The algebraic subschemes are isomorphic to $\mathcal{P}_1$, ..., $\mathcal{P}_n$, which may not be isomorphic to each other. The quotient scheme is isomorphic to $\mathcal{Q}_1$.

Since $\mathcal{P}_1$, ..., $\mathcal{P}_n$ all have the same character table, the character table of $\mathcal{Q}_1/[\mathcal{P}_1; \mathcal{P}_2; \ldots; \mathcal{P}_n]$ is the same as the character table of $\mathcal{Q}_1/\mathcal{P}_1$, which is given in Theorem 3.9.

10.6.2 Crested products

We have seen the bijection between inherent partitions and ideal partitions of a single association scheme, and the link with quotient schemes. Here we use two unrelated partitions, one of each type, on different association schemes, to define a product which generalizes both crossing and nesting.

Theorem 10.12 *Let $\mathcal{Q}_1$ be an association scheme on Ω_1 with associate classes $\mathcal{C}_i$ with adjacency matrices A_i for i in $\mathcal{K}_1$. Let F be an inherent partition in $\mathcal{Q}_1$ with relation matrix R such that $R = \sum_{i \in \mathcal{L}} A_i$. Let $\mathcal{Q}_2$ be an association scheme on Ω_2 with associate classes $\mathcal{D}_j$ with adjacency matrices B_j for j in $\mathcal{K}_2$. Let $\mathcal{P}$ be an ideal partition for $\mathcal{Q}_2$ with classes $\mathcal{G}_m$ with adjacency matrices G_m for m in $\mathcal{M}$. Then the following set of subsets forms an association scheme on $\Omega_1 \times \Omega_2$:*

$$\{\mathcal{C}_i \times \mathcal{G}_m : i \in \mathcal{K}_1 \setminus \mathcal{L},\ m \in \mathcal{M}\} \cup \{\mathcal{C}_i \times \mathcal{D}_j : i \in \mathcal{L},\ j \in \mathcal{K}_2\}.$$

Proof The adjacency matrices are

$$A_i \otimes G_m \qquad \text{for } i \text{ in } \mathcal{K}_1 \setminus \mathcal{L} \text{ and } m \text{ in } \mathcal{M}$$

and

$$A_l \otimes B_j \qquad \text{for } l \text{ in } \mathcal{L} \text{ and } j \text{ in } \mathcal{K}_2.$$

These are symmetric matrices with all entries in $\{0, 1\}$. Moreover, $\mathcal{L}$ contains the 0 index in $\mathcal{K}_1$, so one class is $\mathcal{C}_0 \times \mathcal{D}_0$, which is $\mathrm{Diag}(\Omega_1 \times \Omega_2)$. Also,

$$\begin{aligned} \sum_{i\in\mathcal{K}_1\setminus\mathcal{L}} \sum_{m\in\mathcal{M}} A_i \otimes G_m &= \sum_{i\in\mathcal{K}_1\setminus\mathcal{L}} A_i \otimes \sum_{m\in\mathcal{M}} G_m \\ &= (J_{\Omega_1} - R) \otimes J_{\Omega_2} \end{aligned}$$

and

$$\begin{aligned} \sum_{l\in\mathcal{L}} \sum_{j\in\mathcal{K}_2} A_l \otimes B_j &= \sum_{l\in\mathcal{L}} A_l \otimes \sum_{j\in\mathcal{K}_2} B_j \\ &= R \otimes J_{\Omega_2} \end{aligned}$$

so these matrices do sum to $J_{\Omega_1} \otimes J_{\Omega_2}$.

Let $\mathcal{A}$ be the set of real linear combinations of these adjacency matrices. We need to show that the product of every pair of adjacency matrices is in $\mathcal{A}$. First note that if $m \in \mathcal{M}$ then G_m is a sum of some of the B_j: hence $A_i \otimes G_m \in \mathcal{A}$ for all i in $\mathcal{K}_1$. Therefore $\mathcal{A} = \mathcal{A}_1 \otimes V_{\mathcal{P}} + \mathcal{A}_1[\mathcal{L}] \otimes \mathcal{A}_2$, where $\mathcal{A}_t$ is the Bose–Mesner algebra of $\mathcal{Q}_t$ for $t = 1$, 2, and $\mathcal{A}_1[\mathcal{L}]$ is the subalgebra of $\mathcal{A}_1$ spanned by $\{A_l : l \in \mathcal{L}\}$.

Each of $\mathcal{A}_1 \otimes V_{\mathcal{P}}$ and $\mathcal{A}_1[\mathcal{L}] \otimes \mathcal{A}_2$ is closed under multiplication. If $i \in \mathcal{K}_1$, $l \in \mathcal{L}$, $m \in \mathcal{M}$ and $j \in \mathcal{K}_2$, then $A_iA_l \in \mathcal{A}_1$ because $\mathcal{Q}_1$ is an association scheme and $G_mB_j \in V_{\mathcal{P}}$ because $\mathcal{P}$ is an ideal partition: hence $(A_i \otimes G_m)(A_l \otimes B_j) \in \mathcal{A}_1 \otimes V_{\mathcal{P}} \subseteq \mathcal{A}$. ■

If F is the universal partition U_1 of Ω_1 or $\mathcal{P} = \mathcal{Q}_2$ then the new association scheme is $\mathcal{Q}_1 \times \mathcal{Q}_2$; if F is the equality partition E_1 of Ω_1 and $\mathcal{P}$ is the universal partition of $\Omega_2 \times \Omega_2$ then it is $\mathcal{Q}_1/\mathcal{Q}_2$. The new association scheme is called the *crested product* of $\mathcal{Q}_1$ and $\mathcal{Q}_2$ with respect to F and $\mathcal{P}$. The word *crested* indicates that the crossed and nested products are two special cases; also 'crest' is cognate with a victor's wreath, which is the meaning of 'wreath' in 'wreath product'.

Example 10.10 Let $\mathcal{Q}_2 = \circledR{8}$ and let $\mathcal{P}$ be the ideal partition of $\mathbb{Z}_8 \times \mathbb{Z}_8$ which is derived from the partition $\{0, 4\}$, $\{2, 6\}$, $\{1, 3, 5, 7\}$ of $\mathbb{Z}_8$ as in Theorem 8.1. Let $\mathcal{Q}_1$ be $\underline{\underline{2}}$ on the set $\{x, y\}$ and let $F = E_1$. Then the classes of the crested product are

$$\{(z, a), (z, a) : z \in \{x, y\},\ a \in \mathbb{Z}_8\}$$
$$\{(z, a), (z, b) : z \in \{x, y\},\ a - b = \pm 1\}$$
$$\{(z, a), (z, b) : z \in \{x, y\},\ a - b = \pm 2\}$$

$$\{(z,a),(z,b) : z \in \{x,y\},\ a-b=\pm 3\}$$
$$\{(z,a),(z,b) : z \in \{x,y\},\ a-b=4\}$$
$$\{(x,a),(y,b) : a-b=0 \text{ or } 4\} \cup \{(y,a),(x,b) : a-b=0 \text{ or } 4\}$$
$$\{(x,a),(y,b) : a-b=2 \text{ or } 6\} \cup \{(y,a),(x,b) : a-b=2 \text{ or } 6\}$$
$$\{(x,a),(y,b) : a-b \in \{1,3,5,7\}\} \cup \{(y,a),(x,b) : a-b \in \{1,3,5,7\}\}. \quad \blacksquare$$

Theorem 10.13 *Let $\mathcal{Q}$ be the crested product defined in Theorem 10.12. For $t = 1, 2$, suppose that the strata for $\mathcal{Q}_t$ are $W_e^{(t)}$, for e in $\mathcal{E}_t$, with stratum projectors $S_e^{(t)}$, and let C_t be the character table of $\mathcal{Q}_t$. In $\mathcal{Q}_1$, put $e \approx f$ if $C_1(i,e) = C_1(i,f)$ for all i in $\mathcal{L}$, and put*

$$U_{[[e]]} = \bigoplus_{f \approx e} W_f^{(1)},$$

whose projector $T_{[[e]]}$ is equal to $\sum_{f\approx e} S_f^{(1)}$. In $\mathcal{Q}_2$, let $\mathcal{H}$ be the subset of $\mathcal{E}_2$ with the property that $R_{\vartheta(\mathcal{P})} = k_{\vartheta(\mathcal{P})} \sum_{e\in\mathcal{H}} S_e^{(2)}$. Then the strata for $\mathcal{Q}$ are $W_e^{(1)} \otimes W_f^{(2)}$, for (e,f) in $\mathcal{E}_1 \times \mathcal{H}$, and $U_{[[e]]} \otimes W_f^{(2)}$, for equivalence classes $[[e]]$ of $\approx$ and f in $\mathcal{E}_2 \setminus \mathcal{H}$. The eigenvalues are as follows, where $\bar{C}_2$ is the character table of the quotient of $\mathcal{Q}_2$ by $\vartheta(\mathcal{P})$.

	$W_e^{(1)} \otimes W_f^{(2)} \quad (f \in \mathcal{H})$	$U_{[[e]]} \otimes W_f^{(2)} \quad (f \notin \mathcal{H})$
$A_i \otimes G_m \quad (i \notin \mathcal{L})$	$C_1(i,e)k_{\vartheta(\mathcal{P})}\bar{C}_2(m,f)$	0
$A_i \otimes B_j \quad (i \in \mathcal{L})$	$C_1(i,e)C_2(j,f)$	$C_1(i,e)C_2(j,f)$

Proof The named subspaces are mutually orthogonal with direct sum $\mathbb{R}^{\Omega_1\times\Omega_2}$.

For $t = 1, 2$, let s_t be the number of associate classes of $\mathcal{Q}_t$. Put $x = |\mathcal{L}|$ and $y = |\mathcal{H}|$. Then Theorem 10.7 shows that $\mathcal{P}$ has y classes, so the number of associate classes of $\mathcal{Q}$ is

$$(s_1 + 1 - x)y + x(s_2 + 1) - 1.$$

Hence the number of strata for $\mathcal{Q}$ is

$$(s_1 + 1)y + x(s_2 + 1 - y),$$

which is equal to the number of named subspaces, because $\approx$ has x equivalence classes. Thus it suffices to show that each named subspace is a sub-eigenspace of each adjacency matrix, with the eigenvalue shown.

The proof of Theorem 10.7 shows that $G_m S_f^{(2)} = O$ if $f \notin \mathcal{H}$ while

$G_m S_f^{(2)} = k_{\vartheta(\mathcal{P})}\bar{C}_2(m,f)S_f^{(2)}$ if $f \in \mathcal{H}$. Thus $U_{[[e]]} \otimes W_f^{(2)}$ is a sub-eigenspace of $A_i \otimes G_m$ with eigenvalue 0 if $f \notin \mathcal{H}$, while $W_e^{(1)} \otimes W_f^{(2)}$ is a sub-eigenspace of $A_i \otimes G_m$ with eigenvalue $C_1(i,e)k_{\vartheta(\mathcal{P})}\bar{C}_2(m,f)$ if $f \in \mathcal{H}$. Evidently, $W_e^{(1)} \otimes W_f^{(2)}$ is a sub-eigenspace of $A_i \otimes B_j$ with eigenvalue $C_1(i,e)C_2(j,f)$. If $i \in \mathcal{L}$ then $C_1(i,e') = C_1(i,e)$ for all e' in $[[e]]$, so $A_i T_{[[e]]} = C_1(i,e)T_{[[e]]}$ and therefore $U_{[[e]]} \otimes W_f^{(2)}$ is a sub-eigenspace of $A_i \otimes B_j$ with eigenvalue $C_1(i,e)C_2(j,f)$. ■

Crested products of orthogonal block structures have a particularly attractive presentation. For $t = 1, 2$, let $\mathcal{F}_t$ be an orthogonal block structure on a set Ω_t and let F_t be a partition in $\mathcal{F}_t$. Define the set $\mathcal{G}$ of partitions of $\Omega_1 \times \Omega_2$ as follows:

$$\mathcal{G} = \{G_1 \times G_2 : G_1 \in \mathcal{F}_1,\ G_2 \in \mathcal{F}_2,\ G_1 \preccurlyeq F_1 \text{ or } F_2 \preccurlyeq G_2\}.$$

Then $\mathcal{G}$ contains E and U and its elements are uniform and mutually orthogonal, as shown in Lemmas 6.10 and 6.11. Let $G_1 \times G_2$ and $H_1 \times H_2$ be in $\mathcal{G}$. If $G_1 \wedge H_1 \not\preccurlyeq F_1$ then $G_1 \not\preccurlyeq F_1$ and $H_1 \not\preccurlyeq F_1$ so $F_2 \preccurlyeq G_2$ and $F_2 \preccurlyeq H_2$ and therefore $F_2 \preccurlyeq G_2 \wedge H_2$: thus $(G_1 \wedge H_1) \times (G_2 \wedge H_2) \in \mathcal{G}$. Likewise, if $F_2 \not\preccurlyeq G_2 \vee H_2$ then $G_1 \preccurlyeq F_1$ and $H_1 \preccurlyeq F_1$ so $G_1 \vee H_1 \preccurlyeq F_1$: thus $(G_1 \vee H_1) \times (G_2 \vee H_2) \in \mathcal{G}$. Therefore $\mathcal{G}$ is an orthogonal block structure on $\Omega_1 \times \Omega_2$.

Let $\mathcal{A}_t$ be the Bose–Mesner algebra of the association scheme $\mathcal{Q}_t$ defined by $\mathcal{F}_t$, for $t = 1, 2$. Let $\mathcal{L} = \{G \in \mathcal{F}_1 : G \preccurlyeq F_1\}$. Then $\mathcal{A}_1[\mathcal{L}]$ is the subalgebra of $\mathcal{A}_1$ isomorphic to the Bose–Mesner algebras of the algebraic subschemes of the inherent partition F_1 of $\mathcal{Q}_1$. Let $\mathcal{P}$ be the ideal partition $\varrho(F_2)$ of $\mathcal{Q}_2$, so that $V_{\mathcal{P}} = \operatorname{span}\{R_G : G \in \mathcal{F}_2,\ F_2 \preccurlyeq G\}$. Then the span of the relation matrices of $\mathcal{G}$ is $\mathcal{A}_1[\mathcal{L}] \otimes \mathcal{A}_2 + \mathcal{A}_1 \otimes V_{\mathcal{P}}$, which is the Bose–Mesner algebra of the crested product of $\mathcal{Q}_1$ and $\mathcal{Q}_2$ with respect to F_1 and $\mathcal{P}$: hence this crested product is the association scheme defined by the orthogonal block structure $\mathcal{G}$ and there is no ambiguity in calling $\mathcal{G}$ the crested product of $\mathcal{F}_1$ and $\mathcal{F}_2$ with respect to F_1 and F_2.

Example 10.11 Let $\mathcal{F}_1$ be $\underline{\underline{b}}/\underline{\underline{k}}$ with non-trivial partition B, and let $\mathcal{F}_2$ be the orthogonal block structure defined by a Latin square with non-trivial partitions R, C and L. The crested product of $\mathcal{F}_1$ and $\mathcal{F}_2$ with respect to B and L is the orthogonal block structure whose Hasse diagram is shown in Figure 10.2. ■

Suppose that, for $t = 1, 2$, $\mathcal{F}_t$ is a poset block structure defined by the poset $(M_t, \sqsubseteq_t)$ and that $F_t = F(L_t)$, where L_t is an ancestral subset

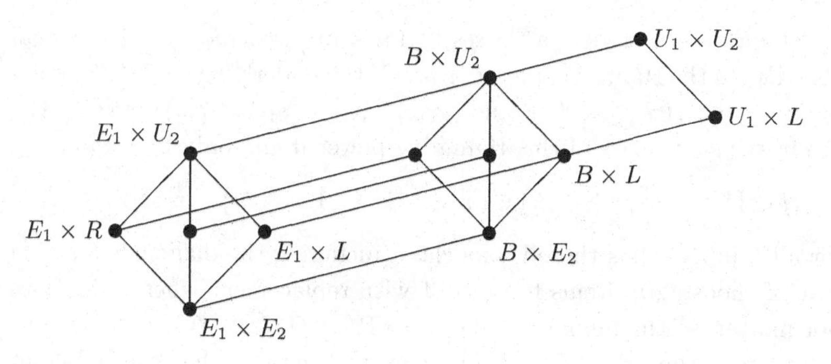

Fig. 10.2. Crested product of orthogonal block structures in Example 10.11

of M_t. Also suppose that $M_1 \cap M_2 = \varnothing$. Then the partitions in the crested product are the $F(K_1 \cup K_2)$, where K_t is an ancestral subset of M_t and either $L_1 \subseteq K_1$ or $K_2 \subseteq L_2$. These subsets are precisely the ancestral subsets of the poset $(M_1 \cup M_2, \sqsubseteq)$ where $i \sqsubseteq j$ if $\{i,j\} \subseteq M_1$ and $i \sqsubseteq_1 j$ or $\{i,j\} \subseteq M_2$ and $i \sqsubseteq_2 j$ or $i \in M_2 \setminus L_2$ and $j \in L_1$. Thus every crested product of two poset block structures is a poset block structure. We can say more than this: *every* poset can be made from smaller posets in this way, for if j is a maximal element of a poset $(M, \sqsubseteq)$ we can take $M_1 = L_1 = \{j\}$, $M_2 = M \setminus M_1$ and $L_2 = M_2 \setminus \{i \in M : i \sqsubseteq j\}$. This proves the following.

Proposition 10.14 *Every poset block structure can be obtained from trivial poset block structures by iterated crested products.*

In turn this poses the following very interesting question: which orthogonal block structures cannot be obtained from smaller orthogonal block structures as crested products? One example is the orthogonal block structure defined by a set of one or more mutually orthogonal Latin squares; another is the group block structure in Example 8.8.

10.6.3 Hamming powers

Repeated crossing of an association scheme $\mathcal{Q}$ gives the mth power $\mathcal{Q}^m$. When $\mathcal{Q} = \underline{n}$ we obtain $(\underline{\underline{n}})^m$. By amalgamating certain classes in $(\underline{\underline{n}})^m$ we obtain the Hamming scheme $\mathrm{H}(m,n)$. We do this by considering not *in which* coordinates $(\alpha_1, \ldots, \alpha_m)$ and $(\beta_1, \ldots, \beta_m)$ differ but *in how many* coordinates they differ.

This idea can be generalized. Let $\mathcal{Q}$ be an association scheme on a set Ω with $s+1$ associate classes $\mathcal{C}_k$ for k in $\mathcal{K}$, including the diagonal class $\mathcal{C}_0$. In the Hamming power $\mathrm{H}(m,\mathcal{Q})$ the underlying set is Ω^m. Pairs $((\alpha_1,\ldots,\alpha_m),(\beta_1,\ldots,\beta_m))$ and $((\gamma_1,\ldots,\gamma_m),(\delta_1,\ldots,\delta_m))$ in $\Omega^m\times\Omega^m$ are in the same class of this Hamming power if and only if

$$|\{i\in\{1,\ldots,m\}:(\alpha_i,\beta_i)\in\mathcal{C}_k\}| = |\{i\in\{1,\ldots,m\}:(\gamma_i,\delta_i)\in\mathcal{C}_k\}|$$

for all k in $\mathcal{K}$. Thus there is one class (including the diagonal) for each way of choosing m items from $s+1$ with replacement when order does not matter, so the number of classes in $\mathrm{H}(m,\mathcal{Q})$ is ${}^{m+s}\mathrm{C}_m$.

To prove that $\mathrm{H}(m,\mathcal{Q})$ is indeed an association scheme, we cannot appeal to distance-regular graphs as we did in Exercise 1.25. Instead we use properties of the ordinary power $\mathcal{Q}^m$ and of the symmetric group S_m of all permutations of $\{1,\ldots,m\}$.

We know that $\mathcal{Q}^m$ is an association scheme, and that its classes are labelled by $\mathcal{K}^m$. Let ϕ be any permutation of $\{1,\ldots,m\}$. Then ϕ acts on $\mathcal{Q}^m$ and $\mathcal{K}^m$ by permuting coordinates; that is,

$$\phi(\alpha)_i = \alpha_{\phi^{-1}(i)} \qquad \text{for } \alpha \text{ in } \Omega^m \text{ and } 1\leqslant i\leqslant m$$
$$\phi(x)_i = x_{\phi^{-1}(i)} \qquad \text{for } x \text{ in } \mathcal{K}^m \text{ and } 1\leqslant i\leqslant m.$$

If $\phi(j)=i$ and $(\alpha_j,\beta_j)\in\mathcal{C}_{x_j}$ in $\mathcal{Q}$ then $(\phi(\alpha)_i,\phi(\beta)_i)\in\mathcal{C}_{x_i}$ in $\mathcal{Q}$. Hence ϕ is a weak automorphism of $\mathcal{Q}^m$. The classes of $\mathrm{H}(m,\mathcal{Q})$ are formed by merging $\mathcal{C}_x$ and $\mathcal{C}_y$ whenever there is a permutation ϕ in S_m with $\phi(x)=y$. Theorem 8.17 shows that $\mathrm{H}(m,\mathcal{Q})$ is an association scheme.

Example 10.12 Let $\mathcal{Q}$ be the association scheme defined by the Petersen strongly regular graph (see Example 1.4) with classes 0 (diagonal), e (edge) and n (non-edge) and corresponding adjacency matrices I, E and N. Then the Hamming power $\mathrm{H}(3,\mathcal{Q})$ has ten associate classes (including the diagonal), with adjacency matrices as follows.

$$\begin{aligned}
B_{[(0,0,0)]} &= I\otimes I\otimes I\\
B_{[(e,e,e)]} &= E\otimes E\otimes E\\
B_{[(n,n,n)]} &= N\otimes N\otimes N\\
B_{[(0,e,e)]} &= I\otimes E\otimes E+E\otimes I\otimes E+E\otimes E\otimes I\\
B_{[(0,0,e)]} &= E\otimes I\otimes I+I\otimes E\otimes I+I\otimes I\otimes E\\
B_{[(0,n,n)]} &= I\otimes N\otimes N+N\otimes I\otimes N+N\otimes N\otimes I\\
B_{[(0,0,n)]} &= N\otimes I\otimes I+I\otimes N\otimes I+I\otimes I\otimes N\\
B_{[(e,n,n)]} &= E\otimes N\otimes N+N\otimes E\otimes N+N\otimes N\otimes E\\
B_{[(e,e,n)]} &= N\otimes E\otimes E+E\otimes N\otimes E+E\otimes E\otimes N
\end{aligned}$$

$$\begin{aligned} B_{[(0,e,n)]} &= I\otimes E\otimes N + N\otimes I\otimes E + E\otimes N\otimes I \\ &\quad + E\otimes I\otimes N + N\otimes E\otimes I + I\otimes N\otimes E \end{aligned}$$

Here is the evaluation of one product:

$$\begin{aligned} & B_{[(0,e,n)]}B_{[(0,e,e)]} \\ &= I\otimes E^2\otimes NE + N\otimes E\otimes E^2 + E\otimes NE\otimes E \\ &\quad + E\otimes E\otimes NE + N\otimes E^2\otimes E + I\otimes NE\otimes E^2 \\ &\quad + E\otimes E\otimes NE + NE\otimes I\otimes E^2 + E^2\otimes N\otimes E \\ &\quad + E^2\otimes I\otimes NE + NE\otimes E\otimes E + E\otimes N\otimes E^2 \\ &\quad + E\otimes E^2\otimes N + NE\otimes E\otimes E + E^2\otimes NE\otimes I \\ &\quad + E^2\otimes E\otimes N + NE\otimes E^2\otimes I + E\otimes NE\otimes E \\ &= (I\otimes E^2\otimes NE + NE\otimes I\otimes E^2 + E^2\otimes NE\otimes I \\ &\quad + E^2\otimes I\otimes NE + NE\otimes E^2\otimes I + I\otimes NE\otimes E^2) \\ &\quad + (N\otimes E\otimes E^2 + E^2\otimes N\otimes E + E\otimes E^2\otimes N \\ &\quad + E\otimes N\otimes E^2 + E^2\otimes E\otimes N + N\otimes E^2\otimes E) \\ &\quad + 2(E\otimes NE\otimes E + E\otimes E\otimes NE + NE\otimes E\otimes E). \end{aligned}$$

Now, $E^2 = N + 3I$ and $NE = 2(E+N)$ so

$$\begin{aligned} & B_{[(0,e,n)]}B_{[(0,e,e)]} \\ &= (2B_{[(0,n,e)]} + 12B_{[(0,0,e)]} + 4B_{[(0,n,n)]} + 12B_{[(0,0,n)]}) \\ &\quad + (2B_{[(n,e,n)]} + 3B_{[(n,e,0)]}) + 2(6B_{[(e,e,e)]} + 2B_{[(e,e,n)]}) \\ &= 12B_{[(e,e,e)]} + 12B_{[(0,0,e)]} + 4B_{[(0,n,n)]} + 12B_{[(0,0,n)]} \\ &\quad + 2B_{[(e,n,n)]} + 4B_{[(e,e,n)]} + 5B_{[(0,e,n)]}. \end{aligned}$$

■

Exercises

10.1 Is the Johnson scheme $\mathrm{J}(m,n)$ a subscheme of $\mathrm{J}(m+1,n)$ or $\mathrm{J}(m,n+1)$ or $\mathrm{J}(m+1,n+1)$?

10.2 Show that $\mathrm{J}(2n,n)$ has an inherent partition with classes of size 2. Describe the quotient scheme when $n=3$ and $n=4$. Find the character table of the quotient scheme when $n=4$.

10.3 Show that the quotient of $\mathrm{Pair}(n)$ by the mirror partition is $\mathrm{T}(n)$.

10.4 Find all inherent partitions on the cube-edge association scheme in Exercise 1.16. Identify the orthogonal block structure which these form. Also identify all the algebraic subschemes and all the quotient schemes.

10.5 Repeat the previous question for the icosahedral and dodecahedral schemes in Exercises 1.20 and 1.21.

10.6 Verify that if $\mathcal{F}$ is an orthogonal block structure and $F \in \mathcal{F}$ and Γ is a class of F then $\left\{G^{\Gamma} : G \in \mathcal{F},\ G \preccurlyeq F\right\}$ is an orthogonal block structure on Γ and $\left\{\bar{G} : G \in \mathcal{F},\ F \preccurlyeq G\right\}$ is an orthogonal block structure on the set of F-classes.

10.7 Find an association scheme, other than T(5), which has no formal dual.

10.8 Find an association scheme, other than trivial schemes and those of Latin square type, which is formally dual to itself.

10.9 Use Exercises 8.7 and 8.8 to show that every Abelian-group scheme is formally dual to another Abelian-group scheme.

10.10 Show that every poset block structure is formally dual to another poset block structure.

10.11 Investigate whether the definition of relaxed wreath product can be applied to blueprints.

10.12 Identify all crested products of $\underline{\underline{n_1}}/\underline{\underline{n_2}}$ with $\underline{\underline{n_3}}/\underline{\underline{n_4}}$.

10.13 Let Δ be a blueprint for the group Ω with classes Δ_k for k in $\mathcal{K}$. By analogy with Section 10.6.3, the m-th Hamming power $\mathrm{H}(m, \Delta)$ is defined to be the partition of Ω^m in which $(\alpha_1, \ldots, \alpha_m)$ and $(\beta_1, \ldots, \beta_m)$ are in the same class if and only if

$$|\{i \in \{1, \ldots, m\} : \alpha_i \in \Delta_k\}| = |\{i \in \{1, \ldots, m\} : \beta_i \in \Delta_k\}|$$

for all k in $\mathcal{K}$. Prove that $\mathrm{H}(m, \Delta)$ is a blueprint for Ω^m.

10.14 Show that the design in Example 7.7 is still partially balanced if the association scheme on the plots is changed to the Hamming square $\mathrm{H}(2, \underline{\underline{2}}/\underline{\underline{3}})$. Find its canonical efficiency factors.

11

Association schemes on the same set

11.1 The partial order on association schemes

So far in this book, we have usually dealt with one association scheme at a time. When there have been two association schemes, either for combining (as in Chapters 3, 9 and 10) or for describing a designed experiment (as in Chapters 5 and 7), they have been defined on different sets. In this chapter we investigate families of association schemes defined on the same set.

In Section 1.1 we defined $\mathcal{Q}$ to be an association scheme on Ω if $\mathcal{Q}$ is a partition of $\Omega \times \Omega$ satisfying certain conditions. Now we can apply the ideas about partitions from Sections 6.1–6.2 to association schemes themselves.

Let $\mathcal{Q}_1$ and $\mathcal{Q}_2$ be association schemes on Ω. Then $\mathcal{Q}_1 \preccurlyeq \mathcal{Q}_2$ if $\mathcal{Q}_2$ is obtained from $\mathcal{Q}_1$ by amalgamation of classes. In particular, if $\underline{\underline{\Omega}}$ denotes the trivial association scheme on Ω then $\mathcal{Q} \preccurlyeq \underline{\underline{\Omega}}$ for every association scheme $\mathcal{Q}$ on Ω.

Example 11.1 Let $\mathcal{Q}_1$ be the rectangular association scheme $\mathrm{R}(n,m)$, and let $\mathcal{Q}_2$ be the partition of $\Omega \times \Omega$ formed from $\mathcal{Q}_1$ by merging the classes 'same row' and 'different row and different column'. Then $\mathcal{Q}_2$ is the group-divisible association scheme $\mathrm{GD}(m,n)$ whose groups are the m columns, and $\mathcal{Q}_1 \prec \mathcal{Q}_2$.

We may write $\mathrm{R}(n,m) \prec \mathrm{GD}(m,n)$, but must be careful not to confuse an association scheme with the isomorphism class of association schemes to which it belongs. Strictly speaking, $\mathrm{GD}(m,n)$ indicates the isomorphism class of group-divisible association schemes with m groups of size n. When we are discussing a single association scheme, there is no harm in identifying it with its isomorphism type. Here, Ω is a given set of size nm and $\mathcal{Q}_1$ is a given association scheme on Ω of type $\mathrm{R}(n,m)$.

Among the many association schemes on Ω of type $\mathrm{GD}(m,n)$, there is only one (if $n \neq m$) which is coarser than $\mathcal{Q}_1$ – the one corresponding to the partition into m columns.

Of course, if $\mathcal{Q}_3$ is the group-divisible association scheme $\mathrm{GD}(n,m)$ on Ω whose groups are the n rows, then we also have $\mathcal{Q}_1 \prec \mathcal{Q}_3$. So we have the Hasse diagram on association schemes shown in Figure 11.1. ■

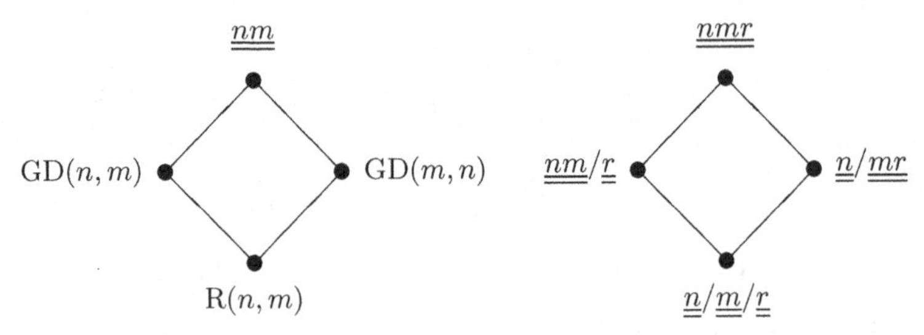

Fig. 11.1. Hasse diagram for Example 11.1

Fig. 11.2. Hasse diagram for Example 11.2

Example 11.2 In the association scheme $\underline{n}/\underline{m}/\underline{r}$ we may merge two classes to obtain $\underline{nm}/\underline{r}$, or we may merge a different two classes to obtain $\underline{n}/\underline{mr}$. The Hasse diagram is shown in Figure 11.2. ■

Proposition 11.1 *Let $\mathcal{P}$ and $\mathcal{Q}$ be cyclic association schemes defined by the blueprints Δ and Γ of $\mathbb{Z}_n$. Then $\mathcal{P} \preccurlyeq \mathcal{Q}$ if and only if $\Delta \preccurlyeq \Gamma$. In particular, $\mathcal{Q}$ is a cyclic association scheme on $\mathbb{Z}_n$ if and only if* ⓝ $\preccurlyeq \mathcal{Q}$.

Example 11.3 Let $\mathcal{Q}$ be an association scheme on a set Ω, and let F be an inherent partition in $\mathcal{Q}$, with $R_F = \sum_{i \in \mathcal{L}} A_i$. Then the matrices $J_\Omega - R_F$ and A_i, for i in $\mathcal{L}$, are the adjacency matrices of an association scheme $\mathcal{P}$ on Ω, and $\mathcal{Q} \preccurlyeq \mathcal{P}$. ■

Example 11.4 Let $\mathcal{F}$ be the set of partitions inherent in an association scheme $\mathcal{Q}$. Theorem 10.1 shows that $\mathcal{F}$ is an orthogonal block structure, so it defines an association scheme $\mathcal{P}$. Then $\mathcal{Q} \preccurlyeq \mathcal{P}$. ■

Example 11.5 Given any association scheme $\mathcal{Q}$ and any integer n with $n \geqslant 2$, the n-fold cross of $\mathcal{Q}$ with itself is strictly finer than the n-th Hamming power of $\mathcal{Q}$; that is, $\mathcal{Q}^n \prec \mathrm{H}(n, \mathcal{Q})$. ■

The characteristic functions of the associate classes in $\mathcal{Q}$ are precisely the adjacency matrices, so the space $V_\mathcal{Q}$ (in $\mathbb{R}^{\Omega\times\Omega}$) is just the Bose–Mesner algebra $\mathcal{A}$. If $\mathcal{A}_1$ and $\mathcal{A}_2$ are the Bose–Mesner algebras of association schemes $\mathcal{Q}_1$ and $\mathcal{Q}_2$ on Ω then Lemma 6.2 shows that $\mathcal{Q}_1 \preccurlyeq \mathcal{Q}_2$ if and only if $\mathcal{A}_1 \geqslant \mathcal{A}_2$. In particular, every Bose–Mesner algebra contains the Bose–Mesner algebra $\mathcal{I}$ of $\underline{\underline{\Omega}}$, which is spanned by I_Ω and J_Ω.

11.2 Suprema

If $\mathcal{Q}_1$ and $\mathcal{Q}_2$ are two association schemes on the same set, we may form their supremum $\mathcal{Q}_1 \vee \mathcal{Q}_2$ by merging any classes of $\mathcal{Q}_1$ which meet the same class of $\mathcal{Q}_2$, and vice versa, and continuing in this way until every class is both a union of $\mathcal{Q}_1$-classes and a union of $\mathcal{Q}_2$-classes. Of course, if $\mathcal{Q}_1 \preccurlyeq \mathcal{Q}_2$ then $\mathcal{Q}_1 \vee \mathcal{Q}_2 = \mathcal{Q}_2$. But what does $\mathcal{Q}_1 \vee \mathcal{Q}_2$ look like in general?

Theorem 11.2 *If $\mathcal{Q}_1$ and $\mathcal{Q}_2$ are both association schemes on Ω then $\mathcal{Q}_1 \vee \mathcal{Q}_2$ is also an association scheme on Ω.*

Proof The diagonal class Diag(Ω) is a class in both $\mathcal{Q}_1$ and $\mathcal{Q}_2$, so it remains a class in $\mathcal{Q}_1 \vee \mathcal{Q}_2$. All classes in $\mathcal{Q}_1$ and $\mathcal{Q}_2$ are symmetric, so the same is true of the merged classes in $\mathcal{Q}_1 \vee \mathcal{Q}_2$.

Let $\mathcal{A}_1$ and $\mathcal{A}_2$ be the Bose–Mesner algebras of $\mathcal{Q}_1$ and $\mathcal{Q}_2$, and let $\mathcal{A}$ be the subspace of $\mathbb{R}^{\Omega\times\Omega}$ spanned by the adjacency matrices of the classes in $\mathcal{Q}_1 \vee \mathcal{Q}_2$. Lemma 6.3 shows that

$$\mathcal{A} = V_{\mathcal{Q}_1 \vee \mathcal{Q}_2} = V_{\mathcal{Q}_1} \cap V_{\mathcal{Q}_2} = \mathcal{A}_1 \cap \mathcal{A}_2. \tag{11.1}$$

Since $\mathcal{A}_1$ and $\mathcal{A}_2$ are both closed under multiplication, $\mathcal{A}$ is also closed under multiplication. Hence $\mathcal{Q}_1 \vee \mathcal{Q}_2$ is an association scheme on Ω. ■

Corollary 11.3 *If $(\Omega, \mathcal{G}, \Theta, \psi)$ is a partially balanced block design with treatment set Θ then there is a unique coarsest association scheme on Θ with respect to which $(\Omega, \mathcal{G}, \Theta, \psi)$ is partially balanced.*

Proof Let $\mathcal{H}$ be the partition of $\Theta \times \Theta$ determined by the concurrences in the design in the blocks defined by the group-divisible association scheme $\mathcal{G}$ on Ω; that is, (θ_1, η_1) and (θ_2, η_2) are in the same class of $\mathcal{H}$ if and only if $\Lambda(\theta_1, \eta_1) = \Lambda(\theta_2, \eta_2)$. Put

$$\mathcal{R} = \{\mathcal{Q} : \mathcal{Q} \preccurlyeq \mathcal{H} \text{ and } \mathcal{Q} \text{ is an association scheme on } \Theta\}.$$

Then $\mathcal{R}$ is not empty, because the design is partially balanced. If $\mathcal{Q}_1$ and $\mathcal{Q}_2$ are in $\mathcal{R}$ then $\mathcal{Q}_1 \vee \mathcal{Q}_2 \in \mathcal{R}$, by Theorem 11.2 and by property (ii) of suprema on page 149. Put $\mathcal{P} = \bigvee \mathcal{R}$. Then $\mathcal{P} \in \mathcal{R}$, because $\mathcal{R}$ is finite. Therefore $\mathcal{P}$ is an association scheme on Θ and $\mathcal{P} \preccurlyeq \mathcal{H}$, so the design is partially balanced with respect to $\mathcal{P}$. Moreover, $\mathcal{P}$ is the coarsest association scheme with this property, because if the design is partially balanced with respect to any association scheme $\mathcal{Q}$ then $\mathcal{Q} \in \mathcal{R}$ so $\mathcal{Q} \preccurlyeq \mathcal{P}$. ■

It is tempting to think that, in the notation of the foregoing proof, the Bose–Mesner algebra of $\mathcal{P}$ must be the algebra generated by the adjacency matrices for the classes of $\mathcal{H}$; that is, the smallest algebra which contains all those matrices. However, this is not true in general.

Example 11.6 Consider the rectangular lattice design for $n(n-1)$ treatments with replication 2, in $2n$ blocks of size $n-1$, which is described in Section 5.5.4. This is partially balanced with respect to the association scheme Pair(n), whose adjacency matrices are I, A, M, B and G: see page 134. These have valency 1, $2(n-2)$, 1, $2(n-2)$ and $(n-2)(n-3)$ respectively. Assume that $n \geqslant 4$ so that $\dim \mathcal{A}_1 = 5$, where $\mathcal{A}_1$ is the Bose–Mesner algebra of Pair(n).

For this design, $\Lambda = 2I + A$, so the adjacency matrices of $\mathcal{H}$ are I, A and $J - A - I$. Let $\mathcal{B}$ be the algebra generated by these matrices.

The information matrix L is equal to $(n-1)^{-1}[2(n-2)I - A]$, so its eigenspaces are those of A. The design is connected, so Corollary 4.10 shows that U_0 is one of these eigenspaces. Hence J is a polynomial in A and $\mathcal{B}$ is spanned by powers of A. Therefore $\dim \mathcal{B}$ is equal to the number of eigenspaces of L, which is equal to the number of distinct eigenvalues of L, which is 4 (from Section 5.5.4).

Now let $\mathcal{P}$ be any association scheme with respect to which the design is partially balanced. Then the Bose–Mesner algebra $\mathcal{A}_2$ of $\mathcal{P}$ must include I and A. But

$$A^2 = 2(n-2)I + (n-3)A + B + 2G,$$

so $\mathcal{A}_2$ must also contain B and G. Hence $\mathcal{A}_2$ contains M, because $M = J - I - A - B - G$. Therefore $\mathcal{A}_2 \geqslant \mathcal{A}_1$ and $\mathcal{P} \preccurlyeq$ Pair(n), so $\dim(\mathcal{A}_2) \geqslant 5$. In particular, $\mathcal{B}$ is not the Bose–Mesner algebra of any association scheme. ■

11.3 Infima

On the other hand, the infimum of two association schemes is not, in general, an association scheme.

Example 11.7 Let $\mathcal{Q}_1$ and $\mathcal{Q}_2$ be the two strongly regular graphs whose edge sets are

$$\{\{1,2\},\{2,3\},\{3,4\},\{4,5\},\{5,1\}\}$$

and

$$\{\{1,3\},\{3,2\},\{2,5\},\{5,4\},\{4,1\}\}$$

respectively. Then one class of $\mathcal{Q}_1 \wedge \mathcal{Q}_2$ consists of the double edges $(4,5)$, $(5,4)$, $(2,3)$ and $(3,2)$: the vertex 1 is not contained in any pair in this class, and so $\mathcal{Q}_1 \wedge \mathcal{Q}_2$ cannot be an association scheme. ■

It is this bad behaviour of infima that causes the problems, noted in Chapter 7, with identifying partial balance in designs with many strata. Suppose that $\mathcal{F}$ is an orthogonal block structure on Ω and that $\psi: \Omega \to \Theta$ is a design map. It may happen that, for each partition F in $\mathcal{F}$, the block design defined by the classes of F is partially balanced with respect to an association scheme $\mathcal{Q}_F$ on Θ. If $\bigwedge_{F \in \mathcal{F}} \mathcal{Q}_F$ were an association scheme $\mathcal{Q}$ then the whole design would be partially balanced with respect to $\mathcal{Q}$. However, in general that infimum is not itself an association scheme.

What does make $\mathcal{Q}_1 \wedge \mathcal{Q}_2$ into an association scheme? It is tempting to explore orthogonality. But orthogonality cannot be necessary, for in Example 11.1 we have $\mathcal{Q}_2 \wedge \mathcal{Q}_3 = \mathcal{Q}_1$, but $\mathcal{Q}_2$ is not orthogonal to $\mathcal{Q}_3$, because the intersections of the classes of $\mathcal{Q}_2$ and $\mathcal{Q}_3$ have the sizes shown in Table 11.1.

			$\mathcal{Q}_3$		
			diagonal	same row	other
			nm	$nm(m-1)$	$nm^2(n-1)$
	diagonal	nm	nm	0	0
$\mathcal{Q}_2$	same column	$nm(n-1)$	0	0	$nm(n-1)$
	other	$n^2m(m-1)$	0	$nm(m-1)$	$nm(n-1)(m-1)$

Table 11.1. *Intersection numbers in Example 11.1*

Nor is orthogonality sufficient, as the next example shows.

Example 11.8 Considered as partitions of the same set of vertices, the two strongly regular graphs in Figure 11.3 are orthogonal to each other, as shown by Corollary 6.5: Table 11.2 shows the sizes of the intersections of their classes. However, starting at vertex 1 and walking first along a double edge and then a single solid one brings us to vertices 2 and 3, while a single solid edge followed by a double one brings us to 4 or 5. So the classes 'same vertex', 'double edge', 'single solid edge', 'single dotted edge' and 'other' do not form an association scheme. ■

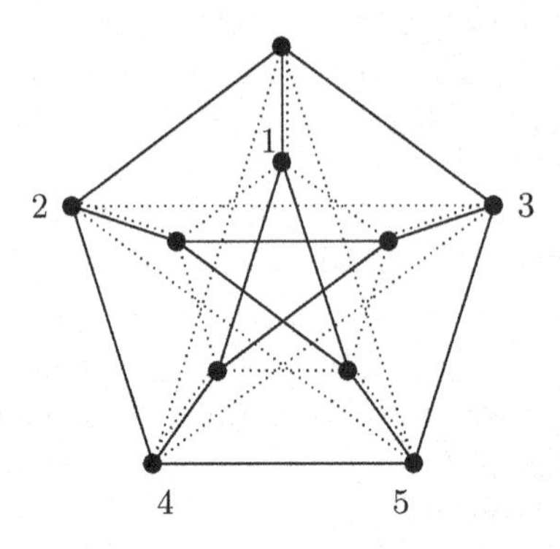

Fig. 11.3. Two Petersen graphs

			first Petersen graph (solid edges)		
			diagonal	edge	non-edge
			10	30	60
second Petersen graph (dotted edges)	diagonal	10	10	0	0
	edge	30	0	10	20
	non-edge	60	0	20	40

Table 11.2. *Intersection numbers in Example 11.8*

So orthogonality is not the key to the infimum being an association scheme. The problem in Example 11.8 is that the adjacency matrices of the classes 'double edge' and 'single solid edge' do not commute. In fact, the adjacency matrices of the two strongly regular graphs do not commute with each other.

Theorem 11.4 *Let $\mathcal{Q}_1$ and $\mathcal{Q}_2$ be association schemes on Ω with Bose–Mesner algebras $\mathcal{A}_1$ and $\mathcal{A}_2$. If $\mathcal{Q}_1 \wedge \mathcal{Q}_2$ is also an association scheme then every matrix in $\mathcal{A}_1$ commutes with every matrix in $\mathcal{A}_2$.*

Proof Suppose that $\mathcal{Q}_1 \wedge \mathcal{Q}_2$ is an association scheme, and let $\mathcal{A}$ be its Bose–Mesner algebra. Then $\mathcal{Q}_1 \wedge \mathcal{Q}_2 \preccurlyeq \mathcal{Q}_1$ and so $\mathcal{A} \geqslant \mathcal{A}_1$. Similarly

$\mathcal{A} \geqslant \mathcal{A}_2$. But $\mathcal{A}$ is commutative, so every matrix in $\mathcal{A}_1$ commutes with every matrix in $\mathcal{A}_2$. ■

However, even commutativity is not sufficient.

Example 11.9 Let Ω consist of the twenty stars in Figure 11.4. Let $\mathcal{Q}_1$ and $\mathcal{Q}_2$ be the group-divisible association schemes on Ω whose groups are the rows and columns, respectively, of Figure 11.4: they both have type GD$(5,4)$. Let A_1 and A_2 be the within-group adjacency matrices of the two schemes. Then

$$(I + A_1)(I + A_2) = (I + A_2)(I + A_1) = \begin{bmatrix} J_{16} & 0 \\ 0 & 4J_4 \end{bmatrix}$$

so $\mathcal{A}_1$ commutes with $\mathcal{A}_2$.

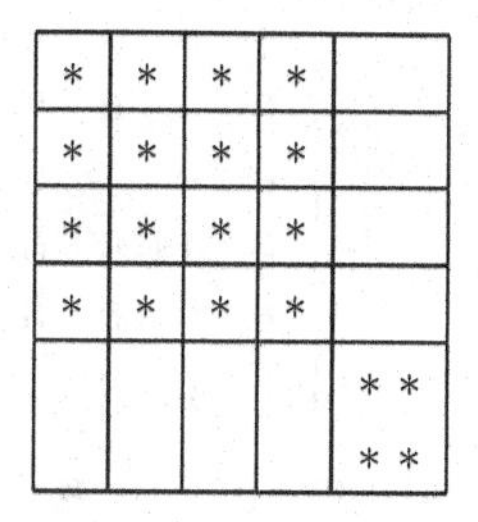

Fig. 11.4. Two group-divisible association schemes in Example 11.9

However, $\mathcal{Q}_1 \wedge \mathcal{Q}_2$ is not an association scheme. One of its classes consists of all distinct pairs among the four stars at the bottom right. None of the remaining sixteen stars is contained in any pair in this class. ■

In Examples 11.7 and 11.9 we found that there is a class $\mathcal{C}$ of $\mathcal{Q}_1 \wedge \mathcal{Q}_2$ such that $|\mathcal{C}(\omega)| = 0$ for some ω in Ω. This implies that if $\mathcal{Q} \preccurlyeq \mathcal{Q}_1 \wedge \mathcal{Q}_2$ and if $V_\mathcal{Q}$ is closed under multiplication then Diag(Ω) splits into at least two classes in $\mathcal{Q}$ and so $\mathcal{Q}$ cannot be an association scheme. Thus there is *no* association scheme finer than $\mathcal{Q}_1 \wedge \mathcal{Q}_2$.

Some of the bad behaviour of infima can be explained by using the Hadamard product of matrices.

Definition Let N and M be two $\Omega \times \Omega$ matrices. Their *Hadamard product* $N \circ M$ is the $\Omega \times \Omega$ matrix whose (α, β)-entry is equal to $N(\alpha,\beta)M(\alpha,\beta)$.

Since we regard $\Omega \times \Omega$ matrices as functions from $\Omega \times \Omega$ to some field, the Hadamard product is just the pointwise product.

Lemma 11.5 *The Bose–Mesner algebra of any association scheme is closed under Hadamard product.*

Proof If $i \neq j$ then $A_i \circ A_j = O_\Omega$, while $A_i \circ A_i = A_i$. ■

Proposition 11.6 *Let $\mathcal{Q}_1$ and $\mathcal{Q}_2$ be association schemes on Ω with sets of adjacency matrices $\{A_i : i \in \mathcal{K}_1\}$ and $\{B_j : j \in \mathcal{K}_2\}$. Then the matrices which are the characteristic functions of the classes in $\mathcal{Q}_1 \wedge \mathcal{Q}_2$ are precisely the Hadamard products $A_i \circ B_j$ for i in $\mathcal{K}_1$ and j in $\mathcal{K}_2$. Consequently, if $\mathcal{Q}_1 \wedge \mathcal{Q}_2$ is an association scheme then each Hadamard product $A_i \circ B_j$ has constant row-sums.*

Example 11.7 revisited Let A_1 and B_1 be the adjacency matrices for the 'edge' class in the two graphs. Then

$$A_1 \circ B_1 = \begin{array}{c} \\ 1 \\ 2 \\ 3 \\ 4 \\ 5 \end{array} \begin{array}{c} \begin{array}{ccccc} 1 & 2 & 3 & 4 & 5 \end{array} \\ \left[\begin{array}{ccccc} 0 & 0 & 0 & 0 & 0 \\ 0 & 0 & 1 & 0 & 0 \\ 0 & 1 & 0 & 0 & 0 \\ 0 & 0 & 0 & 0 & 1 \\ 0 & 0 & 0 & 1 & 0 \end{array}\right] \end{array},$$

which does not have constant row-sums. ■

For the problem of finding partially balanced designs for an orthogonal block structure $\mathcal{F}$ with many strata it is not actually necessary that $\bigwedge_{F \in \mathcal{F}} \mathcal{Q}_F$ be an association scheme. For partial balance of the whole design it is sufficient for there to be an association scheme $\mathcal{Q}$ such that $\mathcal{Q} \preccurlyeq \mathcal{Q}_F$ for all F in $\mathcal{F}$.

Definition Let $\mathcal{Q}_1$ and $\mathcal{Q}_2$ be association schemes on the same set. If there is an association scheme $\mathcal{Q}_3$ such that $\mathcal{Q}_3 \preccurlyeq \mathcal{Q}_1$ and $\mathcal{Q}_3 \preccurlyeq \mathcal{Q}_2$ then $\mathcal{Q}_1$ and $\mathcal{Q}_2$ are *compatible* with each other.

Lemma 11.7 *Let $\mathcal{Q}_1$ and $\mathcal{Q}_2$ be compatible association schemes. Then*

(i) there is a unique coarsest association scheme $\mathcal{P}$ such that $\mathcal{P} \preccurlyeq \mathcal{Q}_1$ and $\mathcal{P} \preccurlyeq \mathcal{Q}_2$ (then $\mathcal{P} \preccurlyeq \mathcal{Q}_1 \wedge \mathcal{Q}_2$ but $\mathcal{P}$ may not be actually equal to $\mathcal{Q}_1 \wedge \mathcal{Q}_2$);
(ii) the Bose–Mesner algebras of $\mathcal{Q}_1$ and $\mathcal{Q}_2$ commute with each other;
(iii) if A and B are adjacency matrices for $\mathcal{Q}_1$ and $\mathcal{Q}_2$ then $A \circ B$ has constant row-sums.

Proof (i) Mimic the proof of Corollary 11.3.
(ii) Mimic the proof of Theorem 11.4.

(iii) The Hadamard product $A \circ B$ is in $V_{\mathcal{Q}_1 \wedge \mathcal{Q}_2}$, which is contained in a Bose–Mesner algebra. ■

Example 11.10 (Example 6.1 continued) Let $\mathcal{Q}_3 = \underline{\underline{3}}/(\underline{\underline{2}} \times \underline{\underline{4}})$, with adjacency matrices I, $R - I$, $C - I$, $B - R - C + I$ and $J - B$. Let $\mathcal{Q}_1$ and $\mathcal{Q}_2$ be the group-divisible association schemes whose groups are the rows and columns respectively. Then the adjacency matrices of $\mathcal{Q}_1$ are I, $R - I$ and $J - R$, while those of $\mathcal{Q}_2$ are I, $C - I$ and $J - C$. Thus $\mathcal{Q}_3 \prec \mathcal{Q}_1$ and $\mathcal{Q}_3 \prec \mathcal{Q}_2$. The adjacency matrices of the classes in $\mathcal{Q}_1 \wedge \mathcal{Q}_2$ are I, $R - I$, $C - I$ and $J - R - C + I$. These do not form an association scheme because $(R - I)(C - I) = B - R - C - I$. However, $\mathcal{Q}_1$ is compatible with $\mathcal{Q}_2$. ■

The questions about whether $\mathcal{Q}_1 \wedge \mathcal{Q}_2$ is an association scheme and about whether $\mathcal{Q}_1$ is compatible with $\mathcal{Q}_2$ seem to be quite hard in general. The next three sections provide partial answers for three classes of association schemes.

11.4 Group schemes

Let Ω be a finite group. Chapter 8 described how to make a group association scheme $\mathcal{Q}(\Delta)$ on Ω from a blueprint Δ for Ω. Blueprints for Ω are partitions of Ω, so they are themselves ordered by $\preccurlyeq$. How is the partial order on the blueprints related to the partial order on the association schemes?

Proposition 11.8 *Let Δ and Γ be partitions of Ω. Then $\Delta \preccurlyeq \Gamma$ if and only if $\mathcal{Q}(\Delta) \preccurlyeq \mathcal{Q}(\Gamma)$.*

Proof Suppose that $\Delta \preccurlyeq \Gamma$. Let Δ_i and Γ_j be classes of Δ and Γ, and let $\mathcal{C}_i^{\Delta}$ and $\mathcal{C}_j^{\Gamma}$ be the corresponding associate classes in $\mathcal{Q}(\Delta)$ and $\mathcal{Q}(\Gamma)$ respectively. Suppose that $\Delta_i \subseteq \Gamma_j$ and that $(\alpha, \beta) \in \mathcal{C}_i^{\Delta}$. Then $\alpha^{-1}\beta \in \Delta_i$ so $\alpha^{-1}\beta \in \Gamma_j$ so $(\alpha, \beta) \in \mathcal{C}_j^{\Gamma}$. Hence $\mathcal{C}_i^{\Delta} \subseteq \mathcal{C}_j^{\Gamma}$.

Conversely, if $\mathcal{Q}(\Delta) \preccurlyeq \mathcal{Q}(\Gamma)$ and $\mathcal{C}_i^{\Delta} \subseteq \mathcal{C}_j^{\Gamma}$ then $\Delta_i = \mathcal{C}_i^{\Delta}(1_\Omega) \subseteq \mathcal{C}_j^{\Gamma}(1_\Omega) = \Gamma_j$. ■

As in Section 8.7, for each ω in Ω define the permutation π_ω of Ω by $\pi_\omega(\alpha) = \omega\alpha$. Lemma 8.18 shows that π_ω is a strong automorphism of every group scheme on Ω.

Lemma 11.9 *Let Δ and Γ be any partitions of Ω. Then $\mathcal{Q}(\Delta) \vee \mathcal{Q}(\Gamma) = \mathcal{Q}(\Delta \vee \Gamma)$ and $\mathcal{Q}(\Delta) \wedge \mathcal{Q}(\Gamma) = \mathcal{Q}(\Delta \wedge \Gamma)$.*

Proof For every ω in Ω, π_ω preserves every class of $\mathcal{Q}(\Delta)$ and $\mathcal{Q}(\Gamma)$ and so it also preserves every class of $\mathcal{Q}(\Delta)\vee\mathcal{Q}(\Gamma)$ and $\mathcal{Q}(\Delta)\wedge\mathcal{Q}(\Gamma)$. By Lemma 8.18, there are partitions Ψ and Φ of Ω such that $\mathcal{Q}(\Delta)\vee\mathcal{Q}(\Gamma) = \mathcal{Q}(\Psi)$ and $\mathcal{Q}(\Delta)\wedge\mathcal{Q}(\Gamma) = \mathcal{Q}(\Phi)$.

Now, $\mathcal{Q}(\Delta) \preccurlyeq \mathcal{Q}(\Delta)\vee\mathcal{Q}(\Gamma) = \mathcal{Q}(\Psi)$ so $\Delta \preccurlyeq \Psi$, by Proposition 11.8. Similarly, $\Gamma \preccurlyeq \Psi$, so $\Delta\vee\Gamma \preccurlyeq \Psi$. Also, $\Delta \preccurlyeq \Delta\vee\Gamma$ so $\mathcal{Q}(\Delta) \preccurlyeq \mathcal{Q}(\Delta\vee\Gamma)$ and similarly $\mathcal{Q}(\Gamma) \preccurlyeq \mathcal{Q}(\Delta\vee\Gamma)$ so $\mathcal{Q}(\Psi) = \mathcal{Q}(\Delta)\vee\mathcal{Q}(\Gamma) \preccurlyeq \mathcal{Q}(\Delta\vee\Gamma)$. Therefore $\Psi \preccurlyeq \Delta\vee\Gamma$ and so $\Psi = \Delta\vee\Gamma$.

The dual argument shows that $\Phi = \Delta\wedge\Gamma$. ■

Now suppose that Δ and Γ are blueprints for Ω, so that $\mathcal{Q}(\Delta)$ and $\mathcal{Q}(\Gamma)$ are association schemes. Theorem 11.2 shows that $\mathcal{Q}(\Delta)\vee\mathcal{Q}(\Gamma)$ is an association scheme. In fact, it is also a group scheme.

Theorem 11.10 *Let Δ and Γ be blueprints for Ω. Then $\Delta\vee\Gamma$ is also a blueprint for Ω and so $\mathcal{Q}(\Delta)\vee\mathcal{Q}(\Gamma)$ is the group scheme $\mathcal{Q}(\Delta\vee\Gamma)$.*

Proof We check the three conditions for a blueprint.

(i)‴ $\{1_\Omega\}$ is a class of both Δ and Γ so it is also a class of $\Delta\vee\Gamma$.

(ii)‴ $\Delta \preccurlyeq \Delta\vee\Gamma$ and every class of Δ is closed under taking inverses, so the same is true of $\Delta\vee\Gamma$.

(iii)‴ By Lemma 6.3, $V_{\Delta\vee\Gamma} = V_\Delta \cap V_\Gamma$. Since V_Δ and V_Γ are both closed under convolution, so is $V_{\Delta\vee\Gamma}$. ■

Although $\mathcal{Q}(\Delta)\wedge\mathcal{Q}(\Gamma)$ may not be an association scheme, there is a better result for infima of group schemes than there is for infima of general association schemes.

Theorem 11.11 *Let Δ and Γ be blueprints for Ω. Then*

(i) if Φ is a blueprint and $\Phi \preccurlyeq \Delta\wedge\Gamma$ then $\mathcal{Q}(\Phi) \preccurlyeq \mathcal{Q}(\Delta)\wedge\mathcal{Q}(\Gamma)$;

(ii) if there is any blueprint Φ finer than $\Delta\wedge\Gamma$ then there is a unique coarsest such blueprint Ξ;

(iii) if $\mathcal{Q}(\Delta)$ is compatible with $\mathcal{Q}(\Gamma)$ then there is a blueprint finer than $\Delta\wedge\Gamma$ and the coarsest association scheme which is finer than $\mathcal{Q}(\Delta)\wedge\mathcal{Q}(\Gamma)$ is $\mathcal{Q}(\Xi)$;

(iv) $\mathcal{Q}(\Delta)\wedge\mathcal{Q}(\Gamma)$ is an association scheme if and only if $\Delta\wedge\Gamma$ is a blueprint.

Proof (i) Use Proposition 11.8.

(ii) This is the same argument as the proof of Corollary 11.3, using Theorem 11.10.

(iii) If there is any association scheme finer than $\mathcal{Q}(\Delta) \wedge \mathcal{Q}(\Gamma)$ then Lemma 11.7(i) shows that there is a coarsest one $\mathcal{P}$.

If $\omega \in \Omega$ then $\pi_\omega(\mathcal{P})$ is an association scheme (see page 246) and $\pi_\omega(\mathcal{P}) \preccurlyeq \pi_\omega(\mathcal{Q}(\Delta)) = \mathcal{Q}(\Delta)$. Similarly, $\pi_\omega(\mathcal{P}) \preccurlyeq \mathcal{Q}(\Gamma)$. Therefore $\pi_\omega(\mathcal{P}) \preccurlyeq \mathcal{Q}(\Delta) \wedge \mathcal{Q}(\Gamma)$ and hence $\pi_\omega(\mathcal{P}) \preccurlyeq \mathcal{P}$. However, $\pi_\omega(\mathcal{P})$ and $\mathcal{P}$ have the same number of classes, so $\pi_\omega(\mathcal{P}) = \mathcal{P}$. Thus $\{\pi_\omega : \omega \in \Omega\}$ forms a group of weak automorphisms of $\mathcal{P}$. By Theorem 8.17, there is an association scheme $\tilde{\mathcal{P}}$ such that $\mathcal{P} \preccurlyeq \tilde{\mathcal{P}}$ and $\{\pi_\omega : \omega \in \Omega\}$ forms a group of strong automorphisms of the scheme $\tilde{\mathcal{P}}$. Each class of $\tilde{\mathcal{P}}$ is formed by merging classes of $\mathcal{P}$ of the form $\mathcal{C}$ and $\pi_\omega(\mathcal{C})$, which lie in the same class of both $\mathcal{Q}(\Delta)$ and $\mathcal{Q}(\Gamma)$. Hence $\tilde{\mathcal{P}} \preccurlyeq \mathcal{Q}(\Delta) \wedge \mathcal{Q}(\Gamma)$ and so $\tilde{\mathcal{P}} = \mathcal{P}$. Then Lemma 8.18 shows that there is a blueprint Φ with $\mathcal{P} = \mathcal{Q}(\Phi)$.

Proposition 11.8 now shows that $\Phi \preccurlyeq \Delta \wedge \Gamma$, so $\Phi \preccurlyeq \Xi$. Hence $\mathcal{Q}(\Phi) \preccurlyeq \mathcal{Q}(\Xi) \preccurlyeq \mathcal{Q}(\Delta) \wedge \mathcal{Q}(\Gamma)$, so the maximality of $\mathcal{P}$ shows that $\mathcal{Q}(\Phi) = \mathcal{Q}(\Xi)$ and so $\Phi = \Xi$ and $\mathcal{P} = \mathcal{Q}(\Xi)$.

(iv) Since $\mathcal{Q}(\Delta) \wedge \mathcal{Q}(\Gamma) = \mathcal{Q}(\Delta \wedge \Gamma)$, this follows from Theorem 8.1. ■

If Ω is Abelian then Inverse is a blueprint nested in every other blueprint, so the infimum of any two group schemes on Ω is always coarser than $\mathcal{Q}(\text{Inverse})$; that is, all group schemes on Ω are compatible. However, if Ω is not Abelian then there may be no association scheme nested in the infimum of two group schemes.

Example 11.11 (Example 8.8 continued) Let

$$\Omega = \langle a, b : a^9 = b^3 = 1,\ b^{-1}ab = a^4 \rangle.$$

The group block structure defined by the lattice of all subgroups of Ω is the group scheme defined by the blueprint Subgroup: this is the blueprint Δ whose classes are $\{1\}$, $\{a^3, a^6\}$, $\{b, b^2\}$, $\{a^3b, a^6b^2\}$, $\{a^3b^2, a^6b\}$, $\{a, a^2, a^4, a^5, a^7, a^8\}$, $\{ab, a^8b^2, a^4b, a^2b^2, a^7b, a^5b^2\}$, and $\{a^2b, a^7b^2, a^5b, ab^2, a^8b, a^4b^2\}$.

Exercise 8.10 shows that there is also a blueprint Γ on Ω such that $\mathcal{Q}(\Gamma) \cong \underline{3}/⑨$: it classes are $\{1\}$, $\{a^3, a^6\}$, $\{a, a^8\}$, $\{a^2, a^7\}$, $\{a^4, a^5\}$ and $\Omega \setminus \langle a \rangle$.

Now, $\Delta \wedge \Gamma$ contains the classes $\{a, a^8\}$ and $\{b, b^2\}$, which do not commute as sets, because

$$\{a, a^8\}\{b, b^2\} = \{ab, ab^2, a^8b, a^8b^2\}$$

while

$$\{b, b^2\}\{a, a^8\} = \{ba, b^2a, ba^8, b^2a^8\} = \{a^7b, a^4b^2, a^2b, a^5b^2\}.$$

By Corollary 8.2, these cannot be classes in the same blueprint, or even unions of classes in the same blueprint. Hence there can be no blueprint nested in $\Delta \wedge \Gamma$. ■

11.5 Poset block structures

In order to investigate how the nesting partial order $\preccurlyeq$ works on poset block structures, we need to make a few small changes to the ideas and notation of Chapter 9. Throughout, Ω will be $\prod_{i \in M} \Omega_i$ with $|\Omega_i| \geqslant 2$ for i in M.

The first change is to use letters like ρ in place of notation like $\sqsubseteq$. Thus we regard the partial order $\sqsubseteq$ on M as the subset ρ of $M \times M$, where

$$\rho = \{(i, j) \in M \times M : i \sqsubseteq j\}.$$

We want to compare different partial orders on M, and the notation '$\rho \subset \sigma$' is more readily comprehensible than '$\sqsubseteq_1 \subset \sqsubseteq_2$'.

The second change is to replace partial orders by *pre-orders*, which have to satisfy only the reflexive and transitive laws. Thus ρ is a pre-order if

(i) (**reflexivity**) $\mathrm{Diag}(M) \subseteq \rho$, and
(ii) (**transitivity**) if $(i, j) \in \rho$ and $(j, k) \in \rho$ then $(i, k) \in \rho$.

If $\rho = \mathrm{Diag}(M)$ then ρ is called the *trivial* partial order on M, as noted on page 261. All partial orders are also pre-orders.

If ρ is a pre-order on M then it defines an equivalence relation $\sim_\rho$ on M: here $i \sim_\rho j$ if $(i, j) \in \rho$ and $(j, i) \in \rho$. The Hasse diagram for ρ has a dot for each equivalence class of $\sim_\rho$, rather than a dot for each element of M.

Ancestral subsets are defined for pre-orders just as they are for partial orders. If L is an ancestral subset and $i \in L$ then L contains the whole equivalence class containing i.

Example 11.12 Let $M = \{1, 2, 3\}$ and let

$$\rho = \{(1,1), (1,2), (2,1), (2,2), (3,1), (3,2), (3,3)\}.$$

Then ρ is a pre-order and its Hasse diagram is shown in Figure 11.5. Its lattice of ancestral subsets is in Figure 11.6. ■

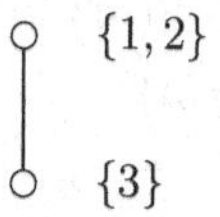

Fig. 11.5. The pre-order in Example 11.12

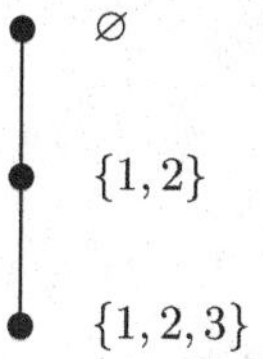

Fig. 11.6. Ancestral subsets in Example 11.12

If K is an equivalence class of $\sim_\rho$, write Ω_K for $\prod_{i\in K}\Omega_K$. All the theory in Chapter 9 carries over to pre-orders; the orthogonal block structure $\mathcal{F}(\rho)$ defined by the pre-order ρ is effectively a poset block structure on the product set $\prod_{\text{classes } K}\Omega_K$. Thus we can call $\mathcal{F}(\rho)$ a poset block structure.

Example 11.12 revisited Put $a = \{1,2\}$ and $b = \{3\}$. Then ρ induces a true partial order σ on $\{a,b\}$:

$$\sigma = \{(a,a),(b,a),(b,b)\}.$$

Put $\Gamma_a = \Omega_1 \times \Omega_2$ and $\Gamma_b = \Omega_3$. The poset block structure $\mathcal{F}(\sigma)$ defined by σ on $\Gamma_a \times \Gamma_b$ is the orthogonal block structure which the pre-order ρ defines on Ω. In fact, $\mathcal{F}(\sigma) = \mathcal{F}_a/\mathcal{F}_b$, where $\mathcal{F}_a$ and $\mathcal{F}_b$ are the trivial orthogonal block structures on $\Omega_1 \times \Omega_2$ and Ω_3 respectively. ■

Denote by $\mathcal{Q}(\rho)$ the association scheme defined by the poset block structure $\mathcal{F}(\rho)$, and let $\mathcal{A}(\rho)$ be its Bose–Mesner algebra. If L is an ancestral subset of M with respect to ρ then its relation matrix $R_{F(L)}$ is given by Equation (9.1) no matter what ρ is, but its adjacency matrix given in Equation (9.3) does depend on ρ so it will be rewritten as A_L^ρ.

Now, pre-orders are subsets of $M \times M$, so they are partially ordered by $\subseteq$. How is the partial order on the pre-orders related to the partial order on the association schemes?

Lemma 11.12 *Let ρ and σ be pre-orders on the set M. If $\rho \subseteq \sigma$ then $\mathcal{Q}(\rho) \preccurlyeq \mathcal{Q}(\sigma)$.*

Proof Suppose that $\rho \subseteq \sigma$. If $L \in \mathcal{S}(\sigma)$ then $L \in \mathcal{S}(\rho)$ so $\mathcal{S}(\sigma) \subseteq \mathcal{S}(\rho)$. Now,

$$\begin{aligned} \mathcal{A}(\sigma) &= \operatorname{span}\{A^\sigma_L : L \in \mathcal{S}(\sigma)\} \\ &= \operatorname{span}\left\{R_{F(L)} : L \in \mathcal{S}(\sigma)\right\} \\ &\leqslant \operatorname{span}\left\{R_{F(L)} : L \in \mathcal{S}(\rho)\right\} \\ &= \operatorname{span}\{A^\rho_L : L \in \mathcal{S}(\rho)\} \\ &= \mathcal{A}(\rho) \end{aligned}$$

and so $\mathcal{Q}(\rho) \preccurlyeq \mathcal{Q}(\sigma)$. ■

Example 11.13 Let $M = \{1, 2, 3\}$ with $n_1 = n$, $n_2 = m$ and $n_3 = r$. The partially ordered family of pre-orders on M gives a partially ordered family of poset block structures on $\Omega_1 \times \Omega_2 \times \Omega_3$. Part of this is shown in Figure 11.7. ■

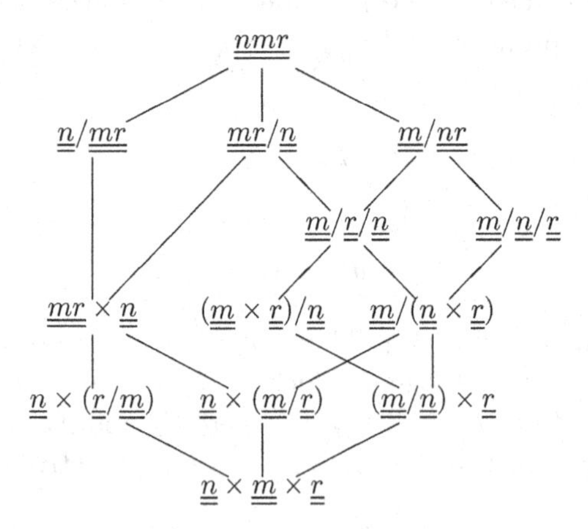

Fig. 11.7. Some of the poset block structures on $\Omega_1 \times \Omega_2 \times \Omega_3$, and their relationships with respect to $\preccurlyeq$

Given pre-orders ρ and σ on M, their union $\rho \cup \sigma$ may not be a pre-order, because it may not be transitive. Write $\rho + \sigma$ for the smallest pre-order containing $\rho \cup \sigma$. It is obtained from $\rho \cup \sigma$ as follows: given i, j in M, if there is any finite sequence $i = i_1, i_2, \ldots, i_r = j$ of elements of M such that $(i_k, i_{k+1}) \in \rho \cup \sigma$ for $k = 1, \ldots, r-1$, then $(i, j) \in \rho + \sigma$; otherwise $(i, j) \notin \rho + \sigma$.

Example 11.14 Let $M = \{1,2,3,4\}$. Suppose that

$$\rho = \{(1,1),(1,2),(1,3),(2,2),(3,3),(4,4)\}$$

and

$$\sigma = \{(1,1),(2,2),(3,1),(3,3),(4,1),(4,3),(4,4)\}.$$

Then

$$\rho + \sigma = \rho \cup \sigma \cup \{(3,2),(4,2)\}$$

and the equivalence classes of $\sim_{\rho+\sigma}$ are $\{1,3\}$, $\{2\}$ and $\{4\}$. ■

Theorem 11.13 *Suppose that ρ and σ are pre-orders on the set M. Then $\mathcal{Q}(\rho) \vee \mathcal{Q}(\sigma) = \mathcal{Q}(\rho + \sigma)$.*

Proof First note that $\mathcal{S}(\rho+\sigma) = \mathcal{S}(\rho) \cap \mathcal{S}(\sigma)$, for both are equal to

$$\{L \subseteq M : i \in L \text{ and } (i,j) \in \rho \cup \sigma \Rightarrow j \in L\}.$$

Now, $\rho \subseteq \rho + \sigma$, so $\mathcal{Q}(\rho) \preccurlyeq \mathcal{Q}(\rho+\sigma)$, by Lemma 11.12. Similarly, $\mathcal{Q}(\sigma) \preccurlyeq \mathcal{Q}(\rho+\sigma)$. Therefore $\mathcal{Q}(\rho) \vee \mathcal{Q}(\sigma) \preccurlyeq \mathcal{Q}(\rho+\sigma)$. Thus

$$\mathcal{A}(\rho+\sigma) = V_{\mathcal{Q}(\rho+\sigma)} \geqslant V_{\mathcal{Q}(\rho)\vee\mathcal{Q}(\sigma)} = V_{\mathcal{Q}(\rho)} \cap V_{\mathcal{Q}(\sigma)} = \mathcal{A}(\rho) \cap \mathcal{A}(\sigma)$$

by Equation (11.1). However, $\mathcal{A}(\rho) = \operatorname{span}\left\{R_{F(L)} : L \in \mathcal{S}(\rho)\right\}$ and $\mathcal{A}(\sigma) = \operatorname{span}\left\{R_{F(L)} : L \in \mathcal{S}(\sigma)\right\}$, so

$$\mathcal{A}(\rho) + \mathcal{A}(\sigma) = \operatorname{span}\left\{R_{F(L)} : L \in \mathcal{S}(\rho) \cup \mathcal{S}(\sigma)\right\}.$$

Corollary 9.8(ii) shows that the matrices $R_{F(L)}$, for all subsets L of M, are linearly independent. Hence

$$\begin{aligned}
\dim(\mathcal{A}(\rho) \cap \mathcal{A}(\sigma)) &= \dim(\mathcal{A}(\rho) + \mathcal{A}(\sigma)) - \dim \mathcal{A}(\rho) - \dim \mathcal{A}(\sigma) \\
&= |\mathcal{S}(\rho) \cup \mathcal{S}(\sigma)| - |\mathcal{S}(\rho)| - |\mathcal{S}(\sigma)| \\
&= |\mathcal{S}(\rho) \cap \mathcal{S}(\sigma)| \\
&= |\mathcal{S}(\rho+\sigma)| \\
&= \dim \mathcal{A}(\rho+\sigma)
\end{aligned}$$

so $\mathcal{A}(\rho) \cap \mathcal{A}(\sigma) = \mathcal{A}(\rho+\sigma)$ and $\mathcal{Q}(\rho) \vee \mathcal{Q}(\sigma) = \mathcal{Q}(\rho+\sigma)$. ■

This shows that the supremum of the association schemes of two poset block structures is itself the association scheme of a poset block structure.

In the other direction, the pre-orders are well behaved, in that $\rho \cap \sigma$ is a pre-order if both ρ and σ are pre-orders, but the association schemes are not well behaved, because $\mathcal{Q}(\rho) \wedge \mathcal{Q}(\sigma)$ may not be an association scheme.

Theorem 11.14 *Let ρ and σ be pre-orders on M. Then $\mathcal{Q}(\rho\cap\sigma)$ is the coarsest association scheme nested in $\mathcal{Q}(\rho)\wedge\mathcal{Q}(\sigma)$; in particular, $\mathcal{Q}(\rho)$ is compatible with $\mathcal{Q}(\sigma)$.*

Proof Lemma 11.12 gives $\mathcal{Q}(\rho\cap\sigma)\preccurlyeq\mathcal{Q}(\rho)$, because $\rho\cap\sigma\subseteq\rho$. Similarly, $\mathcal{Q}(\rho\cap\sigma)\preccurlyeq\mathcal{Q}(\sigma)$. Therefore $\mathcal{Q}(\rho\cap\sigma)\preccurlyeq\mathcal{Q}(\rho)\wedge\mathcal{Q}(\sigma)$. Thus there is at least one association scheme nested in $\mathcal{Q}(\rho)\wedge\mathcal{Q}(\sigma)$.

Let $\mathcal{P}$ be any association scheme with $\mathcal{P}\preccurlyeq\mathcal{Q}(\rho)\wedge\mathcal{Q}(\sigma)$. We need to show that $\mathcal{P}\preccurlyeq\mathcal{Q}(\rho\cap\sigma)$. Let $\mathcal{A}$ be the Bose–Mesner algebra of $\mathcal{P}$, and put $\mathcal{L}=\{L\subseteq M : R_{F(L)}\in\mathcal{A}\}$; that is, $L\in\mathcal{L}$ if $F(L)$ is inherent in $\mathcal{P}$. Theorem 10.1 shows that if K and L are in $\mathcal{L}$ then $F(K)\vee F(L)$ and $F(K)\wedge F(L)$ are inherent in $\mathcal{P}$, so Lemma 9.1 shows that $\mathcal{L}$ is closed under $\cap$ and $\cup$.

For i in M, put

$$\begin{aligned} G_i &= \{j\in M : (i,j)\in\rho\} \\ H_i &= \{j\in M : (i,j)\in\sigma\}. \end{aligned}$$

Then $R_{F(G_i)}\in\mathcal{A}(\rho)\leqslant\mathcal{A}$ because $\mathcal{P}\preccurlyeq\mathcal{Q}(\rho)$; so $G_i\in\mathcal{L}$. Similarly, $H_i\in\mathcal{L}$.

Let $L\in\mathcal{S}(\rho\cap\sigma)$. If $i\in L$ then $i\in G_i\cap H_i\subseteq L$ and so

$$L=\bigcup_{i\in L}(G_i\cap H_i),$$

which is in $\mathcal{L}$. Thus $R_{F(L)}\in\mathcal{A}$. Hence $\mathcal{A}(\rho\cap\sigma)\leqslant\mathcal{A}$ and therefore $\mathcal{P}\preccurlyeq\mathcal{Q}(\rho\cap\sigma)$. ∎

Example 11.10 revisited Put $M=\{1,2,3\}$ and $n_1=3$, $n_2=2$, $n_3=4$. Then $\mathcal{Q}_k=\mathcal{Q}(\rho_k)$ for $k=1, 2, 3$, where

$$\begin{aligned} \rho_1 &= \{(1,1),(1,2),(2,1),(2,2),(3,1),(3,2),(3,3)\} \\ \rho_2 &= \{(1,1),(1,3),(2,1),(2,2),(2,3),(3,1),(3,3)\} \\ \rho_3 &= \{(1,1),(2,1),(2,2),(3,1),(3,3)\} = \rho_1\cap\rho_2. \end{aligned}$$

The Hasse diagrams for these pre-orders are shown in Figure 11.8. Now Theorem 11.14 shows that $\mathcal{Q}_3$ is the coarsest association scheme nested in $\mathcal{Q}_1\wedge\mathcal{Q}_2$, which we have already seen is not itself an association scheme. ∎

Since $\rho\cap\sigma\subseteq\rho$, we have $\mathcal{S}(\rho)\subseteq\mathcal{S}(\rho\cap\sigma)$, and similarly for σ, so we always have $\mathcal{S}(\rho)\cup\mathcal{S}(\sigma)\subseteq\mathcal{S}(\rho\cap\sigma)$. However, equality is not always achieved: in Example 11.10, $\{1\}\in\mathcal{S}(\rho_3)$ but $\{1\}\notin\mathcal{S}(\rho_1)\cup\mathcal{S}(\rho_2)$.

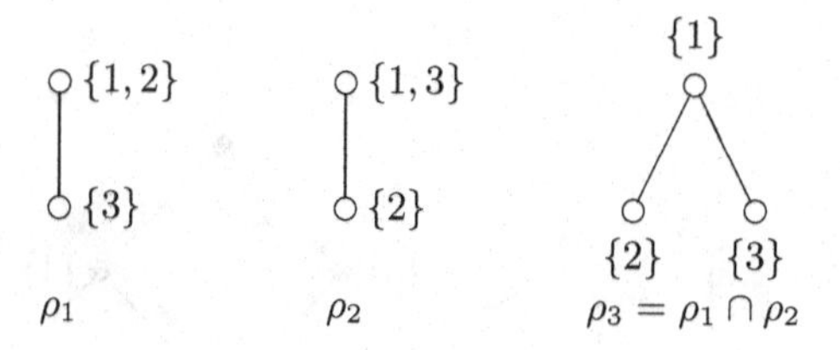

Fig. 11.8. Pre-orders in Example 11.10

Theorem 11.15 *Let ρ and σ be pre-orders on M. If $\mathcal{S}(\rho) \cup \mathcal{S}(\sigma) = \mathcal{S}(\rho \cap \sigma)$ then $\mathcal{Q}(\rho) \wedge \mathcal{Q}(\sigma)$ is an association scheme.*

Proof Since $\mathcal{Q}(\rho \cap \sigma) \preccurlyeq \mathcal{Q}(\rho) \wedge \mathcal{Q}(\sigma) \preccurlyeq \mathcal{Q}(\rho)$, and similarly for σ, we have

$$\mathcal{A}(\rho \cap \sigma) \geqslant V_{\mathcal{Q}(\rho)\wedge\mathcal{Q}(\sigma)} \geqslant \mathcal{A}(\rho) + \mathcal{A}(\sigma). \tag{11.2}$$

Now, $\dim \mathcal{A}(\rho \cap \sigma) = |\mathcal{S}(\rho \cap \sigma)|$ and $\dim(\mathcal{A}(\rho) + \mathcal{A}(\sigma)) = |\mathcal{S}(\rho) \cup \mathcal{S}(\sigma)|$; if these dimensions are equal then all three terms in Equation (11.2) are equal, so $\mathcal{Q}(\rho \cap \sigma) = \mathcal{Q}(\rho) \wedge \mathcal{Q}(\sigma)$. ■

The converse is not true. Note that if K and L are subsets of M then Equation (9.1) and the fact that

$$(N_1 \otimes N_2) \circ (N_3 \otimes N_4) = (N_1 \circ N_3) \otimes (N_2 \circ N_4)$$

show that $R_{F(L)} \circ R_{F(K)} = R_{F(L \cup K)}$.

Example 11.15 Let $M = \{1, 2, 3\}$ and let ρ and σ be the pre-orders on M shown, together with their ancestral subsets, in Figure 11.9. Now,

$$\begin{aligned} V_{\mathcal{Q}(\rho)\wedge\mathcal{Q}(\sigma)} &= \operatorname{span}\left\{R_{F(L)} \circ R_{F(K)} : L \in \mathcal{S}(\rho),\ K \in \mathcal{S}(\sigma)\right\} \\ &= \operatorname{span}\left\{R_{F(L \cup K)} : L \in \mathcal{S}(\rho),\ K \in \mathcal{S}(\sigma)\right\} \\ &= \operatorname{span}\left\{R_{F(H)} : H \subseteq M\right\} \end{aligned}$$

so

$$\mathcal{Q}(\rho) \wedge \mathcal{Q}(\sigma) = \underline{\underline{1}} \times \underline{\underline{2}} \times \underline{\underline{3}} = \mathcal{Q}(\rho \cap \sigma).$$

However, $\{2, 3\} \in \mathcal{S}(\rho \cap \sigma)$ but $\{2, 3\} \notin \mathcal{S}(\rho) \cup \mathcal{S}(\sigma)$. ■

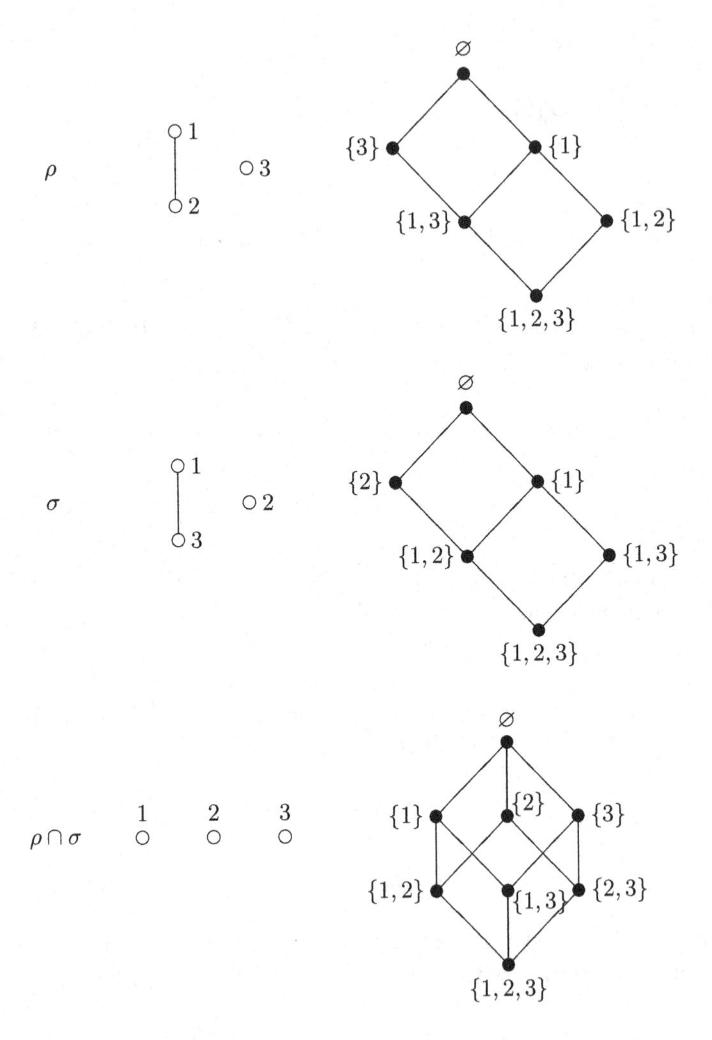

Fig. 11.9. Pre-orders in Example 11.15, and their ancestral subsets

11.6 Orthogonal block structures

In this section, Ω is a fixed set, and we investigate the family of orthogonal block structures on Ω. Let F be a partition on Ω. Then the relation matrix R_F is always given by Equation (6.1), but the adjacency matrix A_F given in Equation (6.4) depends on the orthogonal block structure $\mathcal{F}$, so it will be rewritten as $A_F^{\mathcal{F}}$. Write $\mathcal{Q}(\mathcal{F})$ for the association scheme on Ω defined by the orthogonal block structure $\mathcal{F}$, and $\mathcal{A}(\mathcal{F})$ for its Bose–Mesner algebra.

Orthogonal block structures on Ω are sets of partitions of Ω, so they are partially ordered by $\subseteq$. How is this partial order on the orthogonal block structures related to the partial order on the association schemes which they define?

Lemma 11.16 *Let $\mathcal{F}$ and $\mathcal{G}$ be orthogonal block structures on Ω. If $\mathcal{F} \subseteq \mathcal{G}$ then $\mathcal{Q}(\mathcal{G}) \preccurlyeq \mathcal{Q}(\mathcal{F})$.*

Proof

$$\begin{aligned} \mathcal{A}(\mathcal{F}) &= \operatorname{span}\left\{A_F^{\mathcal{F}} : F \in \mathcal{F}\right\} \\ &= \operatorname{span}\left\{R_F : F \in \mathcal{F}\right\} \\ &\leqslant \operatorname{span}\left\{R_F : F \in \mathcal{G}\right\} \\ &= \operatorname{span}\left\{A_F^{\mathcal{G}} : F \in \mathcal{G}\right\} \\ &= \mathcal{A}(\mathcal{G}) \end{aligned}$$

and so $\mathcal{Q}(\mathcal{G}) \preccurlyeq \mathcal{Q}(\mathcal{F})$. ■

It is clear that $\mathcal{F} \cap \mathcal{G}$ is an orthogonal block structure if $\mathcal{F}$ and $\mathcal{G}$ are.

Lemma 11.17 *Let $\mathcal{F}$ and $\mathcal{G}$ be orthogonal block structures on Ω. Then*

$$\mathcal{Q}(\mathcal{F}) \vee \mathcal{Q}(\mathcal{G}) \preccurlyeq \mathcal{Q}(\mathcal{F} \cap \mathcal{G}).$$

Proof Since $\mathcal{F} \cap \mathcal{G} \subseteq \mathcal{F}$, Lemma 11.16 shows that $\mathcal{Q}(\mathcal{F}) \preccurlyeq \mathcal{Q}(\mathcal{F} \cap \mathcal{G})$. Similarly, $\mathcal{Q}(\mathcal{G}) \preccurlyeq \mathcal{Q}(\mathcal{F} \cap \mathcal{G})$. Hence $\mathcal{Q}(\mathcal{F}) \vee \mathcal{Q}(\mathcal{G}) \preccurlyeq \mathcal{Q}(\mathcal{F} \cap \mathcal{G})$. ■

We have seen that the supremum of two group association schemes is itself a group association scheme and that the supremum of two association schemes derived from poset block structures is itself the association scheme of a poset block structure. Unfortunately, the analogous result does not hold for orthogonal block structures.

Example 11.16 Let Ω be the set of nine cells in the 3×3 square array underlying the pair of mutually orthogonal Latin squares in Figure 11.10. Let R, C, G and L be the partitions of Ω into rows, columns, Latin letters and Greek letters respectively. Put $\mathcal{F} = \{E, R, C, U\}$ and $\mathcal{G} = \{E, G, L, U\}$. Then

$$A_U^{\mathcal{F}} = A_G^{\mathcal{G}} + A_L^{\mathcal{G}}$$

while

$$A_U^{\mathcal{G}} = A_R^{\mathcal{F}} + A_C^{\mathcal{F}}.$$

Thus the adjacency matrices of $\mathcal{Q}(\mathcal{F}) \vee \mathcal{Q}(\mathcal{G})$ are I, $A_U^{\mathcal{F}}$ and $A_U^{\mathcal{G}}$, so $\mathcal{Q}(\mathcal{F}) \vee \mathcal{Q}(\mathcal{G})$ is the Hamming scheme H(2,3), which is not an orthogonal block structure (why not?). On the other hand, $\mathcal{Q}(\mathcal{F} \cap \mathcal{G}) = \mathcal{Q}(\{E, U\})$, which is the trivial association scheme on Ω. ■

A	B	C
C	A	B
B	C	A

α	β	γ
β	γ	α
γ	α	β

Fig. 11.10. Pair of mutually orthogonal Latin squares in Example 11.16

What goes wrong in Example 11.16 is that $\dim \mathcal{A}(\mathcal{F} \cap \mathcal{G})$ is too small, because the analogue of Corollary 9.8(ii) does not hold for all orthogonal block structures.

Definition A set $\mathcal{F}$ of partitions of Ω is *strict* if the set $\{R_F : F \in \mathcal{F}\}$ is linearly independent.

Theorem 11.18 *Let $\mathcal{F}$ and $\mathcal{G}$ be orthogonal block structures on Ω. If $\mathcal{F} \cup \mathcal{G}$ is strict then $\mathcal{Q}(\mathcal{F}) \vee \mathcal{Q}(\mathcal{G}) = \mathcal{Q}(\mathcal{F} \cap \mathcal{G})$.*

Proof Lemma 11.17 gives $\mathcal{Q}(\mathcal{F}) \vee \mathcal{Q}(\mathcal{G}) \preccurlyeq \mathcal{Q}(\mathcal{F} \cap \mathcal{G})$, so

$$\mathcal{A}(\mathcal{F} \cap \mathcal{G}) \geqslant V_{\mathcal{Q}(\mathcal{F}) \vee \mathcal{Q}(\mathcal{G})} = \mathcal{A}(\mathcal{F}) \cap \mathcal{A}(\mathcal{G}),$$

by Equation (11.1). If $\mathcal{F} \cup \mathcal{G}$ is strict then so are $\mathcal{F}$, $\mathcal{G}$ and $\mathcal{F} \cap \mathcal{G}$. Therefore

$$\begin{aligned}
\dim(\mathcal{A}(\mathcal{F}) \cap \mathcal{A}(\mathcal{G})) &= \dim(\mathcal{A}(\mathcal{F}) + \mathcal{A}(\mathcal{G})) - \dim \mathcal{A}(\mathcal{F}) - \dim \mathcal{A}(\mathcal{G}) \\
&= |\mathcal{F} \cup \mathcal{G}| - |\mathcal{F}| - |\mathcal{G}| \\
&= |\mathcal{F} \cap \mathcal{G}| \\
&= \dim \mathcal{A}(\mathcal{F} \cap \mathcal{G})
\end{aligned}$$

and so $\mathcal{A}(\mathcal{F}) \cap \mathcal{A}(\mathcal{G}) = \mathcal{A}(\mathcal{F} \cap \mathcal{G})$ and $\mathcal{Q}(\mathcal{F}) \vee \mathcal{Q}(\mathcal{G}) = \mathcal{Q}(\mathcal{F} \cap \mathcal{G})$. ■

Now we turn to infima. In general, the infimum of two association schemes defined by orthogonal block structures is not an association scheme; indeed, the two association schemes may not be compatible. The slightly surprising result, given that suprema may take us outside the class of orthogonal block structures, is that when the two association schemes *are* compatible then the coarsest association scheme nested in both is also defined by an orthogonal block structure.

Lemma 11.19 *Let $\mathcal{F}$ and $\mathcal{G}$ be orthogonal block structures on Ω. If there is an orthogonal block structure $\mathcal{H}$ on Ω such that $\mathcal{F} \cup \mathcal{G} \subseteq \mathcal{H}$ then $\mathcal{Q}(\mathcal{H}) \preccurlyeq \mathcal{Q}(\mathcal{F}) \wedge \mathcal{Q}(\mathcal{G})$.*

Proof If $\mathcal{F} \cup \mathcal{G} \subseteq \mathcal{H}$ then $\mathcal{F} \subseteq \mathcal{H}$ so Lemma 11.16 shows that $\mathcal{Q}(\mathcal{H}) \preccurlyeq \mathcal{Q}(\mathcal{F})$. Similarly, $\mathcal{Q}(\mathcal{H}) \preccurlyeq \mathcal{Q}(\mathcal{G})$. Hence $\mathcal{Q}(\mathcal{H}) \preccurlyeq \mathcal{Q}(\mathcal{F}) \wedge \mathcal{Q}(\mathcal{G})$. ∎

Theorem 11.20 *Let $\mathcal{F}$ and $\mathcal{G}$ be orthogonal block structures on Ω. If $\mathcal{F} \cup \mathcal{G}$ is an orthogonal block structure then $\mathcal{Q}(\mathcal{F}) \wedge \mathcal{Q}(\mathcal{G}) = \mathcal{Q}(\mathcal{F} \cup \mathcal{G})$.*

Proof If $\mathcal{F} \cup \mathcal{G}$ is an orthogonal block structure then Lemma 11.19 shows that $\mathcal{Q}(\mathcal{F} \cup \mathcal{G}) \preccurlyeq \mathcal{Q}(\mathcal{F}) \wedge \mathcal{Q}(\mathcal{G})$. Hence

$$\mathcal{A}(\mathcal{F} \cup \mathcal{G}) \geqslant V_{\mathcal{Q}(\mathcal{F}) \wedge \mathcal{Q}(\mathcal{G})} \geqslant \mathcal{A}(\mathcal{F}) + \mathcal{A}(\mathcal{G}). \tag{11.3}$$

Also,

$$\begin{aligned} \mathcal{A}(\mathcal{F} \cup \mathcal{G}) &= \operatorname{span}\{R_F : F \in \mathcal{F} \cup \mathcal{G}\} \\ &= \operatorname{span}\{R_F : F \in \mathcal{F}\} + \operatorname{span}\{R_F : F \in \mathcal{G}\} \\ &= \mathcal{A}(\mathcal{F}) + \mathcal{A}(\mathcal{G}). \end{aligned}$$

Therefore equality is forced in Equation (11.3) and so $\mathcal{Q}(\mathcal{F}) \wedge \mathcal{Q}(\mathcal{G}) = \mathcal{Q}(\mathcal{F} \cup \mathcal{G})$. ∎

It is possible for $\mathcal{Q}(\mathcal{F}) \wedge \mathcal{Q}(\mathcal{G})$ to be an association scheme even if $\mathcal{F} \cup \mathcal{G}$ is not an orthogonal block structure.

Example 11.15 revisited Put $\mathcal{F} = \mathcal{F}(\rho)$ and $\mathcal{G} = \mathcal{F}(\sigma)$, where ρ and σ are given in Figure 11.9. Then $\mathcal{Q}(\mathcal{F}) \wedge \mathcal{Q}(\mathcal{G})$ is the association scheme $\underline{\underline{1}} \times \underline{\underline{2}} \times \underline{\underline{3}}$. However,

$$\mathcal{F} = \{F(L) : L \in \mathcal{S}(\rho)\}$$

and

$$\mathcal{G} = \{F(L) : L \in \mathcal{S}(\sigma)\}$$

so $\mathcal{F} \cup \mathcal{G} = \{F(L) : L \subseteq \{1,2,3\},\ L \neq \{2,3\}\}$. Thus $\mathcal{F} \cup \mathcal{G}$ contains $F(\{2\})$ and $F(\{3\})$ but does not contain $F(\{2\}) \wedge F(\{3\})$, so $\mathcal{F} \cup \mathcal{G}$ is not an orthogonal block structure. ∎

Even if $\mathcal{F} \cup \mathcal{G}$ is not itself an orthogonal block structure, it may be contained in one. Since the intersection of orthogonal block structures is itself an orthogonal block structure, we then have a smallest orthogonal block structure containing $\mathcal{F} \cup \mathcal{G}$.

Theorem 11.21 *Let $\mathcal{F}$ and $\mathcal{G}$ be orthogonal block structures on Ω. If $\mathcal{Q}(\mathcal{F})$ is compatible with $\mathcal{Q}(\mathcal{G})$ then $\mathcal{F} \cup \mathcal{G}$ is contained in an orthogonal block structure. Moreover, if $\mathcal{P}$ is the coarsest association scheme nested in $\mathcal{Q}(\mathcal{F}) \wedge \mathcal{Q}(\mathcal{G})$ then $\mathcal{P}$ is the association scheme of the smallest orthogonal block structure containing $\mathcal{F} \cup \mathcal{G}$.*

Proof Let $\mathcal{H}$ be the set of inherent partitions of $\mathcal{P}$, and let $\mathcal{B}$ be the Bose–Mesner algebra of $\mathcal{P}$. If $F \in \mathcal{F}$ then $R_F \in \mathcal{A}(\mathcal{F}) \leqslant \mathcal{B}$ and so $F \in \mathcal{H}$; and similarly for $\mathcal{G}$. Hence $\mathcal{F} \cup \mathcal{G} \subseteq \mathcal{H}$. Theorem 10.1 shows that $\mathcal{H}$ is an orthogonal block structure.

From Lemma 11.19, $\mathcal{Q}(\mathcal{H}) \preccurlyeq \mathcal{Q}(\mathcal{F}) \wedge \mathcal{Q}(\mathcal{G})$. By definition of $\mathcal{H}$, $\mathcal{A}(\mathcal{H}) \leqslant \mathcal{B}$, so $\mathcal{P} \preccurlyeq \mathcal{Q}(\mathcal{H})$. Since $\mathcal{P}$ is the coarsest association scheme nested in $\mathcal{Q}(\mathcal{F}) \wedge \mathcal{Q}(\mathcal{G})$, it follows that $\mathcal{P} = \mathcal{Q}(\mathcal{H})$.

Let $\mathcal{K}$ be any orthogonal block structure containing $\mathcal{F} \cup \mathcal{G}$. Then $\mathcal{Q}(\mathcal{K}) \preccurlyeq \mathcal{Q}(\mathcal{F}) \wedge \mathcal{Q}(\mathcal{G})$ so $\mathcal{Q}(\mathcal{K}) \preccurlyeq \mathcal{P} = \mathcal{Q}(\mathcal{H})$. Therefore $\mathcal{H} \subseteq \mathcal{K}$, and so $\mathcal{H}$ is the smallest orthogonal block structure containing $\mathcal{F} \cup \mathcal{G}$. ■

If $\mathcal{A}(\mathcal{F})$ does not commute with $\mathcal{A}(\mathcal{G})$ then Lemma 11.7(ii) shows that $\mathcal{Q}(\mathcal{F})$ cannot be compatible with $\mathcal{Q}(\mathcal{G})$. Theorem 11.21 shows that even if $\mathcal{A}(\mathcal{F})$ commutes with $\mathcal{A}(\mathcal{G})$ then $\mathcal{Q}(\mathcal{F})$ is not necessarily compatible with $\mathcal{Q}(\mathcal{G})$.

Example 11.17 Let Ω consist of the n^3 cells of a Latin cube whose coordinate partitions are X, Y and Z. Let L be the partition into letters. Put $\mathcal{F} = \{U, X, Y, X \wedge Y, E\}$ and $\mathcal{G} = \{U, Z, L, Z \wedge L, E\}$. Then $\mathcal{F}$ and $\mathcal{G}$ are both orthogonal block structures. Moreover,

$$\begin{aligned}
P_X P_Z = {} & P_Z P_X &= P_U \\
P_X P_L = {} & P_L P_X &= P_U \\
P_X P_{Z\wedge L} = {} & P_{Z\wedge L} P_X &= P_U \\
P_Y P_Z = {} & P_Z P_Y &= P_U \\
P_Y P_L = {} & P_L P_Y &= P_U \\
P_Y P_{Z\wedge L} = {} & P_{Z\wedge L} P_Y &= P_U \\
P_{X\wedge Y} P_Z = {} & P_Z P_{X\wedge Y} &= P_U \\
P_{X\wedge Y} P_L = {} & P_L P_{X\wedge Y} &= P_U.
\end{aligned}$$

If the Latin cube is constructed from an $n \times n$ Latin square Π in the way described on page 250, then $P_{X\wedge Y}$ also commutes with $P_{Z\wedge L}$, and so $\mathcal{A}(\mathcal{F})$ commutes with $\mathcal{A}(\mathcal{G})$. However, if Π is not itself the Cayley table of a group (for example, if Π is the square in Figure 8.5) then Theorem 8.21

shows that there is no orthogonal block structure containing $\mathcal{F} \cup \mathcal{G}$. Theorem 11.21 then shows that $\mathcal{Q}(\mathcal{F})$ is not compatible with $\mathcal{Q}(\mathcal{G})$. ■

11.7 Crossing and nesting

Fortunately, both crossing and nesting of association schemes behave well with respect to the partial order $\preccurlyeq$. For let $\mathcal{Q}_t$ and $\mathcal{P}_t$ be association schemes on Ω_t for $t = 1, 2$. Let $\mathcal{C}$ be a class of $\mathcal{Q}_1$ and $\mathcal{D}$ a class of $\mathcal{Q}_2$. If $\mathcal{Q}_1 \preccurlyeq \mathcal{P}_1$ then there is a class $\mathcal{C}^*$ of $\mathcal{P}_1$ which contains $\mathcal{C}$, and if $\mathcal{Q}_2 \preccurlyeq \mathcal{P}_2$ then there is a class $\mathcal{D}^*$ of $\mathcal{P}_2$ which contains $\mathcal{D}$. Then $\mathcal{C} \times \mathcal{D} \subseteq \mathcal{C}^* \times \mathcal{D}^*$, $\mathcal{C} \times (\Omega_2 \times \Omega_2) \subseteq \mathcal{C}^* \times (\Omega_2 \times \Omega_2)$ and $\mathrm{Diag}(\Omega_1) \times \mathcal{D} \subseteq \mathrm{Diag}(\Omega_1) \times \mathcal{D}^*$. This proves the following.

Theorem 11.22 *Let $\mathcal{Q}_1$ and $\mathcal{P}_1$ be association schemes on Ω_1 and let $\mathcal{Q}_2$ and $\mathcal{P}_2$ be association schemes on Ω_2. If $\mathcal{Q}_1 \preccurlyeq \mathcal{P}_1$ and $\mathcal{Q}_2 \preccurlyeq \mathcal{P}_2$ then $(\mathcal{Q}_1 \times \mathcal{Q}_2) \preccurlyeq (\mathcal{P}_1 \times \mathcal{P}_2)$ and $(\mathcal{Q}_1/\mathcal{Q}_2) \preccurlyeq (\mathcal{P}_1/\mathcal{P}_2)$.*

Example 11.18 Consider the association scheme $\mathrm{Pair}(n)$ described on page 134. The $n(n-1)$ elements fall into mirror-image pairs (i,j) and (j,i): two elements are M-associates if and only if they form a mirror-image pair. The A-associates of (i,j) are (i,k) and (k,j) for $k \notin \{i,j\}$, and these are precisely the B-associates of (j,i). So if we merge the classes whose adjacency matrices are A and B we obtain the association scheme $\mathrm{T}(n)/\underline{\underline{2}}$. Theorem 11.22 shows that $\mathrm{T}(n)/\underline{\underline{2}} \prec \underline{\underline{m}}/\underline{\underline{2}}$, where $m = n(n-1)/2$. So we have the chain

$$\mathrm{Pair}(n) \prec \mathrm{T}(n)/\underline{\underline{2}} \prec \mathrm{GD}\left(\frac{n(n-1)}{2}, 2\right) \prec \underline{\underline{n(n-1)}}. \quad ■$$

Proposition 11.23 *If $\mathcal{Q}_1$ and $\mathcal{Q}_2$ are association schemes (on different sets) then $(\mathcal{Q}_1 \times \mathcal{Q}_2) \preccurlyeq (\mathcal{Q}_1/\mathcal{Q}_2)$.*

11.8 Mutually orthogonal Latin squares

Finally we note an interesting family of association schemes which is closed under both $\wedge$ and $\vee$. Let $\mathcal{F}$ be the orthogonal block structure defined by $n-1$ mutually orthogonal $n \times n$ Latin squares (if such exist): this defines a square association scheme of type $\mathrm{S}(n+1, n)$. The adjacency matrices are I, A_1, A_2, …, A_{n+1} with valencies 1, $n-1$, $n-1$, …, $n-1$. We have $(I + A_i)(I + A_j) = J$ for all i, j with $i \neq j$, and $(I + A_i)^2 = n(I + A_i)$. Consequently, any merging at all of the non-diagonal associate classes gives an association scheme. That is, if $\mathcal{P}$ is any partition of $\{1,\ldots,n+1\}$ then the matrices $\sum_{i\in P} A_i$ for the classes P of $\mathcal{P}$ are the adjacency matrices for an association scheme $\mathcal{Q}(\mathcal{P})$ on Ω. Any association scheme with this property is called *amorphic*.

The *type* of a partition $\mathcal{P}$ of a set is an ordered list of the sizes of its classes, in ascending order. If $\mathcal{P}$ has type $(1,1,\ldots,1)$ then $\mathcal{Q}(\mathcal{P})$ is the association scheme of the original orthogonal block structure $\mathcal{F}$. If $\mathcal{P}$ has type $(1,1,\ldots,1,n+1-m)$ then $\mathcal{Q}(\mathcal{P})$ is the association scheme of the orthogonal block structure obtained from m of the original non-trivial partitions: it is group-divisible when $m=1$ and rectangular when $m=2$. If $\mathcal{P}$ has type $(m,n+1-m)$ then $\mathcal{Q}(\mathcal{P})$ is of Latin-square type $\mathrm{L}(m,n)$: if $m=2$ then $\mathcal{Q}(\mathcal{P})$ is also $\mathrm{H}(2,n)$, while if $m=1$ then $\mathcal{Q}(\mathcal{P})$ is also $\mathrm{GD}(n,n)$. If $\mathcal{P}$ has type $(n+1)$ then $\mathcal{Q}(\mathcal{P})$ is trivial.

When $n=4$ this construction gives 52 association schemes on a set Ω of size 16. A few of these are shown in Figure 11.11 as partitions of $\{1,2,3,4,5\}$.

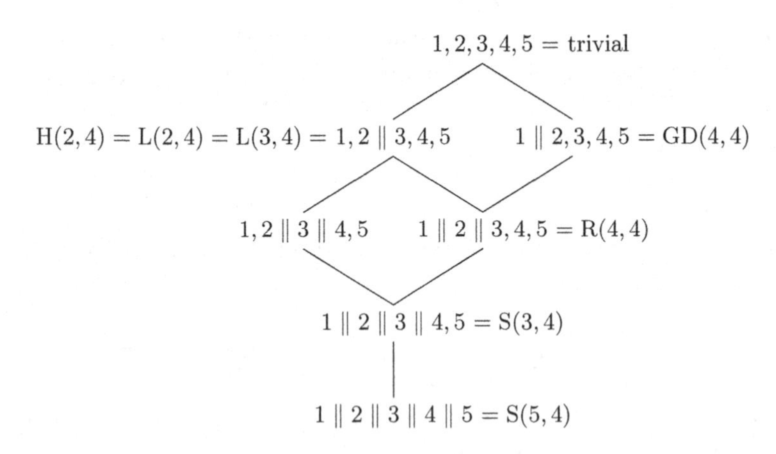

Fig. 11.11. Some of the association schemes formed from a set of three mutually orthogonal 4×4 Latin squares.

In fact, this merging of classes works even if we start with a smaller number of mutually orthogonal Latin squares. If there are $c-2$ squares then the adjacency matrices of the non-diagonal classes are $A_1, \ldots, A_c$, each with valency $n-1$, and A_E, with valency $(n+1-c)(n-1)$. Then $A_E = J-\sum_{i=1}^{c}(I+A_i)+(c-1)I$, so $A_E(I+A_i) = (n-c+1)(J-I-A_i)$ for $1\leqslant i\leqslant c$ and $A_E^2\in \operatorname{span}\{I, A_E, J\}$. So once again each partition of $\{1,\ldots,c,E\}$ gives an association scheme: now the type of the association scheme depends both on the type of the partition and on which class of the partition contains E.

Example 11.19 (Example 6.2 continued) In the square association scheme of type $S(3,n)$ defined by a Latin square of size n there are four non-diagonal associate classes if $n \geqslant 3$: these correspond to the partitions rows, columns, letters and E. Merging $\mathcal{C}_{\text{letters}}$ and $\mathcal{C}_E$ gives a rectangular association scheme $\mathrm{R}(n,n)$, but merging $\mathcal{C}_{\text{rows}}$ and $\mathcal{C}_{\text{columns}}$ gives a new type of association scheme with $s = 3$ and valencies $2(n-1)$, $n-1$ and $(n-2)(n-1)$.

There are 15 partitions of a set of size four. These give the following 15 association schemes on a set Ω of size n^2: three rectangular schemes, three schemes of the new type just described, three Hamming schemes, three group-divisible schemes, one scheme of Latin-square type $\mathrm{L}(3,n)$, the original square association scheme, and the trivial scheme. ■

Exercises

11.1 Suppose that $|\Omega| = nm$. How many association schemes are there on Ω of type $\mathrm{GD}(n,m)$?

11.2 Find all cyclic association schemes for $\mathbb{Z}_8$. Draw the Hasse diagram to show which is coarser than which.

11.3 Find two group schemes for S_3 which are not compatible.

11.4 For each given pair of pre-orders ρ and σ, identify $\mathcal{Q}(\rho) \vee \mathcal{Q}(\sigma)$ and decide whether $\mathcal{Q}(\rho) \wedge \mathcal{Q}(\sigma)$ is an association scheme.

(a) The underlying set M of subscripts is $\{1,2,3\}$ and the pre-orders are as follows.

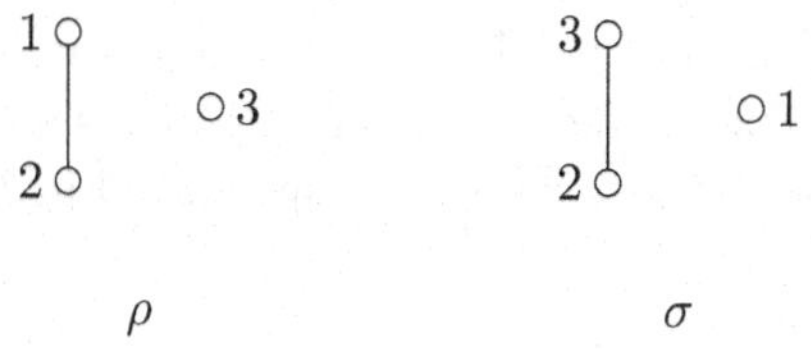

1 ○ ○ 3 ○ ○ 1
| |
2 ○ 2 ○

ρ σ

(b) The underlying set M of subscripts is $\{1,2,3,4\}$ and the pre-orders are as follows.

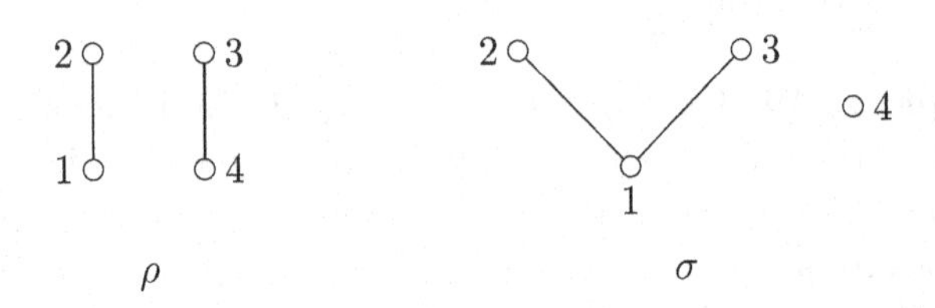

(c) The underlying set M of subscripts is again $\{1, 2, 3, 4\}$ but the pre-orders are as follows.

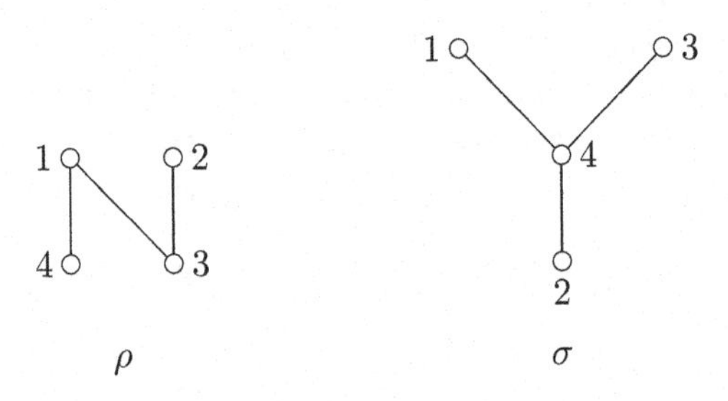

11.5 Draw the Hasse diagrams for the pre-orders ρ, σ and $\rho + \sigma$ in Example 11.14.

11.6 Is it true that the supremum of two group block structures (for the same group) is a group block structure? Give a proof or counter-example.

11.7 Consider the Latin rectangle in Exercise 6.6. The non-trivial partitions are rows, columns and letters. Put $\mathcal{F} = \{\text{rows}, \text{columns}, E, U\}$ and $\mathcal{G} = \{\text{letters}, \text{columns}, E, U\}$.

(a) Find $\mathcal{Q}(\mathcal{F}) \vee \mathcal{Q}(\mathcal{G})$.
(b) When are $\mathcal{Q}(\mathcal{F})$ and $\mathcal{Q}(\mathcal{G})$ compatible?

11.8 For each of the following association schemes $\mathcal{Q}$, find the set $\mathcal{R}$ of all association schemes coarser than $\mathcal{Q}$. Draw the Hasse diagram of $\mathcal{R}$. For each pair $\mathcal{P}_1$ and $\mathcal{P}_2$ of association schemes in $\mathcal{R}$, (i) find the supremum $\mathcal{P}_1 \vee \mathcal{P}_2$; (ii) decide whether $\mathcal{P}_1 \wedge \mathcal{P}_2$ is an association scheme; (iii) find the coarsest association scheme nested in $\mathcal{P}_1 \wedge \mathcal{P}_2$.

(a) $\mathcal{Q} = \underline{\underline{\underline{4}}} \times \underline{\underline{\underline{4}}}$;
(b) $\mathcal{Q} = (\underline{\underline{\underline{4}}} \times \underline{\underline{\underline{4}}})/\underline{\underline{\underline{3}}}$;
(c) $\mathcal{Q} = (\underline{\underline{\underline{2}}}/\underline{\underline{\underline{3}}}) \times \underline{\underline{\underline{4}}}$;

(d) $\mathcal{Q} = \underline{\underline{3}}/(\underline{\underline{2}} \times \underline{\underline{4}})$;

(e) $\mathcal{Q} = \mathcal{L}/\underline{\underline{4}}$, where $\mathcal{L}$ is the square association scheme defined by the orthogonal block structure $\{E, R, C, L, G, U\}$ on the 3×3 square in Example 11.16;

(f) $\mathcal{Q} = \mathrm{J}(6, 3)$;

(g) $\mathcal{Q} = \mathrm{J}(7, 3)$;

(h) $\mathcal{Q} = \mathcal{Q}(\mathsf{Inverse})$ for the Abelian group $\mathbb{Z}_2 \times \mathbb{Z}_2 \times \mathbb{Z}_2$ generated by elements a, b, c of order 2.

11.9 Does Theorem 11.22 generalize to all poset operators? Give a proof or counter-example.

11.10 Is it true that $\mathrm{T}(n) \times \underline{\underline{2}} \prec \mathrm{Pair}(n)$?

11.11 Let ρ and σ be pre-orders on a finite set M. For i in M, let $\mathcal{Q}_i$ be an association scheme on a set with more than one element. Let $\mathcal{Q}(\rho)$ and $\mathcal{Q}(\sigma)$ be the ρ-product and σ-product respectively of the $\mathcal{Q}_i$ for i in M. What relationship between ρ and σ ensures that $\mathcal{Q}(\rho) \preccurlyeq \mathcal{Q}(\sigma)$?

11.12 The labelling in Figure 11.11 uses notation cavalierly. While it is true that one of the association schemes shown is of type $\mathrm{H}(2, 4)$, $\mathrm{L}(2, 4)$ and $\mathrm{L}(3, 4)$, these are not labels for a single isomorphism class of association schemes. Explain what is going on here: in particular, find an association scheme which is of type $\mathrm{L}(3, 4)$ but not of type $\mathrm{L}(2, 4)$.

11.13 In the Hamming association scheme $\mathrm{H}(n, 2)$, let $\mathcal{C}_i$ be the class consisting of pairs of n-tuples which differ in i coordinates, for $0 \leqslant i \leqslant n$.

(a) Put $\mathcal{D}_0 = \mathcal{C}_0$ and $\mathcal{D}_i = \mathcal{C}_i \cup \mathcal{C}_{n+1-i}$ for $1 \leqslant i \leqslant n/2$. Prove that the $\mathcal{D}_i$ form an association scheme. (Hint: consider the folded cube of $\mathrm{H}(n + 1, 2)$.)

(When $n = 4$ this gives a strongly regular graph with 16 vertices and valency 5, which is sometimes called the *Clebsch* graph.)

(b) Now put $\mathcal{D}_0 = \mathcal{C}_0$, $\mathcal{D}_i = \mathcal{C}_{2i-1} \cup \mathcal{C}_{2i}$ for $1 \leqslant i \leqslant n/2$, and, if n is odd, $\mathcal{D}_m = \mathcal{C}_n$ for $m = (n + 1)/2$. Prove that the $\mathcal{D}_i$ form an association scheme. (Hint: consider the half cube of $\mathrm{H}(n+1, 2)$.)

(c) Find all association schemes $\mathcal{Q}$ such that $\mathrm{H}(5, 2) \preccurlyeq \mathcal{Q}$.

11.14 Let Θ consist of all 2-subsets of the set $\{1, 2, 3, 4, 5, 6\}$. Let $\mathcal{Q}_1$ be the triangular association scheme $\mathrm{T}(6)$ on Θ and let $\mathcal{Q}_2$ be the group-divisible association scheme on Θ whose groups are $\{1, 2\}, \{3, 4\}, \{5, 6\} \parallel \{1, 3\}, \{2, 5\}, \{4, 6\} \parallel \{1, 4\}, \{2, 6\}, \{3, 5\} \parallel \{1, 5\}, \{2, 4\}, \{3, 6\} \parallel \{1, 6\}, \{2, 3\}, \{4, 5\}$. Put $\mathcal{Q} = \mathcal{Q}_1 \wedge \mathcal{Q}_2$.

(a) Show that $\mathcal{Q}$ is an association scheme, and find the multiplication table for its adjacency matrices.

(b) Find the character table for $\mathcal{Q}$.

(c) Show that the following design for 15 treatments in the orthogonal block structure $(\underline{\underline{6}} \times \underline{\underline{5}})/\underline{\underline{3}}$ is partially balanced with respect to $\mathcal{Q}$. Find its table of canonical efficiency factors.

$\{1,2\}\ \{3,4\}\ \{5,6\}$	$\{1,3\}\ \{2,5\}\ \{4,6\}$	$\{1,4\}\ \{2,6\}\ \{3,5\}$	$\{1,5\}\ \{2,4\}\ \{3,6\}$	$\{1,6\}\ \{2,3\}\ \{4,5\}$
$\{1,2\}\ \{3,4\}\ \{5,6\}$	$\{1,6\}\ \{2,4\}\ \{3,5\}$	$\{1,5\}\ \{2,3\}\ \{4,6\}$	$\{1,3\}\ \{2,6\}\ \{4,5\}$	$\{1,4\}\ \{2,5\}\ \{3,6\}$
$\{1,6\}\ \{2,4\}\ \{3,5\}$	$\{1,3\}\ \{2,5\}\ \{4,6\}$	$\{1,2\}\ \{3,6\}\ \{4,5\}$	$\{1,4\}\ \{2,3\}\ \{5,6\}$	$\{1,5\}\ \{2,6\}\ \{3,4\}$
$\{1,5\}\ \{2,3\}\ \{4,6\}$	$\{1,2\}\ \{3,6\}\ \{4,5\}$	$\{1,4\}\ \{2,6\}\ \{3,5\}$	$\{1,6\}\ \{2,5\}\ \{3,4\}$	$\{1,3\}\ \{2,4\}\ \{5,6\}$
$\{1,3\}\ \{2,6\}\ \{4,5\}$	$\{1,4\}\ \{2,3\}\ \{5,6\}$	$\{1,6\}\ \{2,5\}\ \{3,4\}$	$\{1,5\}\ \{2,4\}\ \{3,6\}$	$\{1,2\}\ \{3,5\}\ \{4,6\}$
$\{1,4\}\ \{2,5\}\ \{3,6\}$	$\{1,5\}\ \{2,6\}\ \{3,4\}$	$\{1,3\}\ \{2,4\}\ \{5,6\}$	$\{1,2\}\ \{3,5\}\ \{4,6\}$	$\{1,6\}\ \{2,3\}\ \{4,5\}$

(d) Suppose that Π is a symmetric unipotent $2n \times 2n$ Latin square. Let $\mathcal{P}$ be the group-divisible scheme $\mathrm{GD}(2n-1, n)$ defined by the non-diagonal letters of Π. Show that the infimum of $\mathrm{T}(2n)$ and $\mathcal{P}$ is an association scheme.

(e) Investigate when an $m \times m$ Latin square gives a group-divisible scheme on $m(m-1)$ objects which is compatible with $\mathrm{Pair}(m)$.

12
Where next?

Here we give pointers to further work extending the topics in this book.

12.1 Few parameters

One reason for the initial introduction of partially balanced incomplete-block designs was that they typically have a small number of different concurrences. It is widely believed that optimal incomplete-block designs are to be found among those with the smallest range of concurrences. However, in spite of papers like [133, 136, 228], there seems to be little hope of developing any useful theory like that in Chapter 5 by simply imposing the condition that there are few concurrences.

Another reason was the small number of eigenvalues. The Bose–Mesner isomorphism allows us to convert the problem of inverting and diagonalizing $n \times n$ matrices (where $n = |\Omega|$) into the simpler problem of inverting and diagonalizing $(s+1) \times (s+1)$ matrices, where s is the number of non-diagonal associate classes. Therefore another strategy is to concentrate on block designs whose information matrices have $m+1$ distinct eigenvalues, for some small value of m: these designs are said to have *m-th order balance* [132], and are discussed in [70, Chapter 4]. However, having a small number of canonical efficiency factors does not imply a useful pattern in the variances of simple contrasts if the design is not partially balanced.

Second-order balance appears to be useful, because Theorem 5.5 shows that the variance of a simple contrast is a linear function of concurrence in a design with second-order balance. More generally, it is shown in [30] that if the design has m-th order balance then the variance of a simple contrast can be calculated from knowledge of paths of length at most $m-1$ in the treatment-concurrence graph.

Block designs which have second-order balance and one canonical efficiency factor equal to 1 have been studied in [69, 212]. Theorem 4.13 shows that these include the duals of balanced incomplete-block designs which are not symmetric as well as the duals of partially balanced incomplete-block designs with two associate classes and one canonical efficiency factor equal to 1. Most rectangular lattice designs have third-order balance. Further study of designs with second- or third-order balance would be worthwhile.

Two block designs with the same canonical efficiency factors, including multiplicities, are equally good under the A-criterion, but a different pattern of variances among simple contrasts may lead the experimenter to prefer one to the other.

Example 12.1 Compare the following two designs for twelve treatments replicated three times in six blocks of size six. Design ψ_1 has blocks

$$\begin{array}{lll} \{a,b,c,d,e,f\}, & \{g,h,i,j,k,l\}, & \{a,b,c,g,h,i\}, \\ \{d,e,f,j,k,l\}, & \{a,b,c,j,k,l\}, & \{d,e,f,g,h,i\}, \end{array}$$

while design ψ_2 has blocks

$$\begin{array}{lll} \{a,b,d,e,f,j\}, & \{c,g,h,i,k,l\}, & \{b,c,e,f,g,k\}, \\ \{a,d,h,i,j,l\}, & \{a,c,d,f,g,h\}, & \{b,e,i,j,k,l\}. \end{array}$$

The duals of these designs are group-divisible with the same concurrence matrix, so Theorem 4.13 shows that the designs have the same canonical efficiency factors. Design ψ_1 is an inflation of a balanced design for four treatments in blocks of size two, so both designs have second-order balance with canonical efficiency factors 1 and 2/3 (using Example 5.8). Theorem 5.5 shows that the variance of the estimator of $\tau(\theta) - \tau(\eta)$ is equal to $(15 - \Lambda(\theta, \eta))\sigma^2/18$. In design ψ_1 all concurrences are equal to 3 and 1, so the variances are $2\sigma^2/3$ and $7\sigma^2/9$. In design ψ_2 there are concurrences equal to 3, 2, 1 and 0, so variances of some simple contrasts are as high as $5\sigma^2/6$. Thus design ψ_1 would be preferred. ■

12.2 Unequal replication

In some experiments, not all treatments have the same status. Some may be standard treatments, often called *controls*, while others are new. In plant breeding, there may be insufficient seed from the new varieties to replicate them as much as the controls. The experimenter may be more interested in the simple contrasts between one new treatment and one control than in any contrast between treatments of the same type. Alternatively, the controls may be just for demonstration purposes, with contrasts among the new treatments being of most interest.

Suppose that there is a set Θ_1 of t_1 controls and a set Θ_2 of t_2 new treatments, and we seek a binary incomplete-block design in b blocks of size k. Such a design is said to have *supplemented balance* [191] if there are integers r_1, r_2, λ_{11}, λ_{22} and λ_{12} such that each control is replicated r_1 times, each new treatment is replicated r_2 times, and

$$\Lambda(\theta,\eta) = \begin{cases} \lambda_{11} & \text{if } \theta \text{ and } \eta \text{ are distinct controls} \\ \lambda_{22} & \text{if } \theta \text{ and } \eta \text{ are distinct new treatments} \\ \lambda_{12} & \text{if } \theta \in \Theta_1 \text{ and } \eta \in \Theta_2. \end{cases}$$

Counting arguments show that

$$\begin{aligned} t_1 r_1 + t_2 r_2 &= bk, \\ (t_1-1)\lambda_{11} + t_2\lambda_{12} &= r_1(k-1), \\ t_1\lambda_{12} + (t_2-1)\lambda_{22} &= r_2(k-1). \end{aligned}$$

In place of Equation (4.5), the information matrix L is given by

$$L = \mathrm{diag}(r) - k^{-1}\Lambda.$$

Partitioning the rows and columns of L into the two types of treatment, we can write

$$kL = \begin{array}{c} \\ \Theta_1 \\ \Theta_2 \end{array} \begin{array}{c} \begin{array}{cc} \quad\Theta_1 & \qquad\qquad\Theta_2 \end{array} \\ \left[\begin{array}{cc} r_1(k-1)I - \lambda_{11}(J-I) & -\lambda_{12}J \\ -\lambda_{12}J & r_2(k-1)I - \lambda_{22}(J-I) \end{array} \right] \end{array}.$$

Let x be a treatment contrast. If x takes the value 0 on all new treatments then

$$kLx = (r_1(k-1)+\lambda_{11})x = (t_1\lambda_{11} + t_2\lambda_{12})x.$$

Similarly, if x takes the value 0 on all controls then

$$kLx = (t_1\lambda_{12} + t_2\lambda_{22})x.$$

Finally, if $x = t_2\chi_{\Theta_1} - t_1\chi_{\Theta_2}$ then

$$\begin{aligned} kLx &= [(r_1(k-1) - \lambda_{11}(t_1-1))t_2 + t_1t_2\lambda_{12}]\chi_{\Theta_1} \\ &\qquad - [t_1t_2\lambda_{12} + (r_2(k-1) - \lambda_{22}(t_2-1))t_1]\chi_{\Theta_2} \\ &= (t_1+t_2)\lambda_{12}x. \end{aligned}$$

This gives us the eigenvectors and eigenvalues of kL. Canonical efficiency factors are of dubious use for unequally replicated designs, because it is not obvious what replications should be used in the comparator design. However, we obtain the projectors onto the eigenspaces as

$$P_1 = \begin{bmatrix} I - t_1^{-1}J & O \\ O & O \end{bmatrix}, \quad P_2 = \begin{bmatrix} O & O \\ O & I - t_2^{-1}J \end{bmatrix}$$

and

$$P_3 = \begin{bmatrix} \left(\frac{1}{t_1} - \frac{1}{t}\right) J & -\frac{1}{t}J \\ -\frac{1}{t}J & \left(\frac{1}{t_2} - \frac{1}{t}\right) J \end{bmatrix},$$

where $t = t_1 + t_2$. Hence the generalized inverse of L is

$$\frac{k}{t_1\lambda_{11} + t_2\lambda_{12}} P_1 + \frac{k}{t_1\lambda_{12} + t_2\lambda_{22}} P_2 + \frac{k}{t\lambda_{12}} P_3.$$

Equation (4.4) shows that the variance of the estimator of $\tau(\theta) - \tau(\eta)$ is

$$\begin{cases} \dfrac{2k\sigma^2}{t_1\lambda_{11} + t_2\lambda_{12}} & \text{if } \theta \text{ and } \eta \text{ are in } \Theta_1; \\ \dfrac{2k\sigma^2}{t_1\lambda_{12} + t_2\lambda_{22}} & \text{if } \theta \text{ and } \eta \text{ are in } \Theta_2; \\ \dfrac{k\sigma^2(t_1-1)}{(t_1\lambda_{11} + t_2\lambda_{12})t_1} + \dfrac{k\sigma^2(t_2-1)}{(t_1\lambda_{12} + t_2\lambda_{22})t_2} + \dfrac{k\sigma^2}{t_1t_2\lambda_{12}} & \text{otherwise.} \end{cases}$$

Example 12.2 In an experiment on the possible side-effects of new medicines, twelve healthy volunteers tried four medicines each. There were six medicines in the trial; three of these (a, b, c) were well developed and promising, while the other three (d, e, f) were new ideas and of less immediate interest to the drug company. The design consisted of four copies of each of the following blocks.

$$\{a, b, c, d\} \quad \{a, b, c, e\} \quad \{a, b, c, f\}$$

Thus it had supplemented balance with $b = 12$, $k = 4$, $t_1 = t_2 = 3$, $r_1 = 12$, $r_2 = 4$, $\lambda_{11} = 12$, $\lambda_{12} = 4$ and $\lambda_{22} = 0$. The three different variances of simple contrasts were $\sigma^2/6$, $2\sigma^2/3$ and $7\sigma^2/18$ respectively. ■

For given sets Θ_1 and Θ_2, the information matrices of all designs with supplemented balance have the same eigenvectors and therefore commute. Thus two such designs may be juxtaposed without changing the eigenvectors. Moreover, in a multi-stratum design, supplemented balance for each system of blocks implies supplemented balance overall;

if random effects are assumed then information from different strata can be combined because the information matrices commute.

However, if we try to generalize supplemented balance to supplemented partial balance, we run into difficulties.

Example 12.3 Alternatives to the design in Example 12.2 are ψ_1, with blocks

$$\{a,b,c,d\},\ \{a,b,c,d\},\ \{a,b,c,e\},\ \{a,b,c,e\},\ \{a,b,c,f\},\ \{a,b,c,f\},$$
$$\{b,c,d,f\},\ \{b,c,d,e\},\ \{a,c,e,f\},\ \{a,c,d,e\},\ \{a,b,e,f\},\ \{a,b,d,f\},$$

and ψ_2, with blocks being two copies of

$$\{a,b,c,d\},\ \{a,b,c,e\},\ \{a,b,c,f\},\ \{b,c,e,f\},\ \{a,c,d,f\},\ \{a,b,d,e\}.$$

The corresponding information matrices L_1 and L_2 satisfy

$$4L_1 = \begin{bmatrix} 30 & -8 & -8 & -4 & -5 & -5 \\ -8 & 30 & -8 & -5 & -4 & -5 \\ -8 & -8 & 30 & -5 & -5 & -4 \\ -4 & -5 & -5 & 18 & -2 & -2 \\ -5 & -4 & -5 & -2 & 18 & -2 \\ -5 & -5 & -4 & -2 & -2 & 18 \end{bmatrix}$$

and

$$4L_2 = \begin{bmatrix} 30 & -8 & -8 & -6 & -4 & -4 \\ -8 & 30 & -8 & -4 & -6 & -4 \\ -8 & -8 & 30 & -4 & -4 & -6 \\ -6 & -4 & -4 & 18 & -2 & -2 \\ -4 & -6 & -4 & -2 & 18 & -2 \\ -4 & -4 & -6 & -2 & -2 & 18 \end{bmatrix}.$$

Now, L_1 and L_2 do not commute with each other, so they cannot have the same eigenvectors. However,

$$L_1^- = \frac{1}{42 \times 759} \begin{bmatrix} 2999 & -361 & -361 & -871 & -703 & -703 \\ -361 & 2999 & -361 & -703 & -871 & -703 \\ -361 & -361 & 2999 & -703 & -703 & -871 \\ -871 & -703 & -703 & 5015 & -1369 & -1369 \\ -703 & -871 & -703 & -1369 & 5015 & -1369 \\ -703 & -703 & -871 & -1369 & -1369 & 5015 \end{bmatrix}$$

(simply verify that $L_1 L_1^- = L_1^- L_1 = I - (1/6)J$), which has the same pattern of entries as L_1. Writing the six treatments in the rectangular array

a	b	c
d	e	f

,

we see that concurrences and variances in ψ_1 are as follows.

θ, η	$\Lambda(\theta,\eta)$	$\mathrm{Var}(\hat{\tau}(\theta)-\hat{\tau}(\eta))$
both in top row	8	$160\sigma^2/759$
both in bottom row	2	$304\sigma^2/759$
in same column	4	$1626\sigma^2/(7\times 759)$
other	5	$1570\sigma^2/(7\times 759)$

Thus ψ_1 has some features of partial balance. So does ψ_2, which exhibits a similar pattern of concurrences and variances. ■

12.3 Generalizations of association schemes

Given a partition $\mathcal{Q}$ of $\Omega\times\Omega$, let $\mathcal{A}$ be the set $V_\mathcal{Q}$ of real linear combinations of the adjacency matrices of $\mathcal{Q}$. In Chapter 1 we defined $\mathcal{Q}$ to be an association scheme if it satisfies the following three conditions:

(A) $\mathrm{Diag}(\Omega)$ is a class of $\mathcal{Q}$;
(B) if C is a class of $\mathcal{Q}$ then $C = C'$;
(C) $\mathcal{A}$ is an *algebra* in the sense that if A and B are in $\mathcal{A}$ then so is AB.

We could weaken each condition as follows:

(a) if C is a class of $\mathcal{Q}$ and $C\cap\mathrm{Diag}(\Omega)\neq\varnothing$ then $C\subseteq\mathrm{Diag}(\Omega)$;
(b) if C is a class of $\mathcal{Q}$ then so is C';
(c) $\mathcal{A}$ is a *Jordan algebra* in the sense that if A and B are in $\mathcal{A}$ then so is $AB+BA$.

Definition A partition $\mathcal{Q}$ of $\Omega\times\Omega$ is

reflective	if it satisfies (a) and (b);
homogenous	if it satisfies (A);
symmetric	if it satisfies (B);
a *coherent configuration*	if it is reflective and satisfies (C);
regular	if each class C_i has a valency a_i such that $\lvert C_i(\omega)\rvert = a_i$ for all ω in Ω;
commutative	if $AB = BA$ for all A and B in $\mathcal{A}$.

We shall consider only reflective partitions. If $\mathcal{Q}$ is reflective, then its *symmetrization* $\mathcal{Q}^{\mathrm{sym}}$ is the partition whose classes are $\mathcal{C}\cup\mathcal{C}'$ for classes $\mathcal{C}$ of $\mathcal{Q}$. The next proposition formalizes and extends the argument on page 8 which showed that condition (v)′ for the second definition of an association scheme is redundant.

Proposition 12.1 *If the partition $\mathcal{Q}$ is reflective then commutativity implies regularity and regularity implies homogeneity.*

Proof The matrix J_Ω is always in $\mathcal{A}$. For each adjacency matrix A_i of $\mathcal{Q}$, and for all α, β in Ω,

$$(A_iJ_\Omega)(\alpha,\beta) = |\mathcal{C}_i(\alpha)| \quad \text{and} \quad (J_\Omega A_i)(\alpha,\beta) = |\mathcal{C}_i'(\beta)|\,.$$

Commutativity therefore implies that $|\mathcal{C}_i(\alpha)| = |\mathcal{C}_i'(\beta)|$ for all α and β, which implies that $|\mathcal{C}_i(\omega)|$ is independent of ω.

Suppose that $\mathcal{Q}$ is regular and $\mathcal{C}$ contains (α,α). Then $|\mathcal{C}(\alpha)| > 0$, so $|\mathcal{C}(\beta)| > 0$ for all β in Ω. By condition (a), $\mathcal{C} = \mathrm{Diag}(\Omega)$. ■

Proposition 12.2 *(i) If $\mathcal{A}$ is an algebra then it is a Jordan algebra.*

(ii) If $\mathcal{Q}$ is commutative and $\mathcal{A}$ is a Jordan algebra then $\mathcal{A}$ is an algebra.

Proposition 12.3 *(i) If $\mathcal{Q}$ is symmetric and $\mathcal{A}$ is an algebra then $\mathcal{Q}$ is commutative.*

(ii) If $\mathcal{Q}$ is symmetric and homogeneous and $\mathcal{A}$ is a Jordan algebra then $\mathcal{Q}$ is regular.

(iii) If $\mathcal{Q}$ is homogeneous and $\mathcal{A}$ is an algebra then $\mathcal{Q}$ is regular.

Proof (i) This is the same as the proof of Lemma 1.3.

(ii) Let A_i be an adjacency matrix of $\mathcal{Q}$. If $\mathcal{A}$ is a Jordan algebra then it contains $J_\Omega A_i + A_iJ_\Omega$, whose (ω,ω) entry is equal to $|\mathcal{C}_i(\omega)| + |\mathcal{C}_i'(\omega)|$. If $\mathcal{Q}$ is symmetric then $\mathcal{C}_i = \mathcal{C}_i'$ and so this diagonal entry is equal to $2\,|\mathcal{C}_i(\omega)|$. If $\mathcal{Q}$ is homogeneous then all these diagonal entries must be the same.

(iii) Similar to (ii). ■

Proposition 12.4 *If $\mathcal{Q}$ is a coherent configuration then $\mathcal{Q}^{\mathrm{sym}}$ satisfies condition (c).*

Proof The set of linear combinations of adjacency matrices of $\mathcal{Q}^{\mathrm{sym}}$ is just $\{A \in \mathcal{A} : A \text{ is symmetric}\}$. Let A and B be symmetric matrices in $\mathcal{A}$. Since $\mathcal{Q}$ is a coherent configuration, $\mathcal{A}$ contains $AB + BA$. Now, $(AB+BA)' = B'A' + A'B' = BA + AB$, so $AB+BA$ is symmetric. ■

	homogeneous			inhomogeneous	
	regular		irregular		
	commutative	not commutative			
algebra	A C				symmetric
algebra	C directed circuit	C $\mathcal{Q}(E)$ for S_3		C $\mathcal{S}$ $\mathcal{V}$	not symmetric
Jordan algebra but not algebra		$\mathcal{Q}$(Inverse) for S_3		$\mathcal{S}^{\mathrm{sym}}$ $\mathcal{U}^{\mathrm{sym}}$	symmetric
Jordan algebra but not algebra		$\mathcal{W}$	$\mathcal{T}$	$\mathcal{U}$	not symmetric

$\mathcal{S}$ the coherent configuration for supplemented balance, with classes $\mathrm{Diag}(\Theta_1)$, $\mathrm{Diag}(\Theta_2)$, $(\Theta_1 \times \Theta_1) \setminus \mathrm{Diag}(\Theta_1)$, $(\Theta_2 \times \Theta_2) \setminus \mathrm{Diag}(\Theta_2)$, $\Theta_1 \times \Theta_2$ and $\Theta_2 \times \Theta_1$

$\mathcal{T}$ obtained from $\mathcal{S}$ by merging the two diagonal classes and by merging the two other symmetric classes

$\mathcal{U}$ see Example 12.4

$\mathcal{V}$ see Example 12.4

$\mathcal{W}$ $\mathcal{Q}(\Delta)$ in the quaternion group $\langle a, b : a^4 = 1,\ b^2 = a^2,\ b^{-1}ab = a^3 \rangle$, where Δ is the partition with one class $\{a, a^3\}$ and all the rest singletons

Fig. 12.1. Possible combinations of conditions

Propositions 12.1–12.3 show that the only possibilities for a partition $\mathcal{Q}$ satisfying (a)–(c) are those shown in Figure 12.1 in squares which are not crossed out. The square marked A denotes association schemes; those marked C coherent configurations. Some examples of partitions satisfying the given conditions are shown in the relevant squares.

Condition (C) ensures that $\mathcal{A}$ contains the powers of every matrix in it. Condition (a) ensures that $I \in \mathcal{A}$, and hence the argument at the start of Section 2.2 shows that $\mathcal{A}$ contains the Moore–Penrose generalized inverse of every diagonalizable matrix which it contains. Hence we retain Theorem 5.2 so long as $\mathcal{Q}$ is a coherent configuration: that is, if the concurrence of θ and η depends only on the class of $\mathcal{Q}$ containing (θ, η) then so does the variance of the estimator of $\tau(\theta) - \tau(\eta)$.

So coherent configurations give a worthwhile generalization of association schemes. However, it is not useful if there are too many classes: the partition $E_{\Omega\times\Omega}$ into singleton ordered pairs is a coherent configuration and every incomplete-block design is 'partially balanced' with respect to it. A reasonable compromise is to generalize association schemes to homogeneous coherent configurations. Proposition 12.3(iii) shows that homogeneous coherent configurations are regular, so there are at most n classes, where $n = |\Omega|$. Nair [178] recommended generalizing association schemes to homogeneous coherent configurations; Sinha [228] suggested using the special case defined by transitive permutation groups.

Coherent configurations have become an important topic in their own right. See Section 13.2.1 for references.

If all we require is that $\mathcal{A}$ contain the generalized inverse of the information matrix, then we can weaken condition (C) to condition (c), as the following proposition shows.

Proposition 12.5 *Suppose that $I \in \mathcal{A}$. Then conditions (i) and (ii) below are equivalent. If all matrices in $\mathcal{A}$ are symmetric then these conditions are also equivalent to condition (iii).*

(i) $\mathcal{A}$ is a Jordan algebra.
(ii) If $A \in \mathcal{A}$ then $A^m \in \mathcal{A}$ for all non-negative integers m.
(iii) If $A \in \mathcal{A}$ then $A^- \in \mathcal{A}$.

Proof First we suppose only that $I \in \mathcal{A}$.

(i)⇒(ii) Suppose that $A \in \mathcal{A}$. By assumption, A^0 and A^1 are in $\mathcal{A}$. Assume for induction that $A^m \in \mathcal{A}$. Then $\mathcal{A}$ contains the matrix $(AA^m + A^m A)/2$, which is A^{m+1}.

(ii)⇒(i) Suppose that A and B are in $\mathcal{A}$. Then $\mathcal{A}$ contains A^2, B^2 and $(A+B)^2$. However, $(A+B)^2 = A^2 + AB + BA + B^2$, so $\mathcal{A}$ contains $AB + BA$.

Now we assume that every matrix in $\mathcal{A}$ is symmetric and hence diagonalizable.

(ii)⇒(iii) Suppose that $A \in \mathcal{A}$. Then every eigenprojector of A is a polynomial in A, and hence is in $\mathcal{A}$. Now, A^- is a linear combination of these eigenprojectors, so it is in $\mathcal{A}$.

(iii)⇒(ii) Suppose that $A \in \mathcal{A}$. Let the distinct eigenvalues of A be λ_1, ..., λ_r, with eigenprojectors $P_1, \ldots, P_r$. We shall use induction on r to show that $P_i \in \mathcal{A}$ for $1 \leqslant i \leqslant r$. If $r = 1$ then $A = \lambda_1 P_1$ so $P_1 \in \mathcal{A}$ and the hypothesis is true. Now suppose that $r \geqslant 2$. Put $B_1 = A - \lambda_1 I$ and $B_2 = A - \lambda_2 I$. Then $B_1^- = (\lambda_2 - \lambda_1)^{-1} P_2 + \sum_{i=3}^r (\lambda_i - \lambda_1)^{-1} P_i$ and $B_2^- = (\lambda_1 - \lambda_2)^{-1} P_1 + \sum_{i=3}^r (\lambda_i - \lambda_2)^{-1} P_i$. Now, B_1^- and B_2^- are both in $\mathcal{A}$, so $\mathcal{A}$ contains $B_1^- - B_2^-$, which has $r - 1$ eigenprojectors $P_1 + P_2$, P_3, P_4, ..., P_r. By inductive hypothesis, $\mathcal{A}$ contains $P_1 + P_2$ and P_i for $3 \leqslant i \leqslant r$. Since $A = \sum_{i=1}^r \lambda_i P_i$ and $A \in \mathcal{A}$ and $\lambda_1 \neq \lambda_2$, $\mathcal{A}$ also contains P_1 and P_2.

For non-negative integers m, $A^m = \sum_{i=1}^r \lambda_i^m P_i$, which is in $\mathcal{A}$. ∎

Shah [219] proposed generalizing association schemes to homogeneous symmetric partitions for which $\mathcal{A}$ is Jordan algebra. Theorem 5.2 is retained. Proposition 12.4 shows that if $\mathcal{Q}$ is a homogeneous coherent configuration then $\mathcal{Q}^{\mathrm{sym}}$ satisfies Shah's conditions. All the examples of Shah schemes in the literature seem to be of this type. It is easier to calculate in an algebra than in a Jordan algebra, so there seems to be no point in using Shah's generalization unless there are some Shah schemes that are not symmetrizations of homogeneous coherent configurations.

Neither the Nair schemes (homogeneous coherent configurations) nor the Shah schemes are necessarily commutative. However, much of this book has used commutativity in an essential way. What do we lose without it?

For partially balanced incomplete-block designs, we lose the fact that basic contrasts are independent of the design. Thus Techniques 5.2 and 5.3 are not available. Each information matrix must be inverted individually. This is an inconvenience.

For building new block designs, we lose the fact that juxtaposition preserves basic contrasts. Thus we lose the simple method, given in Equation (4.9), of calculating canonical efficiency factors for the new design. This is somewhat more than an inconvenience: we can no longer create good juxtaposed designs purposefully by matching high canonical efficiency factors in one component design to low canonical efficiency factors in another.

For designs on structured sets, the loss is serious. The transition from Equation (7.6) to Equation (7.7) is impossible if the partition $\mathcal{P}$ of $\Omega \times \Omega$ is not commutative. Indeed, we cannot define strata in $\mathbb{R}^{\Omega}$ at all unless $\mathcal{P}$ is commutative. Moreover, the method of combining information, given in Section 7.4, requires that all of the information matrices have the same eigenvectors; in other words, that they commute with each other. Thus the partition $\mathcal{Q}$ of $\Theta \times \Theta$ must also be commutative.

If $\mathcal{P}$ is not commutative then treatment effects must be estimated by an alternative method, which uses the inverse of $\mathrm{Cov}(Y)$. So long as $\mathcal{P}$ is reflective and satisfies (c), the inverse of $\mathrm{Cov}(Y)$ has the form $\sum_F \delta_F B_F$; now the coefficients δ_F have to be estimated before treatment effects can be estimated. Jordan algebras were originally introduced into statistics to help the study of patterned covariance matrices, not patterned concurrence matrices: see [168].

To retain the theory and methods of Chapter 7, we need both $\mathcal{P}$ and $\mathcal{Q}$ to be commutative. Figure 12.1 shows that they must be commutative coherent configurations. Concurrence matrices and covariance matrices are both symmetric, so there is no loss in working inside $\mathcal{P}^{\mathrm{sym}}$ and $\mathcal{Q}^{\mathrm{sym}}$, which are association schemes, by Propositions 12.4 and 12.2. Thus association schemes really are the right level of generality for this theory.

If $\mathcal{Q}$ is a coherent configuration, it is possible for $\mathcal{Q}^{\mathrm{sym}}$ to be commutative even if $\mathcal{Q}$ is not. Since we are concerned primarily with symmetric matrices, we define $\mathcal{Q}$ to be *stratifiable* [21] if $\mathcal{Q}^{\mathrm{sym}}$ is commutative: if $\mathcal{Q}$ is stratifiable then there are strata (common eigenspaces) for the symmetric matrices in $\mathcal{Q}$. For example, in the quaternion group Q_8, $\mathcal{Q}(E)$ is stratifiable but not commutative.

The commutativity shown by supplemented balance in Section 12.2 seems to contradict Proposition 12.1. This paradox is explained by the following result. All of the information matrices (except the one for the U stratum) have zero row-sums, and we want them to commute with each other. Each such matrix L commutes with J, because $LJ = JL = O$. Every symmetric matrix with constant row-sums is a linear combination of J and a symmetric matrix with zero row-sums. So we want all the symmetric matrices in $\mathcal{A}$ with constant row-sums to commute with each other.

Theorem 12.6 *Let $\mathcal{Q}$ be a reflective inhomogeneous partition of $\Omega \times \Omega$ which satisfies (c). Let $\mathcal{B}$ be the set of symmetric matrices in $\mathcal{A}$ with constant row-sums. Then $\mathcal{B}$ is commutative if and only if (i) there are* *two diagonal classes* $\mathrm{Diag}(\Omega_1)$ *and* $\mathrm{Diag}(\Omega_2)$, *(ii)* $(\Omega_1 \times \Omega_2) \cup (\Omega_2 \times \Omega_1)$ *is a class of* $\mathcal{Q}^{\mathrm{sym}}$, *and (iii) the remaining classes of* $\mathcal{Q}$ *give stratifiable partitions of* $\Omega_1 \times \Omega_1$ *and* $\Omega_2 \times \Omega_2$ *separately.*

Proof Suppose that the diagonal classes are $\mathrm{Diag}(\Omega_i)$ for $1 \leqslant i \leqslant m$. Write I_i and J_{ij} for I_{Ω_i} and J_{Ω_i,Ω_j} for $1 \leqslant i,j \leqslant m$. Condition (c) shows that $\mathcal{A}$ contains $I_iJ + JI_i$, which is equal to $\sum_j (J_{ij} + J_{ji})$. Since $\mathcal{A}$ is closed under Hadamard product, each class of $\mathcal{Q}$ is contained in $(\Omega_i \times \Omega_j) \cup (\Omega_j \times \Omega_i)$ for some i and j.

Let $|\Omega_i| = n_i$ for $1 \leqslant i \leqslant m$. If $m \geqslant 3$ then there are scalars λ and μ such that $\mathcal{B}$ contains B_1 and B_2, where

$$B_1 = \begin{bmatrix} O & J_{12} & O \\ J_{21} & \lambda J_{22} & J_{23} \\ O & J_{32} & O \end{bmatrix} \quad \text{and} \quad B_2 = \begin{bmatrix} O & O & J_{13} \\ O & O & J_{23} \\ J_{31} & J_{32} & \mu J_{33} \end{bmatrix}.$$

Then

$$B_1B_2 = \begin{bmatrix} O & O & n_2J_{13} \\ n_3J_{21} & n_3J_{22} & (n_1 + \lambda n_2 + \mu n_3)J_{23} \\ O & O & n_2J_{33} \end{bmatrix}.$$

Since B_1B_2 is not symmetric, $\mathcal{B}$ is not commutative.

Now suppose that $m = 2$ and that the class $\mathcal{C}$ of $\mathcal{Q}$ is contained in $(\Omega_1 \times \Omega_2) \cup (\Omega_2 \times \Omega_1)$. Let A be the adjacency matrix of $(\mathcal{C} \cup \mathcal{C}') \cap (\Omega_1 \times \Omega_2)$. The proof of Proposition 12.3(ii) shows that the non-zero rows of A have constant sum, say a_1, and the non-zero columns have constant sum, say a_2, where $n_1a_1 = n_2a_2$. Then $\mathcal{B}$ contains C_1 and C_2, where

$$C_1 = \begin{bmatrix} n_1I_1 & J_{12} \\ J_{21} & J_{22} \end{bmatrix} \quad \text{and} \quad C_2 = \begin{bmatrix} (a_2 - a_1)I_1 & A \\ A' & O \end{bmatrix}.$$

Now,

$$C_1C_2 = \begin{bmatrix} n_1(a_2 - a_1)I_1 + a_1J_{11} & n_1A \\ a_2J_{21} & a_2J_{22} \end{bmatrix}$$

while

$$C_2C_1 = \begin{bmatrix} n_1(a_2 - a_1)I_1 + a_1J_{11} & a_2J_{12} \\ n_1A' & a_2J_{22} \end{bmatrix}$$

so if $\mathcal{B}$ is commutative then $n_1A = a_2J_{12}$ so $A = J_{12}$. Therefore $\mathcal{C} \cup \mathcal{C}' = (\Omega_1 \times \Omega_2) \cup (\Omega_2 \times \Omega_1)$.

Finally, if $\mathcal{B}$ is commutative then the classes contained in each $\Omega_i \times \Omega_i$ must form a stratifiable partition of $\Omega_i \times \Omega_i$. Thus if $\mathcal{B}$ is commutative then $\mathcal{Q}$ satisfies (i), (ii) and (iii).

Conversely, if $\mathcal{Q}$ satisfies (i)–(iii) then it induces stratifiable partitions $\mathcal{Q}_i$ of Ω_i for $i = 1, 2$, and typical matrices D_1 and D_2 in $\mathcal{B}$ have the form

$$D_1 = \begin{bmatrix} A_1 & \lambda J_{12} \\ \lambda J_{21} & A_2 \end{bmatrix} \quad \text{and} \quad D_2 = \begin{bmatrix} B_1 & \mu J_{12} \\ \mu J_{21} & B_2 \end{bmatrix},$$

where, for $i = 1, 2$, A_i and B_i are in the Bose–Mesner algebra of $\mathcal{Q}_i^{\mathrm{sym}}$ with row sums a_i and b_i; moreover, $a_1 + \lambda n_2 = \lambda n_1 + a_2$ and $b_1 + \mu n_2 = \mu n_1 + b_2$. Then

$$D_1 D_2 = \begin{bmatrix} A_1 B_1 + \lambda\mu n_2 J_{11} & (\mu a_1 + \lambda b_2) J_{12} \\ (\lambda b_1 + \mu a_2) J_{21} & \lambda\mu n_1 J_{22} + A_2 B_2 \end{bmatrix}.$$

Now, A_1B_1 and A_2B_2 are symmetric because $\mathcal{Q}_1^{\mathrm{sym}}$ and $\mathcal{Q}_2^{\mathrm{sym}}$ are association schemes. Also

$$\mu a_1 + \lambda b_2 = \mu a_2 + \mu\lambda(n_1 - n_2) + \lambda b_1 + \mu\lambda(n_2 - n_1) = \lambda b_1 + \mu a_2,$$

so D_1D_2 is symmetric and therefore $D_1D_2 = D_2D_1$. Hence $\mathcal{B}$ is commutative. ■

We can extend all of the theory in Chapters 4–7 to partitions which either are stratifiable or satisfy conditions (i)–(iii) of Theorem 12.6. This is a little more general than association schemes and supplemented balance. For example, it covers the design in Exercise 4.1(d).

Example 12.4 Let $\Theta_1 = \{a, b\}$ with the trivial association scheme and let $\Theta_2 = \{c, d, e, f\}$ with the group-divisible association scheme with groups $c, d \parallel e, f$. Put $\Theta = \Theta_1 \cup \Theta_2$ and let $\mathcal{U}$ be the partition of $\Theta_1 \times \Theta_2$ defined by these two association schemes together with the adjacency matrices A and A', where

$$A = \begin{array}{c} \\ a \\ b \\ c \\ d \\ e \\ f \end{array} \begin{array}{c} \begin{array}{cccccc} a & b & c & d & e & f \end{array} \\ \begin{bmatrix} 0 & 0 & 1 & 1 & 0 & 0 \\ 0 & 0 & 0 & 0 & 1 & 1 \\ 0 & 1 & 0 & 0 & 0 & 0 \\ 0 & 1 & 0 & 0 & 0 & 0 \\ 1 & 0 & 0 & 0 & 0 & 0 \\ 1 & 0 & 0 & 0 & 0 & 0 \end{bmatrix} \end{array}.$$

A little checking shows that $\mathcal{A}$ is a Jordan algebra. Thus the set $\mathcal{B}$ in Theorem 12.6 is commutative even though $\mathcal{U}$ satisfies none of conditions (A)–(C).

For applications in designed experiments we may as well either replace

$\mathcal{U}$ by $\mathcal{U}^{\mathrm{sym}}$, which satisfies (B) but not (C), or repartition the classes represented by A and A' into $\Theta_1 \times \Theta_2$ and $\Theta_2 \times \Theta_1$, which gives a coherent configuration $\mathcal{V}$. ■

In view of Theorem 12.6, the following definition seems useful.

Definition Let ψ be a block design for treatment set Θ with concurrence matrix Λ. Suppose that Θ is partitioned into subsets Θ_1 and Θ_2 with association schemes $\mathcal{Q}_1$ and $\mathcal{Q}_2$ whose associate classes are indexed by $\mathcal{K}_1$ and $\mathcal{K}_2$ respectively. Then ψ has *supplemented partial balance* with respect to $\mathcal{Q}_1$ and $\mathcal{Q}_2$ if there are integers λ_i, for i in $\mathcal{K}_1 \cup \mathcal{K}_2$, and λ such that

$$\Lambda(\theta,\eta) = \begin{cases} \lambda_i & \text{if } (\theta,\eta) \in \mathcal{C}_i, \text{ for } i \text{ in } \mathcal{K}_1 \cup \mathcal{K}_2, \\ \lambda & \text{if } (\theta,\eta) \in (\Theta_1 \times \Theta_2) \cup (\Theta_2 \times \Theta_1). \end{cases}$$

Such a design has equal replication r_1 for treatments in Θ_1 and equal replication r_2 for treatments in Θ_2. If there are b blocks of size k, and if there are t_1, t_2 treatments in Θ_1, Θ_2 respectively, then

$$\begin{aligned} t_1 r_1 + t_2 r_2 &= bk, \\ \sum_{i\in\mathcal{K}_1} a_i\lambda_i + t_2\lambda &= r_1 k \\ t_1\lambda + \sum_{i\in\mathcal{K}_2} a_i\lambda_i &= r_2 k. \end{aligned}$$

A straightforward way of constructing a block design with supplemented partial balance is to take a partially balanced design on the new treatments and adjoin every control treatment to every block. When there are relatively few control treatments, with no limit on their replication, this approach is sensible.

Example 12.4 revisited Suppose that a and b are controls while c, d, e and f are new treatments with replication limited to four, and that we seek a design in eight blocks of size four. The following set of blocks achieves supplemented partial balance.

$$\begin{array}{llll} \{a,b,c,d\}, & \{a,b,c,d\}, & \{a,b,c,e\}, & \{a,b,c,f\}, \\ \{a,b,d,e\}, & \{a,b,d,f\}, & \{a,b,e,f\}, & \{a,b,e,f\} \end{array}$$ ■

However, sometimes an alternative design with supplemented partial balance is more efficient for the contrasts between new treatments and controls. There is scope for further work here.

13
History and references

13.1 Statistics

13.1.1 Basics of experimental design

The statistical theory of design for experiments was developed initially by R. A. Fisher during his time at Rothamsted Experimental Station (1919–1933). This theory was published in two famous books: *Statistical Methods for Research Workers* and *The Design of Experiments* [96, 97]. These books focussed on three main designs: (i) trivial structure on both the experimental units and the treatments; (ii) complete-block designs; (iii) Latin squares. He also introduced factorial experiments, that is, experiments where the treatments have the structure $\underline{\underline{n_1}} \times \underline{\underline{n_2}} \times \cdots \times \underline{\underline{n_m}}$.

F. Yates joined Fisher at Rothamsted in 1930, and remained there until his retirement in 1965. Through annual involvement with real agricultural experiments, he extended Fisher's ideas in three important ways: factorial experiments in blocks; 'complex experiments', that is, experiments where the structure on the experimental units is more complicated than a single system of blocks; and designs in incomplete blocks. The long works [252, 255] contain many worked examples of all three, which feed independently into the theory of association schemes in this book.

13.1.2 Factorial designs

In a factorial experiment there may be more treatments than can fit into each block. Fisher and Yates developed the technique of *confounding*, described in Section 8.6, to deal with this. Examples are in [252, 255]. In the 1940s, Fisher systematized the method in [99, 100] in terms of elementary Abelian groups and their characters. Almost immediately,

R. C. Bose [53] and O. Kempthorne [146] rephrased this method in terms of affine geometries, and the group theoretic approach was lost until J. A. John and A. M. Dean [88, 135] resurrected it in 1975 for Abelian groups other than elementary Abelian. This was then related to the group characters in the series of papers by R. A. Bailey and colleagues [15, 16, 18, 20, 33], which also implicitly introduced the association scheme defined by the blueprint Subgroup. This association scheme was discovered independently by P. Delsarte [89] and W. G. Bridges and R. A. Mena [62].

An alternative approach to factorial designs was given by R. L. Plackett and J. Burman [195] in 1946. This led C. R. Rao to introduce orthogonal arrays (Section 6.2) in [203, 204]. A comprehensive recent reference on orthogonal arrays is [115].

A different class of factorial designs, those with *factorial balance*, was introduced by Yates [252] in 1935. The same concept was given by Bose [53] and K. R. Nair and C. R. Rao [181] in the late 1940s and by B. V. Shah [218, 220] a decade later. In 1966, A. M. Kshirsagar [154] showed that factorial balance is partial balance with respect to a factorial association scheme $\underline{\underline{n_1}} \times \cdots \times \underline{\underline{n_m}}$. Much work on this association scheme and its Bose–Mesner algebra has been done outside the general framework of association schemes: for example, see [109, 155, 156].

13.1.3 Incomplete-block designs

More obviously related to the invention of association schemes was the work by Yates on incomplete-block designs, which he introduced in 1936 [253, 254]. These included balanced incomplete-block designs, which he originally called *symmetrical* incomplete-block designs. Fisher proved his famous inequality in [98]. Some cyclic balanced incomplete-block designs were given in [101]. Yates [254] and Bose [51] independently showed the link between complete sets of mutually orthogonal Latin squares and the balanced designs which geometers were already calling projective planes.

Of course, balanced incomplete-block designs have an aesthetic appeal quite apart from designed experiments. They have been attributed to Kirkman [150] and Steiner [232], amongst others. R. D. Carmichael [77] called them *complete configurations*, having used the term *tactical configuration* for a binary equi-replicate incomplete-block design, following E. H. Moore [173]. (See H. Gropp [108] for configurations in the sense that no concurrence is greater than one.) In 1962 D. R. Hughes [128] renamed them 2-designs on the suggestion of his colleague D. G. Higman[*] because every 2-subset of treatments is contained in the same number of blocks: more generally, a t-design is a collection of blocks of the same size in which every t-subset of the treatments is contained in the same number of blocks. See [46] for a recent overview of the considerable industry in this area.

Yates always told colleagues that balanced incomplete-block designs were an obvious idea that he had invented while in the bath, and there is no suggestion that he did not believe this.[†] However, while at school in Clifton he was awarded a Cay scholarship prize, with which he bought a copy of [209], whose Chapter 10 concerns Kirkman's school-girl problem. There are no marks on the book to indicate whether he read that chapter or used its contents in later life, but it seems likely that he had absorbed the idea of balanced incomplete-block designs from this book.[‡]

Yates also introduced such non-balanced incomplete-block designs as square lattices, cubic lattices and pseudo-factorial designs. In fact, very many of his incomplete-block designs in [252, 253, 254, 255] are partially balanced, often with respect to a factorial association scheme. The design in Example 4.3 is intentionally partially balanced with respect to the association scheme $\underline{\underline{2}} \times \underline{\underline{2}} \times \underline{\underline{2}}$. In fact it is also partially balanced with respect to the coarser association scheme $\mathrm{H}(3,2)$. In presenting this example, I have taken advantage of the isomorphism between $\mathrm{H}(3,2)$ and $\mathrm{R}(2,4)$.

Why then did not Yates take the extra step to inventing association schemes? Apparently, he hated matrices. While he was head of the Statistics Department at Rothamsted (1933–1965) he forbade his staff to use matrices in their writing.[§] The first five chapters of this book show how essential matrices are for the development and understanding of association schemes.

However, Yates did understand very well the importance of variance and efficiency. He introduced the efficiency factor for a balanced incomplete-block design in [254]. The extension to general incomplete-block designs was developed in 1956 by Kempthorne [147], who proved Theorem 4.12. Theorem 4.13 was given by S. C. Pearce [193] in 1968: see also [189]. Canonical efficiency factors and basic contrasts were so named in [132] and [194] respectively in the early 1970s.

[*] Personal communication from D. R. Hughes
[†] Personal communication from D. A. Preece
[‡] Personal communication from G. J. S. Ross
[§] Personal communications from G. V. Dyke and D. A. Preece

The treatment-concurrence graph is in [189]. It is a collapsed version of the treatment-block incidence graph, which was introduced by F. W. Levi [164]. The relationship between variance and distance in the treatment-concurrence graph was queried by R. Mead [169] in 1990. Section 5.3 is based on [30].

A good reference for optimal block designs is [221]. Theorems 5.16, 5.18, 5.19 and 5.21 were proved by Kshirsagar [153], H. D. Patterson and E. R. Williams [189], C. S. Cheng [80] and Cheng and Bailey [82] respectively. The formula in the proof of Theorem 5.18 was given by J. Roy [210]. J. A. John and T. Mitchell [136] found some optimal designs, such as the one in Example 5.16, by exhaustive search.

13.1.4 Partial balance

It was Bose and his students who introduced and developed partial balance and association schemes. In 1939 Bose and Nair [60] defined partially balanced incomplete-block designs. Association schemes were introduced implicitly but not defined. In this paper the authors insisted that concurrences for different associate classes should be different. In 1952 Bose and T. Shimamoto [61] realised that this condition was unnecessary. They also gave the first formal definition of association scheme, and listed four types of association scheme with rank three: group-divisible, Latin-square type, cyclic and triangular. They called the remaining rank three association schemes 'miscellaneous'.

Bose and W. S. Connor studied group-divisible designs in [57], coining the terms *regular*, *semi-regular* and *singular*.

In 1959 Bose and D. M. Mesner [59] took the major step of working out the algebra on which Chapter 2 is based, although Connor and W. H. Clatworthy had already produced the results in Section 2.5 in [84].

The 1950s and 1960s were a busy time for inventing new association schemes. It seems that there was a hope that all association schemes could be classified and all partially balanced incomplete-block designs up to a certain size tabulated.

B. Harshbarger [112, 113, 114] had invented rectangular lattice designs in the late 1940s to augment the series of square lattice designs. In [180], Nair pointed out that Pair(n) is an association scheme and that 2-replicate rectangular lattices are partially balanced with respect to it. G. Ishii and J. Ogawa also gave this association scheme in [130]. (The remainder of Section 5.5.4 is based on the much later papers [39, 85].)

M. N. Vartak introduced the rectangular association scheme in [244, 245]; P. M. Roy the hierarchical group-divisible association scheme in [211]: see also [123, 200]. K. Hinkelmann and Kempthorne reinvented the factorial association scheme as the extended group-divisible scheme in [123], as did K. Kusumoto [158] and P. U. Surendran [233]. With hindsight these can all be seen to be examples of schemes obtainable from trivial schemes by iterated crossing and nesting. So are S. K. Tharthare's generalized right-angular schemes [239, 240], which are $(\underline{\underline{n_1}} \times \underline{\underline{n_2}})/\underline{\underline{n_3}}$; B. Adhikary's [1] are $\underline{\underline{n_1}}/(\underline{\underline{n_2}} \times \underline{\underline{n_3}})$.

The method of inflation was given in [244, 257], together with the observation that it replaced the association scheme $\mathcal{Q}$ by $\mathcal{Q}/\underline{\underline{m}}$. The nested association scheme $\mathrm{L}(r,n)/\underline{\underline{m}}$ appears in [179, 205] and $\mathrm{T}(n)/\underline{\underline{m}}$ in [227]. M. B. Rao gave $\underline{\underline{m}}/\mathcal{Q}$ in [207]. However, the general method of nesting one association scheme in another does not seem to have been given until 1976 [213]. The product construction of designs was in [143, 156, 244], with the fact that it replaced $\mathcal{Q}_1$ and $\mathcal{Q}_2$ by $\mathcal{Q}_1 \times \mathcal{Q}_2$ (orthogonal superposition is in [22], and the proof of Proposition 5.11 in [144]). No systematic attempt to create new association schemes by crossing and nesting seems to have been made in this period. Chapter 3 is based on [230] from 1982. Poset operators in are [31] and crested products in [32].

Other new association schemes in this period were generalizations of those on Bose and Shimamoto's list. Adhikhary [2] generalized blueprints from cyclic groups to Abelian groups.

P. W. M. John generalized triangular schemes to the schemes $\mathrm{J}(n,3)$ in [139]. There might be case for calling these *John* schemes, except that the same generalization was made apparently independently in [58, 157], and M. Ogasawara [188] made the second parameter quite general, giving schemes of *triangular type* $\mathrm{T}_m(n)$. Someone called Johnson did indeed invent a new association scheme in [141], but, ironically, it was not the scheme today called Johnson; it was the unnamed scheme in Example 11.19 with valencies 1, $2(n-1)$, $n-1$ and $(n-1)(n-2)$.

The $\mathrm{L}(2,n)$ scheme is the same as $\mathrm{H}(2,n)$. In [202], D. Raghavarao and K. Chandrasekhararao generalized this to the *cubic* scheme $\mathrm{H}(3,n)$, which was in turn generalized to the *hypercubic* scheme $\mathrm{H}(m,n)$ in [157]. In fact, this association scheme, but not its name, had been given by Shah in [218]. By 1965 S. Yamamoto, Y. Fujii and N. Hamada [251] were able to list six infinite families of association schemes: the cyclic schemes; the factorial schemes F_m (that is $\underline{\underline{n_1}} \times \cdots \times \underline{\underline{n_m}}$), the iterated nested schemes N_m (that is $\underline{\underline{n_1}}/\cdots/\underline{\underline{n_m}}$), the hypercubic schemes C_m (that is, $\mathrm{H}(m,n)$), those of triangular type $\mathrm{T}_m(n)$ (that is, $\mathrm{J}(n,m)$), and the square schemes $\mathrm{S}(c,n)$ made from orthogonal block structures defined by a set of mutually orthogonal Latin squares, as in Section 6.5.

Useful review articles on partial balance and on association schemes respectively were published by Bose [54] in 1963 and S. Kageyama [144] in 1974. By this time Clatworthy's book of tables of partially balanced designs [83] had appeared, as had the books [140, 201], both of which have substantial coverage of partially balanced designs. Cyclic designs were tabulated in [138], and are the main topic of the books [134, 137].

Balance and partial balance both imply equal replication. To extend balance to unequal replication, Pearce defined supplemented balance in [191]. In addition, both Rao [206] and Nigam [187] proposed generalizing partial balance to unequal replication.

13.1.5 Orthogonal block structures

In almost all designed experiments, the structure of the experimental units is a poset block structure. From *Complex Experiments* [252] in 1935 onwards, statisticians sought the right general definition for these. Kempthorne and his colleagues at Ames, Iowa were particularly productive in the 1960s: see [148, 149, 241, 259]. As shown in [25], their *complete balanced response structure* is effectively the same as a poset block structure, but their definition is much harder to work with. In fact, G. Zyskind virtually defined poset block structures in [259].

When he replaced Yates as head of Rothamsted Statistics department in 1965, J. A. Nelder had a vision of a statistical computing program which could analyse many classes of design without needing a separate command for each class: this vision resulted in Genstat [190], building on earlier algorithms of Yates and also of Wilkinson [249] and his colleagues at C.S.I.R.O. Thinking in terms of algorithms, Nelder formalized crossing and nesting in [182] in 1965, and defined a *simple orthogonal block structure* to be any structure obtainable from trivial block structures by crossing and nesting, such as the structures in Section 3.5.

Combining the Ames results with Nelder's in [230, 231] in the 1980s, T. P. Speed and Bailey realized that not only could a single non-recursive definition be given for poset block structures, but that the more general orthogonal block structures were easier both for definition and calculation. Perhaps this was because the earlier work was done without knowledge of the general Möbius function, which therefore had to be calculated explicitly.[†] In [242] T. Tjur showed that even weaker conditions sufficed to define strata and proved Theorem 6.7.

Chapter 4 of [215] is an excellent synthesis of the North American work on orthogonal block structures. The Möbius function is never mentioned.

Section 8.8 is taken from [37].

13.1.6 Multi-stratum experiments

Most early designs for experiments with more than one system of blocks were orthogonal in the sense of Section 7.5.1. An exception was the lattice square (Section 7.5.4), introduced by Yates [256] in 1937 and generalized in [250]. Later came balance in the sense of Section 7.5.2, introduced for nested block designs and nested row-column designs by D. A. Preece [196] in 1967.

If the random effects model is assumed then it is convenient if all strata have the same basic contrasts: indeed, there is no unique best way of combining information otherwise. In 1965 Nelder [183] defined a design to have *general balance* if all the information matrices commute with each other. For a block design this condition is always satisfied, but for more complicated orthogonal block structures the only generally balanced designs that appear to be used in practice are those that have overall partial balance, defined in [127] in 1983. The particular case of designs generated by groups (Sections 7.5.3 and 8.5) is in [20, 36].

The section on valid randomization is based on [13, 35, 172]. Parts of Section 7.7 are in [28, 29, 63, 64].

Apart from the foregoing papers, literature about designs on orthogonal block structures tends to deal with a single structure at a time. Row-column designs balanced in the bottom stratum are given in [152, 197, 198]. Some partially balanced row-column designs are in [4, 5, 24, 165, 223].

For nested blocks, a recent catalogue of balanced designs is [175]. Partial balance was considered in [40, 125, 126]. Optimality properties are in [26, 50, 170, 174].

Nested row-column designs which are balanced in the bottom stratum are constructed in [6, 176, 226, 243]. Cheng clarified the difference between overall balance and balance in the bottom stratum in [81]. Par-

[†] In 1990 O. Kempthorne said to me something like 'This Möbius function really does the job. I wish that we had known about it.'

tial balance is in [7, 8, 110]. The paradox demonstrated in Example 7.15 is explored in [14, 23, 78, 79, 160, 176].

Semi-Latin squares are discussed in [19, 22, 86, 199, 252]. The design in Figure 7.6 is taken from [34], those in Exercise 7.3 from [38]. Some other partially balanced row-column designs with split plots are in [92].

13.1.7 Order on association schemes

The two statistical questions in Chapter 11 are quite old. Partially balanced designs have been presented with as few associate classes as possible ever since they were introduced. In the early 1970s Kageyama [143, 144] said that an association scheme $\mathcal{Q}$ is *reducible* to an association scheme $\mathcal{P}$ if $\mathcal{Q} \prec \mathcal{P}$. He investigated the reducibility of several of the main families of association schemes, but failed to give any results for $\mathrm{J}(n,m)$. However, at about the same time, L. A. Kaluzhnin and M. H. Klin [145] showed that if n is large enough in terms of m and $\mathrm{J}(n,m) \prec \mathcal{P}$ then $\mathcal{P}$ is trivial. Reducibility of group schemes was considered in [3].

In the 1990s E. Bannai and his co-workers [41, 44] said that $\mathcal{P}$ is a *subscheme* of $\mathcal{Q}$, or a *fusion scheme* of $\mathcal{Q}$, or that $\mathcal{Q}$ is a *fission scheme* of $\mathcal{P}$, if $\mathcal{Q} \prec P$. This meaning of 'subscheme' conflicts with the one in Chapter 10.

In the mid 1970s Homel and Robinson conjectured in [125, 126] that nested block designs which were partially balanced with respect to each system of blocks were partially balanced overall. This conjecture, which was shown to be false in [26], leads naturally to the idea of compatible association schemes.

Chapter 11 is an expanded version of [27]. Analogues of Theorem 11.2 for coherent configurations were given in [121, 151].

13.2 Algebra and combinatorics

13.2.1 Permutation groups, coherent configuratons and cellular algebras

The definition of a blueprint given in Chapter 8 can be weakened by replacing the second condition (ii)″ by the requirement that if Δ_i is any class of the blueprint then so is $\{\omega^{-1} : \omega \in \Delta_i\}$. This weaker form was effectively given by I. Schur in 1933 [214]. The analogue of the Bose–Mesner algebra is called a *Schur ring*. The corresponding theory, which assumes a permutation group that is regular but not necessarily Abelian, is explained in Chapter IV of the influential book on permutation groups [248] published by H. Wielandt in 1964 but based on a German version of 1959. Recent work on Schur rings includes [177].

In the 1960s the hunt for finite simple groups sparked a search for interesting combinatorial objects whose automorphism groups might be almost simple. D. G. Higman and C. C. Sims [122] discovered a new simple group by finding a very interesting strongly regular graph on 100 vertices. (According to [131], Mesner had given this graph in his Ph. D. thesis [171] in 1956.) These two authors had been independently developing the technique, already started in [248, Chapter III], of obtaining information about a permutation group G on a set Ω by examining the partition of $\Omega \times \Omega$ into *orbits* of G; that is, sets of the form $\{(g(\alpha), g(\beta)) : g \in G\}$. If the group is transitive then this partition satisfies the first and third conditions for an association scheme; the second condition is also satisfied if the group has an extra property which P. M. Neumann called *generous transitivity* in [185]. Sims concentrated on the graph defined by a single class of the partition, forging an interplay between graph theory and group theory in [224, 225]. In [116, 117], D. G. Higman developed the theory of the partition as a whole, which is quite similar to that for association schemes but can also be seen as a generalization of Schur rings to permutation groups which may not be regular. To extract the basic combinatorial ideas, independent of the action of a permutation group, he defined *coherent configurations* in 1971 [118] and gave further theory in [119, 120, 121].

Non-regular permutation groups which preserve concurrence matrices of designs were used in [36, 228].

Essentially the same concept as coherent configuration was introduced independently by B. Yu. Weisfeiler and A. A. Leman in 1968 [246] and studied in the former Soviet Union under the name *cellular algebra*. An account of this work is in [94], which includes a discussion of amorphic association schemes in [104]. The term 'cellular algebra' has a quite different meaning in recent papers such as [105].

Both of these strands of work appear to be independent of C. R. Nair's [178] effective introduction of homogeneous coherent configurations in 1964.

A good recent reference for permutation groups from this point of view is P. J. Cameron [71].

13.2.2 Strongly regular graphs

Two remarkably closely related papers appeared one after the other in the *Pacific Journal of Mathematics* in 1963. R. H. Bruck had already reinvented square lattice designs in [67], calling them *nets*. In [68], he posed a question equivalent to asking when the association scheme of type $\mathrm{L}(r, n)$ is determined by its parameters. In [55], Bose coined the terms *strongly regular graph* and *partial geometry*: his choice of journal introduced these ideas to pure mathematicians. The time was ripe. A. J. Hoffman and R. R. Singleton had defined Moore graphs in 1960 [124], while W. Feit and G. Higman were using integer conditions similar to those in Section 2.5 in their investigation of generalized polygons [95]. Permutation group theorists were using strongly regular graphs to describe permutation groups of rank three.

Suddenly there was an explosion of interest in strongly regular graphs among pure mathematicians. An accessible account [74] appeared in 1975, and a survey by J. J. Seidel [217] in 1979. It was realised that the task of classifying all strongly regular graphs, let alone all association schemes, was hopeless. Nevertheless, Neumaier [184] showed that, for each given value of the smallest eigenvalue of the adjacency matrix, almost all strongly regular graphs are either group-divisible, of Latin-square type, or of the type constructed in Exercise 5.28; this followed the work of Seidel [216], who had characterized strongly regular graphs with smallest eigenvalue -2.

There is more detail on strongly regular graphs in the books [75, 102, 103, 166, 222]. See [65] for a table of strongly regular graphs with at most 280 vertices, as well as many recent references.

13.2.3 Distance-regular graphs

If the edges and non-edges in a strongly regular graph have equal status then the natural generalization is an association scheme. Otherwise it is a distance-regular graph. These were introduced by N. L. Biggs in [47]. Treatments at book length are in [48, 66].

A. E. Brouwer maintains a database of distance-regular graphs (which, of course, includes strongly regular graphs). To interrogate it, send an email to `aeb@cwi.nl` whose subject line is `exec drg` and whose body contains one or more lines of the form

$$\texttt{drg d=}s\ \ b_0,\ b_1,\ \ldots,\ b_{s-1}\ :\ c_1,\ c_2,\ \ldots,\ c_s$$

in the notation of Section 1.4.4.

13.2.4 Geometry

Influenced by the geometer Levi in Calcutta in the 1930s, Bose naturally used affine and projective geometries over finite fields in his constructions of balanced incomplete-block designs from [52] onwards. Some of the cyclic association schemes discovered in the 1950s and 1960s are based on these geometries. More sophisticated geometry also leads to association schemes and partially balanced designs, as in [56]. However, geometers were largely unaware of the statistical work before the mid-1960s. In what was to become the standard reference on finite geometries, P. Dembowski [90] included sections on association schemes, generalized polygons and balanced incomplete-block designs, thus drawing association schemes to the attention of another section of the mathematical community. He made a valiant attempt to call group-divisible designs simply *divisible* and to rename perfect difference sets as *quotient sets*; he was more successful in introducing the terms *design* and *partial design* for balanced and partially balanced incomplete-block designs respectively.

13.2.5 Delsarte's thesis and coding theory

Delsarte completed his thesis [89] in 1973. His aim was to link association schemes to coding theory. He concentrated on two established classes of association schemes and renamed them to emphasize these links. Thus he renamed the hypercubic schemes as Hamming schemes because the associate classes are defined by Hamming distance [111]. General codes are subsets of Hamming schemes, so Delsarte investigated properties of subsets of general association schemes, but with particular reference to both Hamming schemes and the schemes of triangular type, which he renamed Johnson schemes because S. M. Johnson [142] had worked on constant-weight codes, which are subsets of these schemes. His general framework also linked Hamming schemes to orthogonal arrays and Johnson schemes to t-designs. He described (but did not name) Hamming powers.

For more details about coding theory and its relationship to association schemes, see [45, 75, 76, 167].

13.2.6 Duals

As hinted in Exercises 8.7 and 8.8, there is a well-developed theory of duals of Abelian groups: see [129, Section V.6]. This led O. Tamaschke

[234, 235] to a notion of duality for (some) Schur rings, and Delsarte [89] to effectively the same notion for Abelian-group schemes. In fact, Delsarte gave a more general definition of dual association scheme, but no examples other than Abelian group schemes. By the third edition of their book [75], Cameron and J. van Lint, on the suggestion of Seidel,† had generalized this to the formal duals described in Section 10.5.

13.2.7 Imprimitivity

Cameron, J.-M. Goethals and Seidel recognized the importance of inherent partitions in [73], where they proved Theorem 10.5 and showed that inherent partitions give rise to quotient schemes. They used the term 'subscheme' for what I have called an *algebraic subscheme*; Whelan [247] called them *normal subschemes.*

Parts of Sections 10.1–10.4 are also in [102, 208].

13.2.8 Recent work

The last twenty years have seen important progress in topics related to association schemes which are outside the scope of this book. The polynomial schemes introduced by Delsarte in [89] have been studied extensively, especially by D. A. Leonard [161, 162, 163]. P. Terwilliger developed the subconstituent algebra of an association scheme in [236, 237, 238]. The relationship between the partial order on the association schemes on a given set and the partial order on their automorphism groups was explored by I. A. Faradžev, A. A. Ivanov and M. H. Klin in [93]. Bannai and T. Ito [43] have a project to classify all primitive association schemes. The table algebras of [12] are generalizations of association schemes which lose the underlying set.

It is unfortunate that some of these authors use the term 'association scheme' for 'homogeneous coherent configuration'.

Bannai [42] lists some interesting open problems about homogeneous coherent configurations. He gives a table of these with at most 25 points. An updated version is available at

`http://math.shinshu-u.ac.jp/hanaki/as/as.html`

Cameron [72] shows which of these are association schemes, thus providing answers to some of the exercises in this book. Theorem 8.17 is in [10].

† Personal communication from P. J. Cameron

Glossary of notation

This list excludes notation that is used only within a short proof or example. The right-hand column shows where the notation is used.

a	valency of a strongly regular graph	Chapter 2
a	group element	Chapters 8, 11
a_i	valency of class $\mathcal{C}_i$	throughout
A	harmonic mean of the canonical efficiency factors	Chapters 4–5
A	adjacency matrix for 'same row or column but different' in Pair(n)	Chapters 5, 7, 11
A_i	adjacency matrix for class $\mathcal{C}_i$	throughout
Adj(i)	set of adjacency matrices of association scheme $\mathcal{Q}_i$	Chapter 9
$\mathcal{A}$	Bose–Mesner algebra of an association scheme	throughout
$\mathcal{A}(\mathcal{F})$	Bose–Mesner algebra of the association scheme defined by the orthogonal block structure $\mathcal{F}$	Chapter 11
$\mathcal{A}(\rho)$	Bose–Mesner algebra of the association scheme of the poset block structure defined by the pre-order ρ	Chapter 11
$\mathcal{A}[\mathcal{L}]$	subalgebra of $\mathcal{A}$ spanned by $\{A_i : i \in \mathcal{L}\}$	Chapter 10
b	a natural number, often the size of a set	Chapters 1–3
b	number of blocks	Chapters 4–5
b	group element	Chapters 8, 11
b_i	the constant $p^i_{i+1,1}$ in a distance-regular graph	Chapter 1
b_i	number of copies of the i-th Latin square	Chapter 6

b_x	valency of an associate class in the second association scheme	Chapter 3
$b_{[i]}$	valency of a class in an ideal partition	Chapter 10
$\mathrm{bot}(L)$	$L \setminus \max(L)$	Chapter 9
B	adjacency matrix for black edges of the cube	Chapter 2
B	adjacency matrix for 'same block'	Chapters 4–7
B	adjacency matrix for 'same box'	Examples 5.13, 6.4
B	adjacency matrix AM in $\mathrm{Pair}(n)$	Chapters 5, 7, 11
B	adjacency matrix for 'same rectangle'	Examples 6.1, 11.10
B	an adjacency matrix for $\mathcal{P}$	Chapter 10
B_x	adjacency matrix for the second association scheme	Chapters 3, 10
B_F	adjacency matrix for the association scheme on plots	Chapter 7
c	two more than the number of Latin squares	Chapters 6–7, 11
c	group element	Chapter 8
c_i	the constant $p^i_{i-1,1}$ in a distance-regular graph	Chapter 1
$\mathrm{cov}(Y, Z)$	covariance of random variables Y and Z	Chapters 4, 7
C	adjacency matrix for 'same column'	throughout
C	character table	throughout
$C(i, e)$	eigenvalue of A_i on W_e	throughout
C_n	the graph which is an n-circuit	Chapter 1
${}^n\mathrm{C}_m$	the binomial coefficient which gives the number of m-subsets of an n-subset	throughout
$\mathrm{Cov}(Y)$	covariance matrix of the random vector Y	Chapters 4, 7
$\mathbb{C}$	the complex numbers	throughout
$\mathcal{C}$	subset of $\Omega \times \Omega$	throughout
$\mathcal{C}'$	dual of subset $\mathcal{C}$	Chapter 1
$\mathcal{C}^\Gamma$	restriction $\mathcal{C} \cap (\Gamma \times \Gamma)$	Chapter 10
$\mathcal{C}_i$	an associate class	throughout
$\mathcal{C}_F$	associate class defined by the partition F	Chapter 6
$\mathcal{C}_0$	the diagonal associate class	throughout
$\mathcal{C}_i(\alpha)$	$\{\beta : (\alpha, \beta) \in \mathcal{C}_i\}$	throughout
d_e	dimension of stratum W_e	throughout
d^*	dimension of residual space	Chapters 4, 7
$\det M$	determinant of the matrix M	throughout
$\mathrm{diag}(f)$	diagonal matrix with f on the diagonal	throughout

$\dim W$	dimension of the vector space W	throughout
D	inverse of the character table	throughout
D_ω	$\{(\alpha, \beta) : \beta - \alpha = \omega\}$	Chapter 1
$\mathrm{Diag}(\Omega)$	$\{(\omega, \omega) : \omega \in \Omega\}$	throughout
$\mathcal{D}$	a class of partition $\mathcal{P}$	Chapter 8
$\mathcal{D}_x$	an associate class of the second association scheme	Chapters 3, 10–11
$\mathcal{D}_i$	a block design	Chapter 5
e	edge in a graph	Chapter 1
e	index for a stratum	Chapter 2 onwards
E	the equality partition	Chapter 6 onwards
$\mathcal{E}$	set indexing the strata	throughout
$\mathcal{E}_F$	$\{e \in \mathcal{E} : \varepsilon_{Fe} \neq 0\}$	Chapter 7
$\mathbb{E}(Y)$	expectation of the random vector Y	Chapters 4, 7
f	a function on Ω	throughout
f	index for a stratum	Chapter 2 onwards
F	a field	Chapter 1
F	a partition	Chapter 6 onwards
F^Ω	the set of functions from Ω to F	Chapter 1
$F(\Upsilon)$	partition into left cosets of the subgroup Υ	Chapter 8
$F(L)$	partition of $\prod_{i \in M} \Omega_i$ defined by the subset L of M	Chapter 9 onwards
$\mathcal{F}$	a set of partitions, usually an orthogonal block structure	Chapter 6 onwards
$\mathcal{F}$	set indexing the associate classes on the plots	Section 7.7
$\mathcal{F}(\sqsubseteq)$	poset block structure defined by the partial order $\sqsubseteq$	Chapters 9, 11
$\mathcal{F}_e$	$\{F \in \mathcal{F} : \varepsilon_{Fe} \neq 0\}$	Chapter 7
g	a function on Ω	throughout
g	an element of G	Section 8.8
G	adjacency matrix for 'same group'	Chapters 2, 5–7
G	adjacency matrix for 'same Greek letter'	Chapter 11
G	adjacency matrix for 'the rest' in $\mathrm{Pair}(n)$	Chapters 5, 7, 11
G	a partition	Chapter 6 onwards
G	an abstract finite group	Section 8.8
G	a subset of M	Chapter 9
$\mathcal{G}$	a graph	Chapters 1, 10
$\mathcal{G}$	a strongly regular graph	Chapter 2

$\mathcal{G}$	group-divisible association scheme defined by the blocks	Chapters 4–5, 11
$\mathcal{G}$	an orthogonal block structure	Chapters 10–11
$\mathcal{G}$	set indexing the strata on the plots	Section 7.7
$\mathcal{G}_F$	group-divisible scheme defined by the partition F	Chapters 7–8
$\mathcal{G}_i$	set of pairs of vertices at distance i in a graph $\mathcal{G}$	Chapter 1
$\mathcal{G}_i(\alpha)$	$\{\beta : (\alpha, \beta) \in \mathcal{G}_i\}$	Chapter 1
$\mathcal{G}_m$	a class of the ideal partition $\mathcal{P}$	Chapter 10
$\mathrm{GD}(b, m)$	group-divisible association scheme with b groups of size m	throughout
h	unknown vector of block effects	Chapter 4
h	an element of G	Section 8.8
h_F	unknown vector of effects of F-classses	Chapter 7
$\mathrm{H}(m, n)$	Hamming scheme on the m-th power of an n-set	throughout
H	a partition	Chapters 6, 9–10
H	a subset of M	Chapter 9
$\mathcal{H}$	an orthogonal block structure	Chapters 8, 11
$\mathcal{H}$	a subset of $\mathcal{E}$	Chapter 10
$\mathcal{H}$	the concurrence partition	Chapter 11
i	index for an associate class	throughout
i	element of the poset M	Chapter 9
$[i]$	equivalence class containing i	Chapter 10
i	square root of -1 in $\mathbb{C}$	throughout
I_Ω	identity matrix on Ω	throughout
$\mathrm{Im}\, M$	image of the matrix M, that is, $\{Mx : Mx \text{ is defined}\}$	throughout
Inverse	finest blueprint for an Abelian group	Chapter 8
$\mathcal{I}$	span $\{I, J\}$	Chapter 11
j	index for an associate class	throughout
J_Ω	all-1 matrix on $\Omega \times \Omega$	throughout
$J_{\Delta,\Theta}$	the $\Delta \times \Theta$ matrix with all entries equal to 1	throughout
$\mathrm{J}(n, m)$	Johnson scheme on m-subsets of an n-set	throughout
k	index for an associate class	throughout
k	a natural number, often the size of a set	Chapters 1–3
k	block size	Chapters 4–5, 8

k_F	size of all classes of the uniform partition F	Chapter 6 onwards
$\ker M$	kernel of the matrix M, that is, $\{x : Mx = 0\}$	throughout
K	a subset of the poset M, usually ancestral	Chapters 9–10
K_n	complete undirected graph on n vertices	Chapter 1
$\mathcal{K}$	set indexing the associate classes	throughout
l	a natural number	Chapters 1, 5, 7
l	index of a subset of a group	Chapters 5, 8
L	adjacency matrix for 'same letter'	Examples 1.6, 6.2, 10.11
L	information matrix	Chapters 4–6
L	a subset of the poset M, usually ancestral	Chapter 9 onwards
L_F	information matrix in stratum W_F	Chapter 7
$\mathrm{L}(r, n)$	association scheme of Latin-square type defined by $r - 2$ Latin squares of order n	throughout
$\mathcal{L}$	orthogonal block structure defined by a Latin square	Chapters 6–7
$\mathcal{L}$	subset of $\mathcal{K}$ corresponding to an inherent partition	Chapter 10
m	a natural number, often the size of a set	throughout
m	number of distinct canonical efficiency factors	Chapters 4–5
m_i	A_iR has entries m_i and 0	Chapter 10
$m_\theta(\Upsilon)$	number of ways of making θ as a difference (or quotient) between two elements of Υ	Chapters 5, 8
$\max(L)$	set of maximal elements of L	Chapter 9
$\min(L)$	set of minimal elements of L	Chapter 9
M	a matrix	Chapters 1–4, 7, 11
M	adjacency matrix for mirror-image pairs in $\mathrm{Pair}(n)$	Chapters 5, 7, 11
M	a poset	Chapter 9 onwards
M_ω	adjacency matrix for $\mathcal{D}_\omega$	Chapters 1, 5
$\mathcal{M}$	set of circulant matrices	Chapter 1
$\mathcal{M}$	a subset of $\mathcal{K}$	Chapter 10
$\mathcal{M}$	set indexing the classes of $\mathcal{P}$	Chapter 10
n	a natural number, often the size of Ω	throughout
$\underline{\underline{n}}$	trivial association scheme on n points	throughout
ⓝ	cyclic association scheme defined by the n-circuit	throughout

n_F	number of classes in the partition F	Chapter 6 onwards
n_i	the size of Ω_i	Chapter 9
N	a matrix	Chapters 2–3, 11
N	incidence matrix of a block design	Chapters 4–5
N	a subset of M	Chapter 9
$\mathcal{N}(M)$	the set of antichains in the poset M	Chapter 9
O_Ω	zero matrix on $\Omega \times \Omega$	throughout
p	parameter of a strongly regular graph	Chapter 2
p_{ij}^k	number of ijk triangles through each k edge	throughout
$\mathrm{Pair}(n)$	the association scheme on ordered pairs of distinct elements from an n-set	Chapter 5 onwards
P	an orthogonal projector	Chapters 2–3
P	the orthogonal projector onto V_B	Chapter 4
P_i	matrix in the Bose–Mesner isomorphism	Chapter 2
P_i	the orthogonal projector onto U_i	Chapters 4–5
P_F	the orthogonal projector onto V_F	Chapter 6 onwards
$\mathcal{P}$	a partition of $\Omega \times \Omega$, sometimes an association scheme, sometimes an ideal partition	Chapter 7 onwards
$\mathcal{P}$	partition of $\{1, \ldots, n+1\}$	Chapter 11
q	parameter of a strongly regular graph	Chapter 2
q_{xy}^z	p_{ij}^k for the second assocation scheme	Chapters 3, 8, 10
$q(F; \theta, \eta)$	number of designs in which any given pair of F-th associates are allocated treatments θ and η	Chapter 7
Q	an orthogonal projector	Chapters 2–3
Q	the orthogonal projector onto $V_B^\perp$	Chapter 4
Q_F	stratum projector for the association scheme on plots	Chapter 7
$\mathcal{Q}$	a set of subsets of $\Omega \times \Omega$, usually an association scheme	Chapter 3 onwards
$\mathcal{Q}(\Delta)$	association scheme defined by the blueprint Δ	Chapters 8, 11
$\mathcal{Q}(\mathcal{F})$	association scheme defined by the orthogonal block structure $\mathcal{F}$	Chapters 10–11
$\mathcal{Q}(\rho)$	association scheme of the poset block structure defined by the pre-order ρ	Chapter 11
r	a natural number, often the size of a set	Chapters 1–3
r	replication	Chapter 4 onwards

r_θ	replication of treatment θ	Chapters 4, 7
R	adjacency matrix for 'same row'	Chapters 1, 5–6, 10
R	adjacency matrix for red edges of the cube	Chapter 2
R	orthogonal projector onto the residual space	Chapter 4
R_F	relation matrix for the partition F	Chapter 6 onwards
$\mathrm{R}(n,m)$	rectangular association scheme on the rectangle with n rows and m columns	throughout
$\mathbb{R}$	the real numbers	throughout
$\mathbb{R}^\Omega$	the set of functions from Ω to $\mathbb{R}$	throughout
$\mathbb{R}^{\Omega\times\Omega}$	the set of functions from $\Omega\times\Omega$ to $\mathbb{R}$, that is, the set of $\Omega\times\Omega$ matrices with real entries	throughout
s	number of non-diagonal associate classes	throughout
$\mathrm{S}(c,n)$	square association scheme defined by $c-2$ Latin squares of order n	Chapter 6 onwards
$\mathrm{span}\{\ \}$	the set of all real linear combinations of the matrices shown	throughout
S_e	stratum projector	throughout
S_n	symmetric group of all permutations on n points	Chapters 8, 10
Subgroup	blueprint defined by all subgroups	Chapter 8
$\mathcal{S}(\sqsubseteq)$	set of ancestral subsets	Chapters 9, 11
t	index for association schemes or orthogonal block structures	Chapters 3, 6, 8–11
t	number of treatments	Chapters 4–5, 7–8
$\mathrm{top}(L)$	$L\setminus\min(L)$	Chapter 9
$\mathrm{tr}(M)$	trace of the matrix M	throughout
T	partition of the plots defined by the treatments	Chapter 7
$\mathrm{T}(n)$	triangular association scheme with $n(n-1)/2$ points	throughout
T_f	stratum projector for the second association scheme	Chapter 3
u	a vector	Chapters 2–4, 7
U	a vector space	Chapters 2–3
U	the universal partition	Chapter 6 onwards
U_e	stratum for the first association scheme	Chapter 3
U_e	stratum in the treatments space $\mathbb{R}^\Theta$	Chapter 5
U_i	eigenspace of the information matrix	Chapter 4
U_0	subspace of $\mathbb{R}^\Theta$ spanned by the all-1 vector	Chapters 4–5, 11

v	a vector	Chapters 2–4, 6
V	a vector space	Chapters 2–3
V_f	stratum for the second association scheme	Chapter 3
V_B	vector space of functions constant on each block	Chapter 4
V_F	vector space of functions constant on each class of the partition F	Chapter 6 onwards
$\mathrm{Var}(Y)$	variance of the random variable Y	Chapters 4–5, 7
w	a vector	Chapters 2, 4, 6–7
W	a vector space	Chapter 2
W_e	a stratum	throughout
W_F	subspace defined by a partition F in a set of partitions $\mathcal{F}$, which is a stratum if $\mathcal{F}$ is an orthogonal block structure	Chapter 6 onwards
W_0	stratum spanned by the all-1 vector	throughout
x	a vector, especially a treatment contrast	Chapters 2–7
x	index for an associate class of the second association scheme	Chapters 3, 8, 10
x	shorthand for a specific element defined by 9-th roots of unity	Example 5.16
x	indeterminate in a polynomial	throughout
X	design matrix	Chapters 4–5, 7
X	incidence matrix	Chapter 10
y	a vector	Chapters 2–3
y	index for an associate class of the second association scheme	Chapter 3
y	shorthand for a specific element defined by 9-th roots of unity	Example 5.16
Y	adjacency matrix for yellow edges of the cube	Chapter 2
Y	random vector	Chapters 4–5, 7
z	a vector	Chapters 2–4, 7
z	index for an associate class of the second association scheme	Chapter 3
z	shorthand for a specific element defined by 9-th roots of unity	Example 5.16
Z	a random vector	Chapters 4, 7
$\mathbb{Z}$	the integers	throughout
$\mathbb{Z}_n$	the integers modulo n	throughout

$\mathbb{Z}[x]$	the ring of polynomials with coefficients in $\mathbb{Z}$	Chapter 4
α	an element of Ω	throughout
β	an element of Ω	throughout
γ	an element of Γ, Δ or Ω	throughout
γ_F	covariance of responses on plots which are F-th associates	Chapter 7
Γ	a set	Chapter 1
Γ	a blueprint	Chapters 8, 11
δ	an element of Γ, Δ or Ω	throughout
δ	a block	Chapter 4
Δ	a set	Chapter 1
Δ	the set of blocks	Chapter 4
Δ	a blueprint	Chapters 8, 10–11
Δ_i	a class in a blueprint	Chapters 1–3, 5, 8
ϵ	a complex root of unity	Chapters 2, 5, 7–8
ϵ_n	a primitive n-th root of unity in $\mathbb{C}$	Chapters 2, 5, 7
ε	efficiency factor	Chapters 4–5, 7–8
ζ	a treatment	Chapters 4–5
ζ	zeta function of a poset	Chapter 6 onwards
η	a treatment	Chapters 4–7, 11
θ	a treatment	Chapters 4–7, 11
ϑ	map from ideal partitions to inherent partitions	Chapter 10
ϑ_e	algebra homomorphism from $\mathcal{A}$ to $\mathbb{R}$	Chapter 2
ϑ_L	projection onto the coordinates in L	Theorem 9.13
Θ	set of treatments	Chapter 4 onwards
κ_i	proportional to the variance of the simple contrast between i-th associates	Chapter 5
$\kappa(M)$	scalar function of matrix M	Chapter 7
λ	a scalar	Chapters 1–3
λ	an eigenvalue	Chapters 2–3
λ	unique concurrence of distinct treatments in a balanced incomplete-block design	Chapter 5
λ_i	concurrence of i-th associates in a partially balanced design	Chapter 5
λ_θ	concurrence of 0 and θ in a cyclic design	Chapters 5, 8
λ_{Fi}	concurrence of i-th associates in classes of the partition F	Chapter 7
Λ	concurrence matrix	Chapters 4–5

$\Lambda(\theta, \eta)$	concurrence of treatments θ and η	Chapters 4–5
Λ_F	concurrence matrix in classes of F	Chapter 7
μ	a scalar	Chapters 1–3, 5
μ	an eigenvalue	Chapters 3–4
μ	Möbius function of a poset	Chapter 6 onwards
ν_i	coefficient of A_i in the Drazin inverse of the information matrix	Chapter 5
ξ	an element of Ξ	Chapter 8
ξ_F	stratum variance for stratum W_F	Chapter 7
Ξ	a subgroup	Chapter 8
Ξ	a blueprint	Chapter 11
π	component of an isomorphism between association schemes	Chapters 3, 10
π	bijection between sets of blocks	Chapters 4–5
π	component of an isomorphism between block designs	Chapter 5
π	permutation of Ω	Chapters 7–8
π	component of a homomorphism between association schemes	Chapter 10
π_ω	permutation of Ω obtained by pre-multiplying by ω	Chapter 8
Π	Latin square	throughout
ρ	a design	Section 7.7
ρ	a pre-order	Chapter 11
ϱ	map from inherent partitions to ideal partitions	Chapter 10
σ	permutation of $\{0, \ldots, s\}$	Chapter 8
σ	permutation of $\{1, \ldots, m\}$	Chapter 10
σ	a pre-order	Chapter 11
σ^2	variance of the response on a single plot	Chapters 4–7
τ	unknown vector of treatment constants	Chapters 4–7
υ	an element of Υ	Chapters 5, 7–8
Υ	a subset of a group	Chapters 5, 7–8
ϕ	component of an isomorphism between association schemes	Chapter 3
ϕ	component of an isomorphism between block designs	Chapter 5
ϕ	a design	Section 7.7
ϕ	irreducible character of an Abelian group	Chapter 8
ϕ	group automorphism	Chapter 8

ϕ	component of a homomorphism between association schemes	Chapter 10
φ	algebra isomorphism from $F\mathbb{Z}_n$ to $\mathcal{M}$	Chapters 1, 8
φ	the Bose–Mesner isomorphism	Chapter 2
φ	a particular lattice isomorphism	Theorem 9.13
φ_e	algebra homomorphism from $\mathcal{A}$ to $\mathcal{A}$	Chapter 2
Φ	group of automorphisms	Chapter 8
Φ	a blueprint	Chapter 11
χ_Δ	characteristic function of the set Δ	throughout
χ_ω	characteristic function of the element ω	throughout
ψ	design function	Chapters 4–5, 7, 11
ψ^*	dual design function	Chapter 4
ψ^*	inflated design function	Chapter 4
Ψ	set of designs	Chapter 7
Ψ	a blueprint	Chapter 11
ω	an element of Ω	throughout
Ω	set on which the association scheme is defined	throughout
Ω	set of plots	Chapters 4–5, 7
$\varnothing$	the empty set	throughout
$'$	M' denotes the transpose of the matrix M	throughout
$^-$	M^- denotes the Drazin inverse of the matrix M	Chapter 2 onwards
$\perp$	$W^\perp$ denotes the orthogonal complement of the subspace W	throughout
$\widehat{\ }$	notation for estimator of what is under the hat	Chapter 4
$\lceil x \rceil$	the smallest integer not less than x	throughout
$\uparrow(N)$	$\{i \in M : \exists j \text{ in } N \text{ with } j \sqsubseteq i\}$	Chapter 9
$\not\downarrow(N)$	$\{i \in M : \nexists j \text{ in } N \text{ with } i \sqsubseteq j\}$	Chapter 9
$*$	duality	Chapter 10
$+$	$\gamma + \Delta$ means $\{\gamma + \delta : \delta \in \Delta\}$	Chapters 1, 5
$+$	$U + W$ means $\{u + w : u \in U,\ w \in W\}$	Chapter 2
$+$	$\rho + \sigma$ means the transitive closure of $\rho \cup \sigma$	Chapter 11
$\oplus$	direct sum	throughout
$-$	$-\Delta$ means $\{-\delta : \delta \in \Delta\}$	Chapter 1
$\times$	Cartesian product of sets	throughout
$\times$	crossing operator	Chapter 3 onwards
$\otimes$	tensor product of vectors or matrices	Chapter 3 onwards
$/$	nesting operator	Chapter 3 onwards

$\wedge$	infimum of the partitions on either side	Chapter 6 onwards
$\vee$	supremum of the partitions on either side	Chapter 6 onwards
$*$	convolution	Chapter 8
$\circ$	Hadamard product	Chapter 11
$\langle\ ,\ \rangle$	inner product	Chapters 2–4
$\langle\ :\ \rangle$	group generated by the elements on the left satisfying the relations on the right	Chapters 8, 11
$\perp$	is orthogonal to	throughout
$\cong$	is isomorphic to	throughout
$\leqslant$	the item on the left is contained in the item on the right, and has the same algebraic structure, e.g. vector space	throughout
$\preccurlyeq$	partial order on partitions	Chapter 6 onwards
$\sqsubseteq$	partial order on the set M	Chapter 9
$\sim_L$	equivalence relation defined by the subset L	Chapter 9
$\sim$	equivalence relation on associate classes	Chapter 10
$\approx$	equivalence relation on strata	Chapter 10
$\sim_\rho$	equivalence relation defined by the pre-order ρ	Chapter 11

References

[1] B. ADHIKARY: Some types of m-associate P.B.I.B. association schemes, *Calcutta Statistical Association Bulletin*, **15** (1966), pp. 47–74.

[2] B. ADHIKARY: A new type of higher associate cyclical association scheme, *Calcutta Statistical Association Bulletin*, **16** (1967), pp. 40–44.

[3] B. ADHIKARY & T. K. SAHA: Construction and combinatorial problems of higher associate cyclic PBIB designs, *Calcutta Statistical Association Bulletin*, **40** (1990–91), pp. 163–181.

[4] H. AGRAWAL: Some methods of construction of designs for two-way elimination of heterogeneity-1, *Journal of the American Statistical Association*, **61** (1966), pp. 1153–1171.

[5] H. L. AGRAWAL: Some systematic methods of construction of designs for two-way elimination of heterogeneity, *Calcutta Statistical Association Bulletin*, **15** (1966), pp. 93–108.

[6] H. L. AGRAWAL & J. PRASAD: Some methods of construction of balanced incomplete block designs with nested rows and columns, *Biometrika*, **69** (1982), pp. 481–483.

[7] H. L. AGRAWAL & J. PRASAD: On nested row-column partially balanced incomplete block designs, *Calcutta Statistical Association Bulletin*, **31** (1982), pp. 131–136.

[8] H. L. AGRAWAL & J. PRASAD: Some methods of construction of GD-RC and rectangular-RC designs, *Australian Journal of Statistics*, **24** (1982), pp. 191–200.

[9] M. AIGNER: *Combinatorial Theory*, Springer-Verlag, New York (1979).

[10] P. P. ALEJANDRO, R. A. BAILEY & P. J. CAMERON: Association schemes and permutation groups, *Discrete Mathematics*, **266** (2003), pp. 47–67.

[11] H. ANTON: *Elementary Linear Algebra*, Wiley, New York (2000).

[12] Z. ARAD & H. BLAU: On table algebras and applications to finite group theory, *Journal of Algebra*, **138** (1991), pp. 137–185.

[13] J.-M. AZAÏS: Design of experiments for studying intergenotypic competition, *Journal of the Royal Statistical Society, Series B*, **49** (1987), pp. 334–345.

[14] S. Bagchi, A. C. Mukhopadhyay & B. K. Sinha: A search for optimal nested row-column designs, *Sankhyā, Series B*, **52** (1990), pp. 93–104.

[15] R. A. Bailey: Patterns of confounding in factorial designs, *Biometrika*, **64** (1977), pp. 597–603.

[16] R. A. Bailey: Dual Abelian groups in the design of experiments, in: *Algebraic Structures and Applications* (eds. P. Schultz, C. E. Praeger & R. P. Sullivan), Marcel Dekker, New York (1982), pp. 45–54.

[17] R. A. Bailey: Partially balanced designs, in: *Encyclopedia of Statistical Sciences* (eds. S. Kotz & N. L. Johnson), J. Wiley, New York Volume 6 (1985), pp. 593–610.

[18] R. A. Bailey: Factorial design and Abelian groups, *Linear Algebra and its Applications*, **70** (1985), pp. 349–368.

[19] R. A. Bailey: Semi-Latin squares, *Journal of Statistical Planning and Inference*, **18** (1988), pp. 299–312.

[20] R. A. Bailey: Cyclic designs and factorial designs, in: *Probability, Statistics and Design of Experiments* (ed. R. R. Bahadur), Wiley Eastern, New Delhi (1990), pp. 51–74.

[21] R. A. Bailey: Strata for randomized experiments, *Journal of the Royal Statistical Society, Series B*, **53** (1991), pp. 27–78.

[22] R. A. Bailey: Efficient semi-Latin squares, *Statistica Sinica*, **2** (1992), pp. 413–437.

[23] R. A. Bailey: Recent advances in experimental design in agriculture, *Bulletin of the International Statistical Institute*, **55(1)** (1993), pp. 179–193.

[24] R. A. Bailey: General balance: artificial theory or practical relevance?, in: *Proceedings of the International Conference on Linear Statistical Inference LINSTAT '93* (eds. T. Caliński & R. Kala), Kluwer, Amsterdam (1994), pp. 171–184.

[25] R. A. Bailey: Orthogonal partitions in designed experiments, *Designs, Codes and Cryptography*, **8** (1996), pp. 45–77.

[26] R. A. Bailey: Choosing designs for nested blocks, *Listy Biometryczne*, **36** (1999), pp. 85–126.

[27] R. A. Bailey: Suprema and infima of association schemes, *Discrete Mathematics*, **248** (2002), pp. 1–16.

[28] R. A. Bailey: Balanced colourings of strongly regular graphs, *Discrete Mathematics*, to appear.

[29] R. A. Bailey: Designs on association schemes in:, *Science and Statistics: A Festschrift for Terry Speed* (ed. Darlene R. Goldstein), Institute of Mathematical Statistics Lecture Notes – Monograph Series, 40, IMS, Beachwood, Ohio (2003), pp. 79–102.

[30] R. A. Bailey: Variance, concurrence and distance in block designs, in preparation.

[31] R. A. Bailey: Generalized wreath products of association schemes, in preparation.

[32] R. A. Bailey & P. J. Cameron: Crested products of association schemes, in preparation.

[33] R. A. Bailey, F. H. L. Gilchrist & H. D. Patterson: Identification of effects and confounding patterns in factorial designs, *Biometrika*, **64** (1977), pp. 347–354.

[34] R. A. Bailey & H. Monod: Efficient semi-Latin rectangles: designs for plant disease experiments, *Scandinavian Journal of Statistics*, **28** (2001), pp. 257–270.

[35] R. A. Bailey & C. A. Rowley: Valid randomization, *Proceedings of the Royal Society, Series A*, **410** (1987), pp. 105–124.

[36] R. A. Bailey & C. A. Rowley: General balance and treatment permutations, *Linear Algebra and its Applications*, **127** (1990), pp. 183–225.

[37] R. A. Bailey & C. A. Rowley: Latin cubes, in preparation.

[38] R. A. Bailey & G. Royle: Optimal semi-Latin squares with side six and block size two, *Proceedings of the Royal Society, Series A*, **453** (1997), pp. 1903–1914.

[39] R. A. Bailey & T. P. Speed: Rectangular lattice designs: efficiency factors and analysis, *Annals of Statistics*, **14** (1986), pp. 874–895.

[40] S. Banerjee & S. Kageyama: Methods of constructing nested partially balanced incomplete block designs, *Utilitas Mathematica*, **43** (1993), pp. 3–6.

[41] E. Bannai: Subschemes of some association schemes, *Journal of Algebra*, **144** (1991), pp. 166–188.

[42] E. Bannai: An introduction to association schemes, in: *Methods of Discrete Mathematics* (eds. S. Löwe, F. Mazzocca, N. Melone & U. Ott), Quaderni di Mathematica, 5, Dipartimento di Matematica della Seconda Università di Napoli, Napoli (1999), pp. 1–70.

[43] E. Bannai & T. Ito: *Algebraic Combinatorics. I: Association Schemes*, Benjamin, Menlo Park, California (1984).

[44] E. Bannai & S. Y. Song: Character tables of fusion schemes and fission schemes, *European Journal of Combinatorics*, **14** (1993), pp. 385–396.

[45] A. Barg & S. Litsyn (editors): *Codes and Association Schemes*, DIMACS Series in Discrete Mathematics and Theoretical Computer Science, 56, American Mathematical Society, Providence, Rhode Island (2001).

[46] T. Beth, D. Jungnickel & H. Lenz: Design Theory (2nd edition) Volumes 1 and 2, Cambridge University Press, Cambridge (1999).

[47] N. L. Biggs: Intersection matrices for linear graphs, in: *Combinatorial Mathematics and its Applications* (ed. D. J. A. Welsh), Academic Press, London (1971), pp. 15–23.

[48] N. L. Biggs: *Algebraic Graph Theory*, Cambridge Tracts in Mathematics, 67, Cambridge University Press, Cambridge (1974).

[49] N. L. Biggs: *Discrete Mathematics*, Oxford University Press, Oxford (1985).

[50] B. Bogacka & S. Mejza: Optimality of generally balanced experimental block designs, in: *Proceedings of the International Conference on Linear Statistical Inference LINSTAT '93* (eds. T. Caliński & R. Kala), Kluwer, Amsterdam (1994), pp. 185–194.

[51] R. C. Bose: On the application of the properties of Galois fields to the problem of construction of hyper-Graeco-Latin squares, *Sankhyā*, **3** (1938), pp. 323–338.

[52] R. C. Bose: On the construction of balanced incomplete block designs, *Annals of Eugenics*, **9** (1939), pp. 353–399.

[53] R. C. Bose: Mathematical theory of the symmetrical factorial design, *Sankhyā*, **8** (1947), pp. 107–166.
[54] R. C. Bose: Combinatorial properties of partially balanced designs and association schemes, *Sankhyā*, **25** (1963), pp. 109–136.
[55] R. C. Bose: Strongly regular graphs, partial geometries and partially balanced designs, *Pacific Journal of Mathematics*, **13** (1963), pp. 389–419.
[56] R. C. Bose & I. M. Chakravarti: Hermitian varieties in a finite projective space $\mathrm{PG}(N, q^2)$, *Canadian Journal of Mathematics*, **18** (1966), pp. 1161–1182.
[57] R. C. Bose & W. S. Connor: Combinatorial properties of group divisible incomplete block designs, *Annals of Mathematics Statistics*, **23** (1952), pp. 367–383.
[58] R. C. Bose & R. Laskar: A characterization of tetrahedral graphs, *Journal of Combinatorial Theory*, **2** (1967), pp. 366–385.
[59] R. C. Bose & D. M. Mesner: On linear associative algebras corresponding to association schemes of partially balanced designs, *Annals of Mathematical Statistics*, **30** (1959), pp. 21–38.
[60] R. C. Bose & K. R. Nair: Partially balanced incomplete block designs, *Sankhyā*, **4** (1939), pp. 337–372.
[61] R. C. Bose & T. Shimamoto: Classification and analysis of partially balanced incomplete block designs with two associate classes, *Journal of the American Statistical Association*, **47** (1952), pp. 151–184.
[62] W. G. Bridges & R. A. Mena: Rational circulants with rational spectra and cyclic strongly regular graphs, *Ars Combinatoria*, **8** (1979), pp. 143–161.
[63] C. J. Brien & R. A. Bailey: Multi-tiered experiments: I. Design and randomization, in preparation.
[64] C. J. Brien & R. A. Bailey: Multi-tiered experiments: II. Structure and analysis, in preparation.
[65] A. E. Brouwer: Strongly regular graphs, in: *The CRC Handbook of Combinatorial Designs* (eds. C. J. Colburn & J. H. Dinitz), CRC Press, Boca Raton, Florida (1996), pp. 667–685.
[66] A. E. Brouwer, A. M. Cohen & A. Neumaier: *Distance-Regular Graphs*, Springer, Berlin (1989).
[67] R. H. Bruck: Finite nets. I. Numerical invariants, *Canadian Journal of Mathematics*, **3** (1951), pp. 94–107.
[68] R. H. Bruck: Finite nets. II. Uniqueness and embedding, *Pacific Journal of Mathematics*, **13** (1963), pp. 421–457.
[69] T. Caliński: On some desirable patterns in block designs, *Biometrics*, **27** (1971), pp. 275–292.
[70] T. Caliński & S. Kageyama: *Block Designs: A Randomization Approach. Volume I: Analysis*, Lecture Notes in Statistics, 150, Springer-Verlag, New York (2000).
[71] P. J. Cameron: *Permutation Groups*, London Mathematical Society Student Texts, 45, Cambridge University Press, Cambridge (1999).
[72] P. J. Cameron: Coherent configurations, association schemes and permutation groups, in: *Groups, Combinatorics and Geometry* (eds. A. A. Ivanov, M. W. Liebeck & J. Saxl), World Scientific, Singapore (2003), pp. 55–71.

[73] P. J. CAMERON, J.-M. GOETHALS & J. J. SEIDEL: The Krein condition, spherical designs, Norton algebras and permutation groups, *Proceedings of the Koninklijke Nederlandse Akademie van Wetenschappen, Series A*, **81** (1978), pp. 196–206.

[74] P. J. CAMERON & J. H. VAN LINT: *Graph Theory, Coding Theory and Block Designs*, London Mathematical Society Lecture Note Series, 19, Cambridge University Press, Cambridge (1975).

[75] P. J. CAMERON & J. H. VAN LINT: *Designs, Graphs, Codes and their Links*, London Mathematical Society Student Texts, 22, Cambridge University Press, Cambridge (1991).

[76] P. CAMION: Codes and association schemes, in: *Handbook of Coding Theory* (eds. V. Pless, W. C. Huffman & R. Brualdi), Elsevier, New York (1998).

[77] R. D. CARMICHAEL: *Introduction to the Theory of Groups of Finite Order*, Ginn & Co, Boston (1937).

[78] J. Y. CHANG & W. I. NOTZ: A method for constructing universally optimal block designs with nested rows and columns, *Utilitas Mathematica*, **38** (1990), pp. 263–276.

[79] J. Y. CHANG & W. I. NOTZ: Some optimal nested row-column designs, *Statistica Sinica*, **4** (1994), pp. 249–263.

[80] C.-S. CHENG: Optimality of certain asymmetrical experimental designs, *Annals of Statistics*, **6** (1978), pp. 1239–1261.

[81] C.-S. CHENG: A method for constructing balanced incomplete block designs with nested rows and columns, *Biometrika*, **73** (1986), pp. 695–700.

[82] C.-S. CHENG & R. A. BAILEY: Optimality of some two-associate-class partially balanced incomplete-block designs, *Annals of Statistics*, **19** (1991), pp. 1667–1671.

[83] W. H. CLATWORTHY: *Tables of Two-Associate-Class Partially Balanced Designs*, Applied Mathematics Series, 63, National Bureau of Standards, Washington, D.C. (1973).

[84] W. S. CONNOR & W. H. CLATWORTHY: Some theorems for partially balanced designs, *Annals of Mathematical Statistics*, **25** (1954), pp. 100–112.

[85] L. C. A. CORSTEN: Rectangular lattices revisited, in: *Linear Statistical Inference* (eds. T. Caliński & W. Klonecki), Lecture Notes in Statistics, 35, Springer-Verlag, Berlin (1985), pp. 29–38.

[86] L. A. DARBY & N. GILBERT: The Trojan square, *Euphytica*, **7** (1958), pp. 183–188.

[87] B. A. DAVEY & H. A. PRIESTLEY: *Introduction to Lattices and Order*, Cambridge University Press, Cambridge (1990).

[88] A. M. DEAN & J. A. JOHN: Single replicate factorial arrangements in generalized cyclic designs: II. Asymmetrical arrangements, *Journal of the Royal Statistical Society, Series B*, **37** (1975), pp. 72–76.

[89] P. DELSARTE: *An algebraic approach to the association schemes of coding theory*, Ph. D. thesis, Université Catholique de Louvain (1973). (appeared as Philips Research Reports Supplement, No. 10, 1973)

[90] P. DEMBOWSKI: *Finite Geometries*, Springer-Verlag, Berlin (1968).

[91] A. DEY: *Theory of Block Designs*, Wiley Eastern, New Delhi (1986).

[92] R. N. EDMONDSON: Trojan square and incomplete Trojan square designs for crop research, *Journal of Agricultural Science*, **131** (1998), pp. 135–142.
[93] I. A. FARADŽEV, A. A. IVANOV & M. H. KLIN: Galois correspondence between permutation groups and cellular rings (association schemes), *Graphs and Combinatorics*, **6** (1990), pp. 303–332.
[94] I. A. FARADŽEV, A. A. IVANOV, M. H. KLIN & A. J. WOLDAR: *Investigations in Algebraic Theory of Combinatorial Objects*, Mathematics and its Applications (Soviet Series), 84, Kluwer, Dordrecht (1994).
[95] W. FEIT & G. HIGMAN: The nonexistence of certain generalized polygons, *Journal of Algebra*, **1** (1964), pp. 434–446.
[96] R. A. FISHER: *Statistical Methods for Research Workers*, Oliver and Boyd, Edinburgh (1925).
[97] R. A. FISHER: *The Design of Experiments*, Oliver and Boyd, Edinburgh (1935).
[98] R. A. FISHER: An examination of the different possible solutions of a problem in incomplete blocks, *Annals of Eugenics*, **10** (1940), pp. 52–75.
[99] R. A. FISHER: The theory of confounding in factorial experiments in relation to the theory of groups, *Annals of Eugenics*, **11** (1942), pp. 341–353.
[100] R. A. FISHER: A system of confounding for factors with more than two alternatives, giving completely orthogonal cubes and higher powers, *Annals of Eugenics*, **12** (1945), pp. 282–290.
[101] R. A. FISHER & F. YATES: *Statistical Tables for Biological, Agricultural and Medical Research*, Oliver and Boyd, Edinburgh (1938).
[102] C. D. GODSIL: *Algebraic Combinatorics*, Chapman and Hall, New York (1993).
[103] C. GODSIL & G. ROYLE: *Algebraic Graph Theory*, Graduate Texts in Mathematics, 207, Springer-Verlag, New York (2001).
[104] JA. JU. GOL'FAND, A. V. IVANOV & M. H. KLIN: Amorphic cellular rings, in: *Investigations in Algebraic Theory of Combinatorial Objects* (eds. I. A. Faradžev, A. A. Ivanov, M. H. Klin & A. J. Woldar), Mathematics and its Applications (Soviet Series), 84, Kluwer, Dordrecht (1994), pp. 167–187.
[105] J. J. GRAHAM & G. I. LEHRER: Cellular algebras, *Inventiones Mathematicae*, **123** (1996), pp. 1–34.
[106] G. GRÄTZER: *Lattice Theory. First Concepts and Distributive Lattices*, W. H. Freeman and Company, San Francisco (1971).
[107] G. R. GRIMMETT & D. R. STIRZAKER: *Probability and Random Processes*, Oxford University Press, Oxford (1982).
[108] H. GROPP: Configurations, in: *The CRC Handbook of Combinatorial Designs* (eds. C. J. Colburn & J. H. Dinitz), CRC Press, Boca Raton, Florida (1996), pp. 253–255.
[109] S. GUPTA & R. MUKERJEE: *A Calculus for Factorial Arrangements*, Lecture Notes in Statistics, 59, Springer-Verlag, Berlin (1989).
[110] S. C. GUPTA & M. SINGH: Partially balanced incomplete block designs with nested rows and columns, *Utilitas Mathematica*, **40** (1991), pp. 291–302.

[111] R. W. HAMMING: Error detecting and error correcting codes, *Bell System Technical Journal*, **29** (1950), pp. 147–160.

[112] B. HARSHBARGER: Preliminary report on the rectangular lattices, *Biometrics*, **2** (1946), pp. 115–119.

[113] B. HARSHBARGER: Rectangular lattices, Virginia Agricultural Experiment Station, Memoir 1 (1947).

[114] B. HARSHBARGER: Triple rectangular lattices, *Biometrics*, **5** (1949), pp. 1–13.

[115] A. S. HEDAYAT, N. J. A. SLOANE & J. STUFKEN: *Orthogonal Arrays*, Springer-Verlag, New York (1999).

[116] D. G. HIGMAN: Finite permutation groups of rank 3, *Mathematische Zeitschrift*, **86** (1964), pp. 145–156.

[117] D. G. HIGMAN: Intersection matrices for finite permutation groups, *Journal of Algebra*, **6** (1967), pp. 22–42.

[118] D. G. HIGMAN: *Combinatorial Considerations about Permutation Groups*, Mathematical Institute, Oxford (1971).

[119] D. G. HIGMAN: Coherent configurations I, *Geometriae Dedicata*, **4** (1975), pp. 1–32.

[120] D. G. HIGMAN: Coherent configurations II, *Geometriae Dedicata*, **5** (1976), pp. 413–424.

[121] D. G. HIGMAN: Coherent algebras, *Linear Algebra and its Applications*, **93** (1987), pp. 209–239.

[122] D. G. HIGMAN & C. C. SIMS: A simple group of order 44,352,000, *Mathematische Zeitschrift*, **105** (1968), pp. 110–113.

[123] K. HINKELMANN & O. KEMPTHORNE: Two classes of group divisible partial diallel crosses, *Biometrika*, **50** (1963), pp. 281–291.

[124] A. J. HOFFMAN & R. R. SINGLETON: On Moore graphs of diameter two and three, *IBM Journal of Research and Development*, **4** (1960), pp. 497–504.

[125] R. HOMEL & J. ROBINSON: Nested partially balanced incomplete block designs, in: *Proceedings of the First Australian Conference on Combinatorial Mathematics 1972* (eds. J. Wallis & W. D. Wallis), TUNRA, Newcastle, New South Wales (1972), pp. 203–206.

[126] R. J. HOMEL & J. ROBINSON: Nested partially balanced incomplete block designs, *Sankhyā, Series B*, **37** (1975), pp. 201–210.

[127] A. M. HOUTMAN & T. P. SPEED: Balance in designed experiments with orthogonal block structure, *Annals of Statistics*, **11** (1983), pp. 1069–1085.

[128] D. R. HUGHES: Combinatorial analysis: t-designs and permutation groups, in: *Proceedings of the Symposium in Pure Mathematics* 6, American Mathematical Society, Providence, Rhode Island (1962), pp. 39–41.

[129] B. HUPPERT: *Endlicher Gruppen I*, Springer-Verlag, Berlin (1967).

[130] G. ISHII & J. OGAWA: On the analysis of balanced and partially balanced designs, *Osaka City University Business Review*, **81** (1965), pp. 1–31.

[131] T. B. JAJCAYOVÁ & R. JAJCAY: On the contributions of Dale Marsh Mesner, *Bulletin of the Institute of Combinatorics and its Applications*, **36** (2002), pp. 46–52.

[132] A. T. JAMES & G. N. WILKINSON: Factorization of the residual operator and canonical decomposition of nonorthogonal factors in the analysis of variance, *Biometrika*, **58** (1971), pp. 279–294.

[133] R. G. JARRETT: Definitions and properties for m-concurrence designs, *Journal of the Royal Statistical Society, Series B*, **45** (1983), pp. 1–10.

[134] J. A. JOHN: *Cyclic Designs*, Chapman and Hall, London (1987).

[135] J. A. JOHN & A. M. DEAN: Single replicate factorial arrangements in generalized cyclic designs: I. Symmetrical arrangements, *Journal of the Royal Statistical Society, Series B*, **37** (1975), pp. 63–71.

[136] J. A. JOHN & T. J. MITCHELL: Optimal incomplete block designs, *Journal of the Royal Statistical Society, Series B*, **39** (1977), pp. 39–43.

[137] J. A. JOHN & E. R. WILLIAMS: *Cyclic and Computer-Generated Designs*, Chapman and Hall, London (1995).

[138] J. A. JOHN, F. W. WOLOCK & H. A. DAVID: *Cyclic Designs*, Applied Mathematics Series, 62, National Bureau of Standards, Washington, D.C. (1972).

[139] P. W. M. JOHN: An extension of the triangular association scheme to three associate classes, *Journal of the Royal Statistical Society, Series B*, **28** (1966), pp. 361–365.

[140] P. W. M. JOHN: *Statistical Design and Analysis of Experiments*, Macmillan, New York (1971).

[141] J. D. JOHNSON: *Adding partially balanced incomplete block designs*, Ph. D. thesis, University of California at Davis (1967).

[142] S. M. JOHNSON: A new upper bound for error-correcting codes, *IEEE Transactions on Information Theory*, **8** (1962), pp. 203–207.

[143] S. KAGEYAMA: On the reduction of associate classes for certain PBIB designs, *Annals of Mathematical Statistics*, **43** (1972), pp. 1528–1540.

[144] S. KAGEYAMA: Reduction of associate classes for block designs and related combinatorial arrangements, *Hiroshima Mathematical Journal*, **4** (1974), pp. 527–618.

[145] L. A. KALUZHNIN & M. H. KLIN: On certain maximal subgroups of symmetric and alternating groups, *Mathematics of the USSR, Sbornik*, **16** (1972), pp. 95–123.

[146] O. KEMPTHORNE: A simple approach to confounding and fractional replication in factorial experiments, *Biometrika*, **34** (1947), pp. 255–272.

[147] O. KEMPTHORNE: The efficiency factor of an incomplete block design, *Annals of Mathematical Statistics*, **27** (1956), pp. 846–849.

[148] O. KEMPTHORNE: Classificatory data structures and associated linear models, in: *Statistics and Probability: Essays in Honor of C. R. Rao* (eds. G. Kallianpur, P. R. Krishnaiah & J. K. Ghosh), North-Holland, Amsterdam (1982), pp. 397–410.

[149] O. KEMPTHORNE, G. ZYSKIND, S. ADDELMAN, T. N. THROCKMORTON & R. F. WHITE: *Analysis of Variance Procedures*, Aeronautical Research Laboratory, Ohio Report No. 149 (1961).

[150] T. P. KIRKMAN: On a problem in combinations, *Cambridge and Dublin Mathematical Journal*, **2** (1847), pp. 191–204.

[151] M. Klin, C. Rücker, G. Rücker & G. Tinhofer: Algebraic combinatorics in mathematical chemistry. Methods and algorithms. I. Permutation groups and coherent (cellular) algebras, *MATCH*, **40** (1999), pp. 7–138.

[152] A. M. Kshirsagar: On balancing in designs in which heterogeneity is eliminated in two directions, *Calcutta Statistical Association Bulletin*, **7** (1957), pp. 469–476.

[153] A. M. Kshirsagar: A note on incomplete block designs, *Annals of Mathematical Statistics*, **29** (1958), pp. 907–910.

[154] A. M. Kshirsagar: Balanced factorial designs, *Journal of the Royal Statistical Society, Series B*, **28** (1966), pp. 559–569.

[155] B. Kurkjian & M. Zelen: A calculus for factorial arrangements, *Annals of Mathematical Statistics*, **33** (1962), pp. 600–619.

[156] B. Kurkjian & M. Zelen: Applications of the calculus for factorial arrangements. I: Block and direct product designs, *Biometrika*, **50** (1963), pp. 63–73.

[157] K. Kusumoto: Hyper cubic designs, *Wakayama Medical Reports*, **9** (1965), pp. 123–132.

[158] K. Kusumoto: Association schemes of new types and necessary conditions for existence of regular and symmetrical PBIB designs with those association schemes, *Annals of the Institute of Statistical Mathematics*, **19** (1967), pp. 73–100.

[159] W. Ledermann: *Introduction to Group Characters*, Cambridge University Press, Cambridge (1977).

[160] J. A. Leeming: *Efficiency versus resolvability in block designs*, Ph. D. thesis, University of London (1998).

[161] D. A. Leonard: Orthogonal polynomials, duality and association schemes, *SIAM Journal of Mathematical Analysis*, **13** (1982), pp. 656–663.

[162] D. A. Leonard: Parameters of association schemes that are both P- and Q- polynomial, *Journal of Combinatorial Theory, Series A*, **36** (1984), pp. 355–363.

[163] D. A. Leonard: Rational functions and association scheme parameters, *Journal of Algebraic Combinatorics*, **6** (1997), pp. 269–277.

[164] F. W. Levi: *Geometrische Konfigurationen*, Hirzel, Leipzig (1929).

[165] S. M. Lewis & A. M. Dean: On general balance in row–column designs, *Biometrika*, **78** (1991), pp. 595–600.

[166] J. H. van Lint & R. M. Wilson: *A Course in Combinatorics*, Cambridge University Press, Cambridge (1992).

[167] F. J. MacWilliams & N. J. A. Sloane: *The Theory of Error-Correcting Codes*, North-Holland, Amsterdam (1977).

[168] J. D. Malley: *Optimal Unbiased Estimation of Variance Components*, Lecture Notes in Statistics, 39, Springer-Verlag, Berlin (1986).

[169] R. Mead: The non-orthogonal design of experiments, *Journal of the Royal Statistical Society, Series A*, **153** (1990), pp. 151–178.

[170] S. Mejza & S. Kageyama: On the optimality of certain nested block designs under a mixed effects model, in: *MODA 4 – Advances in Model-Oriented Data Analysis* (eds. C. P. Kitsos & W. G. Müller), Physica-Verlag, Heidelberg (1995), pp. 157–164.

[171] D. M. Mesner: An investigation of certain combinatorial properties of partially balanced incomplete block experimental designs and association schemes, with a detailed study of designs of Latin square and related types, Ph. D. thesis, Michigan State University (1956).

[172] H. Monod & R. A. Bailey: Valid restricted randomization for unbalanced designs, *Journal of the Royal Statistical Society, Series B*, **55** (1993), pp. 237–251.

[173] E. H. Moore: Tactical memoranda, *American Journal of Mathematics*, **18** (1896), pp. 264–303.

[174] J. P. Morgan: Nested designs, in: *Handbook of Statistics 13: Design and Analysis of Experiments* (eds. S. Ghosh & C. R. Rao), North-Holland, Amsterdam (1996), pp. 939–976.

[175] J. P. Morgan, D. A. Preece & D. H. Rees: Nested balanced incomplete block designs, *Discrete Mathematics*, **231** (2001), pp. 351–389.

[176] J. P. Morgan & N. Uddin: Optimality and construction of nested row and column designs, *Journal of Statistical Planning and Inference*, **37** (1993), pp. 81–93.

[177] M. Muzychuk: The structure of rational Schur rings over cyclic groups, *European Journal of Combinatorics*, **14** (1993), pp. 479–490.

[178] C. R. Nair: A new class of designs, *Journal of the American Statistical Association*, **59** (1964), pp. 817–833.

[179] K. R. Nair: Partially balanced incomplete block designs involving only two replication, *Calcutta Statistical Association Bulletin*, **3** (1950), pp. 83–86.

[180] K. R. Nair: Rectangular lattices and partially balanced incomplete block designs, *Biometrics*, **7** (1951), pp. 145–154.

[181] K. R. Nair & C. R. Rao: Confounding in asymmetric factorial experiments, *Journal of the Royal Statistical Society*, **10** (1948), pp. 109–131.

[182] J. A. Nelder: The analysis of randomized experiments with orthogonal block structure. I. Block structure and the null analysis of variance, *Proceedings of the Royal Society of London, Series A*, **283** (1965), pp. 147–162.

[183] J. A. Nelder: The analysis of randomized experiments with orthogonal block structure. II. Treatment structure and the general analysis of variance, *Proceedings of the Royal Society of London, Series A*, **283** (1965), pp. 163–178.

[184] A. Neumaier: Strongly regular graphs with smallest eigenvalue $-m$, *Archiv der Mathematik*, **38** (1979), pp. 89–96.

[185] P. M. Neumann: Generosity and characters of multiply transitive permutation groups, *Proceedings of the London Mathematical Society*, **31** (1975), pp. 457–481.

[186] P. M. Neumann, G. A. Stoy & E. C. Thompson: *Groups and Geometry*, Oxford University Press, Oxford (1994).

[187] A. K. Nigam: Nearly balanced incomplete block designs, *Sankhyā, Series B*, **38** (1976), pp. 195–198.

[188] M. Ogasawara: *A necessary condition for the existence of regular and symmetric PBIB designs of T_m type*, Institute of Statistics Mimeo Series, 418, University of North Carolina at Chapel Hill, North Carolina (1965).

[189] H. D. Patterson & E. R. Williams: Some theoretical results on general block designs, *Congressus Numerantium*, **15** (1976), pp. 489–496.

[190] R. W. Payne, P. W. Lane, P. G. N. Digby, S. A. Harding, P. K. Leech, G. W. Morgan, A. D. Todd, R. Thompson, G. Tunnicliffe Wilson, S. J. Welham, R. P. White, A. E. Ainsley, K. E. Bicknell, M. F. Franklin, J. C. Gower, T. J. Hastie, S. K. Haywood, J. H. Maindonald, J. A. Nelder, H. D. Patterson, D. L. Robinson, G. J. S. Ross, H. R. Simpson, R. J. Tibshirani, L. G. Underhill & P. J. Verrier: *Genstat 5 Release 3 Reference Manual*, Clarendon Press, Oxford (1993).

[191] S. C. Pearce: Supplemented balance, *Biometrika*, **47** (1960), pp. 263–271.

[192] S. C. Pearce: The use and classification of non-orthogonal designs, *Journal of the Royal Statistical Society, Series A*, **126** (1963), pp. 353–377.

[193] S. C. Pearce: The mean efficiency of equi-replicate designs, *Biometrika*, **55** (1968), pp. 251–253.

[194] S. C. Pearce, T. Caliński, & T. F. de C. Marshall: The basic contrasts of an experimental design with special reference to the analysis of data, *Biometrika*, (1974), **61** pp. 449–460.

[195] R. L. Plackett & J. P. Burman: The design of optimum multifactorial experiments, *Biometrika*, **33** (1946), pp. 305–325.

[196] D. A. Preece: Nested balanced incomplete block designs, *Biometrika*, **54** (1967), pp. 479–486.

[197] D. A. Preece: Balanced 6×6 designs for 9 treatments, *Sankhyā, Series B*, **30** (1968), pp. 443–446.

[198] D. A. Preece: A second domain of balanced 6×6 designs for nine equally-replicated treatments, *Sankhyā, Series B*, **38** (1976), pp. 192–194.

[199] D. A. Preece & G. H. Freeman: Semi-Latin squares and related designs, *Journal of the Royal Statistical Society, Series B*, **45** (1983), pp. 267–277.

[200] D. Raghavarao: A generalization of group divisible designs, *Annals of Mathematical Statistics*, **31** (1960), pp. 756–771.

[201] D. Raghavarao: *Constructions and Combinatorial Problems in Design of Experiments*, John Wiley and Sons, New York (1971).

[202] D. Raghavarao & K. Chandrasekhararao: Cubic designs, *Annals of Mathematical Statistics*, **35** (1964), pp. 389–397.

[203] C. R. Rao: Factorial arrangements derivable from combinatorial arrangements of arrays, *Journal of the Royal Statistical Society, Supplement*, **9** (1947), pp. 128–139.

[204] C. R. Rao: On a class of arrangements, *Proceedings of the Edinburgh Mathematical Society*, **8** (1949), pp. 119–125.

[205] C. R. Rao: A general class of quasifactorial and related designs, *Sankhyā*, **17** (1956), pp. 165–174.

[206] M. B. Rao: Partially balanced block designs with two different number of replications, *Journal of the Indian Statistical Association*, **4** (1966), pp. 1–9.

[207] M. B. Rao: Group divisible family of PBIB designs, *Journal of the Indian Statistical Association*, **4** (1966), pp. 14–28.

[208] S. B. Rao, D. K. Ray-Chaudhuri & N. M. Singhi: On imprimitive association schemes, in: *Combinatorics and Applications* (eds. K. S. Vijayan & N. M. Singhi), Indian Statistical Institute, Calcutta (1984), pp. 273–291.

[209] W. W. Rouse Ball: Mathematical Recreations and Essays (8th edition), Macmillan, London (1919).

[210] J. Roy: On the efficiency-factor of block designs, *Sankhyā*, **19** (1958), pp. 181–188.

[211] P. M. Roy: Hierarchical group divisible incomplete block designs with m-associate classes, *Science and Culture*, **19** (1953), pp. 210–211.

[212] G. M. Saha: On Calinski's patterns in block designs, *Sankhyā, Series B*, **38** (1976), pp. 383–392.

[213] G. M. Saha & Gauri Shanker: On a generalized group divisible family of association schemes and PBIB designs based on the schemes, *Sankhyā, Series B*, **38** (1976), pp. 393–404.

[214] I. Schur: Zur Theorie der einfach transitiven Permutationsgruppen, *Sitzungberichte der Preussischen Akademie der Wissenschaften, Physikalische Mathematische Klasse*, **18/20** (1933), pp. 598–623.

[215] S. R. Searle, G. Casella & C. E. McCulloch: *Variance Components*, Wiley, New York (1992).

[216] J. J. Seidel: Strongly regular graphs with $(-1, 1, 0)$ adjacency matrix having eigenvalue 3, *Linear Algebra and its Applications*, **1** (1968), pp. 281–298.

[217] J. J. Seidel: Strongly regular graphs, in: *Surveys in Combinatorics 1979* (ed. B. Bollobás), London Mathematical Society Lecture Note Series, 38, Cambridge University Press, Cambridge (1979), pp. 157–180.

[218] B. V. Shah: On balancing in factorial experiments, *Annals of Mathematical Statistics*, **29** (1958), pp. 766–779.

[219] B. V. Shah: A generalization of partially balanced incomplete block designs, *Annals of Mathematical Statistics*, **30** (1959), pp. 1041–1050.

[220] B. V. Shah: Balanced factorial experiments, *Annals of Mathematical Statistics*, **31** (1960), pp. 502–514.

[221] K. R. Shah & B. K. Sinha: *Theory of Optimal Designs*, Lecture Notes in Statistics, 54, Springer, New York (1989).

[222] M. S. Shrikhande & S. S. Sane: *Quasi-Symmetric Designs*, London Mathematical Society Lecture Note Series, 164, Cambridge University Press, Cambridge (1991).

[223] S. S. Shrikhande: Designs for two-way elimination of heterogeneity, *Annals of Mathematical Statistics*, **22** (1951), pp. 235–247.

[224] C. C. Sims: Graphs and finite permutation groups, *Mathematische Zeitschrift*, **95** (1967), pp. 76–86.

[225] C. C. Sims: Graphs and finite permutation groups. II, *Mathematische Zeitschrift*, **103** (1968), pp. 276–281.

[226] M. Singh & A. Dey: Block designs with nested rows and columns, *Biometrika*, **66** (1979), pp. 321–326.

[227] N. K. Singh & K. N. Singh: The non-existence of some partially balanced incomplete block designs with three associate classes, *Sankhyā A*, **26** (1964), pp. 239–250.

[228] B. K. SINHA: Some aspects of simplicity in the analysis of block designs, *Journal of Statistical Planning and Inference*, **6** (1982), pp. 165–172.

[229] K. SINHA: Generalized partially balanced incomplete block designs, *Discrete Mathematics*, **67** (1987), pp. 315–318.

[230] T. P. SPEED & R. A. BAILEY: On a class of association schemes derived from lattices of equivalence relations, in: *Algebraic Structures and Applications* (eds. P. Schultz, C. E. Praeger & R. P. Sullivan), Marcel Dekker, New York (1982), pp. 55–74.

[231] T. P. SPEED & R. A. BAILEY: Factorial dispersion models, *International Statistical Review*, **55** (1987), pp. 261–277.

[232] J. STEINER: Combinatorische Aufgabe, *Journal für reine und angewandte Mathematik*, **45** (1853), pp. 181–182.

[233] P. U. SURENDRAN: Association matrices and the Kronecker product of designs, *Annals of Mathematical Statistics*, **39** (1968), pp. 676–680.

[234] O. TAMASCHKE: Zur Theorie der Permutationsgruppen mit regulärer Untergruppe. I., *Mathematische Zeitschrift*, **80** (1963), pp. 328–354.

[235] O. TAMASCHKE: Zur Theorie der Permutationsgruppen mit regulärer Untergruppe. II., *Mathematische Zeitschrift*, **80** (1963), pp. 443–465.

[236] P. TERWILLIGER: The subconstituent algebra of an association scheme. I, *Journal of Algebraic Combinatorics*, **1** (1992), pp. 363–388.

[237] P. TERWILLIGER: The subconstituent algebra of an association scheme. II, *Journal of Algebraic Combinatorics*, **2** (1993), pp. 73–103.

[238] P. TERWILLIGER: The subconstituent algebra of an association scheme. III, *Journal of Algebraic Combinatorics*, **2** (1993), pp. 177–210.

[239] S. K. THARTHARE: Right angular designs, *Annals of Mathematical Statistics*, **34** (1963), pp. 1057–1067.

[240] S. K. THARTHARE: Generalized right angular designs, *Annals of Mathematical Statistics*, **36** (1965), pp. 1535–1553.

[241] T. N. THROCKMORTON: *Structures of classification data*, Ph. D. thesis, Ames, Iowa (1961).

[242] T. TJUR: Analysis of variance models in orthogonal designs, *International Statistical Review*, **52** (1984), pp. 33–81.

[243] N. UDDIN & J. P. MORGAN: Some constructions for balanced incomplete block designs with nested rows and columns, *Biometrika*, **77** (1990), pp. 193–202.

[244] M. N. VARTAK: On an application of Kronecker product of matrices to statistical designs, *Annals of Mathematical Statistics*, **26** (1955), pp. 420–438.

[245] M. N. VARTAK: The non-existence of certain PBIB designs, *Annals of Mathematical Statistics*, **30** (1959), pp. 1051–1062.

[246] B. YU. WEISFEILER & A. A. LEMAN: Reduction of a graph to a canonical form and an algebra which appears in the process, *Scientific-Technical Investigations, Series 2*, **9** (1968), pp. 12–16.

[247] M. C. WHELAN: *Imprimitive association schemes*, Ph. D. thesis, University of London (1989).

[248] H. WIELANDT: *Finite Permutation Groups*, Academic Press, New York (1964).

[249] G. N. WILKINSON: A general recursive procedure for analysis of variance, *Biometrika*, **57** (1970), pp. 19–46.

[250] E. R. WILLIAMS & J. A. JOHN: A note on optimality in lattice square designs, *Biometrika*, **83** (1996), pp. 709–713.
[251] S. YAMAMOTO, Y. FUJII & N. HAMADA: Composition of some series of association algebras, *Journal of Science of Hiroshima University, Series A-I*, **29** (1965), pp. 181–215.
[252] F. YATES: Complex experiments, *Journal of the Royal Statistical Society, Supplement*, **2** (1935), pp. 181–247.
[253] F. YATES: A new method for arranging variety trials involving a large number of varieties, *Journal of Agricultural Science*, **26** (1936), pp. 424–455.
[254] F. YATES: Incomplete randomized blocks, *Annals of Eugenics*, **7** (1936), pp. 121–140.
[255] F. YATES: *The Design and Analysis of Factorial Experiments*, Technical Communication, 35, Imperial Bureau of Soil Science, Harpenden (1937).
[256] F. YATES: A further note on the arrangement of variety trials: quasi-Latin squares, *Annals of Eugenics*, **7** (1937), pp. 319–332.
[257] M. ZELEN: A note on partially balanced designs, *Annals of Mathematical Statistics*, **25** (1954), pp. 599–602.
[258] P.-H. ZIESCHANG: *An Algebraic Approach to Association Schemes*, Lecture Notes in Mathematics, 1628, Springer-Verlag, Berlin (1996).
[259] G. ZYSKIND: On structure, relation, sigma, and expectation of mean squares, *Sankyhā, Series A*, **24** (1962), pp. 115–148.

Index

Printed in the United States
By Bookmasters